Geophysical Monograph Series

Including
IUGG Volumes
Maurice Ewing Volumes
Mineral Physics Volumes

Geophysical Monograph Series

146 **The Extreme Proterozoic: Geology, Geochemistry, and Climate** Gregory S. Jenkins, Mark A. S. McMenamin, Christopher P. McKay, and Linda Sohl (Eds.)

147 **Earth's Climate: The Ocean–Atmosphere Interaction** Chunzai Wang, Shang-Ping Xie, and James A. Carton (Eds.)

148 **Mid-Ocean Ridges: Hydrothermal Interactions Between the Lithosphere and Oceans** Christopher R. German, Jian Lin, and Lindsay M. Parson (Eds.)

149 **Continent-Ocean Interactions Within East Asian Marginal Seas** Peter Clift, Wolfgang Kuhnt, Pinxian Wang, and Dennis Hayes (Eds.)

150 **The State of the Planet: Frontiers and Challenges in Geophysics** Robert Stephen John Sparks, and Christopher John Hawkesworth (Eds.)

151 **The Cenozoic Southern Ocean: Tectonics, Sedimentation, and Climate Change Between Australia and Antarctica** Neville Exon, James P. Kennett, and Mitchell Malone (Eds.)

152 **Sea Salt Aerosol Production: Mechanisms, Methods, Measurements, and Models** Ernie R. Lewis and Stephen E. Schwartz

153 **Ecosystems and Land Use Change** Ruth S. DeFries, Gregory P. Anser, and Richard A. Houghton (Eds.)

154 **The Rocky Mountain Region—An Evolving Lithosphere: Tectonics, Geochemistry, and Geophysics** Karl E. Karlstrom and G. Randy Keller (Eds.)

155 **The Inner Magnetosphere: Physics and Modeling** Tuija I. Pulkkinen, Nikolai A. Tsyganenko, and Reiner H. W. Friedel (Eds.)

156 **Particle Acceleration in Astrophysical Plasmas: Geospace and Beyond** Dennis Gallagher, James Horwitz, Joseph Perez, Robert Preece, and John Quenby (Eds.)

157 **Seismic Earth: Array Analysis of Broadband Seismograms** Alan Levander and Guust Nolet (Eds.)

158 **The Nordic Seas: An Integrated Perspective** Helge Drange, Trond Dokken, Tore Furevik, Rüdiger Gerdes, and Wolfgang Berger (Eds.)

159 **Inner Magnetosphere Interactions: New Perspectives From Imaging** James Burch, Michael Schulz, and Harlan Spence (Eds.)

160 **Earth's Deep Mantle: Structure, Composition, and Evolution** Robert D. van der Hilst, Jay D. Bass, Jan Matas, and Jeannot Trampert (Eds.)

161 **Circulation in the Gulf of Mexico: Observations and Models** Wilton Sturges and Alexis Lugo-Fernandez (Eds.)

162 **Dynamics of Fluids and Transport Through Fractured Rock** Boris Faybishenko, Paul A. Witherspoon, and John Gale (Eds.)

163 **Remote Sensing of Northern Hydrology: Measuring Environmental Change** Claude R. Duguay and Alain Pietroniro (Eds.)

164 **Archean Geodynamics and Environments** Keith Benn, Jean-Claude Mareschal, and Kent C. Condie (Eds.)

165 **Solar Eruptions and Energetic Particles** Natchimuthukonar Gopalswamy, Richard Mewaldt, and Jarmo Torsti (Eds.)

166 **Back-Arc Spreading Systems: Geological, Biological, Chemical, and Physical Interactions** David M. Christie, Charles Fisher, Sang-Mook Lee, and Sharon Givens (Eds.)

167 **Recurrent Magnetic Storms: Corotating Solar Wind Streams** Bruce Tsurutani, Robert McPherron, Walter Gonzalez, Gang Lu, José H. A. Sobral, and Natchimuthukonar Gopalswamy (Eds.)

168 **Earth's Deep Water Cycle** Steven D. Jacobsen and Suzan van der Lee (Eds.)

169 **Magnetospheric ULF Waves: Synthesis and New Directions** Kazue Takahashi, Peter J. Chi, Richard E. Denton, and Robert L. Lysal (Eds.)

170 **Earthquakes: Radiated Energy and the Physics of Faulting** Rachel Abercrombie, Art McGarr, Hiroo Kanamori, and Giulio Di Toro (Eds.)

171 **Subsurface Hydrology: Data Integration for Properties and Processes** David W. Hyndman, Frederick D. Day-Lewis, and Kamini Singha (Eds.)

172 **Volcanism and Subduction: The Kamchatka Region** John Eichelberger, Evgenii Gordeev, Minoru Kasahara, Pavel Izbekov, and Johnathan Lees (Eds.)

173 **Ocean Circulation: Mechanisms and Impacts—Past and Future Changes of Meridional Overturning** Andreas Schmittner, John C. H. Chiang, and Sidney R. Hemming (Eds.)

174 **Post-Perovskite: The Last Mantle Phase Transition** Kei Hirose, John Brodholt, Thorne Lay, and David Yuen (Eds.)

175 **A Continental Plate Boundary: Tectonics at South Island, New Zealand** David Okaya, Tim Stem, and Fred Davey (Eds.)

176 **Exploring Venus as a Terrestrial Planet** Larry W. Esposito, Ellen R. Stofan, and Thomas E. Cravens (Eds.)

177 **Ocean Modeling in an Eddying Regime** Matthew Hecht and Hiroyasu Hasumi (Eds.)

178 **Magma to Microbe: Modeling Hydrothermal Processes at Oceanic Spreading Centers** Robert P. Lowell, Jeffrey S. Seewald, Anna Metaxas, and Michael R. Perfit (Eds.)

179 **Active Tectonics and Seismic Potential of Alaska** Jeffrey T. Freymueller, Peter J. Haeussler, Robert L. Wesson, and Göran Ekström (Eds.)

180 **Arctic Sea Ice Decline: Observations, Projections, Mechanisms, and Implications** Eric T. DeWeaver, Cecilia M. Bitz, and L.-Bruno Tremblay (Eds.)

Geophysical Monograph 181

Midlatitude Ionospheric Dynamics and Disturbances

Paul M. Kintner Jr.
Anthea J. Coster
Tim Fuller-Rowell
Anthony J. Mannucci
Michael Mendillo
Roderick Heelis
Editors

American Geophysical Union
Washington, DC

Published under the aegis of the AGU Books Board

Kenneth R. Minschwaner, Chair; Gray E. Bebout, Joseph E. Borovsky, Kenneth H. Brink, Ralf R. Haese, Robert B. Jackson, W. Berry Lyons, Thomas Nicholson, Andrew Nyblade, Nancy N. Rabalais, A. Surjalal Sharma, Darrell Strobel, Chunzai Wang, and Paul David Williams, members.

Library of Congress Cataloging-in-Publication Data

Midlatitude ionospheric dynamics and disturbances / Paul M. Kintner, Jr. ... [et al.], editors.
    p. cm. — (Geophysical monograph, ISSN 0065-8448 ; 181)
   ISBN 978-0-87590-446-7
 1. Ionosphere—Research. 2. Ionospheric storms. 3. Sudden ionospheric disturbances. 4. Space environment. I. Kintner, Paul M.
   QC807.5.M53 2008
   551.51′45—dc22

2008045394

ISBN: 978-0-87590-446-7
ISSN: 0065-8448

Cover Photo: Isocontours of electron density during the 30 October 2003 storm from University of Bath MIDAS GPS tomography (courtesy of Cathryn Mitchell and Paul Spencer).

Copyright 2008 by the American Geophysical Union
2000 Florida Avenue, N.W.
Washington, DC 20009

Figures, tables and short excerpts may be reprinted in scientific books and journals if the source is properly cited.

Authorization to photocopy items for internal or personal use, or the internal or personal use of specific clients, is granted by the American Geophysical Union for libraries and other users registered with the Copyright Clearance Center (CCC) Transactional Reporting Service, provided that the base fee of $1.50 per copy plus $0.35 per page is paid directly to CCC, 222 Rosewood Dr., Danvers, MA 01923. 0065-8448/08/$01.50+0.35.

This consent does not extend to other kinds of copying, such as copying for creating new collective works or for resale. The reproduction of multiple copies and the use of full articles or the use of extracts, including figures and tables, for commercial purposes requires permission from the American Geophysical Union.

Printed in the United States of America.

CONTENTS

Preface
Paul M. Kintner Jr. ix

Midlatitude Ionospheric Dynamics and Disturbances: Introduction
*Paul M. Kintner Jr., Anthea J. Coster, Tim Fuller-Rowell, Anthony J. Mannucci,
Michael Mendillo, and Roderick Heelis*...1

Section I: Characterization of Midlatitude Storms

Review and Overview

Ionospheric Storms at Mid-Latitude: A Short Review
Gerd W. Prölss 9

The Mid-Latitude Trough—Revisited
Alan Rodger 25

**Assimilation of Observations With Models to Better Understand Severe Ionospheric Weather
at Mid-Latitudes**
Jan J. Sojka, R. W. Schunk, D. C. Thompson, L. Scherliess, and M. David 35

Low- and Middle-Latitude Ionospheric Dynamics Associated With Magnetic Storms
R. A. Heelis 51

A Data-Model Comparative Study of Ionospheric Positive Storm Phase in the Midlatitude F Region
G. Lu, L. P. Goncharenko, A. J. Coster, A. D. Richmond, R. G. Roble, N. Aponte, and L. J. Paxton 63

Recent Results

**High-Resolution Observations of Subauroral Polarization Stream-Related Field Structures During
a Geomagnetic Storm Using Passive Radar**
Melissa G. Meyer 77

Ionization Dynamics During Storms of the Recent Solar Maximum
C. N. Mitchell, P. Yin, P. S. J. Spencer, and D. Pokhotelov 83

Mapping the Time-Varying Distribution of High-Altitude Plasma During Storms
G. S. Bust and Geoff Crowley 91

Section II: Electric Field Coupling From the Heliosphere and Inner Magnetosphere

Review and Overview

Interplanetary Causes of Middle Latitude Ionospheric Disturbances
Bruce T. Tsurutani, Ezequiel Echer, Fernando L. Guarnieri, and Olga P. Verkhoglyadova................................99

Ionospheric-Magnetospheric-Heliospheric Coupling: Storm-Time Thermal Plasma Redistribution
John C. Foster...121

Recent Results

The Linkage Between the Ring Current and the Ionosphere System
P. C. Brandt, Y. Zheng, T. S. Sotirelis, K. Oksavik, and F. J. Rich..135

Storm Phase Dependence of Penetration of Magnetospheric Electric Fields to Mid and Low Latitudes
Takashi Kikuchi, Kumiko K. Hashimoto, and Kenro Nozaki..145

Relating the Interplanetary-Induced Electric Fields With the Low-Latitude Zonal Electric Fields Under Geomagnetically Disturbed Conditions
Adela Anghel, David Anderson, Jorge Chau, Kiyohumi Yumoto, and Archana Bhattacharyya.....................157

Simulation of PPEF Effects in Dayside Low-Latitude Ionosphere for the October 30, 2003, Superstorm
Olga P. Verkhoglyadova, Bruce T. Tsurutani, Anthony J. Mannucci, Akinori Saito, Tohru Araki, David Anderson, M. Abdu, and J. H. A. Sobral..169

Impact of the Neutral Wind Dynamo on the Development of the Region 2 Dynamo
T. W. Garner, Geoff Crowley, and R. A. Wolf...179

Section III: Thermospheric Control of the Mid-Latitude Ionosphere

Review and Overview

Global Modeling of Storm-Time Thermospheric Dynamics and Electrodynamics
T. J. Fuller-Rowell, A. D. Richmond, and N. Maruyama 187

Thermospheric Dynamics at Low and Mid-Latitudes During Magnetic Storm Activity
J. W. Meriwether 201

Disturbed O/N_2 Ratios and Their Transport to Middle and Low Latitudes
Geoff Crowley and R. R. Meier..221

Storm Time Energy Budgets of the Global Thermosphere
William J. Burke 235

Recent Results

Sources of *F*-Region Height Changes During Geomagnetic Storms at Mid Latitudes
Mariangel Fedrizzi, T. J. Fuller-Rowell, Naomi Maruyama, Mihail Codrescu, and Hargobind Khalsa 247

Neutral Composition and Density Effects in the October-November 2003 Magnetic Storms
T. J. Immel, Geoff Crowley, J. M. Forbes, R. S. Nerem, and E. K. Sutton..259

Optical and Radio Observations and AMIE/TIEGCM Modeling of Nighttime Traveling Ionospheric Disturbances at Midlatitudes During Geomagnetic Storms
K. Shiokawa, T. Tsugawa, Y. Otsuka, T. Ogawa, G. Lu, A. Saito, and M. Yamamoto..............................271

Section IV: Ionospheric Gradients, Irregularities and User Needs

A Digest of Electrodynamic Coupling and Layer Instabilities in the Nighttime Midlatitude Ionosphere
Roland T. Tsunoda 283

Irregularities Within Subauroral Polarization Stream-Related Troughs and GPS Radio Interference at Midlatitudes
Evgeny Mishin and Natan Blaunstein..291

DEMETER Satellite Observations of Plasma Irregularities in the Topside Ionosphere at Low, Middle, and Sub-Auroral Latitudes and Their Dependence on Magnetic Storms
Robert F. Pfaff Jr., Carmen Liebrecht, Jean-Jacques Berthelier, Michel Malingre, Michel Parrot, and Jean-Pierre Lebreton 297

Optical and Radio Observations of Structure in the Midlatitude Ionosphere: Midlatitude Ionospheric Dynamics and Disturbances
Jonathan J. Makela 311

Section V: Experimental Methods and New Techniques

Global-Scale Observations of the Limb and Disk (GOLD): New Observing Capabilities for the Ionosphere-Thermosphere
R. W. Eastes, W. E. McClintock, M. V. Codrescu, A. Aksnes, D. N. Anderson, L. Andersson, D. N. Baker, A. G. Burns, S. A. Budzien, R. E. Daniell, K. F. Dymond, F. G. Eparvier, J. E. Harvey, T. J. Immel, A. Krywonos, M. R. Lankton, J. D. Lumpe, G. W. Prölss, A. D. Richmond, D. W. Rusch, O. H. Siegmund, S. C. Solomon, D. J. Strickland, and T. N. Woods 319

PREFACE

In September 2002, the NASA-sponsored Living with a Star Geospace Mission Definition Team (GMDT) issued its report entitled "The LWS Geospace Storm Investigations: Exploring the Extremes of Space Weather." This report identified the mid-latitude ionosphere as a critical component of space weather. Preliminary results from John Foster and Anthea Coster combining TEC measurements from networks of GPS receivers across North America demonstrated for the first time that the response of the mid-latitude ionosphere during magnetic storms was organized. Earlier measurements from scattered sites had identified the positive and negative phases of mid-latitude ionospheric storms, but had not appreciated the organization over continent-size scales. This new model showed promise but it had not yet matured beyond a curiosity.

Nearly three years later, J. Grebowsky organized a session of invited speakers for the Spring 2005 AGU General Assembly to address the science of the original GMDT report. R. Pfaff, T. Fuller-Rowell, G. Lu, L. Paxton and I offered separate viewpoints on the science. At this juncture, the accumulated evidence irrefutably confirmed the preliminary results of Foster and Coster and changed the community viewpoint toward the mid-latitude ionosphere as a quiet and uninteresting region. Other experiments, involving dense arrays of GPS TEC receivers in Japan and imaging of ionospheric air glow, revealed even more complex and fascinating structures with regional length scales in the mid-latitude ionosphere. Global models and simulations were utilized to actively investigate the cause of mid-latitude ionospheric storms. After the session, it was obvious that the 90 minutes available for discussing progress in understanding the mid-latitude ionosphere was woefully inadequate.

The organizer, invited speakers, and audience members decided to hold an impromptu meeting late in the afternoon in an empty meeting room. To the small group assembled somewhat haphazardly in New Orleans it was clear that the mid-latitude ionosphere deserved attention as a special region unto itself. A decision was made to propose to AGU a Chapman Conference entitled Mid-latitude Ionospheric Dynamics and Disturbances and I volunteered to lead the effort. A. Coster, T. Fuller-Rowell, A. Mannucci, and M. Mendillo offered to be co-organizers. The conference occurred at Yosemite National Park, California, 3–6 January 2007, and inspired this monograph.

This book is our best attempt to assemble in one place a comprehensive examination of the mid-latitude ionosphere and to convey the spirit of the Chapman Conference. There was substantial debate over the question of what was a mid-latitude ionospheric storm: Was it the auroral zone pushing equatorwards, equatorial convective storms pushing pole ward, or a unique process? In the end, most participants had begun thinking of the ionosphere as responding globally during a magnetic storm. In this view, the daytime mid-latitude ionosphere is the region through which the storm transports huge volumes of ionospheric plasma into the polar caps and onto the night side. It is also the region where inner-magnetospheric electric fields and thermospheric-driven dynamo electric fields and chemistry compete to control the mid-latitude ionosphere. A central theme was that space-based measurements of the ionosphere and thermosphere are required for progress. Resolving these issues is the subject of a future Chapman Conference built upon future measurements of the mid-latitude ionosphere.

Paul M. Kintner Jr.
Cornell University

Midlatitude Ionospheric Dynamics and Disturbances: Introduction

Paul M. Kintner Jr.,[1] Anthea J. Coster,[2] Tim Fuller-Rowell,[3] Anthony J. Mannucci,[4]
Michael Mendillo,[5] and Roderick Heelis[6]

1. OVERVIEW

Recent discoveries have demonstrated that the ionosphere responds over regions extending from the equator to the poles during geomagnetic storms and experiences the most extreme changes at midlatitudes. The midlatitude ionosphere was first studied during the "discovery era" of radio physics and space flight 50 or more years ago, but for the past three decades the polar and tropical ionosphere have dominated scientific activity, resulting in the false impression that the midlatitude ionosphere was an uninteresting region of known morphology and well-understood processes. During the past five years, however, the ability to image the ionosphere and thermosphere with large arrays of ground-based GPS receivers and satellite-borne UV imagers changed this viewpoint dramatically and led to the inception of the Chapman Conference on Mid-Latitude Ionospheric Dynamics and Disturbances (MIDD) and to this monograph.

The most dramatic changes in ionospheric content occur at midlatitudes, not at high or equatorial latitudes. The most extreme examples of ionospheric total electron content (TEC) perturbations occur at midlatitudes during geomagnetic storms, where TEC can change by factors of three to ten over the duration of a magnetic storm. The ionosphere responds to magnetic storms over regions extending from the equator to the poles, where huge volumes of plasma are produced and transported polewards. Sharp gradients in ionospheric content, extending thousands of kilometers, are created by unknown factors. These gradients spawn irregularities that together impact users of RF signals, either transiting across or reflecting from the ionosphere. At higher altitudes, dramatic changes in the ionosphere are accompanied by movement and transport of the plasmasphere.

The midlatitude ionosphere is controlled by two largely unconstrained mechanisms: the inner magnetospheric electric field, originating in the heliosphere, and the dynamic properties of the thermosphere. The recent MIDD Chapman Conference brought together three communities to investigate this control: (1) the ionospheric community, which is characterizing and modeling the midlatitude domain, (2) the magnetospheric and solar wind community, which is investigating how inner magnetospheric electric fields map to and transport the midlatitude ionosphere, and (3) the thermospheric community, which is investigating how thermospheric winds and composition control the midlatitude ionosphere. During geomagnetic storms these fields and winds are strongly driven, yielding ionospheric space weather in the form of gradients and irregularities. When these occur at midlatitudes, where the U.S. taxpayer resides, the effects of ionospheric space weather take on special importance, particularly in the area of GPS and aviation.

At the beginning of this monograph, midlatitude ionospheric storms are defined and the historical record reviewed. On the poleward side, the midlatitude ionosphere is bounded by the auroral ionosphere and on the equatorward side by the equatorial ionosphere. To some extent, processes in these two regions enter the midlatitude ionosphere, but processes unique to the midlatitude ionosphere also appear to exist. Alternately, the midlatitude ionosphere may be a plasma source region for the polar cap and the magnetosphere. Addressing the relative significance of these regions and their

[1] School of Electrical and Computer Engineering, Cornell University, Ithaca, New York, USA.
[2] Haystack Observatory, Massachusetts Institute of Technology, Westford, Massachusetts, USA.
[3] CIRES University of Colorado and NOAA Space Weather Prediction Center, Boulder, Colorado, USA.
[4] Jet Propulsion Laboratory, California Institute of Technology, Pasadena, California, USA.
[5] Center for Space Physics, Boston University, Boston, Massachusetts, USA.
[6] Hanson Center for Space Sciences, University of Texas at Dallas, Dallas, Texas, USA.

Midlatitude Ionospheric Dynamics and Disturbances
Geophysical Monograph Series 181
Copyright 2008 by the American Geophysical Union.
10.1029/181GM02

processes is largely an experimental effort aided by models and visualizations that ingest large quantities of data.

Defining a midlatitude ionospheric storm leads directly to the debate over what causes them. There are two possibilities: external electric fields originating in the solar wind and inner magnetosphere and thermospheric winds. Electric fields within the solar wind are modified by magnetospheric plasma populations as the fields propagate inward and eventually reach low and middle latitudes. The asymptotic behavior consists of shielding the inner magnetosphere from solar wind electric fields, but rapidly changing solar wind electric fields can reach the inner magnetosphere during a period of 1 to 2 hours, before the plasma populations can react. This electric field is referred to as a prompt penetration electric field (PPEF). On the other hand, thermospheric winds also drive the ionosphere, and the equivalent electric fields are referred to as disturbance dynamo (DD) electric fields. Either electric field source could explain midlatitude ionospheric storms, which led at the conference to a spirited debate conducted primarily with models and simulations. The sustaining reason for this controversy is a lack of comprehensive measurements to act as drivers of the models or to distinguish their predictions.

The monograph is organized around the principles discussed above: investigations of midlatitude ionospheric storms, investigations of electric field coupling from the heliosphere and inner magnetosphere, and thermospheric control of the midlatitude ionosphere. In addition, a section on ionospheric gradients and irregularities examines the space weather aspects of the midlatitude ionosphere. Our future understanding of midlatitude ionospheric storms as a part of global ionospheric storms will depend on new experimental techniques based on satellites to resolve the current controversies. This monograph highlights the scientific progress in characterizing the midlatitude ionosphere and the new questions raised by this progress.

2. MIDLATITUDE IONOSPHERIC STORMS

Central to the theme of this monograph is the characterization of midlatitude ionospheric storms. The papers in this section cover a broad spectrum of issues in this area, beginning with a general review of ionospheric storms at midlatitudes by Gerd Prölss. This paper provides a brief history of ionospheric storm research, describing how early observations from either a single observatory or from small subgroups of observatories were used to discover the positive and negative phases of ionospheric storms. The positive phase of an ionospheric storm is when the total electron content, TEC, increases early during the storm period. This is followed by a decrease in the TEC during the storm recovery phase (the negative phase). This paper also discusses different possible definitions of "middle latitudes." One definition has the midlatitudes sandwiched between the equatorward boundary of subauroral phenomena and the edge of the equatorial anomaly peaks. An alternate definition holds that the middle latitudes are a region unaffected by subauroral or equatorial phenomena, at least during quiet or moderately disturbed conditions (e.g., 25 to 55 degree invariant latitude). Prölss concludes his paper with a discussion of the origin of the positive phase of ionospheric storms. He points out that both winds and electric fields are important mechanisms and notes that the question is not which one of these mechanisms is responsible for the positive phase, but rather which of these two mechanisms is more important, especially at middle latitudes.

The next paper, by A. S. Rodger, discusses the midlatitude trough, which is an extremely consequential feature that forms near the boundary of the midlatitudes. The midlatitude ionospheric trough is a region at F-region altitudes, typically a few degrees wide in latitude, where the plasma concentration is usually lower as compared with regions immediately poleward and equatorward. Figure 1 in his paper presents two visualizations of this region. The trough normally lies close to the equatorward boundary of the auroral precipitation and is where the corotation and convection electric fields are oppositely directed (approximately). The trough is a region that is strongly coupled to the spatial and temporal variation of electric fields in its vicinity and thus has important consequences for midlatitude ionospheric storms. During geomagnetically active periods, the detailed morphology of troughs becomes difficult to predict, as the cross-polar cap electric field is usually increasing while the auroral oval is expanding.

The paper by J. Sojka discusses the inadequacies of models in describing the evolution of the distribution of electric fields and neutral winds during ionospheric storm periods in the midlatitudes. In his paper, he presents a simplified convection electric pattern and uses this to drive a physics-based ionospheric model demonstrating how superstorm ionospheric conditions can be generated. He points out that future models that incorporate data assimilation may be able to someday overcome the limitations of present-day empirical and physical models.

The next paper is R. Heelis's review of our current understanding of low and middle latitude ionospheric dynamics and energetics associated with magnetic storms. The DMSP data shown in Figure 2 of his paper is significant. It clearly indicates that during the large geomagnetic storm of 20 November 2003, magnetic latitudes near 50 degrees, normally associated with the midlatitudes, become fully engulfed in the auroral zone. Furthermore, this figure also shows that near dusk the sunward (westward) flows normally associated with the auroral zone now exist at latitudes as low as 30 degrees. These data illustrate that, under these circumstances, middle latitudes, which are usually dominated by corotation and dynamo fields, may be directly influenced by auroral electric fields and particles. Heelis also points out that it appears prudent to consider the formation and evolution of the TEC enhancements at middle and low latitudes as separate features, even if they may at times be collocated. His conclusions, which agree with those of the earlier papers in this section, are that significant questions remain concerning the role that winds and electric fields play in producing the dramatic midlatitude ionospheric density perturbations observed during major geomagnetic storms.

This paper is followed by a discussion by G. Lu et al. of the global modeling of ionospheric TEC and the role of neutral winds and electric fields in producing enhanced TEC during a moderate geomagnetic storm on 10 September 2005. Her paper shows that, by using realistic time-dependent ionospheric convection and auroral precipitation as input, the Thermosphere Ionosphere Electrodynamics General Circulation Model (TIEGCM) is able to reproduce the large-scale storm features in the electron density, electron temperature, and vertical ion drift observed by incoherent scatter radars (ISRs). The model also captures the temporal and spatial TEC variations shown in the global GPS maps. The agreement in the data-model comparison suggests that using the TIEGCM to investigate which mechanisms have played a significant role in generating the observed storm-time features is a valid approach. The result of this model investigation suggests that the primary cause of the dayside positive storm phase during this moderate geomagnetic storm is the storm-enhanced meridional neutral wind.

The final papers in this section concern recent results. They describe new techniques that are being used to monitor the midlatitude storm features. The first of these new-techniques papers is by M. Meyer and includes a report on observations of features associated with storm-time electric fields: the sub-auroral polarization stream (SAPS). These observations were collected using passive, coherent radar facilities at the University of Washington. The receivers are situated in Washington State and have an effective field of view of the sub-auroral region over southwestern Canada. These passive radars use FM radio waves to observe E-region structures at a range resolution of 1.5 km. Further techniques involving interferometry enable the resolution of scattering volumes in the cross-beam dimension and facilitate the localization of echoes within the radar field of view.

Figure 2 in Meyer's paper shows a fine-scale wavelike modulation propagating through the SAPS electric field, which is itself drifting (more slowly) equatorward. The period of the modulation observed over a 3-hour period is between 1 and 3 minutes (corresponding to 5–16 mHz). The electric field sub-structures (referred to as sub-auroral ionization drifts (SAIDs) by some authors) appear to be propagating equatorward at an average phase velocity of 415 m/s while the entire channel drifts equatorward at approximately 140 m/s.

The next new technique paper provides a discussion of 4D ionization dynamics by C. Mitchell et al. This paper describes a new observation technique called GPS imaging where the line-of-sight TEC observations from a global network of receivers are inverted into 3D time-dependent maps of electron density. Three large geomagnetic storm periods were studied. In each case, it was observed that the main phase of the storm involves a sudden uplift in F-layer altitude extending across the entire midlatitude region, and that this uplift propagates westward (from Europe to the United States).

Finally, G. Bust reports on another GPS imaging analysis using the 4D imaging algorithm called Ionospheric Data Assimilation Three-Dimensional (IDA3D). His study addresses the question of whether storm enhanced density (SED) midlatitude plasma is due to high altitude transport of equatorial plasma. His study also examines the similarities and differences between the 2003 October 30 and November 20 storms. His analysis suggests that the enhanced plasma in the midlatitude region is an extension of the equatorial anomaly region. However, again in agreement with the other authors in this section, he points out that the relative roles of electric fields, winds, and composition remain to be understood in the context of the detailed formation of the SED. He also raises the issue of another unknown, which is the question of how plasma within the midlatitude bulge region, which is not moving at a high speed, transformed into a tongue of ionization (TOI) that is moving poleward at large plasma velocities.

The papers in this section raise several questions and identify many unknowns in our understanding of midlatitude ionospheric storms. Combining new models with new data sources and new data analysis techniques are all critical next steps to increasing our understanding of how ionospheric storm features are produced. An understanding of the physical forces and their dynamic interactions is required to predict the space weather effects associated with large midlatitude ionospheric storms.

3. ELECTRIC FIELD COUPLING FROM THE HELIOSPHERE AND INNER MAGNETOSPHERE

Tsurutani et al. begin this section with a review of the origins of global ionospheric storms from outbursts on the sun to solar wind coupling to the ionosphere. This paper firmly establishes on theoretical and empirical grounds that geomagnetic storms causing significant midlatitude disturbances originate with conditions in the solar wind. Solar coronal mass ejections (CMEs) generally lead to geomagnetic storms if they reach the Earth. The more intense CMEs become "superstorms" with significant midlatitude consequences, especially if the southward magnetic field component has large values (–10 nT or more) that persist for several hours. Increases in solar wind ram pressure (dynamic pressure) are also associated with geo-effectiveness, as are sudden increases in solar irradiance due to solar flares. Intense flares cause short-duration increases in dayside electron density at all sunlit latitudes, known as sudden ionospheric disturbances (SIDs). Tsurutani et al. discuss the impact of the major flare of 28 October 2003, the largest flare ever recorded in the EUV portion of the spectrum. Uncertainties in measuring the transient spectra and in modeling the thermosphere define a difficult modeling problem. How well understood the physical processes that lead to electron density and TEC increases are remains an open question.

Anghel et al. emphasize the important role of electric fields in linking solar wind conditions to ionospheric disturbances. Electric fields are induced over the Earth's ionosphere by moving plasma clouds. A net B_z southward component is associated with the highest degrees of geo-effectiveness ($E_y = V \times B_z$ in GSM coordinates). Dawn–dusk electric fields induced by the heliosphere can "penetrate" (propagate) from high to low latitudes within minutes of reaching the magnetopause. This direct solar wind–ionosphere link is exploited by Anghel et al., who perform a joint statistical analysis of measured ionosphere electric fields and electric fields induced at Earth by the solar wind. Such a joint statistical analysis provides information on the coupling as well as the frequency components in the heliospheric driver itself.

The third paper in this section emphasizes that midlatitude dynamics are strongly affected by physical coupling between the magnetosphere and ionosphere. The J. C. Foster paper discusses recent research by the Millstone Hill group in M–I coupling during dusk-time, including the important role of subauroral electric fields. This paper focuses on storm-enhanced density (SED) and how the SED structure depends on the relative magnitudes of the Earth's corotation electric field and electric fields in the inner-magnetosphere, which evolve on time scales of tens of minutes during storms. The increased plasma content over a large fraction of Earth is caused by what has been termed the "Dayside Superfountain." The possibility of a "preferred longitude" for high-content SED is also suggested by Foster et al. due to the reduced magnetic field in the American sector (also known as the South Atlantic Anomaly).

Verkhoglyadova et al. present data from the Jicamarca incoherent scatter radar that measures electron density at the F2 layer peak (foF2) as a function of time for the 30 October 2003 superstorm. A striking aspect of this superstorm is the enormous depletion of plasma near the equator as the storm progresses through the main phase. Afternoon peak electron densities reduce by a factor of three in about ~1.5 hours during the storm's main phase. The measured depletion is compared to three separate model runs that produce electron densities that are 50% to >200% larger than measured. A possible reason for the discrepancy between measurements and modeled foF2 is electric field inputs to the model being too low (peak value ~1 mV/m estimated by the dual magnetometer method), since PPEF may attain values of 4–5 mV/m during large storms. Verkhoglyadova et al. conclude with a forward-looking discussion of PPEF modeling and its consequences and call for more sophisticated modeling of PPEF at local dawn.

Brandt et al. explore several elements contributing to the SAPS electric fields and more generally to magnetosphere–ionosphere coupling. This paper uses observations and modeling to study M–I coupling via Region 2 currents from a unique end-to-end perspective, combining solar wind, magnetospheric, and ionospheric data. The Energetic Neutral Atom (ENA) images clearly show a highly asymmetric ring current during the main "driven" phase of the storm. Plasma pressure within the ring current is the driver for closure currents into the ionosphere, and these currents close through a medium with finite conductivity. Brandt et al. present data showing a correlation between high-speed flow in the ionosphere and HENA image intensity. Then they show that ionospheric conductivity distributions strongly modify magnetospheric electric fields and influence particle motions, creating a feedback loop between ionospheric conductivity and electric fields with magnetospheric currents and electric fields.

The next paper in this volume returns to more global considerations. Kikuchi et al. analyze three magnetic storms to further understand a fundamental issue of global M–I coupling: How do electric fields penetrate globally into the ionosphere? They find that *equatorial* DP2 currents respond nearly instantaneously to increased *auroral* electrojet activity (*AE* index). This global-scale coupling is viewed as being due to an electric field propagation in the zeroth-order transverse mode of the Earth–ionosphere waveguide. Kikuchi et al. show that the magnitude and sign of equatorial DP2 currents are sensitive indicators of the relative roles of penetration versus shielding effectiveness and that, in the tightly coupled M–I–T system, magnetospheric current systems will generate ionospheric electric fields mapping back out to the magnetosphere, suppressing or enhancing the ring current.

The final paper in this section, by Garner et al., is a reminder that the ionosphere is not coupled solely to the solar wind and magnetosphere—it is part of the thermosphere. During geomagnetic disturbances, thermospheric modification is known to occur via high-latitude Joule heating; global circulation changes as a consequence. Ion-neutral drag generates electric fields via the disturbance dynamo mechanism. Garner et al. investigate the "second-order" impact of the neutral-wind dynamo electric fields on the Region-2 current system. The significance of this science question is perhaps highest during periods with multiple storms, when geo-effective solar wind conditions influence a previously disturbed thermosphere (so-called "preconditioning"). A model is used to map the electric fields generated from the disturbance dynamo electric field out to the magnetosphere. The model used is the Thermosphere–Ionosphere–Mesosphere Electrodynamics General Circulation Model (TIME-GCM), a coupled thermosphere–ionosphere model that self-consistently includes ionospheric electrodynamics.

4. THERMOSPHERIC CONTROL OF THE MIDLATITUDE IONOSPHERE

The cause of long-lived depletions in the midlatitude ionosphere during recovery from a geomagnetic storm, the so-called negative phase, has long been thought to be a consequence of neutral composition changes. Apart from the plasma structure associated with the sub-auroral trough or depletions associated with the high-velocity plasma streams associated with SAPS, neutral composition changes still remain the most likely explanation. Crowley and Meier review the neutral composition theory for the negative phase by comparing in detail the numerical modelling results with relatively recent and fairly extensive observations of the O/N_2 ratio from the GUVI instrument onboard the TIMED spacecraft. They show that numerical models simulate the storm-time thermospheric neutral composition changes quite well, except for details in the response and recovery time-scales.

Neutral composition changes arise from adjustment of the global circulation following the impulsive injection of energy at high latitudes. The dynamical changes are reviewed by both Meriwether and Shiokawa et al. from the observational perspective and by Fuller-Rowell and Richmond from the modelling side. Meriwether's review indicates that over the last 20 to 30 years, quite an extensive database of observations has been assembled from the combination of ground-based and space-based observations. Much of this database has been captured in the latest Horizontal Wind Model (HWM) by Emmert et al., which is much improved over the earlier versions and which explicitly includes a Disturbance Wind Model (DWM) component. Although becoming more extensive, thermospheric neutral wind observations still have many holes in their coverage and have rarely been taken in conjunction with the related plasma and electrodynamic components. Consequently, although the numerical results presented by Fuller-Rowell and Richmond appear to capture many of the observed dynamical features, such as the wave features presented by Shiokawa, the realism of some of the thermospheric dynamo characteristics, in particular, the apparent rapid dynamo response, have yet to be confirmed. Thus, ambiguity remains in separating the prompt penetration and disturbance dynamo, particularly on the night side. This ambiguity has prompted the need for coupled models of the thermosphere–ionosphere–plasmasphere and the inner magnetosphere, as was presented by Maruyama at the conference.

The thermospheric dynamics and neutral composition response to geomagnetic storms is strongly influenced by the intensity and time dependence of the energy injection at high latitudes. A good indicator of the amount of energy injected is the increase in neutral temperature and density globally. Burke et al. attempt to quantify this integrated energy source by direct comparison of the geomagnetic indicators, such as Dst and solar wind parameters, with observations of neutral density from the CHAMP satellite. They concluded from energy budget considerations that the solar wind–magnetospheric–ionosphere pathway is the primary route and that the path through the ring current is not the main driver of neutral thermospheric density change.

The relationship between neutral density and composition change is explored by Immel et al. They show that the physics of density change and neutral composition, although ultimately driven by the same high latitude heating, are different; and that although the two are closely related, sometimes due to the common source, at other times the two parameters can be quite different. The difference in the physics is that density responds to heating and thermal expansion of the atmosphere whereas the O/N_2 ratio change observed by a GUVI-type instrument requires a change in the global circulation. Fedrizzi et al. show a second consequence of thermospheric expansion: It can raise the height of the F layer due to the vertical winds associated with expansion. Further, they separate and quantify the relative contribution of expansion and horizontal wind in changing the height of the ionosphere at midlatitude during a storm.

5. IONOSPHERIC GRADIENTS AND IRREGULARITIES

Spatial ionospheric perturbations at midlatitudes fall roughly into several categories. At the longest length scales (50–200 km), features associated with equatorial convective storms can propagate poleward to midlatitudes such as Hawaii. At mesoscales (100 km), traveling ionospheric disturbances (MSTID) with a northwest to southeast orientation dominate and are sometimes associated with magnetic storms. At yet shorter scales (<1 km), irregularities occur on background gradients with several possible origins.

Makela et al. discuss optical and GPS observations of gradients and scintillations at Hawaii and Puerto Rico. Using 630-nm observations of night glow, they demonstrate that the plumes typically associated with equatorial convective storms occur over and to the north of Hawaii. They are primarily oriented along magnetic field lines and tilt westward with time because of shear and conductivity gradients. These structures frequently exhibit braided behavior on their western flanks, which Makela et al. ascribe to the gradient drift instability produced by thermospheric winds blowing across ionospheric gradients. Using the same 630-nm imaging technique, they then examine ionospheric structures over Puerto Rico where MSTID are found. These structures are aligned northwest to southeast and propagate mostly to the southwest, but occasionally in a northeast direction. The peculiar orientation of the MSTID is explained by the Perkins instability, although the local linear growth rate is too small to be significant.

This problem was considered by Tsunoda et al., who considered E-region coupling to the F region. They examine the "E-layer" instability, which is plane wave fluctuation in altitude in the presence of a neutral wind shear. These fluctuations have the largest growth rates at propagation directions, very similar to the Perkins instability but the growth rates are larger. Coupling of the E and F regions, driven by shear in the zonal wind, leads to altitude fluctuations with the correct pattern and scale lengths to explain MSTID.

Pfaff et al. present electric field fluctuation and density data from the Demeter satellite in circular orbit at 710 km altitude. They showed that, during the main phase of magnetic storms, the ionospheric density increased dramatically at midlatitudes by a factor of 10–100 at the altitude of Demeter. Simultaneously, these same regions exhibited electric field fluctuations over the frequency range of 1–500 Hz, which correlated well with regions of depleted ionosphere and ionospheric gradients. Furthermore, this behavior was similar on both the dayside and night side. Pfaff et al. also note that the midlatitude ionospheric disturbances are limited to about 6 hours during the main phase of a magnetic storm. During this same period, the equatorial ionosphere rises above the altitude of Demeter, and Pfaff et al. conclude that an eastward electric field is responsible for the vertical motion.

Mishin et al. propose an alternate explanation for midlatitude irregularities, using DMSP data to note the relationship between electric field and density fluctuations, the inner edge of the ring current, and subauroral polarization streams (SAPS). They call the electric field "supauroral polarization stream wave structures" or SAPSWS. The SAPSWS are found on the inner edge of the ring current and within SAPS regions. They propose that SAPSWS are generated by field-aligned currents on the inner edge of the ring current that are unstable to current convective instabilities. The SAPSWS, acting on plasma density gradients, then produce irregularities responsible for scintillations at UHF and L-band frequencies.

6. CONCLUSIONS

J. Foster has summarized the controversy over the origin of midlatitude dynamics and disturbances with the statement, "What we see depends on how we look." In "looking" at the midlatitude ionosphere, two big challenges exist. First, there are no continuous global scale measurements characterizing the state of the midlatitude ionosphere and its obvious connections to the low-latitude and high-latitude ionosphere. Second, there are no global or *in situ* measurements characterizing thermospheric winds and composition. Ground-based measurements can be helpful. For example, daytime electric fields in the tropics can be inferred from magnetometers, but midlatitude electric field measurements will require *in situ* electric field measurements with an accuracy of better than 1 mV/m. Dual frequency GPS receiver networks have also been helpful in identifying the global nature of ionospheric storms, especially at midlatitudes, but gaps in coverage over the oceans, the tropics, and polar regions and the limitation of only sensing TEC leave major questions unaddressed.

At regional and shorter-length scales, our understanding of the generation of midlatitude ionospheric gradients and irregularities is less well advanced, with characterization still being an important issue. MSTID observed in the Japanese sector with dense GPS arrays and observed in Puerto Rico with ground-based airglow cameras have revealed this phenomenon. Questions about their global distribution, the importance of E-region coupling through thermospheric wind shear, the implications of conjugate behavior, the origin of TEC fluctuations, and the origin of Fresnel scale irregularities are just now being addressed, with no consensus on the basic physical process involved.

Ionospheric storms unquestionably originate in the heliosphere, driven by solar processes. The details of how the ionosphere responds depend critically on magnetospheric properties and the thermospheric history. Yet, perhaps one of the most significant issues addressed by this monograph is that the ionosphere does not react just passively to driving forces from above. Changes in ionospheric conductivity react back on ring current pressure gradients and modulate ring current development. SED structures flowing poleward through the midlatitude may be a reservoir, providing heavy ions to the magnetosphere.

Controversy over the origin of ionospheric storms can only be resolved by measurements that provide continuous global-scale characterization complemented by *in situ* measurements of quantities that cannot be imaged. For example, ionospheric and thermospheric imaging from geostationary orbit (GEO), proposed by R. Eastes et al., has several major advantages. GEO continuously stares at a full hemisphere, measuring the evolution of the ionosphere and thermosphere, which is required for comparison with models and simulations. Next, it provides context for other experimental techniques, especially ground-based methods yielding only local results. Finally, it complements *in situ* satellite experiments within the ionosphere and thermosphere that cannot resolve space-time ambiguities on their own. However, GEO imaging also has its limitations, namely, the inability to measure ionospheric electric fields and thermospheric winds. Measuring these properties requires in situ instruments on low-earth orbiting satellites. Some of the issues raised in this monograph can be addressed by increasing the density of inexpensive, ground-based instrumentation, as suggested in the NSF Distributed Arrays of Small Instruments (DASI) program, but this technique cannot address the required global characterization or the measurement of ionospheric electric fields and thermospheric winds. For in situ measurements, new technology leading to greatly reduced satellite size and decreased launch costs may open new research opportunities. Exploiting the opportunities will be required if the prediction of midlatitude ionospheric storms is to contribute to the national space weather program.

Ionospheric Storms at Mid-Latitude: A Short Review

Gerd W. Prölss

Argelander Institut für Astronomie, Universität Bonn, Bonn, Germany

In this contribution, four different aspects of ionospheric storms are discussed: (1) the early history of ionospheric storm research; (2) the publication statistics in this field; (3) the definition of middle latitudes; and (4) the origin of positive ionospheric storms. To illustrate the history of ionospheric storm research, early publications in this field are listed in Table 1. It is remarkable that most of the salient features of ionospheric storms were discovered before 1950. Since then, the number of publications on ionospheric F-region storms has increased dramatically, and this is documented in Figure 4 and also by the reference list compiled in the Appendix. Conspicuous fluctuations in the rate of publication are attributed to various trends including new observation techniques, space weather research activities, and the unprecedented number of superstorms observed in recent years. With regard to the definition of middle latitudes, we suggest that they extend only up to the subauroral region. Here, the equatorward edge of the subauroral trough is used to define this poleward boundary. Positive ionospheric storms remain an ill-understood phenomenon, and various disturbance scenarios are discussed. Clearly, more precise and more comprehensive measurements, especially of winds and electric fields, are needed to solve this long-standing puzzle.

1. INTRODUCTION

The term "ionospheric storm" is used in this work to designate a large-scale perturbation of the ionospheric density caused by the strongly enhanced dissipation of solar wind energy in the space environment of the Earth. A typical example of such an event is shown in Figure 1. Using the AE and Dst indices, the first two panels of this figure describe the level of geomagnetic activity during a 3-day interval in May 1972. A moderately strong storm occurred on 15 May. The ionospheric response to this disturbance is shown in the lower part of the figure. Plotted are the total electron content (TEC) and the maximum ionization density of the F_2 layer (Nmax) as observed at two neighboring stations. The lines without data points indicate the normal daily variations, the lines marked by data points the storm time variations. On the first storm day, a significant increase in both the TEC and maximum ionization density is observed. This constitutes the positive phase of an ionospheric storm. On the second storm day, both quantities exhibit an anomalous decrease; this is the negative phase of an ionospheric storm.

Ionospheric storms are important for two reasons. First, they constitute an important link in the complex chain of solar–terrestrial relations. Second, they are of great practical interest since sub- and transionospheric radio communications may be severely degraded or even disrupted during such events. In the present communication, four rather diverse aspects of this phenomenon are addressed. In section 2, some annotations to the early history of ionospheric storm research are made. Next, the publication statistics in this

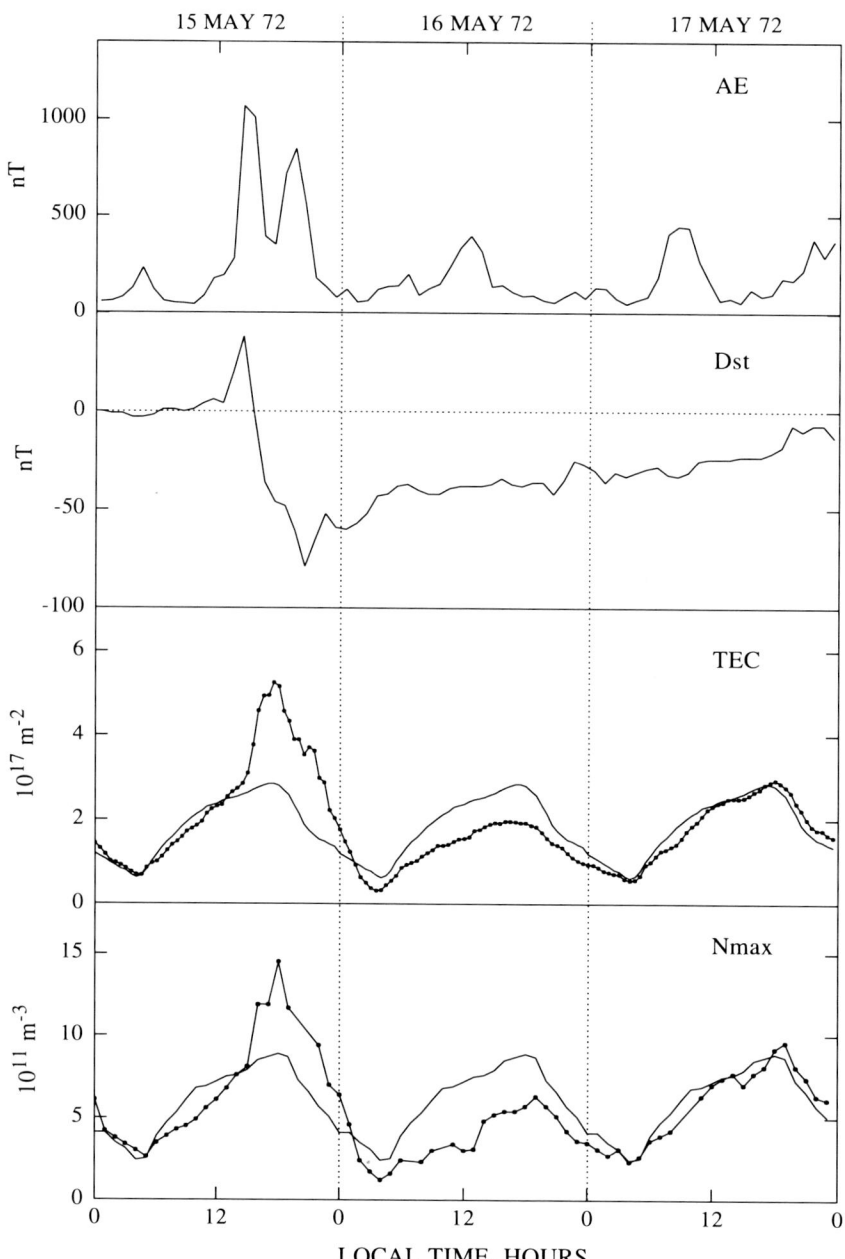

Figure 1. Ionospheric response to the magnetic storm of 15 May 1972. The *AE* and *Dst* indices are used to indicate the level of geomagnetic activity during this event. The last two panels show the storm-associated changes in the total electron content (TEC) and maximum electron density (Nmax) of the ionosphere, respectively. The curves marked by data points indicate the storm-time variations, whereas the curves without data points serve as quiet-time references. These reference curves correspond to the mean variations observed on the 7 days before the storm event. The TEC data were obtained at Hamilton, MA, using the Faraday polarization twist of VHF radio waves transmitted from the geostationary satellite ATS-3. At an altitude of 420 km, the subionospheric coordinates were 38.9°N, 70.7°W. The maximum ionization density measurements were recorded at the ionosonde station Wallops Island (37.9°N, 75.5°W). The ionospheric data shown in this figure are taken from *Mendillo and Klobuchar* [1974].

field are discussed (section 3). What is meant by "middle latitudes" is investigated in section 4. Finally, in section 5, a few annotations to a somewhat controversial topic, namely, the origin of the positive phase of ionospheric storms, are made.

2. EARLY HISTORY OF IONOSPHERIC STORM RESEARCH

The beginning of ionospheric storm research can be traced back to the early 1920s when experiments in one-way radio telephone transmissions were conducted from Rocky Point in the United States to New Southgate in England. During these experiments, a definite correlation was found between abnormal radio transmissions and disturbances of the Earth's magnetic field [*Espenschied et al.*, 1925a, 1925b]. The observed effect was a significant decrease in the nighttime signal strength and a moderate increase in the daytime value (see Figure 2). Subsequent studies confirmed this effect, and early publications in this field are listed in Table 1. Note that in some of these publications no reference is made to the Kennelly–Heaviside layer, as the ionosphere was called at that time. Evidently, the existence of such a layer was by no means generally accepted, especially among radio engineers.

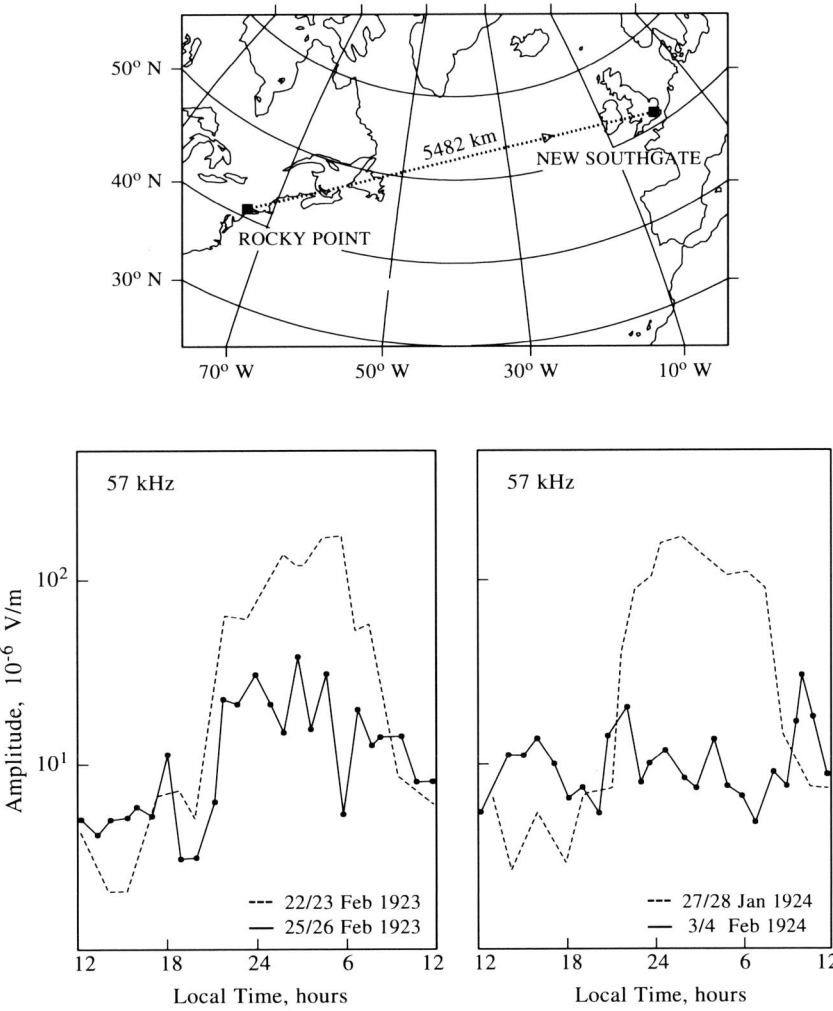

Figure 2. Disturbed radio transmissions following magnetic storms. The radio signals were transmitted from Rocky Point (Long Island) to New Southgate (near London) over a distance of 5482 km (top). The transmission frequency was 57 kHz. In the lower two panels, the received signal (electric field) strength is plotted as a function of local time at the receiving station. The lines marked by data points indicate the storm-time variations observed on 25–26 February 1923 and 3–4 February 1924, respectively. The dashed lines serve as quiet time references. [After *Espenschied et al.*, 1925a, 1925b.]

Table 1. Magnetic Activity Associated Perturbations of the Ionosphere: Early Publications

Author(s)	Year	Disturbance Effects and Their Interpretations
Espenschied et al.	1925a, 1925b	Observe abnormal transatlantic radio telephone transmissions during magnetically disturbed conditions
Maurain	1926	Does not find any correlation between magnetic activity and quality of radio communication over short distances at middle latitudes
Larmor	1926	Suggests ray path twisting as cause of communication blackout during disturbed conditions
Espenschied et al.	1926	Same as *Espenschied et al.* [1925a, 1925b]
Pickard	1927a, 1927b, 1927c	Investigates correlation of radio reception with magnetic and solar activity
Eckersley and Appleton	1927 1927	Suggest that reduction in signal strength is due to absorption of wave energy by additional ionization in the lower Kennelly–Heaviside layer produced by electrified streams of particles emitted from the sun
Sreenivasan	1927	Attributes differences in long- and short-wave disturbances to height variations of storm effects
Anderson	1928	Presents statistics of correlation between quality of radio transmissions and magnetic activity
Dahl and Gebhardt	1928	Receive no return signal from Heaviside layer during increased solar activity
Eckersley	1928	Again attributes long fades (>1/2 h) of short waves to absorption caused by storm-induced increase in ionization density
Mesny	1929	Reviews influence of solar and magnetic activity on radio wave propagation
Sreenivasan	1929	Observes no correlation between magnetic activity and long-wave radio communication at low latitudes
Wymore	1929	Investigates increase in signal strength of long-wave transmissions during daytime following magnetic activity
Anderson	1929	Documents the different responses of long- and short-wave transmissions to magnetic storms during daytime conditions
Eckersley	1929	Presents further evidence for his theory of short wave attenuation during magnetic storms
Hafstad and Tuve	1929a, 1929b	Observe increase in height of Kennelly–Heaviside layer during a magnetic storm
Maris and Hulburt	1929	Attribute disturbance of Kennelly–Heaviside layer to flash of UV radiation emitted by sun
Mögel	1930	Differentiates between solar activity associated short-duration and magnetic storm associated long-duration pertubations of short wave radio transmissions[a]
Appleton et al.	1933	Report on absorption effects and large increases in ionospheric E layer density at high latitudes during disturbed conditions
Berkner and Wells	1934	Observe decrease in F_1 layer critical frequency at magnetic equator during disturbed conditions[b]
Schafer and Goodall	1935	Document correlation between decrease in F_1 layer critical frequency and magnetic activity
Appleton and Ingram	1935	Report on both increases and decreases of the F_1 and F_2 layer critical frequencies at noon and midnight during disturbed conditions. Increases are attributed to particle precipitation, decreases to heating and subsequent expansion of the upper atmosphere
Kirby et al.	1935	Document magnetic storm associated decrease in noon time F_2 layer critical frequency and concurrent increase in layer height. Disturbance effects are attributed to thermal expansion of F layer caused by abnormal heating
Harang	1936	Studies increases in E layer and decreases in F layer densities, and absorption effects at high latitudes during disturbed conditions
Kirby et al.	1936	Investigate severe ionospheric disturbance during which the F_2 layer critical frequency fell below that of the F_1 layer (G condition). Note that ionospheric response to magnetic storms depends on onset time of the disturbance
Seaton	1936	Reports on large decrease in F layer density (including G conditions) during magnetically active period

Table 1. (continued)

Author(s)	Year	Disturbance Effects and Their Interpretations
Appleton	1937	Summarizes his results concerning the geomagnetic activity effect on the ionosphere. Considers solar eruption associated radio wave fade-outs a different phenomenon
Appleton et al.	1937	Observe inverse correlation between magnetic activity and F layer critical frequency at middle and high latitudes except during winter and/or weak magnetic activity. As before, increase in ionization density is attributed to particle precipitation, and decrease to heating and subsequent thermal expansion of upper atmosphere
Kirby et al.	1937	Note that increases in the F layer density are preferentially observed during the beginning, and decreases during the main phase of a magnetic disturbance. Largest ionospheric perturbations are observed during equinox and smallest during winter conditions
Harang	1937	Finds that noontime depressions in F layer density correlate best with magnetic activity observed during previous 24 h (delayed effect). Finds positive correlation between magnetic activity and virtual height of F layer
Gilliland et al.	1938	Document G condition during two "ionosphere storms" (First to use this latter term)
Kirby et al.	1938	Distinguish between two phases of an ionospheric storm: an initial violent phase in the auroral zone followed by a moderate phase at middle latitudes caused by the spreading out of the disturbance effects
Berkner et al.	1939	Thorough case study of intense ionospheric storm. Effects observed include depletion of F region density, increase in layer height, scatter of nighttime echoes, and seasonal variations. Attribute spread F to breakup of smooth ionospheric stratification. At equatorial latitudes, large height changes, sudden disappearance of F layer ionization density, as well as absorption effects are observed during nighttime
Berkner and Seaton	1940a	Document sudden disappearance of daytime F_2 layer at equatorial latitudes during another severe storm event. At middle latitudes and almost simultaneously, a large depression in the electron density and an increase in layer height are observed in the nighttime ionosphere
Berkner and Seaton	1940b	Demonstrate statistically that at equatorial latitudes, the peak electron density increases with increasing geomagnetic activity, independent of season. At middle latitudes, the peak electron density decreases with increasing magnetic activity except during winter, when an increase in density is observed during moderately disturbed conditions
Eckersley	1942	Electric fields ($E \times B$ drifts) caused by charge separation in solar particle streams penetrating into the upper atmosphere are made responsible for the depression in ionization density observed during magnetic storms
Menzel and Salisbury	1948	Consider heating of the upper atmosphere by solar radio waves a possible explanation for some phases of ionospheric storms
Yumura	1948	Reports that the correlation between magnetic activity and ionospheric perturbations depends on season and local time
Nagata et al.	1949	Find that the correlation between changes in the H component of the geomagnetic field and F layer density perturbations is positive in summer and negative in winter. Suggests that ionospheric disturbance effects propagate from high to low latitudes
Fukushima	1950	Notes that a more extensive study does not support his disturbance propagation hypothesis
Appleton and Piggott	1950	Derive mean variation of noontime F layer critical frequency at Slough during disturbed conditions; observe initial positive and subsequent negative phase; and emphasize differences in the ionospheric disturbance effects recorded at high, middle, and low latitudes. Also discuss longitudinal and local time variations
Burkard	1950	First global synopsis of an ionospheric storm. Documents latitudinal variations of disturbance effects. Speculates that negative and positive storm phases are due to abnormally low and abnormally high solar particle precipitation, respectively
Appleton	1950	Super-posed epoch study of mean ionospheric storm behavior at Slough. Observe time delay between magnetic and ionospheric storms. Report on dependence of disturbance effects on latitude, longitude, and local time

[a] Last study on magnetic storm associated perturbations of radio communication considered in this compilation.
[b] Beginning with this reference, only studies dealing with F layer storm effects are considered.

Also note that at a transmission frequency of 57 kHz, radio waves are reflected (and absorbed) in the ionospheric D region. Accordingly, the storm effects documented in Figure 2 refer to the lowermost ionosphere.

The first to present direct evidence for magnetic storm associated perturbations of the ionospheric F layer were *Hafstad and Tuve* [1929]. As illustrated in Figure 3, these authors observed a marked increase in the (virtual) reflection height. Later, it was discovered that not only the height but also the density of the ionosphere is affected by magnetic activity [e.g., *Appleton et al.*, 1933; *Berkner and Wells*, 1934; *Schafer and Goodall*, 1935; *Appleton and Ingram*, 1935; *Kirby et al.*, 1935]. It was also found that these density perturbations exhibited a number of systematic changes. In fact, by the end of the 1940s the following properties of ionospheric storms had been identified:

- Both abnormal increases and decreases in the ionization density are observed.
- The magnitude of these effects depends on the strength of the magnetic activity.
- Perturbations are much larger in the F_2 layer than in the F_1 layer. During severe storms, the maximum F_2 layer density may even drop below that of the F_1 layer (G condition).
- Decreases in the ionization density predominate at higher latitudes, and increases at equatorial latitudes.
- At middle latitudes, increases in the ionization density are primarily observed in winter and/or during periods of weaker magnetic activity. During equinox and summer conditions, and especially during stronger magnetic storms, decreases in the ionization density are the dominant feature.

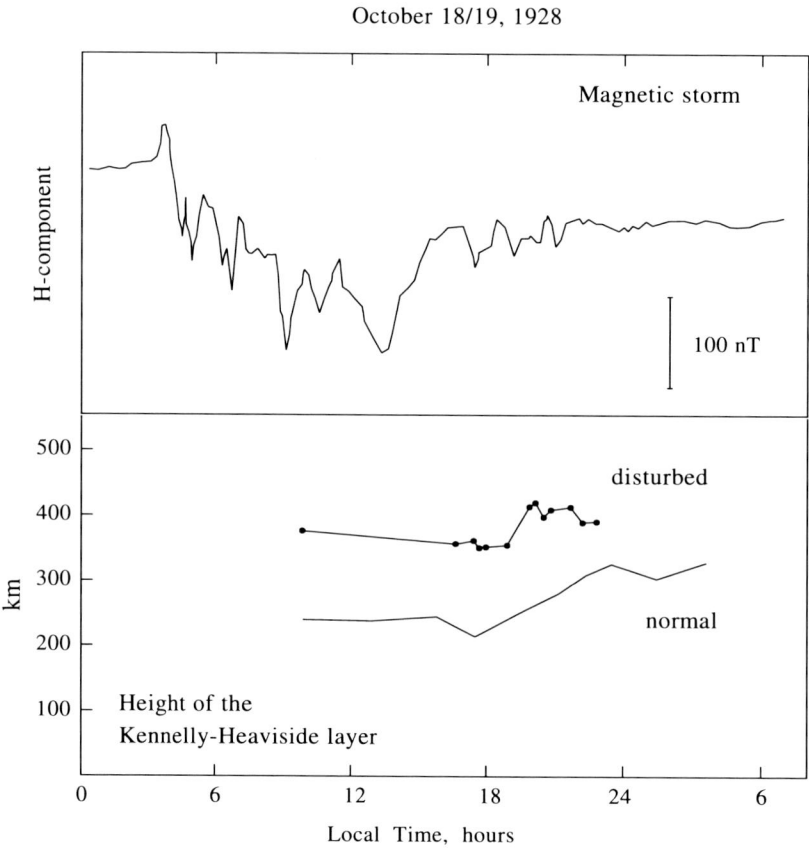

Figure 3. Increase in the virtual height of the Kennelly–Heaviside layer during a magnetic storm. The upper part shows the horizontal-intensity magnetogram obtained at Cheltenham, MD, on 18–19 October 1928. The height of the "equivalent reflecting layer" at 4.435 MHz observed on 18 October is indicated by the line marked by data points. Measurements obtained on 7–8 October serve as a quiet time reference (line without data points; after *Hafstad and Tuve* [1929a, 1929b]).

- The type of ionospheric disturbance effect also depends on the local time of the magnetic storm onset.
- Both increases and decreases in the ionization density are accompanied by an increase in the apparent layer height.
- During large magnetic storms, dramatic but brief increases in layer height and decreases in ionization density are observed. This is especially true for the equatorial ionosphere, where veritable holes in the density may be created.
- At middle latitudes, the nighttime ionosphere loses its smooth stratification and becomes rather irregular due to dynamic effects.
- At higher latitudes, even weak magnetic activity causes blanketing and absorption, rendering a continuous monitoring of the F layer impossible.
- At times, disturbance effects appear to propagate from high to low latitudes.

Today, these early publications are only of historical interest. This is because most of them are based on rather limited data sets and also use somewhat crude analysis methods. The very careful and instructive interpretations of storm ionograms presented by *Berkner et al.* [1939] and *Berkner and Seaton* [1940a] are exceptions that are still worth being studied.

As for the early theories of ionospheric storms, some do have a somewhat exotic ring. This should not come as a surprise since so little was known about the physics of the ionosphere at that time. For example, people were still debating whether ionization is formed by radiation or particle precipitation. Also, little was known about the composition and density of the ionizable gases of the upper atmosphere. Moreover, there was no agreement on whether the ionosphere is made up of electrons and positive ions or negative and positive ions. In addition, theoreticians had to struggle with the complexity and diversity of the observations. Radio transmissions were affected differently at long and short waves, during day and night, and at high and low latitudes. Later, it was discovered that the ionization density both increases and decreases during disturbed conditions. And then there were the widely differing time scales to be considered. Whereas some fadings lasted for an hour or less, others continued for days.

In this situation, it is surprising that the explanation offered for the first documented ionospheric storm effects remains valid up to the present. In 1927, Eckersley and Appleton independently suggested that the reduction in signal strength documented in Figure 2 is caused by the absorption of wave energy by additional ionization in the lower ionosphere. This additional ionization was thought to be produced by electrified streams of particles emitted from the Sun. If, besides the Sun, the magnetosphere is considered a source of precipitating particles, the above explanation is entirely correct.

Another hypothesis that should be mentioned here is the so-called thermal expansion theory. In 1935, Appleton and Ingram and Kirby et al. independently suggested that the negative phase of ionospheric storms is caused by the thermal expansion of the upper atmosphere. Thus, for a production-loss equilibrium situation, the maximum ionization density of the F_2 layer was found to be proportional to $1/\sqrt{T}$ or $1/T$, depending on the form of the loss term used in the calculation. Here, T denotes the neutral gas temperature. If it is assumed that the upper atmosphere is heated during geomagnetic storms, this would explain the observed decrease of the ionization density. The observed increase in layer height (see Figure 3) and the seasonal anomaly of the F_2 layer density seemingly supported this disturbance scenario. Accordingly, the thermal expansion theory was the preferred explanation for the negative phase of ionospheric storms during the next two decades. In fact, even after this explanation was shown to be incomplete at best [e.g., *Lepechinsky*, 1951; *Martyn*, 1953; *Maeda and Sato*, 1959], it was revived several times when new upper atmospheric heating mechanisms such as magnetohydrodynamic waves [*Dessler*, 1969; *Matuura*, 1963, 1972] or heat conduction waves [*Volland*, 1967] were discussed in the literature.

3. PUBLICATION STATISTICS

In the early 1950s, the number of publications on ionospheric storms increased significantly; see Figure 4. This histogram is based on more than 800 publications on ionospheric F region storms found in various journals and books. For a complete listing of these articles, see the CD-ROM accompanying this volume. Note that not all of these publications deal exclusively with ionospheric storm effects. In fact, in some cases only short subsections are dedicated to this subject. Accordingly, these papers contribute only partly to the publication statistics. This explains why the number of papers shown in the histogram is less than the total number of publications it is based upon.

As it turns out, almost 80% of these papers deal with the morphology of ionospheric storms, although the number of theoretical studies steadily increased over the years (see the dotted line in Figure 4). Also, most of these studies deal with storm effects at middle latitudes, and only relatively few with disturbance effects at low (ca. 12%) or high (ca. 9%) latitudes.

A prominent feature of the histogram shown in Figure 4 is the pronounced fluctuations in the publication rate (see also Figure 30 of *Mendillo* [2006]). Up to now, three major peaks can be identified, each with a different background. For example, the increase in the early 1950s is mainly due to the diligence of Japanese authors, who contributed more

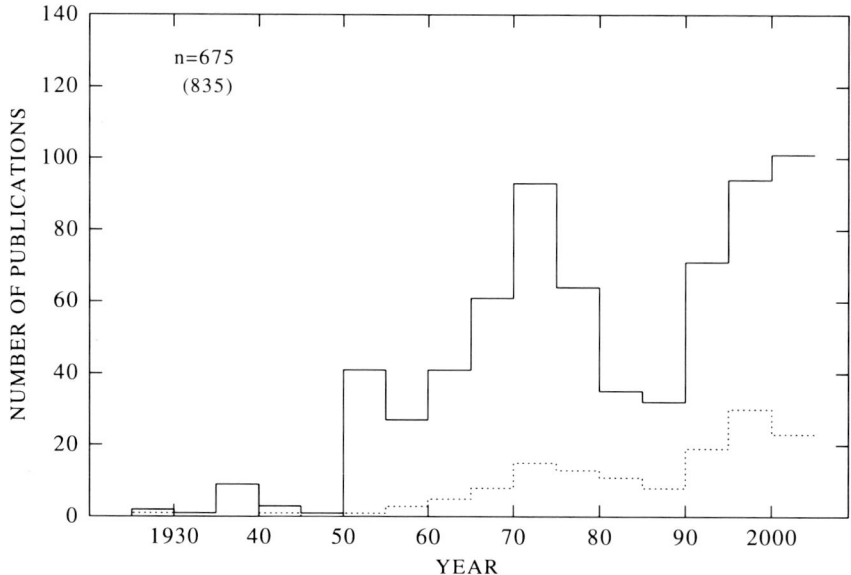

Figure 4. Number of papers on ionospheric F region storms published within each 5-year interval between 1925 and 2004. This histogram is based on a total of 835 publications that appeared in various journals and books. A complete listing of all these papers can be found in the Appendix (see the CD-ROM accompanying this volume). Note that this reference list also contains publications that only partly deal with ionospheric storm effects. In this case, only a certain percentage of these papers contributes to the histogram. This explains why the number of papers shown in the histogram (n = 675) is less than the total number of publications it is based upon (n = 835). The dotted line indicates the number of theoretical studies (including storm simulations).

than half of the papers published within this time interval. An exciting new topic was the so-called electron drift theory introduced by *Martyn* [1951, 1953]. This theory assumes that the electric field associated with the intense current system set up in the auroral zone during geomagnetic storms spreads all over the world. Together with the geomagnetic field, it causes up- and downward directed $\vec{E} \times \vec{B}$ drifts, which are made responsible for negative and positive ionospheric storms, respectively. A variant of this theory assumes the electric fields to be produced locally in the dynamo region of the upper atmosphere by storm-induced winds [*Maeda*, 1953]. A quantitative assessment of these ideas proved to be difficult, and some of the assumptions made at that time are no longer acceptable. Nevertheless, some elements of these mechanisms have survived up to the present day and are important ingredients of modern storm theories (see section 5).

Another disturbance mechanism first discussed in the 1950s is the so-called composition change theory. Prompted by *Lepechinsky* [1951], who found it necessary to invoke increased recombination to explain larger negative ionospheric storms, *Seaton* [1956] suggested storm-induced turbulence and mixing in the lower ionosphere. This would increase the molecular oxygen abundance and therefore the ionization loss rate in the F region. Again, some aspects of this disturbance mechanism are still important and are part of modern storm theories.

In the early 1970s, a second peak in the publication rate is observed. This may partly be attributed to the increased availability of satellite-supported measurements. In particular, measurements of the TEC of the ionosphere based on radio transmissions from geostationary satellites greatly contributed to this surge in publication rate. In this context, the following question arises, "What is the relationship between storm-induced changes in the TEC of the ionosphere and storm-induced changes in the maximum electron density of the F_2 layer, the key parameter of previous storm studies?" A first answer to this question is given in Figure 1. Thus, storm-induced changes in both quantities look very much alike. A quantitative confirmation of this close correlation is presented in Figure 5.

Storm-induced changes in the total electron content (dTEC) are plotted as a function of the concurrently observed changes in the maximum electron density dNmax. The data used in this figure were taken from the excellent atlas of ionospheric storms compiled by *Mendillo and Klobuchar* [1974]. To avoid overcrowding the correlogram with more than 6000 data points, the following procedure was adopted. For each 10%-wide interval of dNmax, the median and the

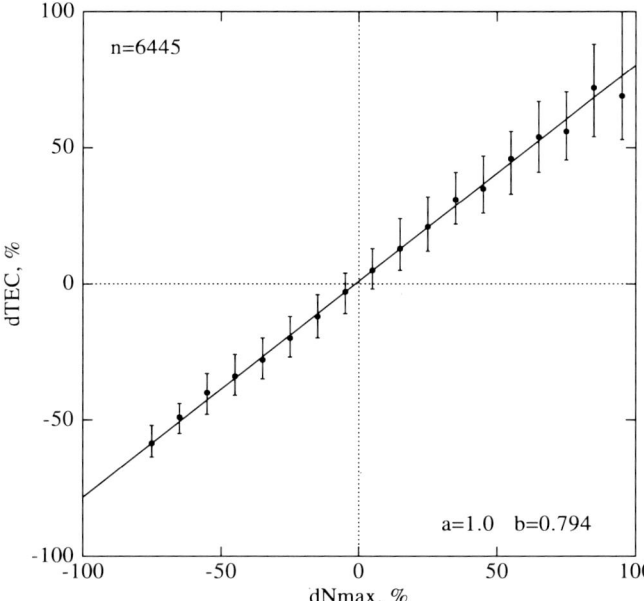

Figure 5. Correlation between storm-induced changes in the maximum electron density of the F_2 layer (dNmax) and the associated changes in the TEC of the ionosphere (dTEC). These parameters were taken from the atlas of ionospheric storms compiled by *Mendillo and Klobuchar* [1974]. Sixty-nine storm periods, each lasting 5 days, were selected for the present study. This resulted in a total of 6445 pairs of hourly values of dNmax and dTEC. To document their correlation, all dTEC values were sorted into 10%-wide intervals of dNmax. For each of these intervals, the median and the upper and lower quartiles of the dTEC values were determined. They are indicated by the dots and bars. To emphasize the close linear correlation between dNmax and dTEC, a linear regression line was fitted to the medians. The associated line parameters are given in the lower right-hand corner.

upper and lower quartiles of the associated dTEC values were determined. They are indicated by the dots and bars. As can be seen, an excellent linear relationship exists between dNmax and dTEC, and this is emphasized by the regression line fitted to the median values. Accordingly, storm studies based on either quantity should produce the same results, at least in a statistical sense. It is also clear that the storm-induced changes in the maximum ionization density are somewhat larger than those in the TEC, on average by about 20%. If we distinguish between day- and nighttime data, a tighter correlation is observed during daytime. The corresponding correlation coefficients are $r_{day} = 0.86$ and $r_{night} = 0.82$.

Returning once more to the publication statistics shown in Figure 4, we note that the most recent surge in the publication rate is based on at least three different trends. First, space weather research activities in general; second, TEC measurements based on GPS; and third, an unprecedented number of exceptionally large geomagnetic storms. To illustrate this last point, the storm classification scheme shown in Table 2 is used.

This storm classification scheme is based on a total of 1085 geomagnetic storms identified within the time interval 1957–1993 [*Loewe and Prölss*, 1997]. Here, the minimum *Dst* index observed during each storm serves as a convenient indicator of the storm intensity. Based on the distribution of these storm intensities, five different storm classes are introduced. For example, class 1 storms are weak storms with a minimum *Dst* index between −30 and −50 nT. And class 5 storms are superstorms with a minimum *Dst* index of less than −350 nT.

How many of these superstorms have been observed in recent times? Between 1957 and 1988, that is, within a 32-year time interval, only four of these superstorms occurred. However, within the much shorter time interval between 1989 and 2004 (16-year time interval), no less than six superstorms were observed. This corresponds to a factor 3 increase in the occurrence rate. And, of course, many people feel inspired to report on unusual observations made during such events, which in turn increases the number of publications.

Faced with the large number of papers on ionospheric F region storms, a newcomer in this field must feel overwhelmed. Fortunately, numerous review articles on various aspects of this phenomenon are available to help in this situation. Representative examples include those by *Maeda and Sato* [1959], *Obayashi* [1964], *Matuura* [1972], *Rishbeth* [1975], *Prölss* [1995], *Fuller-Rowell et al.* [1997], *Buonsanto* [1999], *Mikhailov* [2000], *Förster and Jakowski* [2000], *Danilov and Lastovicka* [2001], *Abdu et al.* [2006], *Mendillo* [2006], and *Burns et al.* [2007].

4. DEFINING MIDDLE LATITUDES

How do we define middle latitudes? Answering this seemingly simple question turns out to be more difficult than expected. In this work, we approach this topic by first considering some data.

Table 2. Classification of Geomagnetic Storms

Storm Class	(*Dst*)min, Range	Percentage	Designation
1	−30 to −50	44%	weak
2	−50 to −100	32%	moderate
3	−100 to −200	19%	strong
4	−200 to −350	4%	very strong
5	<−350	<1%	great

This classification scheme is based on the intensity distribution of 1085 geomagnetic storms identified within the time interval 1957–1993. Here, the minimum *Dst* index serves as an indicator of the storm intensity [*Loewe and Prölss*, 1997].

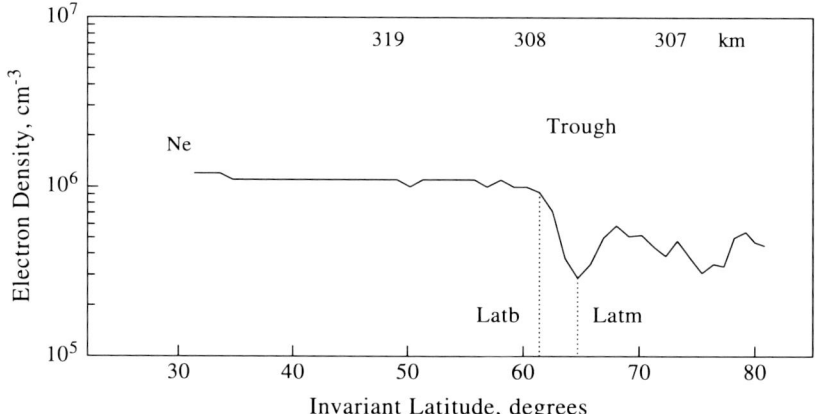

Figure 6. Electron density variation at middle and subauroral latitudes. The data were obtained by the DE 2 satellite in the northern hemisphere on 9 December 1981 at about 0736 UT. Solar and magnetic local times were approximately 1800 and 1700, respectively. The observation heights are indicated at the top of the figure. Vertical dotted lines mark the equatorward (*latb*) and poleward (*latm*) boundaries of the density drop in the trough wall.

In Figure 6, we have plotted the ionospheric electron density as a function of invariant latitude. These data were obtained in the evening sector of the northern hemisphere during moderately disturbed conditions. As can be seen, the electron density varies rather smoothly up to about 60° invariant latitude. There, a sudden drop in density is observed. This density drop is, of course, part of the ionospheric trough [e.g., *Rodger*, this volume]. Nowadays, most people call this a "mid-latitude trough," implying that it is a mid-latitude phenomenon and also that middle latitudes extend all the way to the auroral oval. Here, we prefer to call it a "subauroral trough," implying that we consider subauroral latitudes a separate region. Among other things, this region is characterized by phenomena such as density troughs, electron temperature peaks, and polarization streams. If mapped into the magnetosphere, it corresponds to the plasmasphere–plasmasheet transition region, which also contains the plasmaspheric boundary layer. If this partition is accepted, middle latitudes extend only up to subauroral latitudes. Accordingly, the poleward boundary of middle latitudes is fixed by the equatorward boundary of subauroral phenomena. Here, the equatorward edge of the subauroral trough is used to determine the approximate location of this boundary.

In a first step, the latitude *latb*, at which the density begins to decrease, and the latitude *latm*, at which the first density minimum is observed, were determined for a larger number of density profiles (see, for example, Figure 6). The latitude range in between then defines the location of the equatorward trough wall. This location depends on local time and the level of geomagnetic activity, as is illustrated in Figure 7.

The locations of *latb*, *latm*, and the trough wall in between are plotted as functions of magnetic local time for two levels

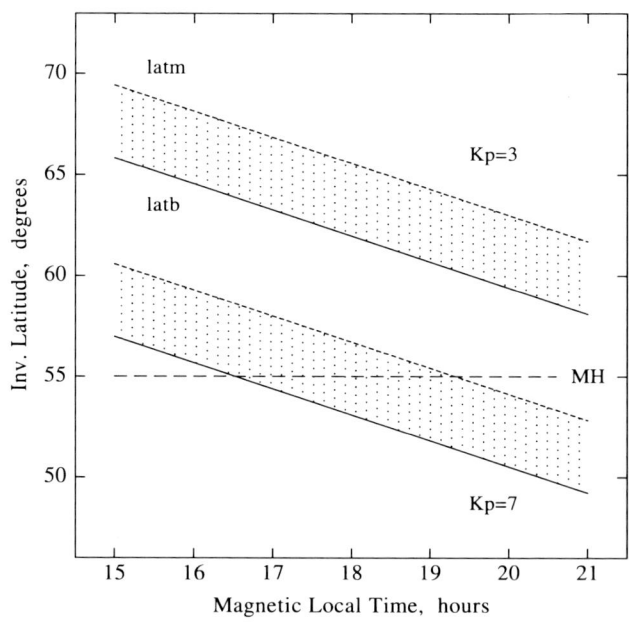

Figure 7. Poleward boundary of the mid-latitude region as a function of local time and geomagnetic activity. The beginning of the trough-associated density drop (*latb*, solid lines) serves as an indicator of this boundary. Also shown is the location of the first density minimum within the trough (*latm*, dashed lines). The shaded area in between indicates the location of the equatorward trough wall. This plot is based on results presented by *Prölss* [2007]. Here, *Kp* indices of 3 and 7 correspond to *AE6* indices of about 270 and 760 nT, respectively. Also indicated is the location of the incoherent backscatter facility Millstone Hill (MH). To be consistent with the trough data presented in Figures 7 and 8, this location refers to the early 1980s. Currently, the invariant latitude of Millstone Hill at *F* region heights is closer to 53–54°.

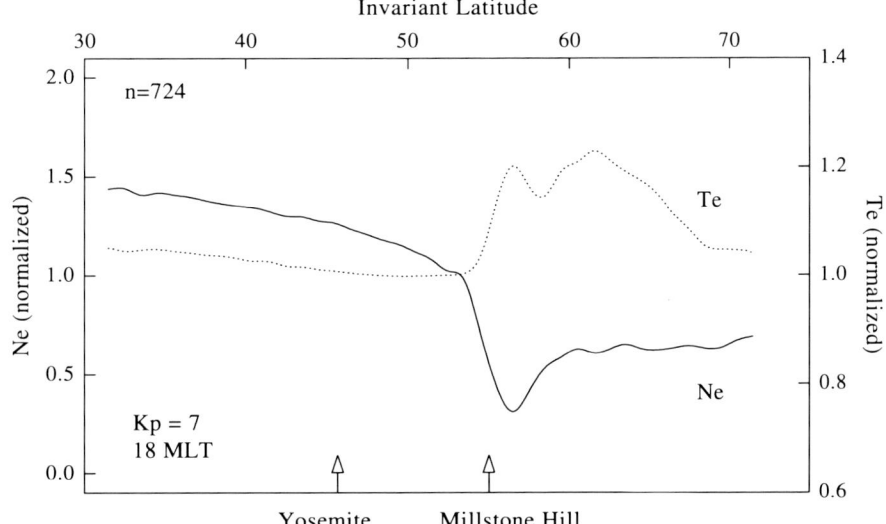

Figure 8. Mean latitudinal profiles of the electron density (solid line) and the electron temperature (dotted line) in the neighborhood of the equatorward trough wall. Plotted are the medians of 724 density and temperature profiles. These profiles have been superimposed in such a way that the latitude of the inner boundary of the trough wall *latm* serves as a common reference location. In addition, all density and temperature profiles have been normalized to the respective density and temperature values observed at the equatorward boundary of the trough wall, *latb*. The latitudes given on the topside abscissa refer to a *Kp* index of 7 and a magnetic local time of 1800. Also indicated are the locations of Millstone Hill and Yosemite. [After *Prölss*, 2007.]

of geomagnetic activity. Within the local time sector considered (1500–2100 magnetic local time), all quantities move equatorward with progressing time in a quasi-linear fashion. Also, all quantities are significantly displaced toward lower latitudes during disturbed conditions. Therefore, a typical observing station such as Millstone Hill is located at middle latitudes during moderately disturbed conditions, but becomes a subauroral station during storm conditions, at least in the evening sector. This is illustrated once more in a different manner in Figure 8.

Here, the mean latitudinal profiles of the subauroral trough and electron temperature enhancement are shown. The invariant latitude is indicated on the upper abscissa and refers to a *Kp* index of 7 and a magnetic local time of 18 h. Under these conditions, Millstone Hill is located right beneath the electron temperature enhancement and equatorward trough wall. Yosemite, on the other hand, remains a mid-latitude location.

On the equatorward side, middle latitudes are limited by the equatorial anomaly. Here, the footpoints of the anomaly crests may be used to define such an equatorward boundary. An interesting situation arises during the initial phase of geomagnetic superstorms. Because of the equatorward motion of its poleward boundary and the poleward motion of its equatorward boundary, the mid-latitude region will shrink considerably and may even disappear altogether.

Alternatively, middle latitudes may be defined as a region that is not affected by subauroral or equatorial phenomena, at least during quiet or moderately disturbed conditions (e.g., 25–55° invariant latitude). In this case, storm effects at middle latitudes contain two components. The first is due to the displacement of subauroral and auroral phenomena toward lower latitudes; and the poleward expansion of the equatorial anomaly. The second is due to the propagation of disturbance effects from high to low latitudes, and includes changes induced by prompt penetration electric fields and traveling atmospheric disturbances.

5. POSITIVE IONOSPHERIC STORMS

Whereas the origin of the negative phase of ionospheric storms is thought to be well understood [see, e.g., *Prölss and Werner*, 2002], this is by no means the case for the positive phase. In fact, people have been trying to understand this latter phenomenon for more than 70 years without arriving at a generally accepted explanation. In this context, the following observation is of special significance.

With the *AE* index serving as an indicator, the upper panel of Figure 9 shows an isolated burst of substorm activity. In response to this activity and with a certain time delay, a positive ionospheric storm develops at middle latitudes (bottom panel). The crucial point is that the increase in the ionization

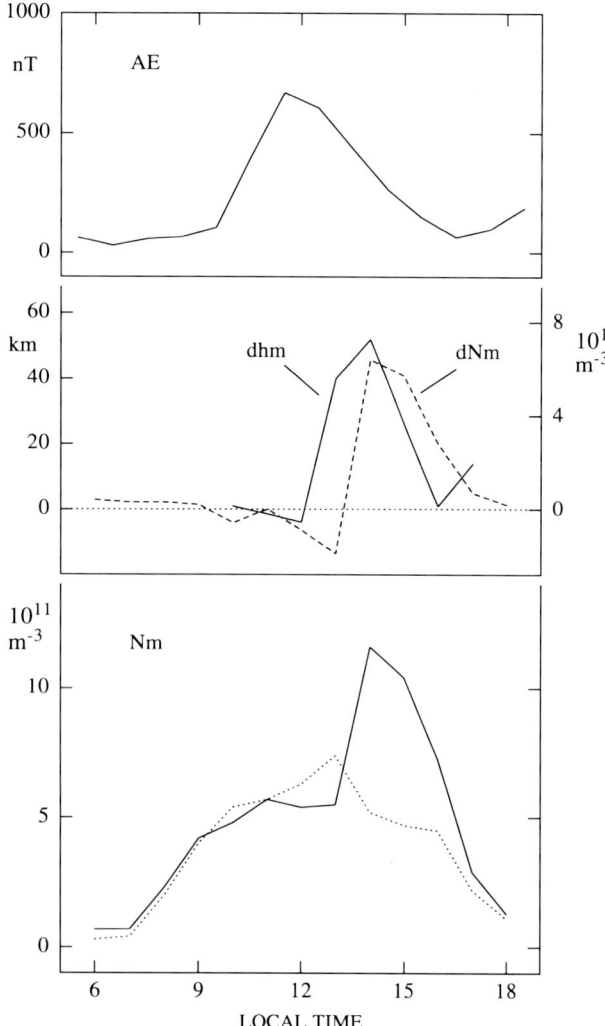

Figure 9. Short-duration positive ionospheric storm in the local afternoon sector. In response to an isolated burst of substorm activity on 23 January 1973 (hourly averaged *AE* index, top), the ionosonde at Slough (50° invariant magnetic latitude) first observes an impulse-like uplifting of the F_2 layer (dhm, middle), and subsequently an impulse-like increase in the ionization density (dNm, middle; Nm, bottom). Data recorded on 22 January serve as a quiet-time reference (dotted lines). [From *Prölss*, 1993.]

directed electric fields, and advection of high-density plasma. All these mechanisms are probably important and will contribute to the positive phase of ionospheric storms. However, neither the first mechanism (neutral composition changes) nor the fourth mechanism (advection) can be the dominant one since they do not explain the observed increase in layer height. On the other hand, mechanisms two and three are actually based on an increase in layer height. This is illustrated in Figure 11.

In the case of an equatorward-directed wind, the ions and electrons will feel a frictional force. Now, charged particles can only move freely along and parallel to the geomagnetic field. Accordingly, the field-aligned component of the frictional force will push the ionization up the inclined magnetic field lines. Obviously, this motion results in an uplifting of the F_2 layer, as is observed. That an increase in layer height will lead to an increase in the ionization density (at least during daytime) is well understood and has to do with the reduced losses at higher altitudes.

In the case of the electric field mechanism, the height increase is caused by an $\vec{E} \times \vec{B}$ drift. Since this drift is perpendicular to the inclined geomagnetic field, it also leads to an uplifting of the ionosphere. At the same time, a poleward drift is observed, in contrast to the wind mechanism that is associated with an equatorward-directed drift component.

Clearly, both these mechanisms are important. Remember that electrons and ions are but trace constituents of the upper atmosphere. Thus, whatever the much denser neutral atmosphere does during disturbed conditions, the ions and electrons will be affected. On the other hand, field-perpendicular drifts are observed, especially at higher and equatorial latitudes. This shows that electric fields must play an important role. Therefore, the question is not whether positive ionospheric storms are caused by winds *or* electric fields. Rather, the question should be, "Which of these two mechanisms is more important, especially at middle latitudes?" Here, opinions are divided. Good examples for this difference in opinion are the two review papers by *Prölss* [1995] and *Mendillo* [2006]. Whereas the former emphasizes wind-induced changes, the latter strongly favors the electric field mechanism. Note that the first author to contrast these two mechanisms was *Evans* [1970].

Even if wind-induced changes are preferred, one is still faced with the question as to the origin of these winds. Are they primarily caused by changes in the large-scale circulation, as first suggested by *Jones and Rishbeth* [1971]? Or are they part of so-called traveling atmospheric disturbances (TADs), as first suggested, implicitly by *Roble et al.* [1978] and explicitly by *Prölss and Jung* [1978]? In the first case, the whole thermosphere is in motion; in the second case, a perturbation is moving through a thermosphere more or less at rest.

density is preceded by a significant increase in the height of the F_2 layer (see the center panel of Figure 9). This prior uplifting of the ionosphere is typical and is almost always observed. Therefore, any explanation of positive ionospheric storms must be consistent with this observation.

In the second column of Figure 10, four of the more popular of these explanations are listed. They include changes in the neutral gas composition, equatorward winds, eastward-

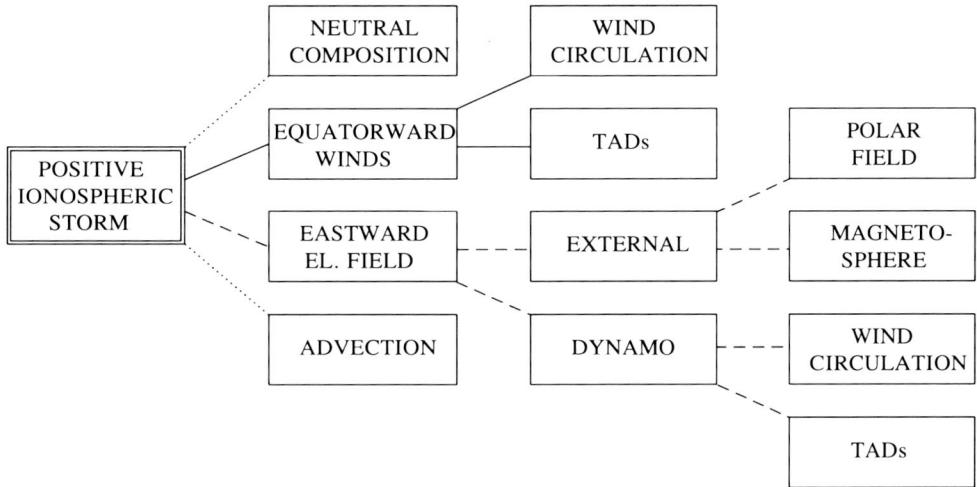

Figure 10. Mechanisms contributing to the positive phase of ionospheric storms at middle latitudes.

Things become even more complicated if the electric field mechanism is preferred. Is the electric field imposed on the ionosphere from the outside, or is it produced internally in the dynamo region of the upper atmosphere? And, if one favors external electric fields, do these propagate from polar to middle latitudes or are they created within the inner magnetosphere? Finally, if dynamo electric fields are considered more important, are they generated by changes in the large-scale wind circulation or by TADs? For a more detailed discussion of these various possibilities see, for example, the contributions by *Anghel et al.* [this volume], *Brandt et al.* [this volume], *Foster* [this volume], *Garner et al.* [this volume], *Heelis* [this volume], and *Kikuchi et al.* [this volume].

Evidently, there are many questions and few answers. Missing are suitable measurements that allow us to single out the correct explanation(s). Therefore, progress in this

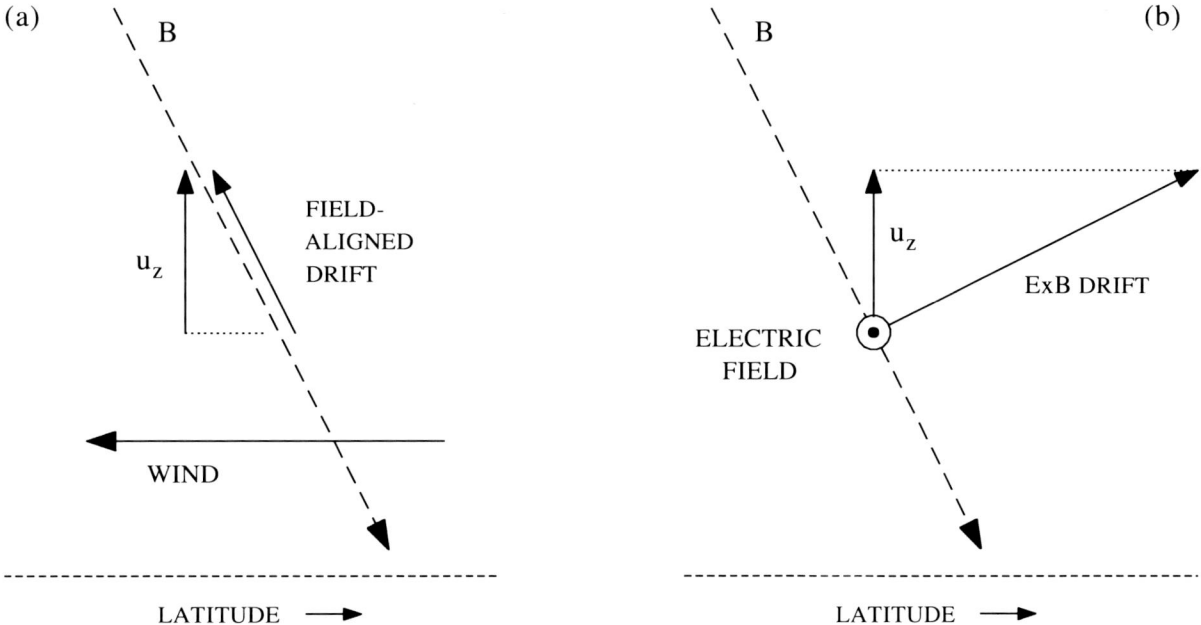

Figure 11. Uplifting of the ionospheric *F* layer by (a) an equatorward directed wind and (b) by an eastward directed electric field. The situation illustrated applies to the northern hemisphere at a magnetic latitude of 45°. B is the Earth's magnetic field, and u_z is the upward directed component of the drift velocity.

field will critically depend on better and more comprehensive data sets. Thus, positive ionospheric storms represent a real challenge for experimentalists. At the same time, theoretical simulation studies such as those described in the companion papers by *Crowley and Meier* [this volume], *Fedrizzi et al.* [this volume], *Fuller-Rowell et al.* [this volume], *Lu et al.* [this volume], *Sojka et al.* [this volume], and *Verkhoglyadova et al.* [this volume] should help to uncover the complex chain of events leading to the observed disturbance effects.

Acknowledgments. I thank the meeting organizers for the opportunity to prepare this contribution. I am also grateful to M. Hanussek, K. Schrüfer, and M. Weigand for their help in preparing this manuscript.

REFERENCES

Abdu, M. A., J. R. de Souza, J. H. A. Sobral, and I. S. Batista (2006), Magnetic storm associated disturbance dynamo effects in the low and equatorial latitude ionosphere, in *Recurrent Magnetic Storms: Corotating Solar Wind Streams, Geophys. Monogr. Ser.*, vol. 167, edited by B. Tsurutani et al., pp. 283–304, AGU Washington, D. C.

Anderson, C. N. (1928), Correlation of long wave transatlantic radio transmission with other factors affected by solar activity, *Proc. Inst. Radio Eng., 16*, 297–347.

Anderson, C. N. (1929), Notes on the effect of solar disturbances on transatlantic radio transmission, *Proc. Inst. Radio Eng., 17*, 1528–1535.

Anghel, A., D. Anderson, J. Chau, K. Yumoto, and A. Bhattacharyya (2008), Relating the interplanetary-induced electric fields with the low-latitude zonal electric fields under geomagnetically disturbed conditions, this volume.

Appleton, E. V. (1927), Magnetic storms and wireless transmission, *Electrician, 98*, 256–257.

Appleton, E. V. (1937), Regularities and irregularities in the ionosphere—I, *Proc. R. Soc. London, A162*, 451–479.

Appleton, E. V. (1950), Magnetic and ionospheric storms, *Arch. Met. Geophys. Biokl., A3*, 113–119.

Appleton, E. V., and L. J. Ingram (1935), Magnetic storms and upper-atmospheric ionisation, *Nature, 136*, 548–549.

Appleton, E. V., and W. R. Piggott (1950), World morphology of ionospheric storms, *Nature, 165*, 130–131.

Appleton, E. V., R. Naismith, and G. Builder (1933), Ionospheric investigations in high latitudes, *Nature, 132*, 340–341.

Appleton, E. V., R. Naismith, and L. J. Ingram (1937), British radio observations during the second international polar year 1932–33, *Philos. Trans. R. Soc, A 236*, 191–259.

Berkner, L. V., and S. L. Seaton (1940a), Ionospheric changes associated with the magnetic storm of March 24, 1940, *Terr. Magn. Atmos. Electr., 45*, 393–418.

Berkner, L. V., and S. L. Seaton (1940b), Systematic ionospheric changes associated with geomagnetic activity, *Terr. Magn. Atmos. Electr., 45*, 419–423.

Berkner, L. V., and H. W. Wells (1934), F-region ionosphere-Investigations at low latitudes, *Terr. Magn. Atmos. Electr., 39*, 215–230.

Berkner, L. V., H. W. Wells, and S. L. Seaton (1939), Ionospheric effects associated with magnetic disturbances, *Terr. Magn. Atmos. Electr., 44*, 283–311.

Blanc, M., and A. D. Richmond (1980), The ionospheric disturbance dynamo, *J. Geophys. Res., 85*, 1669–1686.

Brandt, P. C., Y. Zheng, T. S. Sotirelis, K. Oksavik, and F. J. Rich (2008), The linkage between the ring current and the ionosphere system, this volume.

Buonsanto, M. J. (1999), Ionospheric storms—A review, *Space Sci. Rev., 88*, 563–601.

Burkard, O. (1950), Studie zur weltweiten Ionosphärenstörung vom 15. März 1948 (in German with English abstract), *Arch. Met. Geophys. Biokl., A2*, 315–324.

Burns, A. G., S. C. Solomon, W. Wang, and T. L. Killeen (2007), The ionospheric and thermospheric response to CMEs: Challenges and successes, *J. Atmos. Sol. Terr. Phys., 69*, 77–85.

Crowley, G., and R. Meier (2008), Disturbed O/N_2 ratios and their transport to middle and low latitudes, this volume.

Dahl, O., and L. A. Gebhardt (1928), Measurements of the effective heights of the conducting layer and the disturbances of August 19, 1927, *Proc. Inst. Radio Eng., 16*, 290–296.

Danilov, A. D., and J. Lastovicka (2001), Effects of geomagnetic storms on the ionosphere and atmosphere, *Int. J. Geomagn. Aeron., 2*, 209–224.

Dessler, A. J. (1959), Ionospheric heating by hydromagnetic waves, *J. Geophys. Res., 64*, 397–401.

Eckersley, T. L. (1927), Short-wave wireless telegraphy, *J. Inst. Electr. Eng., London, 65*, 600–644.

Eckersley, T. L. (1928), A discussion on short wave fading, *Marconi Rev., 1*, 23–28.

Eckersley, T. L. (1929), An investigation of short waves, *J. Inst. Elec. Eng., 67*, 992–1032.

Eckersley, T. L. (1942), Holes in the ionosphere and magnetic storms, *Nature, 150*, 177.

Espenschied, L., C. N. Anderson, and A. Bailey (1925a), Transatlantic radio telephone transmission, *Bell Syst. Tech. J., 4*, 459–507.

Espenschied, L., C. N. Anderson, and A. Bailey (1925b), Transatlantic radio telephone transmission, *Electr. Commun., 4*, 7–23

Espenschied, L., C. N. Anderson, and A. Bailey (1926), Transatlantic radio telephone transmission, *Proc. Inst. Radio Eng., 14*, 7–56.

Evans, J. V. (1970), The June 1965 magnetic storm: Millstone Hill observations, *J. Atmos. Terr. Phys., 32*, 1629–1640.

Fedrizzi, M., T. J. Fuller-Rowell, N. Maruyama, M. Codrescu, and H. Khalsa (2008), Sources of F-region height changes during geomagnetic storms at mid latitudes, this volume.

Förster, M., and N. Jakowski (2000), Geomagnetic storm effects on the topside ionosphere and plasmasphere: A compact tutorial and new results, *Surv. Geophys., 21*, 47–87.

Foster, J. C. (2008), Ionospheric–magnetospheric–heliospheric coupling: Storm-time thermal plasma redistribution, this volume.

Fukushima, N., Propagation of ionospheric disturbance in F_2-layer, *Rep. Ionos. Res. Jpn.*, *4*, 47.

Fuller-Rowell, T. J., M. V. Codrescu, R. G. Roble, and A. D. Richmond (1997), How does the thermosphere and ionosphere react to a geomagnetic storm?, in *Magnetic Storms, Geophys. Monogr. Ser.*, vol. 98, edited by B. T. Tsurutani et al., pp. 203–225, AGU, Washington, D. C.

Fuller-Rowell, T. J., A. D. Richmond, and N. Maruyama (2008), Global modeling of storm-time thermospheric dynamics and electrodynamics, this volume.

Garner, T. W., G. Crowley, and R. A. Wolf (2008), Impact of the neutral wind dynamo on the development of the region 2 dynamo, this volume.

Gilliland, T. R., S. S. Kirby, and N. Smith (1938), Characteristics of the ionosphere at Washington, D. C., April 1938, *Proc. Inst. Radio Eng.*, *26*, 781–785.

Hafstad, L. R., and M. A. Tuve (1929a), Note on Kennelly-Heaviside layer observations during a magnetic storm, *Terr. Magn. Atmos. Electr.*, *34*, 39–44.

Hafstad, L. R., and M. A. Tuve (1929b), Further studies of the Kennelly-Heaviside layer by the echo-method, *Proc. Inst. Radio Eng.*, *17*, 1513–1522.

Harang, L. (1936), Änderungen der Ionisation der höchsten Atmosphärenschichten während der Nordlichter und erdmagnetischer Störungen (in German with English abstract), *Gerlands Beitr. Geophys.*, *46*, 438–454.

Harang, L. (1937), Further studies on the vertical movements of the air in the upper atmosphere, *Terr. Magn. Atmos. Electr.*, *42*, 55–72.

Heelis, R. A. (2008), Low- and middle-latitude ionospheric dynamics associated with magnetic storms, this volume.

Jones, K. L., and H. Rishbeth (1971), The origin of storm increases of mid-latitude F-layer electron concentration, *J. Atmos. Terr. Phys.*, *33*, 391–401.

Kikuchi, T., K. K. Hashimoto, and K. Nozaki (2008), Storm phase dependence of penetration of magnetospheric electric fields to mid and low latitudes, this volume.

Kirby, S. S., T. R. Gilliland, E. B. Judson, and N. Smith (1935), The ionosphere, sunspots, and magnetic storms, *Phys. Rev.*, *48*, 849.

Kirby, S. S., T. R. Gilliland, N. Smith, and S. E. Reymer (1936), The ionosphere, solar eclipse and magnetic storm, *Phys. Rev.*, *50*, 258–259.

Kirby, S. S., N. Smith, T. R. Gilliland, and S. E. Reymer (1937), The ionosphere and magnetic storms, *Phys. Rev.*, *51*, 992–993.

Kirby, S. S., N. Smith, and T. R. Gilliland (1938), The nature of the ionospheric storm, *Phys. Rev.*, *54*, 23.

Larmor, J. (1926), Magnetic storms and wireless communication, *Nature*, *118*, 662.

Lepechinsky, D. (1951), Effects of temperature variations of the upper atmosphere on the formation of ionospheric layers, *J. Atmos. Terr. Phys.*, *1*, 278–285.

Loewe, C. A., and G. W. Prölss (1997), Classification and mean behavior of magnetic storms, *J. Geophys. Res.*, *102*, 14,209–14,213.

Lu, G., L. P. Goncharenko, A. J. Coster, A. D. Richmond, R. G. Roble, N. Aponte, and L. J. Paxton (2008), A data-model comparative study of ionospheric positive storm phase in the mid-latitude F region, this volume.

Maeda, K.-I. (1953), A theory of distribution and variation of the ionospheric F_2 layer, *Rep. Ionos. Res. Jpn.*, *7*, 81–107.

Maeda, K.-I., and T. Sato (1959), The F region during magnetic storms, *Proc. Inst. Radio Eng.*, *47*, 232–239.

Maris, H. B., and E. O. Hulburt (1929), Wireless telegraphy and magnetic storms, *Proc. Inst. Radio Eng.*, *17*, 494–500

Martyn, D. F. (1951), The theory of magnetic storms and auroras, *Nature*, *167*, 92–94.

Martyn, D. F. (1953), Geo-morphology of F-region ionospheric storms, *Nature*, *171*, 14–16.

Matuura, N. (1963), Thermal effect on the ionospheric F region disturbance, *J. Radio Res. Lab. Japan*, *10*, 1–35.

Matuura, N. (1972), Theoretical models of ionospheric storms, *Space Sci. Rev.*, *13*, 124–189.

Maurain, Ch. (1926), Sur la recherche d'une correspondance entre les perturbations magnétique et les perturbations dans la propagation des ondes électromagnétiques, *Onde Electr.*, *5*, 483–487.

Mendillo, M. (2006), Storms in the ionosphere: Patterns and processes for total electron content, *Rev. Geophys.*, *44*, RG4001, doi:10.1029/2005RG000193.

Mendillo, M., and J. A. Klobuchar (1974), An atlas of the mid-latitude F-region response to geomagnetic storms, *Tech. Rep. 74-0065*, 267 pp., Air Force Cambridge Res. Lab., Cambridge, Mass.

Menzel, D. H., and W. W. Salisbury (1948), Audio-frequency radio waves from the sun, *Nature*, *161*, 91.

Mesny, R. (1929), Activité solaire et propagation, *Onde Electr.*, *8*, 103–110

Mikhailov, A. V. (2000), Ionospheric F_2-layer storms, *Fis. Tierra*, *12*, 223–262.

Mögel, H. (1930), Über die Beziehung zwischen Empfangsstörungen bei Kurzwellen und den Störungen des magnetischen Feldes der Erde, *Telefunken Z.*, *11*, 14–31.

Nagata, T., N. Fukushima, and M. Sugiura (1949), Geomagnetic disturbances and ionospheric storms (in Japanese with English abstract), *Rep. Ionos. Res. Jpn.*, *3*, 41–72.

Obayashi, T. (1964), Morphology of storms in the ionosphere, in *Research in Geophysics, Vol. 1*, edited by H. Odishaw, pp. 335–366, MIT Press, Cambridge, Mass.

Pickard, G. W. (1927a), The correlation of radio reception with solar activity and terrestrial magnetism, *Proc. Inst. Radio Eng.*, *15*, 83–97.

Pickard, G. W. (1927b), The correlation of radio reception with solar activity and terrestrial magnetism II, *Proc. Inst. Radio Eng.*, *15*, 749–766.

Pickard, G. W. (1927c), The relation of radio reception to sunspot position and area, *Proc. Inst. Radio Eng.*, *15*, 1004–1012.

Prölss, G. W. (1993), Common origin of positive ionospheric storms at middle latitudes and the geomagnetic activity effect at low latitudes, *J. Geophys. Res.*, *98*, 5981–5991.

Prölss, G. W. (1995), Ionospheric F-region storms, in *Handbook of Atmospheric Electrodynamics*, 2, edited by H. Volland, pp. 195–248, CRC Press, Boca Raton, Fla.

Prölss, G. W. (2007), The equatorward wall of subauroral troughs in the afternoon/evening sector, *Ann. Geophys.*, *25*, 645–659.

Prölss, G. W., and M. J. Jung (1978), Travelling atmospheric disturbances as a possible explanation for daytime positive storm effects of moderate duration at middle latitudes, *J. Atmos. Terr. Phys.*, *40*, 1351–1354.

Prölss, G. W., and S. Werner (2002), Vibrationally excited nitrogen and oxygen and the origin of negative ionospheric storms, *J. Geophys. Res.*, *107*(A2), 1016, doi:10.1029/2001JA900126.

Rishbeth, H. (1975), F-region storms and thermospheric circulation, *J. Atmos. Terr. Phys.*, *37*, 1055–1064.

Roble, R. G., A. D. Richmond, W. L. Oliver, and R. M. Harper (1978), Ionospheric effects of the gravity wave launched by the September 18, 1974, sudden commencement, *J. Geophys. Res.*, *83*, 999–1009.

Rodger, A. S. (2008), The mid-latitude trough—Revisited, this volume.

Schafer, J. P., and W. M. Goodall (1935), Diurnal and seasonal variations in the ionosphere during the years 1933 and 1934, *Proc. Inst. Radio Eng.*, *23*, 670–681.

Seaton, M. J. (1956), A possible explanation of the drop in F-region critical densities accompanying major ionospheric storms, *J. Atmos. Terr. Phys.*, *8*, 122–124.

Seaton, S. L. (1936), Note on ionospheric disturbance of November 3 and 4, 1936, *Terr. Magn. Atmos. Electr.*, *41*, 407.

Sojka, J. J., R. W. Schunk, D. C. Thompson, L. Scherliess, and M. David (2008), Assimilation of observations with models to better understand severe ionospheric weather at mid-latitudes, this volume.

Sreenivasan, K. (1927), Magnetic storms and wireless transmission, *Electrician*, *98*, 496.

Sreenivasan, K. (1929), On the relation between long-wave reception and certain terrestrial and solar phenomena, *Proc. Inst. Radio Eng.*, *17*, 1793–1814.

Tsurutani, B. T., A. J. Mannucci, B. A. Iijima, A. Komjathy, A. Saito, T. Tsuda, O. P. Verkhoglyadova, W. D. Gonzalez, and F. L. Guarnieri (2006), Dayside ionospheric (GPS) response to corotating solar wind streams, in *Recurrent Magnetic Storms: Corotating Solar Wind Streams*, Geophys. Monogr. Ser., vol. 167, edited by B. Tsurutani et al., pp. 245–270, AGU, Washington D. C.

Verkhoglyadova, O. P., B. T. Tsurutani, A. J. Mannucci, A. Saito, T. Araki, D. Anderson, M. Abdu, and J. H. A. Sobral (2008), Simulation of PPEF effects in dayside low-latitude ionosphere for the October 30, 2003, superstorm, this volume.

Volland, H. (1967), Heat conduction waves in the upper atmosphere, *J. Geophys. Res.*, *72*, 2831–2841.

Wymore, I. J. (1929), The relation of radio propagation to disturbances in terrestrial magnetism, *Proc. Inst. Radio Eng.*, *17*, 1206–1213.

Yumura, T. (1948), Statistical investigation on the relation between the magnetic variations and ionospheric and solar phenomena, *Rept. Ionos. Res. Jpn.*, *2*, 74–78.

G. W. Prölss, Argelander Institut für Astronomie, Universität Bonn, Auf dem Hügel 71, 53121 Bonn, Germany. (gproelss@astro.uni-bonn.de)

The Mid-Latitude Trough—Revisited

Alan Rodger

British Antarctic Survey, Cambridge, UK

The mid-latitude trough is a major feature of the *F*-region ionosphere that forms at the boundary between the mid-latitude and auroral ionospheres. The trough region has a major influence on the propagation of radio waves because of the large gradients in electron concentration, and the irregularities often embedded in the equatorward and poleward walls of the trough. The formation processes of the trough have been reasonably well understood for some time, and physically based models of the trough can now reproduce the climatology of the trough quite well. Prediction of the trough shape and dynamics for individual events, "trough weather," is still under development. Recent scientific progress on major topics affecting trough morphology is described here, together with some suggestions on how the remaining uncertainties can be addressed.

1. INTRODUCTION

The mid-latitude (or main) ionospheric trough is a region at *F*-region altitudes, typically a few degrees wide in latitude, where the plasma concentration is usually lower compared with regions immediately poleward and equatorward. It normally lies close to the equatorward boundary of the auroral precipitation. There is no agreed quantitative definition of a trough and hence rigorous intercomparisons between studies can be challenging.

Many observational, theoretical, and modeling studies of the mid-latitude trough were carried out in the 1970s and 1980s. These established the major characteristics of the trough and a good understanding of trough formation processes (see the review by *Rodger et al.* [1992]).

The mid latitude trough has major effects on the propagation of radio waves because of the strong gradients in plasma concentration, and the trough edges are often regions of high plasma irregularity occurrence being particularly sensitive to the gradient-drift instability. This is important for many practical activities, such as the accuracy of global positioning and radar observations of the Earth from space. Thus, it is important to know precisely the location of the trough and its shape. This has led to resurgence in trough studies to take the general understanding of the trough to a level where accurate prediction may be possible. This short review has three objectives: (1) to summarize the major characteristics of the trough, (2) to highlight areas of recent scientific progress, and (3) to outline the key requirements for further study that will lead to more accurate prediction.

2. MID-LATITUDE TROUGH MORPHOLOGY

The mid-latitude trough is primarily a phenomenon that occurs in darkness, and is thus seen mainly in winter and equinox and is less common in summer. As a result, the local time extent of the trough is small in summer, centered about midnight, and extends further toward dawn and dusk with progression toward winter. In some longitudes, and under the quietest geomagnetic conditions, the trough can be observed at all local times around midwinter.

During steady geomagnetic conditions, the trough moves progressively to lower latitudes with later magnetic local times (MLT), and is observed at lower latitudes at the same MLT as geomagnetic activity increases. There are several empirical equations that describe this motion reasonably well [*Dudeney et al.*, 1983]. Most formulae use the *Kp* magnetic index, which has a number of limitations, such as its 3-h cadence. An index related to the solar wind conditions might lead to improved representations. As solar activity increases, the electron concentration in the trough minimum and both edges of the trough increases.

Two illustrations (Plate 1) derived from different techniques give representations of the trough. The top panel illustrates the three major elements of a trough—the equatorward edge extending up to about 71°N, the poleward edge that lies above about 75°N, and the intervening region that constitutes the trough minimum region. In the bottom image, the trough minimum can be observed near ~73° at 1800 MLT, and at progressively lower latitudes at later times. The equatorward edge initially has a concentration exceeding 5×10^{11} m^{-3} but this diminishes with time as the station moves further into darkness. By ~2300 MLT, the equatorward edge and the trough minimum become almost indistinguishable. The poleward edge is initially outside the field of view near 2000 MLT, but by 2300 MLT it is well defined above about 67°N magnetic latitude.

3. FORMATION PROCESSES OF THE MID-LATITUDE TROUGH

3.1. Steady Geomagnetic Conditions

The equatorward edge is formed by mid-latitude plasma that decays steadily as it corotates further into darkness with time, as illustrated in Plate 1.

There are two major causes of the enhancement of plasma concentration in the poleward edge of the trough in the evening sector:

1. Local energetic particle precipitation causing additional ionization both at *E*- and *F*-region altitudes [*Rodger et al.*, 1986].
2. Transport of dayside plasma, through the cusp, polar cap, and nightside oval.

The mid-latitude trough is the low plasma concentration region between the two edges. It is where the corotation and convection electric fields are approximately oppositely directed (i.e., quasi stagnation), as illustrated in Plate 1. As a result, the plasma residence time at a particular MLT is greater than at lower latitudes. Provided the plasma is in darkness, *F*-region recombination processes occur and plasma concentrations fall to low levels.

In the morning sector, the trough minimum is the extension of the evening sector trough that is slowly convecting and/or corotating toward dawn. This convection can cause the trough to be observed slightly (~1 h) beyond the day–night terminator. The equatorward edge is formed by corotating plasma that continues to decay with time becoming progressively less distinct from the trough minimum (Plate 1, bottom panel). The poleward edge is formed by plasma that has drifted across the polar cap from the dayside ionosphere with some enhancement in the plasma concentration as the flux tubes drift through the nightside oval and then to lower latitudes. The poleward edge occurs equatorward of the auroral oval and, in this respect, is different from the evening sector trough [*Rodger et al.*, 1986].

Thermosphere–ionosphere general circulation models reproduce the steady-state conditions reasonably well, i.e., the climatology of the trough. This includes the variations that occur as a function of season, Universal Time (UT) (including the effects of the offset of the geographic and geographic poles), and the variations of the Bz and By components of the interplanetary magnetic field [e.g., *Bowline et al.*, 1996]. The UT effects lead to particularly strong seasonal variations in trough shape, longitude extent, and gradients of the equatorward and poleward walls of the trough [e.g., *Sojka and Schunk*, 1989]. The importance of convected dayside plasma into the nightside oval has also been demonstrated [e.g., *Foster et al.*, 2005], and many of these features also vary as a function of solar activity [*Sojka et al.*, 2006].

3.2. Formation Processes of the Mid-Latitude Trough— Geomagnetically Active Periods

During geomagnetically active periods, the detailed morphology of troughs becomes much more difficult to predict, although the physics remains essentially the same. The basic differences are that the cross-polar cap electric field is usually increasing, and the auroral oval is expanding. The fundamental problem is that the spatial and temporal variations of critical processes, particularly the electric fields, are not sufficiently well known to allow accurate modeling to be undertaken. For example, the trough forms where the balance between corotation and convection occurs, yet this continually changes both in space and time.

When the plasma velocity (**V**) exceeds about 500 m s^{-1} in the rest frame of the neutral wind (**U**), the plasma loss process becomes highly nonlinear (Figure 1) [*Schunk et al.*, 1975]. When (**V**–**U**) increases from 1 to 2 km s^{-1}, the recombination rate rises by 1 order of magnitude. The neutral wind is often assumed to be zero, but this too can lead to serious errors both in the loss rate, and in the wind-induced field-aligned plasma motion. Plasma velocity enhancements can occur in narrow regions (a few kilometers) associated with auroral arcs. Thus, deep narrow troughs can be formed in

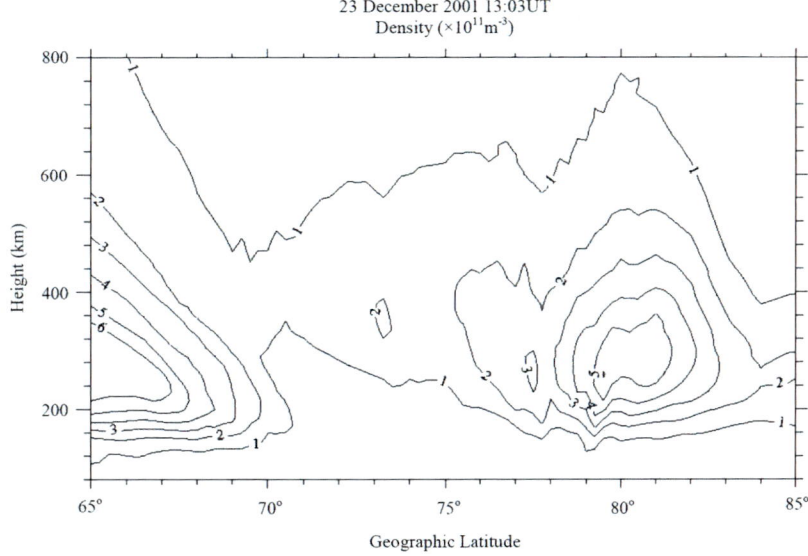

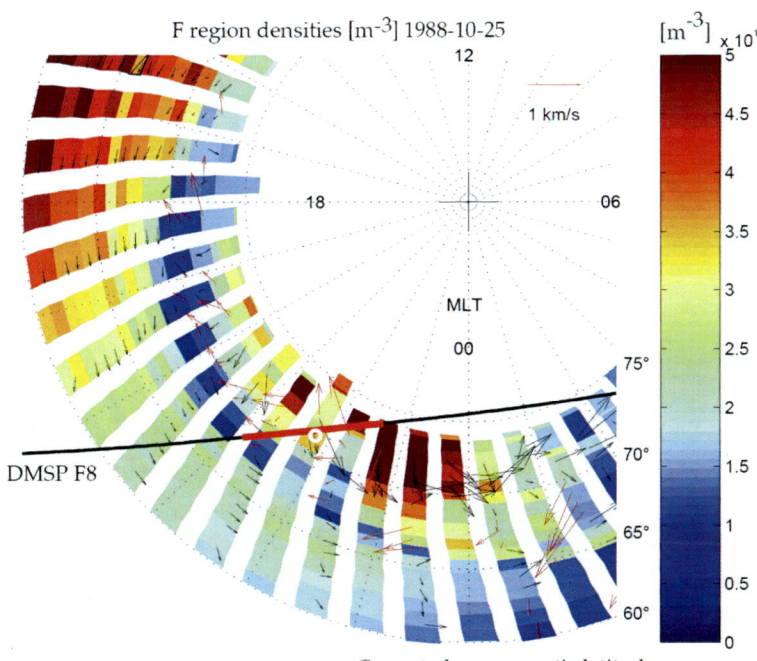

Plate 1. (top) A latitude cross section of electron concentration ($\times 10^{11}$ m^{-3}) near 15°E derived from a tomography reconstruction of a satellite pass at 1303 UT on 23 December 2001 [from *Pryse et al.*, 2005] illustrating the trough and its edges. (bottom) Maximum *F*-region electron concentration (m^{-3}) in the 300- to 500-km altitude interval from European Incoherent Scatter on 25 October 1988 using corrected geomagnetic latitude of the magnetic footpoint at 100 km and MLT. Superimposed are plasma velocity estimates with the arrow length corresponding to the velocity as 250 m s^{-1} per degree latitude length. The black line shows the track of the DMSP F8 satellite mapped along the field line to 100-km altitude. The thick red line shows the region where the satellite observed proton precipitation. The thick white circle shows the 100 km footpoint where the radar was making *F*-region observations at the time corresponding to a red line along the satellite track [from *Nilsson et al.*, 2005].

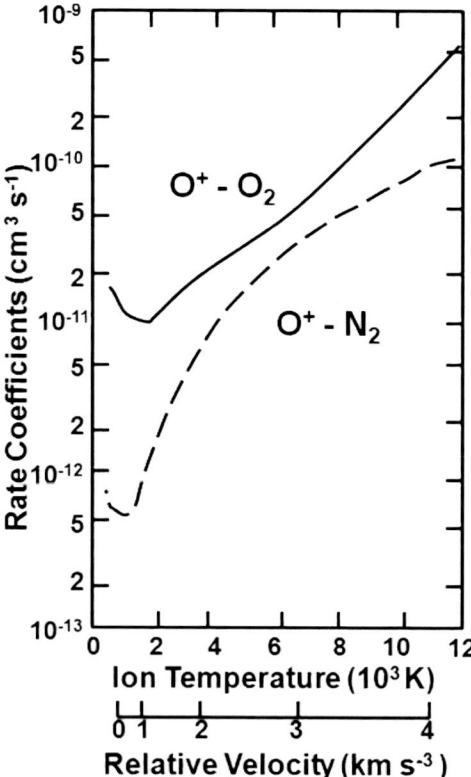

Figure 1. Rate coefficients for the two major loss reactions at F-region altitudes as a function of ion temperature and velocity in the rest frame of the neutral particles. The graph assumes that the neutral temperature is 1000 K.

minutes, which are not detectable by many types of instruments, and are subgrid-scale in general circulation models. However, they can still affect plasma concentrations and irregularity occurrence, and hence radio wave propagation.

4. KEY ISSUES

Successful prediction of the trough morphology and dynamics rely on a detailed understanding of physics and chemistry. The summary above indicates that this is in good order and as a result, the "climatology" of the trough can be predicted with moderate accuracy. The shaded boxes in Figure 2 summarize the main process responsible for the formation of the three components of the trough and its edges. However, predicting the "weather" of the trough region is considerably less secure. The fundamental reasons for this are that sufficiently realistic treatments of the energy input, boundary conditions, and parameterization of subgrid-scale processes are missing from trough models. These are illustrated in the unshaded boxes of Figure 2, and the next sections discuss these uncertainties and how some of them may be addressed.

4.1. Energy Inputs—The Electric Field and Magnetosphere–Ionosphere Coupling

The spatial and temporal variation of electric fields in the vicinity of the trough is of fundamental importance in determining the shape and depth of the trough. A variety of terminology has evolved to describe electric field variations in the trough region. Initially, the term polarization jet [*Galperin et al.*, 1973], as detected from low-altitude satellite observations, was used. *Spiro et al.* [1979] then set a threshold for the polarization jet (a plasma velocity that exceeded 1 km s^{-1}) and termed such occasions as sub-auroral ion drift (SAID) events. *Freeman et al.* [1992] then inferred from magnetometer data that there were broader regions of enhanced flow in the evening sector and termed these substorm-associated radar auroral surges (SARAHs). Finally, *Foster and Burke* [2002], using incoherent scatter radar measurements of plasma velocity, coined the term sub-auroral polarization streams (SAPS) to describe the broad region of enhanced plasma flow equatorward of the auroral oval.

The characteristics SAIDs and polarization jets include:

1. regions of narrow westward convection just equatorward of the auroral oval, typically 1–2° wide in latitude;
2. electric fields that exceed 50 mV m^{-1};
3. observed mainly in the evening sector;
4. smaller-scale intense electric field structures sometimes embedded within SAIDs [*Mishin and Mishin*, 2007];
5. encompass up to about one third of the cross polar cap potential.

SAPs and SARAHs have the following characteristics [*Foster and Vo*, 2002]:

1. observed between ~1800 and 0300 MLT equatorward of the auroral oval;
2. 3–5° wide in latitude;
3. observed at progressively lower latitudes with increasing geomagnetic activity;
4. sometimes have SAIDs embedded within them.

SAPSs/SARAHs arise from changes in the global electrical current circuit and feedback effects. The sequence involves enhancement of the region 2 field-aligned currents (FACs) driven by ring current pressure gradients, with closure via poleward Pedersen currents in the ionosphere and region 1 currents back to the magnetosphere [*Southwood and Wolf*,

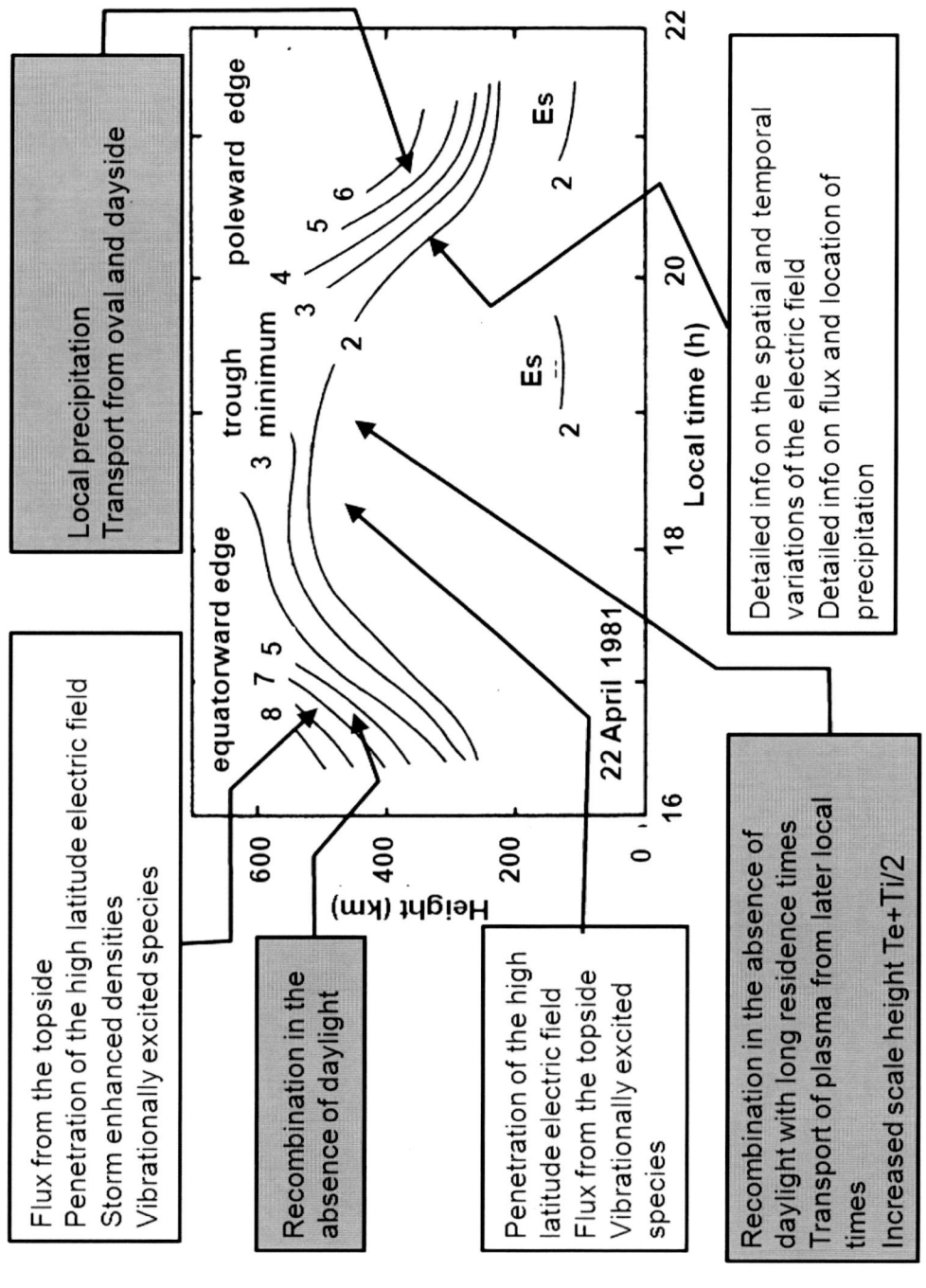

Figure 2. A schematic diagram of the mid-latitude trough in the evening sector for 22 April 1981 [after Dudeney et al., 1982]. The shaded boxes indicate the primary mechanisms for the formation of each of the three component parts of the trough. The unshaded boxes summarize some of the key processes that require further quantification for more accurate modeling of the trough.

1978]. Because of the low ionospheric conductivity at sub-auroral latitudes, the poleward Pedersen currents generate intense poleward electric fields between the region 2 FACs and the low-latitude edge of the electron aurora. The presence of the poleward-directed electric field further reduces the ionospheric conductivity [*Pintér et al.*, 2006], intensifying the field and sometimes forming SAIDs. Coupled ionosphere–magnetosphere models are beginning to reproduce some of the key properties of the electric field at the equatorward edge of the auroral oval [*Goldstein et al.*, 2005].

In the morning sector, westward drift of plasma with speeds of ~300 m s^{-1} has been observed at sub-auroral latitudes during substorm expansion phases [e.g., *Huang et al.*, 2001]. The cause of this additional poleward-directed electric field has not been uniquely identified. Suggestions include the effects of an ionospheric dynamo [*Kamide et al.*, 1994], a polarization electric field [*Huang et al.*, 2001], or a feature associated with the eastward end of the substorm current wedge [*Liang et al.*, 2006].

Large N–S oscillations of the electric field (~30 mV m^{-1}) within SAIDs have recently been identified with scales ~10s of kilometers. Their frequencies are of the order 0.1–0.3 Hz [e.g., *Mishin et al.*, 2003] and are associated with soft (30 keV) proton precipitation. Such waves are not only important for energy deposition in the thermosphere, but are a potential driving mechanism for plasma irregularities in the trough [*Keskinen et al.*, 2004].

The expansion of the cross-polar cap electric field to lower latitudes is particularly important in the formation of storm-enhanced densities (SEDs) in the afternoon sector [*Foster et al.*, 2002]. SEDs lie near the inner edge of the SAPS electric field [*Foster et al.*, 2007]. SEDs are important as high plasma concentration *F*-region mid-latitude plasma initially forms the equatorward edge of the trough region and is entrained into the high-latitude convection pattern, transported into the polar cap, and subsequently into the nightside auroral oval where it contributes to the formation of the poleward edge of the trough.

Thus the most important unknown is the spatial and temporal variation of the electric field. The precise nature of the trough morphology depends critically on the time history of the convection pattern and hence the trajectory of plasma in the sub-auroral and high-latitude ionospheres. As the lifetime of *F*-region plasma is many hours, especially in darkness, and plasma only takes 1–2 h to convect from the cusp into the nightside auroral oval, the electric field over the entire high-latitude region is required, probably with a temporal resolution of a few minutes, to be capable of accurately predicting the electron concentrations and gradients in the trough region. This is clearly not possible, and hence an empirical approach is required (see section 4.4).

4.2. Neutral Winds and Composition of the Thermosphere

The temporal variation of O, N$_2$, and O$_2$ in the thermosphere is never precisely known and hence the detailed loss rate of plasma (section 3.2) is not accurately known. This is further complicated because vibrationally excited species distributions are also not known and these enhance the loss rate. *Pavlov and Foster* [2001] have shown that inclusion of excited species can reduce the peak electron concentration by up to a factor of 3. Although a series of satellite missions have allowed climatology models of composition to be derived, there are no significant real-time measurements that can be assimilated into models.

Neutral winds are very important in field-aligned plasma transport at *F*-region altitudes. In the trough minimum region, Joule heating is not important because the ionospheric conductivity is so low; the Pedersen conductivity usually peaks well into the *F*-region. In the poleward edge of the trough, it cannot be ignored because auroral oval precipitation enhances conductivity at *E*-region altitudes and the heating results in enhanced winds blowing away from localized "hotspots."

Again, there are climatological models of neutral winds, but these have very great uncertainties when applied to specific events. Low-Earth orbit satellites with accelerometers (e.g., CHAMP) are giving new insight into the high-spatial structure of neutral winds patterns but there are insufficient real-time data for addressing trough "weather" requirements.

4.3. Plasma Heating

In the trough region, plasma scale height [proportional to $(T_i + T_e)/2$] is very important in determining the vertical distribution of the *F*-region electron concentration, and the flow of plasma into the top-side ionosphere. The topside plasma forms a reservoir that eventually cools and falls, helping to maintain *F*-region concentrations at night [*Anderson et al.*, 1998].

There are several processes that contribute to raising T_i and T_e in the vicinity of the trough. For example, the electric field is of fundamental importance in determining the ion temperature, T_i. Although the plasma velocity in the trough region is typically ~200–500 m s^{-1} giving T_i ~1000 K, within SAIDs the electric field can exceed 100 mV m^{-1}, and thus T_i can exceed 3000 K.

In the mid-latitude trough region, transfer of energy from ions to electrons enhances the electron temperature, T_e. Another contribution to T_e enhancement results from heat flux from the plasmasphere. Given the relatively low thermal heat capacity of the electron gas, and the inefficient heat conduction in the bottomside of the *F* region, T_e can be in excess of 6000 K. If the heat transfer exceeds a critical threshold, stable auroral red arcs (SARarcs) are formed. Modeling of

SARarc emissions reproduces observations well [*Baumgardner et al.*, 2007], but no consensus has been reached yet on the mechanisms that drive the heat flux downward; those proposed all depend on ring current interaction with the plasmasphere [*Gurgiolo et al.*, 2005]. As with much of geophysics, perhaps several mechanisms may be occurring at the same time, but their relative importance probably changes between events.

Energy deposition as a function of space and time during substorms is a further major unknown. Although some empirical relationships relate energy deposition in each substorm to the energy input from the solar wind at the same time, the spatial distribution of the substorm energy deposition remains unknown. Critical questions include "At what longitude does the substorm start?" and "What controls the longitude extent of the substorm current wedge?" Recent studies have demonstrated that onset is not invariably at the Harang discontinuity as previously thought [*Weygand et al.*, 2007]. The THEMIS spacecraft mission with the very comprehensive, complementary ground-based measurements should help address the problem, and hence allow more complete understanding of the poleward edge of the midlatitude trough.

4.4. Ionosphere–Plasmasphere Coupling

Statistically, the plasmapause maps close to the equatorward edge of the trough (see *Rodger et al.* [1992] for references). During storm onset, the ionospheric footprint of plasmaspheric tails have been shown to be collocated with SEDs regions [e.g., *Foster et al.*, 2002] as one might expect during the erosion of the plasmasphere. However, during storm recovery times, this close relationship breaks down as the timescale for recovery/filling of the trough is typically of the order of hours (i.e., when the trough location rotates into daylight), whereas the plasmapause near $L = 4$ can take ~10 days to refill.

Yizengaw et al. [2005, 2006], using a combination of GPS total electron current data and global mapping of the plasmapause from the IMAGE satellite, have developed very powerful techniques to study the relationship between the trough and the plasmapause and to determine the ionospheric ouflow at storm times. Applying these methods to a complete range of geomagnetic conditions would allow long-standing questions concerning the coupling between the ionosphere in the vicinity of the trough and the equatorial plane to be resolved.

4.5. Magnetosphere–Ionosphere Coupling

Time-dependent coupling of the magnetosphere to the ionosphere is included in some general circulation models. *Leimohn et al.* [2006] carried out an assessment of the accuracy of one such model for two storm intervals. They found that empirical models of the electric field could determine the general morphology quite well but coupling was required to be able to reproduce critical features for trough morphology, such as over- and under-shielding. However, many of the subtleties could not be derived from the model, partly because the conductivity-electric field feedback was not explicitly included in the model runs. Another key feature that was not included was the feedback of the time dependence of the ring current on the dipole magnetic field. *Zaharia et al.* [2006] have demonstrated how important this can be in generating small-scale, intense plasma features; without this, much smoother plasma concentrations are produced by models. Increased structuring of the plasma leads to an increased probability of irregularity occurrence.

4.6. Modeling Subgrid-Scale and Short Timescale Processes

Small-scale, short-lived processes can have important global effects in geospace. For example, substorms are initiated at the most equatorward auroral arc, a feature that is typically 1 km in latitude width. *Rodger et al.* [2001] demonstrated the large differences in the global F-region electron concentration when 6-min field variations of the high-latitude electric field were used in a modeling study compared with the then normal 1-h averages. Also, structuring of the plasma that forms the poleward edge of the trough occurs when changes to the orientation of the interplanetary magnetic field occur [*Rodger et al.*, 1994a, 1994b]. However, because it is not possible to capture such spatial and temporal scales in general circulation modeling, alternative approaches are required. Several options are given below.

Abel et al. [2006] have shown that both electric field variations in the polar cap and the auroral oval have power-law distributions, but interestingly with different exponents. It is suggested that the polar cap variations acquire the characteristic turbulence more directly from the solar wind. Conductivity and FACs are also likely to show power-law characteristics. Therefore, such approaches to parameterize multiscale and subgrid-scale structures in numerical models should improve the description and quantification of energy deposition.

Nested grids, where the resolution of the model is altered to match the scale size of phenomena in different regions of geospace, have been applied to the coupled ionosphere–thermosphere system [*Wang et al.*, 2006]. They demonstrated that increased structuring of the plasma is obtained compared with lower resolution models.

Empirical approaches usually give much better predictions than those determined from physically based models, at least until the latter achieve a significant degree of maturity. Examples are the models of T_e and electron concentration in the trough region developed by *Prölss* [2006, 2007]. Data assimilation is also a powerful way of improving prediction, and large arrays of magnetometers and SuperDARN radars [*Chisham et al.*, 2007] can provide key inputs and nudge thermosphere–ionosphere circulation models in near real-time.

5. SUMMARY AND CONCLUSIONS

1. The mid-latitude trough comprises three regions, the equatorward and poleward edges, separated by the trough minimum.
2. The major physical and chemical processes that are responsible for trough formation are well understood, for example, the subtle balance between production, recombination, and horizontal and vertical transport of plasma (see Figure 2, shaded boxes). As a result, the climatology of the trough can be reasonably accurately modeled by modern thermosphere–ionosphere general circulation models.
3. Accurate prediction of the characteristics of the trough in specific events has yet to be achieved because the necessary spatial and temporal quantification of many critical processes is not yet possible, the two most important of which are the electric field and the field-aligned transport of plasma (see Figure 2, unshaded boxes).
4. Some suggestions about how these uncertainties may be determined and modeled have been provided.

REFERENCES

Abel, G. A., M. P. Freeman, and G. Chisham (2006), Spatial structure of ionospheric convection velocities in regions of open and closed magnetic field topology, *Geophys. Res. Lett.*, *33*(24), L24103, doi:10.1029/2006GL027919.

Anderson, D. N., et al. (1998), Intercomparison of physical models and observations of the ionosphere, *J. Geophys. Res.*, *103*, 2179–2192.

Baumgardner, J., J. Wroten, J. Semeter, J. Kozyra, M. Buonsanto, P. Erickson, and M. Mendillo (2007), A very bright SAR arc: Implications for extreme magnetosphere–ionosphere coupling, *Ann. Geophys.*, *25*, 2593–2608.

Bowline, M. D., J. J. Sojka, and R.W. Schunk (1996), Relationship of theoretical patch climatology to polar cap patch observations, *Radio Sci.*, *31*, 635–644.

Chisham, G., et al. (2007), A decade of the Super Dual Auroral Radar Network (SuperDARN): Scientific achievements, new techniques and future directions, *Surv. Geophys.*, *28*, 33–109.

Dudeney, J. R., M. J. Jarvis, R. I. Kressman, M. Pinnock, A. S. Rodger, and K. H. Wright (1982), Ionospheric troughs in Antarctica, *Nature*, *295*, 307–308.

Dudeney, J. R., A. S. Rodger, and M. J. Jarvis (1983), Radio studies of the main F-region trough in Antarctica, *Radio Sci.*, *18*, 927–936.

Foster, J. C., and W. J. Burke (2002), SAPS: A new categorization of sub-auroral electric fields, *Eos Trans. AGU*, *83*, 393–394.

Foster, J. C., and H. B. Vo (2002), Average characteristics and activity dependence of the subauroral polarization stream, *J. Geophys. Res.*, *107*(A12), 1475, doi:10.1029/2002JA009409.

Foster, J. C., A. J. Coster, P. J. Erickson, J. Goldstein, and F. J. Rich (2002), Ionospheric signatures of plasmaspheric tails, *Geophys. Res. Lett.*, *29*(13), 1623, doi:10.1029/2002GL015067.

Foster, J. C., et al. (2005), Multiradar observations of the polar tongue of ionization, *J. Geophys. Res.*, *110*, A09S31, doi:10.1029/2004JA010928.

Foster, J. C., W. Rideout, B. Sandel, W. T. Forrester and F. J. Rich (2007), On the relationships of SAPS to storm-enhanced density, *J. Atmos. Sol. Terr. Phys.*, *69*, 303–313.

Freeman, M. P., D. J. Southwood, M. Lester, T. K. Yeoman, and G. D Reeves (1992), Substorm-associated radar auroral surges, *J. Geophys. Res.*, *97*, 12,173–12,185.

Galperin, Y. I., V. L. Khalipov, and A. G. Zosimova (1973), Direct measurement of ion drift velocity in the upper ionosphere during a magnetic storm: 2. Results of measurements during the November 3, 1967 magnetic storm, (In Russian), *Kosm. Issled.*, *11*, 273.

Goldstein, J., J. L. Burch, and B. R. Sandel (2005), Magnetospheric model of subauroral polarization stream, *J. Geophys. Res.*, *110*, A09222, doi:10.1029/2005JA011135.

Gurgiolo, C., B. R. Sandel, J. D. Perez, D. G. Mitchell, C. J. Pollock, and B. A. Larsen (2005), Overlap of the plasmasphere and ring current: Relation to subauroral ionospheric heating, *J. Geophys. Res.*, *110*, A12217, doi:10.1029/2004JA010986.

Huang, C.-S., J. C. Foster, and J. M. Holt (2001), Westward plasma drift in the mid-latitude ionospheric F region in the midnight-dawn sector, *J. Geophys. Res.*, *106*, 30,349–30,362.

Kamide, Y., et al. (1994), Ground-based studies of ionospheric convection associated with substorm expansion phase, *J. Geophys. Res.*, *99*, 19,451–19,466.

Keskinen, M. J., S. Basu, and S. Basu (2004), Midlatitude sub-auroral ionospheric small-scale structure during a magnetic storm, *Geophys. Res. Lett.*, *31*, L09811, doi:10.1029/2003GL019368.

Liang, J., G. J. Sofko, and H. U. Frey (2006), Postmidnight convection dynamics during substorm expansion phase, *J. Geophys. Res.*, *111*, A04205, doi:10.1029/2005JA011483.

Liemohn, M. W., A. J. Ridley, J. U. Kozyra, D. L. Gallagher, M. F. Thomsen, M. G. Henderson, M. H. Denton, P. C. Brandt, and J. Goldstein (2006), Analyzing electric field morphology through data-model comparisons of the Geospace Environment Modeling Inner Magnetosphere/Storm Assessment Challenge events, *J. Geophys. Res.*, *111*, A11S11, doi:10.1029/2006JA011700.

Mishin, E. V., and W. M. Mishin (2007), Prompt response of SAPS to stormtime substorms, *J. Atmos. Sol. Terr. Phys.*, *69*, 1222–1240.

Mishin, E. V., W. J. Burke, C. Y. Huang, and F. J. Rich (2003), Electromagnetic wave structures within subauroral polarization streams, *J. Geophys. Res.*, *108*(8), 1309, doi:10.1029/2002JA009793.

Nilsson, H., T. I. Sergienko, Y. Ebihara, and M. Yamauchi (2005), Quiet-time mid-latitude trough: Influence of convection, field-aligned currents and proton precipitation, *Ann. Geophys.*, *23*, 3277–3288.

Pavlov, A. V. and J. C. Foster (2001), Model/data comparison of F region ionospheric perturbation over Millstone Hill during the severe geomagnetic storm of July 15–16, 2000, *J. Geophys. Res.*, *106*, 29,051–29,069.

Pintér, B., S. D. Thom, R. Balthazor, H. Vo, and G. J. Bailey (2006), Modeling subauroral polarization streams equatorward of the plasmapause footprints, *J. Geophys. Res.*, *111*, A10306, doi:10.1029/2005JA011457.

Prölss, G. W. (2006), Subauroral electron temperature enhancement in the nighttime ionosphere, *Ann. Geophys.*, *24*, 1871–1885.

Prölss, G. W. (2007), The equatorward wall of the ionospheric trough in the afternoon/evening sector, *Ann. Geophys.*, *25*, 645–659.

Pryse, S. E., K. L. Dewis, R. L. Balthazor, H. R. Middleton, and M. H. Denton (2005), The dayside high-latitude trough under quiet geomagnetic conditions: Radio tomography and the CTIP model, *Ann. Geophys.*, *23*, 1199–1206.

Rodger, A. S., L. H. Brace, W. T. Hoegy, and J. D. Winningham (1986), The poleward edge of the mid-latitude trough—Its formation, orientation and dynamics, *J. Atmos. Terr. Phys.*, *48*, 715–728.

Rodger, A. S., R. J. Moffett, and S. Quegan (1992), The role of ion drift in the formation of ionisation troughs in the mid- and high-latitude ionosphere—A review, *J. Atmos. Terr. Phys.*, *54*, 1–30.

Rodger, A. S., M. Pinnock, J. R. Dudeney, K. B. Baker, and R. A. Greenwald (1994a), A new mechanism for polar patch formation, *J. Geophys. Res.*, *99*, 6425–6436.

Rodger, A. S., M. Pinnock, J. R. Dudeney, J. Watermann, O. de la Beaujardiere, and K. B. Baker (1994b), Simultaneous two-hemisphere observations of the presence of polar patches in the night-side ionosphere, *Ann. Geophys.*, *12*, 642–648.

Rodger, A. S., G. D. Wells, R. J. Moffett, and G. J. Bailey (2001), The variability of Joule heating, and its effects on the ionosphere and thermosphere, *Ann. Geophys.*, *19*, 773–781.

Schunk, R. W., P. M. Banks, and W. J. Raitt (1975), Effects of electric fields and other processes upon the night-time high latitude F layer, *J. Geophys. Res.*, *80*, 3121–3130.

Southwood, D. J., and R. A. Wolf (1978), An assessment of the role of precipitation in magnetospheric convection, *J. Geophys. Res.*, *83*, 5227–5232.

Spiro, R. W., R. A. Heelis, and W. B. Hanson (1979), Rapid subauroral ion drift events observed by Atmospheric Explorer C, *Geophys. Res. Lett.*, *6*, 660–663.

Sojka, J. J., and R. W. Schunk (1989), Theoretical study of the seasonal behavior of the global ionosphere at solar maximum, *J. Geophys. Res.*, *94*, 6739–6749.

Sojka, J. J., C. Smithtro, and R. W. Schunk (2006), Recent developments in ionosphere–thermosphere modelling with an emphasis on solar variability, *Adv. Space Res.*, *37*, 369–379.

Wang, W., A. G. Burns, S. Solomon, and T. L. Killeen (2005), High-resolution, coupled thermosphere–ionosphere models for space weather applications, *Adv. Space Res.*, *36*, 2486–2491.

Weygand, J. M., R. L. McPherron, K. Kirsti, O. Amm, A. Viljanen, and H. U. Frey (2007), Relation of substorm onset to Harang discontinuity, Greenland Space Sciences Symposium, 4–9 May (abstract only http://www.gsss-2007.org/).

Yizengaw, E., and M. B. Moldwin (2005), The altitude extension of the mid-latitude trough and its correlation with plasmapause position, *Geophys. Res. Lett.*, *32*, L09105, doi:10.1029/2005GL022854.

Yizengaw, E., M. B. Moldwin, P. L. Dyson, B. J. Fraser, and S. Morley (2006), First tomographic image of ionospheric outflows, *Geophys. Res. Lett.*, *33*, L20102, doi:10.1029/2006GL027698.

Zaharia, S., V. K. Jordanova, M. F. Thomsen, and G. D. Reeves (2006), Self-consistent modeling of magnetic fields and plasmas in the inner magnetosphere: Application to a geomagnetic storm, *J. Geophys. Res.*, *111*, A11S14, doi:10.1029/2006JA011619.

A. Rodger, British Antarctic Survey, Madingley Road, Cambridge, CB3 0ET, UK. (ASRO@bas.ac.uk)

Assimilation of Observations With Models to Better Understand Severe Ionospheric Weather at Mid-Latitudes

Jan J. Sojka, R. W. Schunk, D. C. Thompson, L. Scherliess, and M. David

Center for Atmospheric and Space Sciences, Utah State University, Logan, Utah, USA

For many decades, the mid-latitude ionosphere was regarded as well characterized even if not well modeled. As a result, the Federal Aviation Authority's Wide Area Augmentation System (WAAS) was developed to provide augmented GPS positioning information to correct for ionospheric variability. However, over the past 5 years, recurrent superstorms in the ionosphere have forced the WAAS system to go offline for many hours at a time. This report discusses present-day knowledge regarding these conditions and how they are associated with unexpectedly steep horizontal gradients in the mid-latitude ionosphere total electron content (TEC). In a general sense, the possible physical mechanisms are understood, but during a storm the distribution and evolution of the driving forces for these mechanisms are neither understood nor adequately observed, the two main driving forces being the convection electric field and the neutral wind. In this paper a simplified convection electric field pattern is presented and used to drive a physics-based ionospheric model. This demonstrates how the superstorm ionospheric condition could be generated. Data assimilation is a new approach that could exceed present-day empirical and physical model limitations. There are three main expectations for data assimilation: (1) combined with a good ionospheric background model, the data assimilation must provide realistic global specification of the ionosphere; (2) it must also provide additional information about the ionosphere that is not already evident in the observation, that is, altitude profiles of the electron density when only slant TEC integrals of the electron density are available; and (3) with full physics-based models in the assimilation procedure, data assimilation models must also provide the drivers, that is, the neutral wind and electric field patterns. This paper examines the current status of these three expectations with regard to the future for the scientist and the space weather forecaster.

1. INTRODUCTION

Compared to the ionospheric weather at high and low latitudes, the mid-latitude ionosphere was thought to be relatively well understood until the Federal Aviation Authority's (FAA) Wide Area Augmentation System (WAAS) encountered its first mid-latitude geomagnetic storm. These unexpected WAAS problems occurred in the early years of the 21st century. As a result, ionospheric scientists have, for the past few years, returned to studies of the mid-latitude ionosphere with urgency and passion. Given that it was the frequent episodes of WAAS being unavailable during geomagnetic storms that revealed a major shortcoming in our understanding of the mid-latitude ionosphere, this introduction will describe WAAS from an ionospheric science perspective. The authors apologize for a lack of technical depth in describing WAAS itself.

Over the past three decades, positioning on the Earth, in the atmosphere, and even in space has become relatively affordable, accurate, and universally accessible with the availability of L-band radio signals from the US Air Force Global Positioning Satellites (GPS). The accuracy of a geolocation depends on many technical issues and the type of environment the L-band radio signals pass through. Most of the technical issues can be overcome with more sophisticated hardware and software or the next development in GPS, for example, going from two L-band frequencies to three L-band frequencies. Currently, the largest uncontrolled accuracy problem arises in the ionosphere through which L-band waves must propagate. The ionosphere, which includes all its layers as well as the plasmasphere, introduces delays, phase shifts, and—under extreme conditions—scintillation that leads to the introduction of "noise" to the extent that receivers can lose the ability to lock on to the signal. In geopositioning calculations, a receiver combines signals from many satellites and hence from a broad spread of arrival angles. Each satellite signal "samples" a different ionosphere. If unexpected gradients exist in the ionosphere, these map back to geoposition uncertainty. As will be described shortly, this is the major source of the WAAS problem.

For the purpose of this report, it is unnecessary to describe the myriad applications of the GPS system. In the context of ionospheric research, however, it is important to acknowledge that the very worst GPS implementation problem experienced by users has become an ionospheric researcher's greatest asset. GPS signals are collected from receivers distributed around the globe. The international GPS community has adopted a standard format for recording these signals, called RINEX. This format contains not only all the information that users of GPS need, but also sufficient information for scientists to infer a relative measure of the total electron content (TEC) between each GPS satellite transmitter and receiver. With additional knowledge of GPS satellite biases and receiver biases, the relative TEC can, in fact, become an absolute TEC. Given that currently, many thousands of receivers are operating worldwide and their RINEX files are made publicly available, ionospheric scientists have for the first time been able to produce "snapshots" of the global ionospheric TEC distributions. One of the first global ionospheric mapping studies was carried out at the NASA Jet Propulsion Laboratory (JPL) in their Global Ionospheric Mapping (GIM) project beginning in 1991 [*Wilson et al.*, 1992, 1995; *Mannucci et al.*, 1998].

The FAA considered the possibility of using GPS geopositioning on aircraft, especially those flying across the United States as an additional independent geolocation system. They realized that the variability in the ionosphere was a problem, but innovatively developed a solution to this problem. Their solution involved monitoring the ionospheric TEC over the United States and hence modeling the TEC spatial distribution. This, in turn, provided the required correction for the individual satellite to receiver L-band links. By doing this mapping in real-time and via an additional relay satellite providing these corrections to the aircraft, the WAAS was created. In principle, the concept is appropriate. The ionospheric knowledge at the time the system was developed expected that less than 30 ground GPS TEC monitors distributed across the United States would suffice. Given that the FAA bears the responsibility for millions of passengers and aircraft crew, their systems must meet extremely tight design criteria. Initially, FAA had hoped to make the WAAS system a 7-9 system such that it would be available 99.99999% of the time. This corresponds to being unavailable for less than 4 s/yr. The adopted criteria falls short of this and when the service referred to as precision approach (PA) started in July 2003, the following requirements defined the WAAS PA service [*Doherty et al.*, 2004]. PA services are available ≥95% of the time over 75% of the contiguous United States. The Halloween storms that occurred in 2003, only 4 months after WAAS PA went online, rendered the PA services unavailable for 15 and 11.3 h on 29 and 30 October 2003, respectively [*Doherty et al.*, 2004]. This unavailability problem has recurred over the past few years whenever storms and superstorms occur.

Scientists at Millstone Hill, Massachusetts Institute of Technology, and at the Institute for Scientific Research, Boston College, came up with the cause of the problem. Plate 1, provided by Dr. Anthea Coster, shows a TEC map over the United States on 20 November 2003, during an episode of WAAS unavailability. At this particular universal time (UT), the United States is in sunlight and would normally be viewed to have a uniform unstructured TEC distribution. Clearly, Plate 1 does not comply with this expectation. An intense TEC structure runs from the Caribbean, east of Florida toward the Great Lakes, and into Canada over Hudson Bay. Coster and her coworkers were able to show that the real-time operational WAAS systems ground-based GPS TEC receivers that alerted the WAAS system of geopositioning problems were, in fact, colocated with this TEC feature. Today, it is understood that associated with this TEC feature, a ridge of high TEC, there are in places extremely steep spa-

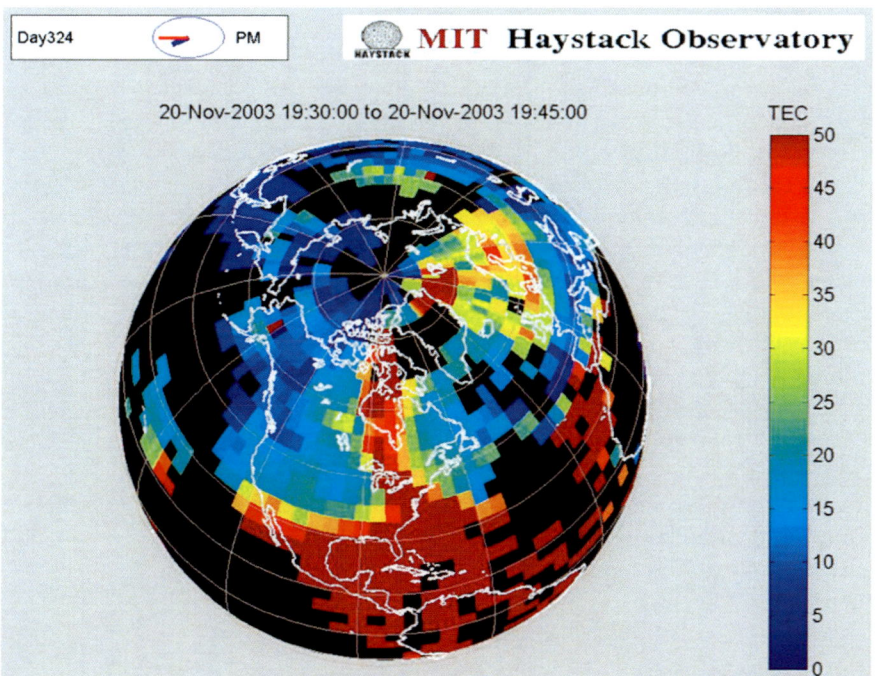

Plate 1. The ionospheric TEC over North America on 20 November 2003, at 19:30 UT. The TEC is color-coded in TEC units (equal to 10^{16} el m^{-2}). TEC is projected onto the Earth's surface (provided by Dr. Anthea Coster, 2007).

tial gradients in TEC. It is these steep spatial gradients that are the source of WAAS being rendered "unavailable." The triangulation algorithms are unable to keep the uncertainty below a WAAS threshold when these steep unexpected gradients occur. In fact, the 30 or so ground GPS TEC monitors are spaced too far apart to capture these gradients, unless the gradient is in the field of view of a monitor.

The present-day ionospheric research tool of observation, theory, and modeling has not provided a forecast solution to this very public FAA WAAS problem. Observations of ionospheric TEC provide an excellent real-time specification but no glimpse into the future. WAAS is based on using such measurements. Theory at this time probably understands the complex interplay processes that occur locally in the ionosphere–thermosphere. It does not understand either the dynamic evolution of neutral dynamics of electrodynamics on regional scales during geomagnetic disturbed times. Models, both empirical and physics-based, contain good descriptions of the ionosphere as observed or theoretically expected, but are very limited in forecast capability during these storms. This is the point at which assimilation techniques have been introduced to the ionospheric community to provide a means to go forward. Specifically, will it be possible to use these techniques to extract information on the neutral dynamics and electrodynamics that drive the ionosphere during storms? If this can be achieved, our knowledge will advance significantly and forecasting will become a possibility. At present, this assimilation capability is only being developed. At this time early versions of assimilation provide excellent real-time specification of the ionosphere. These initial efforts are akin to tomographic reconstruction or direct mapping based on having available in real-time large numbers of measurements covering the area of interest.

In Section 2, a relatively short history focusing on the mid-latitude positive storm phase is given to indicate how we arrived at the WAAS era. The probable driver mechanisms of the storm positive phase are described in Section 3, and one of these is focused on and modeled in Section 4 as a demonstration that theoretically present-day ionospheric physics on a local scale appears to be sufficient. In Section 5, an assimilation model is described and a real-time specification example is used to show that progress is being made and how future assimilation models will provide new knowledge. These hopes are summarized in Section 6.

2. WHY WAS THE WAAS PROBLEM NOT PREDICTED?

Today with hindsight and the old adage that "a picture is worth a thousand words," it is possible to piece together the answer to this question. The observation of geomagnetic storms in the mid-latitude ionosphere, its positive and negative phases, goes back to the earliest days of ionosonde observations [*Hafstad and Tuve*, 1929]. By the early 1970s, a significantly complete description of the storm was pieced together from single station observations and early satellite in situ observations. This included the use of Faraday rotation effects on radio signals from geosynchronous satellites and ground-based receivers to measure the TEC of the ionosphere plus plasmasphere. *Mendillo* [1973] provided an excellent summary of this and took on the task of searching for a forecast algorithm for the storm's positive phase in the ionosphere in the vicinity of Millstone Hill, MA.

Figure 1 shows Mendillo's average TEC enhancement for 28 geomagnetic storms from November 1967 to December 1969. The ionospheric location is the *F*-layer pierce location for the geosynchronous ATS-3 satellite transmitter to the receiver located at Hamilton, MA. With reference to Plate 1, this mid-latitude location is poleward of the storm TEC ridge, confirming that the storm effects are truly at mid-latitudes. In fact, when comparing the TEC average peak identified in Figure 1, one is tempted to infer this is an LT cross section of the two-dimensional (2-D) ridgelike structure in Plate 1. It probably is. However, an essential fact was unknown in 1973, namely, that the ridge structure was a coherent long-lived structure that extended more than 1000 km. Today, with GPS TEC maps, this is self-evident. Hence, both in 1973 and in earlier periods, much of the emphasis in interpreting these structures was to view them as being temporal in nature. Indeed, traveling ionospheric disturbances (TID) covered a range of dynamic phenomena that first appeared

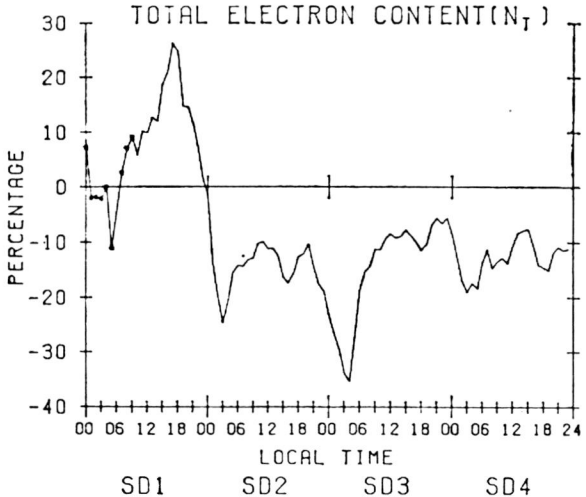

Figure 1. The average disturbed daily variation of TEC at Sagamore Hill Radio Observatory from 28 storm periods. Reprinted from *Mendillo* [1973] with permission from Elsevier.

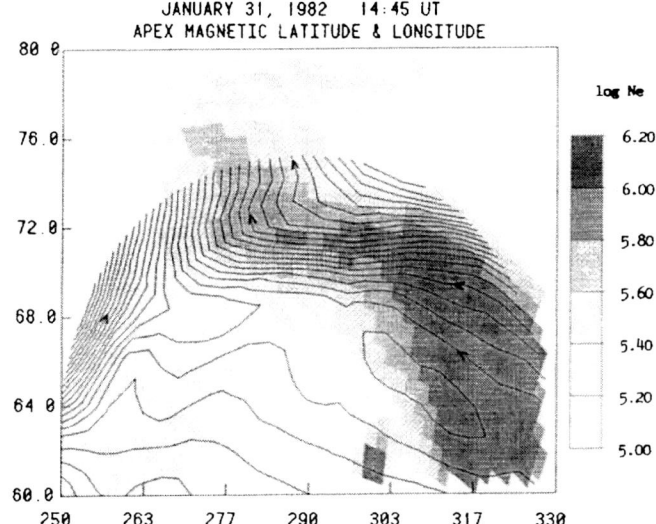

Figure 2. Millstone Hill ISR azimuth scan on 31 January 1982, which shows enhanced density plasma streaming along the observed convection trajectories through the cusp well into the polar cap. From *Foster* [1993].

in Figure 1 were at best informative but not robust. Today, such a challenge could be viewed as analogous to predicting if a specific town in the "hurricane corridor" of the United States will be hit by a specific hurricane based only on a few indices of local and national weather. In both cases, we now know that we are dealing with a complex set of processes, a system-level problem.

Twenty years later, *Foster* [1993] was able to capture the 2-D nature of the ionospheric ridge as it appeared poleward of the Millstone Hill incoherent scatter radar (ISR). Foster described the ridge as a storm-enhanced density (SED) and showed that poleward of this SED was the mid-latitude trough and poleward of that was the auroral oval. The westward–northward extension of the SED was directed toward the cusp and its subsequent entry into the polar cap. Figure 2 shows an example of this SED on 31 January 1982 [from *Foster*, 1993]. Although the SED is shown as an N_mF_2 enhanced structure rather than the TEC structure in Plate 1, the two are indeed synonymous. This 2-D analysis, however, only pertained to regions poleward of the mid-latitude WAAS area. It was not predicted to extend into mid-latitudes under severe storm conditions. The measurements taken in the vicinity of Millstone Hill (see Figure 2) were viewed as the "base" from which the SED extended poleward.

At about the same time as the 1993 Foster SED work was revealing a coherent 2-D ionospheric structure, the first GPS TEC maps were being developed. *Wilson et al.* [1995] described and demonstrated how the fledging International GPS Geodynamics Service (IGS) GPS TEC network could provide global maps of the ionosphere in real-time. Although at that time there were only about 40 IGS receivers globally distributed, Wilson and his coworkers developed algorithms to produce global ionospheric maps with temporal cadences

at higher latitudes and were viewed as the propagation of phenomena in the neutral atmosphere [traveling atmospheric disturbances (TADs)]. Certainly, the predominantly southeast-to-northwest alignment of the TEC ridge (evident in Plate 1), when viewed by individual corotating stations moving under such a structure, gives credence to this TID propagation phenomena. Attempts to find detailed forecast algorithms for the presence of the TEC enhancement shown

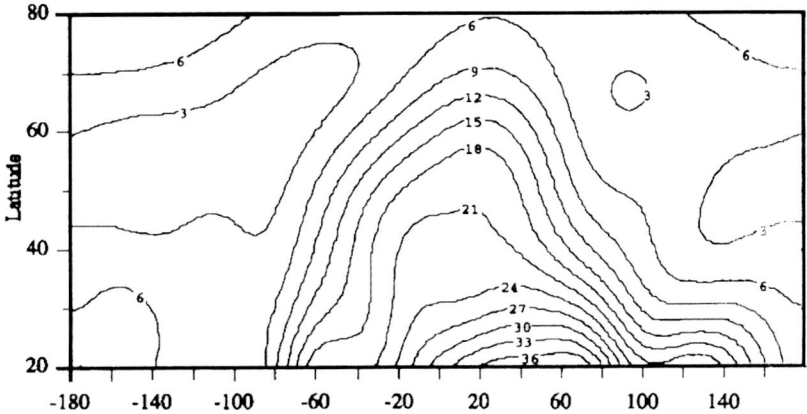

Figure 3. Global ionospheric TEC distribution for 4 January 1993, derived from a 12-h fit. Contours are labeled in units of TEC (equal to 10^{16} el m^{-2}). From *Wilson et al.* [1995].

decreasing for 24, to 12, to 6 h. Figure 3 shows an example of a northern hemisphere 2-D, 12-h TEC map in a Sun-fixed coordinate system. Because the GPS receivers were sparse, the spatial (and temporal) resolution appears very limited compared to what we have today (see Plate 1). However, these maps were useful for the operation of the NASA deep-space network as well as providing ground-based receiver biases. That they ushered in the era of real-time JPL-GIM maps makes this figure particularly significant. However, Figure 3 did not have the resolution in latitude, longitude, or time to foresee the WAAS problem.

This brief history using only three references merely provides a measure of how well the ionospheric community was observing and interpreting *the severe storms*. To an extent, the ionospheric environment during these severe storms was known at the time the WAAS was being developed; however, the major shortcoming lay in the need for WAAS engineers to quantify the very steep latitude and longitude gradients in TEC ridges in order to fully infer their impact on geolocation calculations. The importance of this task was probably not fully appreciated at that time because of the space–time ambiguity inherent in single-point measurements and also the diverse descriptions of these phenomena in the scientific literature.

The three references used to highlight the research community's appreciation of the positive phases of an ionospheric storm comprise only a small selection from the many hundreds of papers on the topic. For a comprehensive review of progress in this field, see *Mendillo* [2006, and references therein]. Returning to our original question why scientists were unable to predict the WAAS problem, it was because scientists were using the same GPS ionospheric measurements to "see" for a first time the mid-latitude storm problem as FAA engineers were implementing WAAS. In September of 2001, *Coster et al.* [2001] gave a presentation on the 15–16 July 2000 geomagnetic SED at a technical meeting of the Satellite Division of the Institute of Navigation while WAAS underwent rigorous testing over a 90-day period prior to July 2003.

3. UNDERSTANDING THE DAYTIME STORM EFFECTS

The mid-latitude dayside ionosphere is relatively homogeneous with well-understood weak latitude and local time gradients. Even the day-to-day variability tends to maintain this homogeneity although the exact local time or magnitude of the peak density shows day-to-day variations. Under nonstorm conditions, the day-to-day variability would have a variance of no more than a few tens of percent. In fact, most of the time, the mid-latitude ionosphere would be in this state and pose no problems to the WAAS methodology. However, over the past 5 years, the ionosphere has been under storm conditions on several occasions per year with each having an adverse impact over periods ranging from minutes to many hours. In this 5-year period, there have also been at least four superstorms whose impacts are significantly longer lived. To understand the impact, it is necessary to have an understanding of the driver mechanisms that can lead to dayside ionospheric density gradients on the small scale that impact GPS geoposition calculations.

A storm is, in fact, a complex set of action and reaction mechanisms that originates at the Sun, propagates through the heliosphere, and when it interacts with the magnetosphere, leads to energy being transferred into geospace. Today, details of the various interactions that constitute these energy transfers are still being studied. The ionospheric storm is then driven by interactions between the ionosphere and magnetosphere, primarily of an electrodynamic nature. The energy and momentum deposited in the ionosphere–thermosphere is due to enhanced particle precipitation, visible and invisible forms of the aurora, and electric field distribution expanded to lower latitudes as the storm intensity increases. Present-day knowledge indicates that auroral and electric fields' input are divided approximately 20% and 80%, respectively, of the total energy deposited. Today, auroral imagery maps the 20% due to auroral deposition, but no equivalent mapping exists for the electric fields at mid-latitudes.

This energy then heats both the ionosphere and thermosphere. As a result, the high-latitude thermosphere, after a period of 30 min to a few hours, will itself become a storm driver because its winds and composition changes move equatorward. Hence, in addition to the magnetospheric direct input via particle precipitation and electric fields, the thermosphere will also cause equatorward winds and composition changes. Imaging the composition changes is becoming a reality, but at this time, the winds cannot be mapped adequately at mid-latitudes.

Of these four drivers, the electric fields and neutral wind have been viewed as the probable source of the mid-latitude dayside ionospheric positive phase. *Jones and Rishbeth* [1971] modeled the effect of a thermospheric heat input at high latitudes causing a wind to flow equatorward through the dayside mid-latitudes. Figure 4, taken from the study of *Jones and Rishbeth* [1971], contrasts the effect of the *F* layer on the peak density. The difference in the storm-enhanced ionospheric density from that of the quiet time (solid line) is well over a factor of 2. Since TEC and the peak density are proportional to first order, this figure demonstrates that the dayside TEC could be doubled by equatorward winds generated after the thermosphere is heated. In a subsequent paper, *Jones* [1971] also modeled how an electric field, primarily

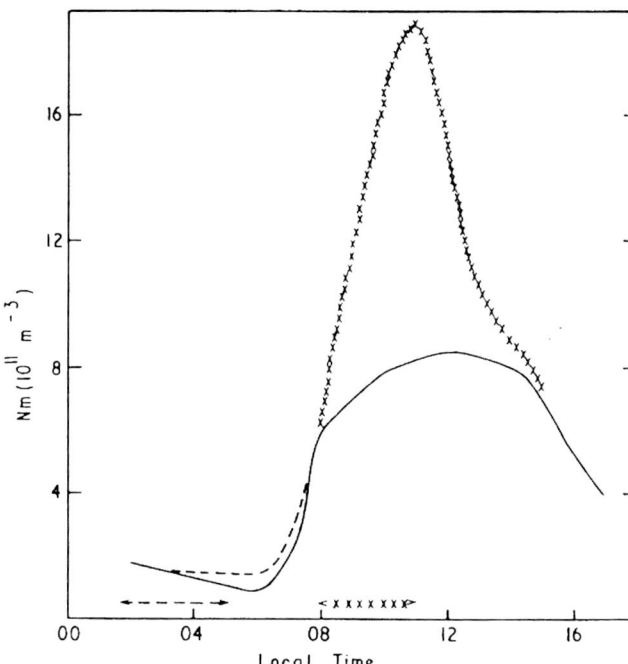

Figure 4. Modeled variation of N_mF_2 at Auckland in winter under quiet conditions (line plot) and for a storm-driven equatorial wind condition (symbols plot). Reprinted from *Jones and Rishbeth* [1971] with permission from Elsevier.

eastward, would enhance the dayside peak density. Figure 5 [from *Jones*, 1971] shows how an oscillating electric field would increase and decrease the density. A more comprehensive model by *Tanaka and Hirao* [1973] studied the relationship between an eastward electric field and both the peak density and height of the F layers. Figure 6 shows the main findings of their parametric study. For cases where the eastward electric field is constant for periods on the order of a few hours, the peak density can be readily doubled or tripled while its layer height is increased by more than 100 km.

These studies discussed the two main mechanisms that can, in general, double TEC in the mid-latitude ionosphere in the daytime, which is conventionally referred to as the storm positive phase. Sequentially, this phase is usually associated with the beginning of a storm, the storm main phase. A specific subset of the positive phase is the dusk effect. As its name implies, at mid-latitudes during storms there are frequent observations of a very "short-lived," very enhanced F layer at dusk as viewed by a corotating ground observatory. The dusk effect is only 1 or 2 h long in local time and hence, is very distinct from the broad local time width of the dayside storm positive phase. *Papagiannis et al.* [1971] specifically addressed how electric fields might evolve to account for this phenomenon.

The relative merits of the wind and electric field drivers have oscillated over the past five decades. *Mendillo* [2006] in his review paper tabulates this oscillation of favor from the first theoretical offerings by *Martyn* [1953], concluding that electrostatic fields were the source. This was followed by various other mechanisms being proposed until, in 1995, a review by *Prolss* [1995] favored the traveling atmospheric disturbances, a thermospheric source. Today, the TEC images showing the static, coherent, ridge of ionization leaves the community struggling for what combination of mechanisms can literally "freeze" this coherent structure for what appears to be many hours. Unfortunately, our understanding, or observations, of the wind or electric field distributions at mid-latitude do not provide insight on how the very large TEC gradients are created. This, unfortunately, is also the key to understanding the WAAS problem origin.

4. A POSSIBLE ELECTRIC FIELD SCENARIO

In the previous section, the two main mechanisms that would generate enhanced dayside electron densities locally during storms were described. Unfortunately, at present, there are no successful models showing how either an electric field distribution or neutral wind distribution would evolve during a storm at mid-latitudes. *Sojka* [2005] and, most recently, *Sojka and Heelis* [2007] presented a scenario by which an electric field distribution would create TEC storm patterns as seen in Plate 1. The scenario is based in part on ground-based photographic observations of the aurora in the southern states (i.e., Texas, New Mexico, and Arizona), which show the auroral zone extending deep into the

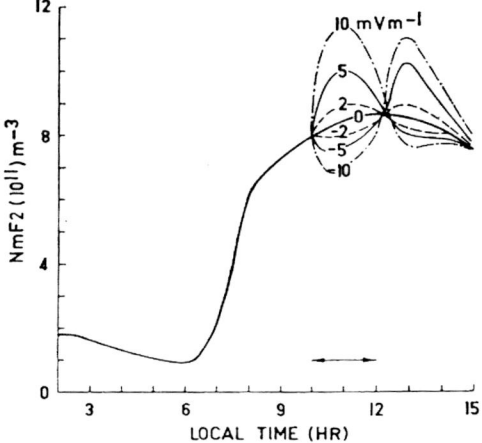

Figure 5. Increases in N_mF_2 due to electrodynamic drifts modeled for the Auckland mid-latitude location. Reprinted from *Jones* [1971] with permission from Elsevier.

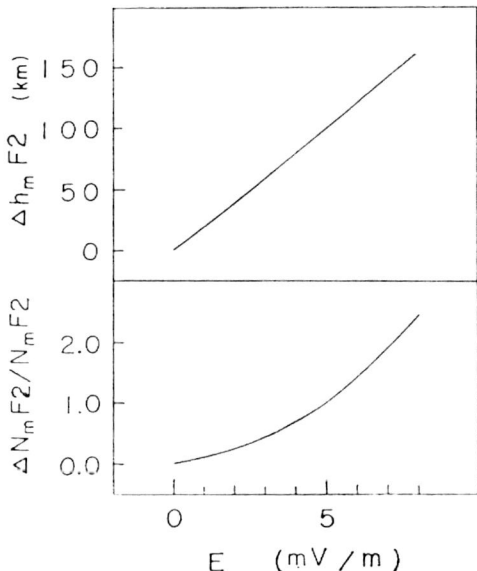

Figure 6. Modeled F-layer (a) peak height and (b) peak density dependence on the amplitude of an eastward electric field at mid-latitudes. Reprinted from *Tanaka and Hirao* [1973] with permission from Elsevier.

can be adjusted, via two parameters, to represent a possible storm-time electric field distribution. These two parameters comprise the polar cap radius and the mid-latitude fall-off factor that enables electric fields to penetrate equatorward. Table 1 lists how these two parameters can be used to create four storm-time electric field distributions at mid-latitude. The electric field and electric potential for these four models are shown in Figure 8. Models 1 and 2 have a quiet time radius, whereas models 3 and 4 have a severe storm radius that has moved from 73° to 60° magnetic latitude. The conventional quiet time mid-latitude fall-off factor is −4, and a storm-like penetration can be captured by a fall-off factor of −2. A dusk sector latitude cross section of the Volland meridional electric field for all four models is shown in Figure 8 (top), whereas their equivalent potential is shown in Figure 8 (bottom). For all four cases, the cross tail (cross polar cap) potential has been maintained at 100 kV. This is a value associated with a geomagnetic storm, but is probably lower than that of a superstorm. When the polar cap radius increases to 30°, the mid-latitude electric fields are significantly increased from less than 1 to more than 5 mV/m. The effect of the fall-off parameter just equatorward of the polar cap boundary reduces the electric field but then at lower lati-

mid-latitude region during storms. In fact, under superstorm conditions experienced during the past decade, the auroral equatorward boundary is at the poleward edge of the equatorial region, almost supporting the argument for no mid-latitudes as a distinct region at these times. From space satellites that image the auroral emissions, this equatorward progression or expansion of the auroral oval is seen. The ground- and space-based assets to measure the electric fields at mid-latitudes are limited. The Sun-synchronous Defense Meteorological Satellite Program (DMSP) carry instruments that infer the electric fields in situ. Dr. Heelis (private communication, 2006) has found that the latitude boundaries of the electric field distributions also move equatorward during storms. Indeed, during superstorms, these equatorward penetration electric fields are observed all the way to the equator.

Figure 7 shows the electric field as an equipotential set of contours in a polar coordinate system for the most symmetric case of a two-cell convection pattern [*Volland*, 1975]. The convection has a polar cap region, poleward of 60° and a mid-latitude region where the potential diminishes. *Volland* [1975] argues that this mid-latitude fall of region can extend further equatorward during times of storm activity. The dashed contours represent examples of this penetration. From the DMSP observations, we know that in addition the polar cap radius also expands. Hence, this Volland pattern

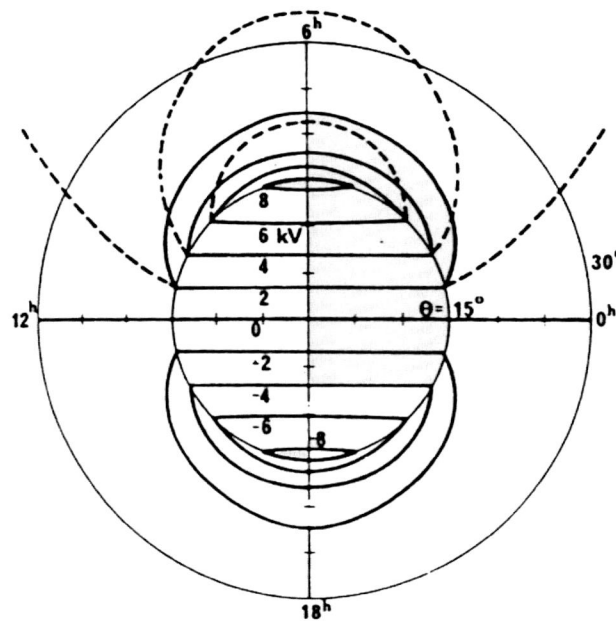

Figure 7. Volland two-cell convection pattern for quiet conditions with a polar cap radius of 15°. Equipotential lines are drawn at 2-kV intervals. Dashed lines represent possible penetration of the electric field to lower latitudes during storm periods. From *Volland* [1975].

Table 1. Volland Model Parameters

Model	Radius	Fall-off
1	17	−4
2	17	−2
3	30	−4
4	30	−2

tudes demonstrates electric fields that are larger than those of the quiet time fall-off. Note that these mid-latitude electric fields are still small compared to those of the auroral or polar cap in Figure 8.

The storm convection models shown in Figure 8 are based on the most symmetric Volland patterns and as such are inherently an oversimplification. However, the question to be raised is: What are their relative effects on the mid-latitude ionosphere? To simulate the ionospheric TEC responses, the Utah State University (USU) Time-Dependent Ionospheric Model (TDIM) has been used.

The model was initially developed as a mid-latitude, multi-ion (NO^+, O_2^+, N_2^+, and O^+) model by *Schunk and Walker* [1973]. The time-dependent ion continuity and momentum equations were solved as a function of altitude for a corotating plasma flux tube, including diurnal variations and all relevant *E*- and *F*-region processes. This model was extended to include high-latitude effects because of convection electric fields and particle precipitation by *Schunk et al.* [1975, 1976]. Flux tubes of plasma were followed as they moved in response to the convection electric fields. The addition of plasma convection and particle precipitation models is described by *Sojka et al.* [1981a, 1981b]. The theoretical development of the TDIM is described by *Schunk* [1988], whereas comparisons with observations are discussed by *Sojka* [1989]. The TDIM requires inputs for the neutral atmosphere, neutral wind, auroral precipitation, and convection electric field. These are represented by the MSIS-86 [*Hedin*, 1987], HWM [*Hedin et al.*, 1991], Hardy oval [*Hardy et al.*, 1987], and the *Volland* [1975] or equivalent convection models. These drivers are empirical models that require the geomagnetic three-hourly *Kp* and *Ap* indices; the MSIS-86 and HWM models require the solar $F_{10.7}$ and $F_{10.7A}$ indices as well.

A midwinter, solar maximum day was chosen and a simulation that began as a quiet day was followed by a storm onset in which *Kp* increased from 1 to 5 in 30 min. The simulation then used the storm convection pattern and continued for several hours. The simulations were timed such that the American sector was in the noon to dusk local time sector during the storm.

Plate 2 shows TEC snapshots of six simulations at 2 h into the storm. The four storm convection models referred to in Figure 8 are the left and right pairs of snapshots with the

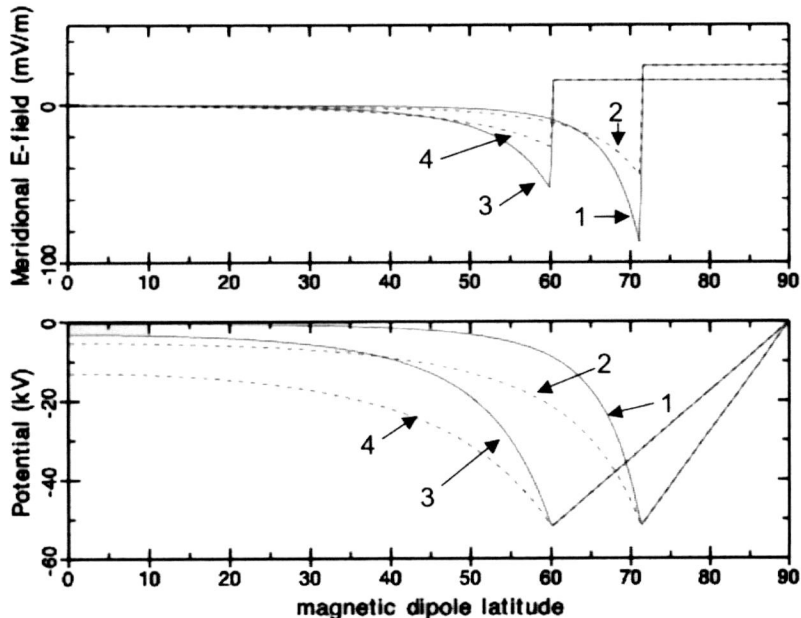

Figure 8. (a) Dusk sector meridional electric field for the four models listed in Table 1. (b) Their corresponding electric potentials.

central pair being the intermediate cases. Plate 2a (left) represents the standard USU-TDIM storm simulation (model 1 in Figure 8 and Table 1). This panel exhibits certain features that can be contrasted with Plate 1 or Figure 2. In the polar diagram, noon is at the top and dusk (1800 MLT) is at the left, whereas magnetic latitude increases from 40° to the pole at the center of the diagram. A just discernible TEC feature extends antisunward across the polar cap from about noon at 75° (the cusp). This is the tongue of ionization (TOI), which enters the polar cap at the cusp. In the afternoon sector before dusk, an enhanced TEC region extends poleward and is funneled toward noon. This is the SED feature described by *Foster* [1993]. In the afternoon lower mid-latitudes, the TEC is higher than the prenoon sector, indicating a positive storm phase. However, for the conditions the simulation is trying to represent, the TEC values are at least a factor of 2 too low.

The sequence of dials from left to right in Plate 2a shows how increasing the polar cap radius significantly addresses this low TEC problem in the standard model 1 simulation. By increasing the radius to 30°, Plate 2a (right), and model 3 in Figure 8 and Table 1, the TEC has been increased by 50%. Now, a significant ridge of higher TEC leads into an SED and TOI. Also note that at mid-latitudes, a strong local time gradient is developed to the east of this SED base region. By increasing the electric field penetration to lower latitudes, that is, changing the fall-off factor, even more dramatic increases in mid-latitude TEC are obtained. This is shown in the set of dials in Plate 2b, where the left and right dials represent the convection models 2 and 4, respectively. In Plate 2b (right), model 4, although still exhibiting the same convection cross polar cap potential as the other simulations, now shows mid-latitude TEC values and gradients that begin to exceed those observed. These simulations do, however, have a density ridge as found in the Plate 1 observations. These simulations, using probably the simplest convection pattern, do generate the storm-enhanced TEC at mid-latitudes, their coherent long-lived ridge of TEC, as well as very large gradients at the edges of the ridges. Yet, this convection pattern does not contain the full richness of electrodynamics imagined to be associated with the mid-latitude magnetosphere, that is, ring current, partial ring currents, over-under shielding, sub-auroral polarization schemes penetration electric fields, storm-time mid- and low-latitude dynamos, etc. These simulations neither include neutral wind nor thermospheric composition changes beyond those included in statistical models as the Kp changes from 1 to 5 during the storm's initial phase.

5. IONOSPHERIC DATA ASSIMILATION

In the previous section, a first-principle ionospheric model (USU-TDIM), using a modified Volland convection pattern, was able to demonstrate that the electric field can be a viable candidate for generating the mid-latitude storm TEC ridges and steep gradients. However, current understanding of how this electric field distribution varies during a storm, how the neutral winds evolve, or the role of other processes remains largely unknown. Since it is extremely difficult to map the drivers by direct measurements either from the ground observations or space platforms, other methods need to be found. Within the past decade, the data assimilation techniques that have worked well in both meteorology and oceanography have been considered for application in the ionosphere, primarily because of the abundances and wide distribution of GPS TEC measurements. It is hoped that the data assimilation will provide several distinct advantages over either empirical or physics-based modeling of the ionosphere: (1) combined with a good ionospheric background model, data assimilation must provide realistic global specification of the ionosphere; (2) it must also provide additional information about the ionosphere that is not already evident in the observation, that is, altitude profiles of the electron density when only slant TEC integrals of the electron density are available; and (3) with full physics-based models in the assimilation procedure, data assimilation must also provide the drivers. An immediate benefit is that data assimilation in near real-time will provide both forecasters and other users, which include scientists, with an optimum real-time specification of the ionosphere. This specification can be either regional or global and is inherently not restricted to the location where observations were made. The validity of filling-in latitude and local time depends on what ionospheric background is used in the data assimilation model.

To show how assimilation of observations with models can produce better ionospheric specifications, the USU Global Assimilation of Ionospheric Measurements (GAIM) will be used. GAIM, in fact, is a suite of different models, and for this paper only, examples from the Gauss–Markov Kalman filter (GMKF) will be shown. This model uses the Ionospheric Forecast Model (IFM) as its ionospheric background representation. The IFM is a streamlined, computationally faster version of the earlier described USU-TDIM. It takes into account NO^+, O_2^+, N_2^+, O^+, and H^+, and covers the E region, F region, and topside ionosphere. In the GMKF version of GAIM, the ionospheric densities obtained from IFM constitute a background ionospheric density field on which perturbations are superimposed based on the available measurements and their errors. The density perturbations and associated errors evolve over time via a statistical Gauss–Markov process. A detailed description of this USU GAIM model and its initial validation is given by *Scherliess et al.* [2006], and the IFM is described by *Schunk et al.* [1997].

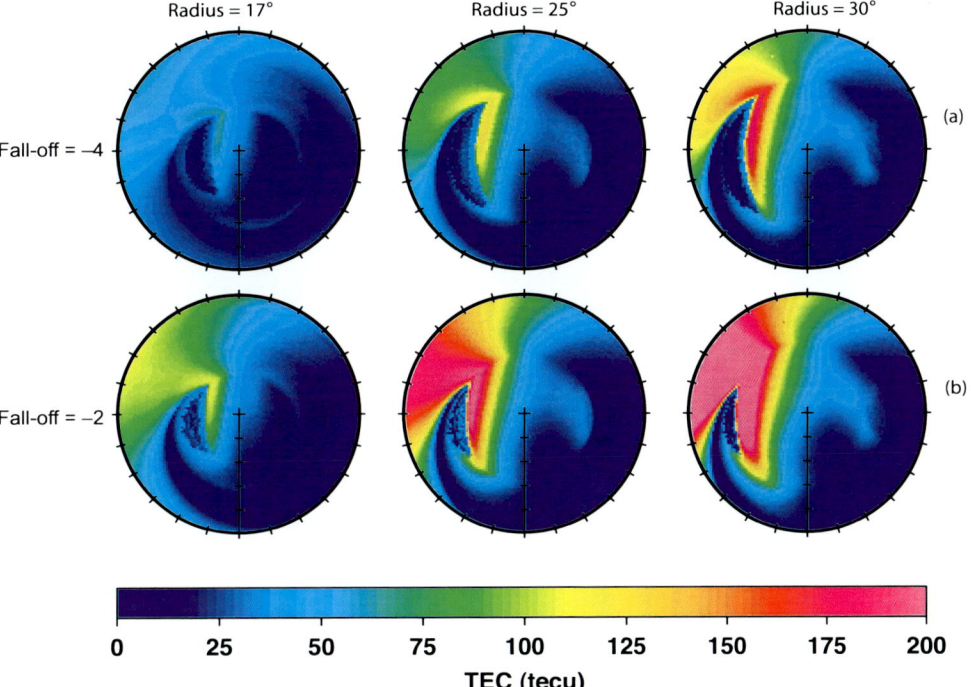

Plate 2. TDIM TEC simulation for six different permutations of storm Volland convection models with each snapshot being taken 2 h after the storm was initiated. Each snapshot is a magnetic latitude–magnetic local time polar plot. Noon is at the top and dusk is at the left. The magnetic latitudes extend from 40° to 90° at the center. TEC is color-coded on a linear scale in units of 10^{16} el m^{-2}.

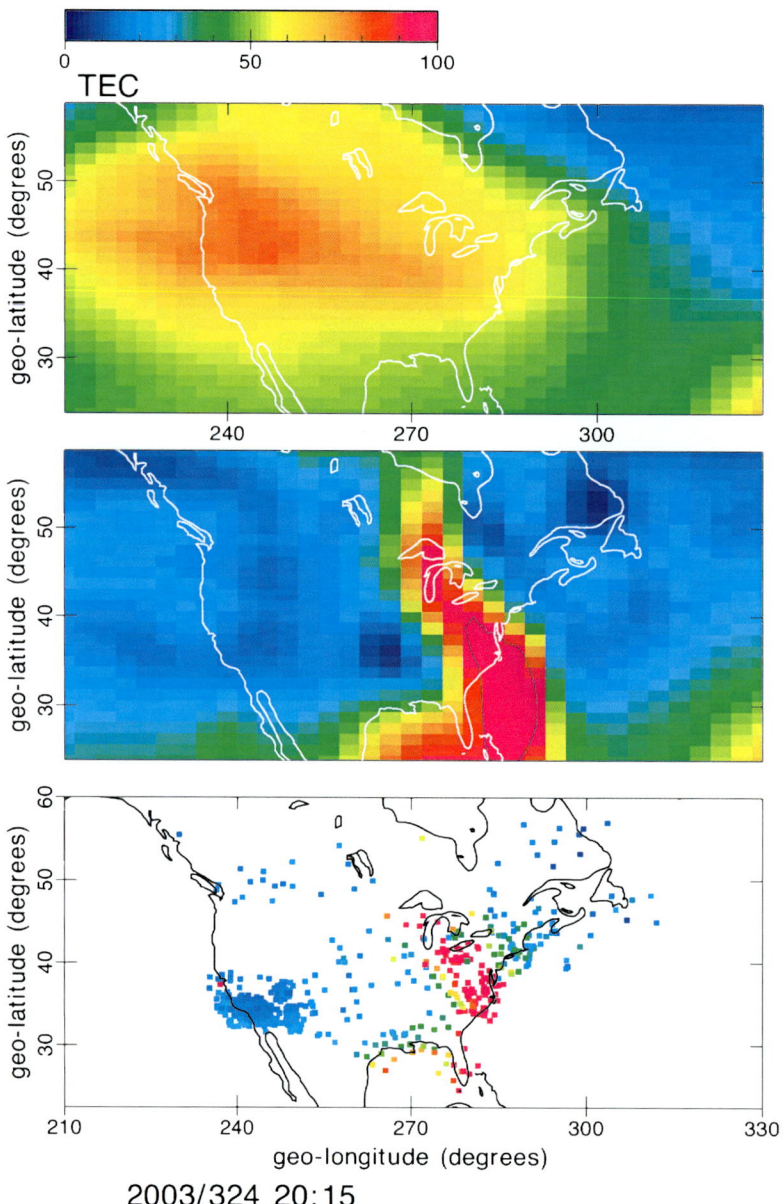

Plate 3. The American sector mid-latitude ionospheric TEC on day 324, 2003, at 20:15 UT as observed by ground GPS receivers (bottom panel), as modeled by the climatological IFM (top panel), and as specified by the USU-GAIM model (middle panel). The TEC is color-coded and given in units of 10^{16} el m^{-2}.

Using IFM, GAIM, and GPS/TEC, it is shown that the first expectation of assimilation is achieved. Plate 3 contrasts the inability of a standard first principles physics model driven by climatology, in this case the IFM, to capture a ridge of TEC across the United States with GAIM's ability not only to do so but also to extend this information to a region devoid of observations. The top panel comprises the IFM TEC, which simply shows a smooth mid-latitude ionosphere. However, on day 324 (2003), at the time when a storm was in progress, the TEC observations (bottom panel) clearly reveal the ridge of high TEC. In the bottom panel, each station's slant TEC values have been converted to an approximate vertical TEC and plotted as a color box at its F-layer pierce point location. The GAIM model has assimilated these slant TEC values as well as data from ionosondes and DMSP in situ plasma measurements. GAIM's data assimilated specification of TEC is shown in the middle panel of Plate 3. In regions where no observations were made, GAIM physics and prior assimilation history fill in the TEC. Plate 3 provides excellent justification for why ionospheric data assimilation is timely. The second expectation of the data assimilation is also partly achieved by GAIM, namely, providing a full 3-D ionospheric specification rather than a 2-D TEC map. Plate 3 gives a reasonable state-of-the-art model specification for present-day ionospheric data assimilation. To compute TEC, the GAIM altitude profiles of ionospheric electron content need to be integrated. The profiles contain an F layer, an h_mF_2, an N_mF_2, etc. However, the present-day capability of these assimilation codes in responding to large changes in layer height is still far from perfect. Figure 6a shows that the layer height change can be more than 100 km for an electric field change of only 5 mV/M [*Tanaka and Hirao*, 1973]. Such changes are observed at mid-latitudes during daytime positive phase storms using ionosondes. Present-day assimilation models usually depend on a climatology-driven physics model that does not contain this extreme range of variability for a background ionosphere. Hence, the assimilation technique cannot readily change layer heights by such a large amount. If measurements are available that contain electron density profile information, other than conventional ground-based GPS, progress can be made to improve the layer height difficulties. At present, there are several research groups working on such models. *Scherliess et al.* [2006] described the USU-GAIM model and their introduction provides references to other data assimilation efforts.

Plate 3, although providing evidence that an improved mid-latitude ionospheric specification is obtained by using data assimilation, does not provide information on the drivers of the ionospheric modifications. The current lack of knowledge on the electric field dynamics and neutral dynamics is also a measure of our lack of understanding of the geomagnetic storm. To successfully forecast storms, we need to understand them. One way to do this is to "observe" their drivers with sufficient spatial and temporal resolution, and infer the dynamics. To date, formal observations have failed to provide sufficient information. Theoretically, first-principles modeling still has a long way to go to get M–I coupling through the whole M–I system operating. Hence, the third expectation of assimilation is that it will provide this information. To address this, a much more sophisticated form of data assimilation is required. Present-day assimilation targets specification of the state variable that is either observed directly or inferred by other measurements. A specific example is the USU GAIM model as represented in Plate 3; its output is the electron density, whereas its inputs are in situ measurements of electron density or line-of-sight electron density integrals that are the slant TEC measurements. The next generation of the USU GAIM, referred to as a full-physics GAIM model, will not only solve for the state variable electron density but also for the dynamic evolution of the physics model drivers. Examples of drivers would be the electric field, neutral wind, thermospheric composition, etc. Clearly, the expectation is that with sufficient observations as well as complementary types of observation and a physics model with robust drivers' dependencies, future data assimilation models would map the drivers as well as the ionospheric density. If this is successful, then even indirectly a method of mapping both the electric field and neutral winds is possible. This would enhance our understanding of the storm evolution and therefore, forecasting the ionosphere during storm periods.

The present-day status of the USU full-physics GAIM work is that for the past 3 years various prototype models of drivers have been tested. These "models" need to be designed to capture the expected dynamics, but also be defined by the least number of variables. Each of these variables becomes an increase in the GAIM state vector and hence rapidly increases CPU needs. Hence, optimization trade-off studies are crucial and under testing. Another aspect of this challenge concerns the self-consistency of the driver impact within the "full-physics." If a process is truncated, preventing a driver's full impact to be determined, then the assimilation may well find inaccurate description for the driver. A good example of this would be the role of meridional winds driving plasma from one hemisphere to another along field lines. If the "full-physics" does not support such flux tube transport, that is, flux tubes are modeled with an inappropriate topside boundary at 1000 km, then an assimilation procedure attempting to extract wind information would not be able to appropriately "weight" physics at the two ionospheric ends of the flux tubes. The result is a meridional wind that is probably inaccurate. The USU researchers, and probably other teams, are prototyping their efforts, which should, in the next 2 or 3 years, begin to provide the community "maps" showing how the "drivers" are evolving. These will have sufficient spatial and temporal resolution that they will be able to provide new insights and constraints on how a mid-latitude ionospheric storm or superstorm evolves. In the same time frame, it is probable that the causal sequences of electrodynamics and neutral dynamics will be uncovered such that possibilities to use forecasting techniques based on real-time assimilation can be discussed.

6. CONCLUSION

This paper has concentrated on how the mid-latitude ionosphere's storm response has produced severe technology impacts on the FAA WAAS system and how this has caught the science community by surprise. Given our understanding of the storm effects, the "damage" is caused by combinations of electric fields and neutral winds, neither of which can presently be mapped with sufficient quality to be useful in either improving specification or forecasting the ionosphere. The present-day success of ionospheric data assimilation lies in producing the best real-time specification. However, it is expected that future efforts involving ionospheric data assimilation based on full-physics models would also provide maps of the electric field and neutral wind drivers. This would then result in a fuller understanding of the storm drivers and the methods of forecasting their evolution during storms. This, in turn, would lead to forecasting the ionosphere while the storm is actually in progress.

Acknowledgments. The research was supported by NSF grant ATM-0408592 and NASA grant NNG04GNG3G to Utah State University. The authors are grateful to the book editor for sharing detailed knowledge about the WAAS PA system and associated technical reference.

REFERENCES

Coster, A. J., J. C. Foster, P. J. Erickson, and F. J. Rick (2001), Regional GPS mapping of storm enhanced density during the 15–16 July 2000 geomagnetic storm, 14th International Technical Meeting of the Satellite Division of the Institute of Navigation (ION GPS 2001), Salt Lake City, UT, 11–14 September 2001, pp. 2531–2539.

Doherty, P., A. J. Coster, and W. Murtagh (2004), Space weather effects of October–November 2003, *GPS Solutions 8*, 267–271, doi:10.1007/s10291-004-0109-3.

Foster, J. C. (1993), Storm time plasma transport at middle and high latitudes, *J. Geophys. Res.*, *98*, 1675–1689.

Hafstad, L. R., and M. A. Tuve (1929), Note on Kennelly-Heaviside layer observations during a magnetic storm, *Terr. Magn. Atmos. Electr.*, *34*, 39–44.

Hardy, D. A., M. S. Gussenhoven, R. Raistrick, and W. J. McNeil (1987), Statistical and function representations of the pattern of auroral energy flux, number flux, and conductivity, *J. Geophys. Res.*, *92*, 12,275–12,294.

Hedin, A. E. (1987), MSIS-86 thermospheric model, *J. Geophys. Res.*, *92*, 4649–4662.

Hedin, A. E., et al. (1991), Revised global model of thermospheric winds using satellite and ground-based observations, *J. Geophys. Res.*, *96*, 7657–7688.

Jones, K. L. (1971), Electrodynamic drift effects in mid-latitude F-region storm phenomena, *J. Atmos. Terr. Phys.*, *33*, 1311–1319.

Jones, K. L., and H. Rishbeth (1971), The origin of storm increases of mid-latitude F-layer electron concentration, *J. Atmos. Terr. Phys.*, *33*, 391–401.

Mannucci, A. J., B. D. Wilson, D. N. Yuan, C. H. Ho, U. J. Lindqwister, and T. F. Runge (1998), A global mapping technique for GPS-derived ionospheric total electron content measurements, *Radio Sci.*, *33*, 565–582.

Martyn, D. F. (1953), The morphology of the ionospheric variations associated with magnetic disturbances, I. Variations at moderately low latitudes, *Proc. R. Soc. London, Ser. A*, *218*, 1–18.

Mendillo, M. (1973), A study of the relationship between geomagnetic storms and ionospheric disturbances at mid-latitudes, *Planet. Space Sci.*, *21*, 349–358.

Mendillo, M. (2006), Storms in the ionosphere: Patterns and processes for total electron content, *Rev. Geophys.*, *44*, RG4001, doi:10.1029/2005RG000193.

Papagiannis, M. D., M. Mendillo, and J. A. Klobucher (1971), Simultaneous Storm-time increases of the ionospheric total electron content and the geomagnetic field in the dusk sector, *Planet. Space Sci.*, *19*, 503–511.

Prolss, G. W. (1995), Ionospheric F-region storms, in *Handbook of Atmospheric Electrodynamics*, vol. 2, edited by H. Volland, pp. 195–248, CRC Press, Boca Raton, Fla.

Scherliess, L., R. W. Schunk, J. J. Sojka, D. C. Thompson, and L. Zhu (2006), Utah State University Global Assimilation of Ionospheric Measurements Gauss–Markov Kalman filter model of the ionosphere: Model description and validation, *J. Geophys. Res.*, *111*, A11315, doi:10.1029/2006JA011712.

Schunk, R. W. (1988), A mathematical model of the middle and high-latitude ionosphere, *Pure Appl. Geophys.*, *127*, 255–303.

Schunk, R. W., and J. C. G. Walker (1973), Theoretical ion densities in the lower ionosphere, *Planet. Space Sci.*, *21*, 1875–1896.

Schunk, R. W., W. J. Raitt, and P. M. Banks (1975), Effect of electric fields on the daytime high-latitude E and F regions, *J. Geophys. Res.*, *80*, 3121–3130.

Schunk, R. W., P. M. Banks, and W. J. Raitt (1976), Effects of electric fields and other processes upon the nighttime high-latitude F layer, *J. Geophys. Res.*, *81*, 3271–3282.

Schunk, R. W., J. J. Sojka, and J. V. Eccles (1997), Expanded capabilities of the Ionospheric Forecast Model, Final Report, AFRL-VS-HA-TR-98-0001, pp. 1–142.

Sojka, J. J. (1989), Global scale, physical models of the *F* region ionosphere, *Rev. Geophys., 27*, 371–403.

Sojka, J. J. (2005), Does the mid-latitude ionosphere exist during superstorms, *Eos Trans. AGU, 86*(18), Jt. Assem. Suppl., Abstract SA12A-02.

Sojka, J. J., and R. Heelis (2007), Discussion of models inadequate spatial distribution of energy into the thermosphere during storms, paper presented at IUGG XXIV General Assembly, Perugia, Italy, 2–13 July 2007.

Sojka, J. J., W. J. Raitt, and R. W. Schunk (1981a), Theoretical predictions for ion composition in the high-latitude winter F-region for solar minimum and low magnetic activity, *J. Geophys. Res., 86*, 2206–2216.

Sojka, J. J., W. J. Raitt, and R. W. Schunk (1981b), A theoretical study of the high-latitude winter F region at solar minimum for low magnetic activity, *J. Geophys. Res., 86*, 609–621.

Tanaka, T., and K. Hirao (1973), Effects of an electric field on the dynamical behavior of the ionospheres and its application to the storm time disturbance of the *F*-layer, *J. Atmos. Terr. Phys., 35*, 1443–1452.

Volland, H. (1975), Models of the global electric fields within the magnetosphere, *Ann. Geophys., 31*, 154–174.

Wilson, B., A. Mannucci, C. Edwards, and T. Roth (1992), Global ionospheric maps using a global network of GPS receivers, *Proceedings of the Beacon Satellite Symposium*, MIT, Cambridge, MA, July 1992.

Wilson, B. D., A. J. Mannucci, and C. D. Edwards (1995), Subdaily northern hemisphere ionospheric maps using an extensive network of GPS receivers, *Radio Sci., 30*, 639–648.

M. David, L. Scherliess, R. W. Schunk, J. J. Sojka, and D. C. Thompson, Center for Atmospheric and Space Sciences, Utah State University, 4405 Old Main Hill, Logan, UT 84322-4405, USA. (jan.sojka@usu.edu)

ns Associated With
Magnetic Storms

R. A. Heelis

Hanson Center for Space Sciences, University of Texas at Dallas, Dallas, Texas, USA

At low and middle latitudes, the dynamics of the ionosphere is affected directly by local neutral wind components parallel to the magnetic field and indirectly through the dynamo action of wind components perpendicular to the magnetic field. Electric fields may also be applied to the region through the application of an external potential at high latitudes, which results from the interaction of the magnetosphere with the interplanetary medium. The application of these fields and the modification of neutral winds by energy inputs at high latitudes produce effects that can dramatically change the usually observed environment at low and middle latitudes. The region over which the high-latitude convection pattern and the associated energy inputs are applied can expand dramatically during a storm, making the normal middle latitudes behave like auroral latitudes. Electric fields can penetrate into the equatorial region with time constants varying from a few minutes to a few hours. These dynamical influences dramatically change the distribution of ionization, with both E×B drifts and neutral winds playing a different role at different latitudes. There is a dynamic interplay between neutral wind transport, electric fields originating from internal processes and electric fields originating from external influences in an inertial frame of reference that must be considered. After examination of these effects, we find that significant questions remain concerning the role that winds and electric field play in producing the ionospheric density perturbations that so dramatically affect the performance of space-based communication and navigation systems.

1. INTRODUCTION

During magnetic quiet times the usual patterns of electric fields in the low-latitude ionosphere are dominated by the dynamo action of neutral winds that result from absorption of solar extreme ultraviolet radiation in the stratosphere and lower thermosphere [*Roble and Ridley*, 1994]. With increasing latitude the neutral winds are increasingly influenced by electromagnetic energy inputs and particle inputs that arise at high latitudes through the interaction of the magnetosphere with the interplanetary medium [*Fesen et al.*, 1993]. Under quasi-steady conditions, the currents produced by the interaction of the magnetosphere with the interplanetary medium are closed through the auroral ionosphere by current systems generated in the magnetosphere. Under these conditions, electric potentials at high latitudes are confined to a region outside the plasmasphere [*Toffoletto et al.*, 2003] and the plasmasphere is "shielded" from high-latitude electric fields. This shielding is never completely effective, but inside the plasmasphere the zonal motion of the plasma is dominated by atmospheric corotation and superimposed on

Midlatitude Ionospheric Dynamics and Disturbances
Geophysical Monograph Series 181
This paper is not subject to U.S. copyright. Published in 2008 by the American Geophysical Union.
10.1029/181GM06

this motion are E×B drifts created by the dynamo action of neutral winds. Near the equator these charged-particle drifts perpendicular to the magnetic field have diurnal variations, being generally westward by day and eastward by night, and upward and poleward by day and downward and equatorward by night [*Fejer et al.*, 1991]. At middle latitudes, a semidiurnal component in the E×B drifts becomes more evident [*Fejer*, 1993].

Quasi-steady conditions rarely prevail and some of the current generated in the magnetosheath and the magnetopause closes through the low- and middle-latitude ionosphere. Thus, some perturbations in the ion drifts at low latitudes can be well correlated with changes in the interplanetary medium [*Kelley et al.*, 2003]. However, during large magnetic storms the electrodynamic configuration of the low- and middle-latitude ionosphere is known to change dramatically, and these changes can be due to interactions between the ionosphere, the thermosphere, and the magnetosphere that remain to be fully understood. Magnetic storms are generally associated with large increases in the electric potential applied across the magnetosphere that occur on timescales of a few hours [*Hairston and Heelis*, 1995]. Under these conditions, the shielding mechanisms operative in the magnetosphere break down and electric potentials applied at high latitudes can result in the appearance of electric fields at low latitudes [*Spiro et al.*, 1988]. A large increase in the electric potential across the magnetosphere may be accompanied by a significant expansion of the polar cap and the auroral zone. Under these circumstances, middle latitudes, which are usually dominated by corotation and dynamo fields, may be directly influenced by auroral electric fields and particles. In addition, energy input from high-latitude electric fields and particles may be applied at latitudes significantly lower than usual. Thus, the dynamo wind systems at low and middle latitudes may be influenced by what are normally regarded as high-latitude energy inputs. In this work we briefly review some of the observed phenomenology in the storm-time electrodynamics of the low- and middle-latitude ionosphere and discuss the physical processes that could be responsible. In so doing, we will also expose some problems that require further observations and interpretation.

2. OBSERVATIONS

Near the equator, the average behavior of E×B drifts has been extensively studied using radar and satellite measurements. These measurements support the role of the E-region tidal wind systems in producing drifts that are upward and westward during the day, and downward and eastward at night. The role of the zonal wind in the F region may also be invoked to produce the so-called prereversal enhancement in the vertical ion drift near sunset [*Eccles*, 1998]. This large-scale behavior of the E×B drift has been extensively modeled using data from the Jicamarca radar [*Fejer et al.*, 1981, 1991] and provides a commonly used baseline from which significant perturbations can be easily recognized.

Figure 1, taken from the work of *Huang et al.* [2005], provides our first perspective that the vertical drift at the equator is directly influenced by changes in the interplanetary electric field. The lowest panel shows the vertical ion drift measured near the dip equator by the Jicamarca incoherent scatter radar and the upper panels show the solar wind dynamic pressure, the north–south component of the interplanetary magnetic field (IMF), and the interplanetary electric field produced by the flowing solar wind embedded in the IMF. Light and heavy traces in each of the panels contrast a nominally quiet day with a more disturbed day, respectively. There are several important points to note from this figure and the associated work. At Jicamarca, the local time lags the universal time by about 5 h. Thus, these data show that in

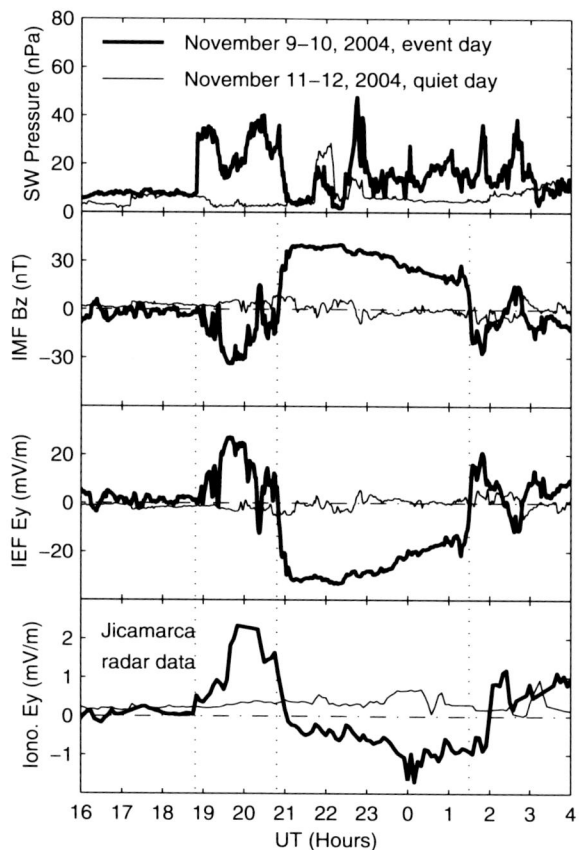

Figure 1. Vertical drift perturbations observed by the Jicamarca radar correlated with changes in the interplanetary electric field [after Huang et al., 2005].

the afternoon and evening hours, increases in the interplanetary electric field associated with southward IMF produce upward drift perturbations, in this case, on the order of 50 m/s. Other studies show that decreases in the interplanetary electric field associated with northward IMF are usually associated with downward drift perturbations in this local time region. We note that the correlations between the interplanetary electric field and the equatorial drift perturbations are not perfect, either in time or magnitude, suggesting that effects other than a penetration field may play a role.

The complex relationships between vertical drift perturbations and variations in the interplanetary electric field and the ring current intensity are emphasized by *Fejer et al.* [2007]. These relationships are further complicated by the fact that changes in the IMF produce drift perturbations at the equator that are different at different local times. A simple picture of equatorial penetration electric fields may be obtained by visualizing the expansion of the auroral convection features to lower latitudes. Increases in the polar cap potential produce drift perturbations at the equator that are predominantly upward during the day and downward at night, westward near dusk, and eastward near dawn. These expectations are in accord with ground-based observations [*Fejer and Scherliess*, 1997], indicating that near 0600 storm-time drift perturbations are very small. However, during large storms significant upward drift perturbations may also be seen near the dusk terminator [*Wolf*, 1970; *Huba et al.*, 2005].

In the evening sector, vertical drift perturbations are poorly described because of the frequent presence of ionospheric irregularities that are observed at this local time [*Basu et al.*, 2005]. These irregularities can be especially severe during storm times because of the previously mentioned upward drift perturbations near dusk. The vertical drift perturbations can drive the *F* region to very high altitudes, where the timescales for gravitational instabilities are very short [*Sultan et al.*, 1996]. Indeed, the work of *Basu et al.* [2007] suggests that upward drift perturbations near dusk will specify the longitude sector in which plasma irregularities will be observed during superstorms. Thus, the local time behavior of storm-time drift perturbations described by *Fejer and Scherliess* [1997] from an analysis of data from Jicamarca may apply at all longitudes, although *Basu et al.* [2007] suggest that the magnitude of the dusk vertical drift perturbations at the magnetic equator may be largest in the longitude sector that includes the South Atlantic anomaly. Rigorous confirmation of this idea will be continuously developed as factors such as intensity of the storm and time history of the event are removed as variables. Furthermore, these drift perturbations will be superimposed on a background drift that certainly will be different at different longitudes [*Kil et al.*, 2007; *Hartman and Heelis*, 2007].

The satellite data shown here later suggest that storm-time drift perturbations are generally short-lived and associated with changes in the IMF, whereas the result shown in Figure 1 suggests that the drift perturbations may be longer lived and span a region of enhanced or reduced interplanetary electric field. In this regard, *Fejer and Scherliess* [1997] note that storm-time drift perturbations appear over two timescales. One is associated almost immediately with the changing IMF and termed a penetration field. The other, occurring with a time delay of order 1 h, is referred to as a disturbance dynamo. For many large storms, the period of enhanced interplanetary electric field may have a duration of several hours, as shown Figure 1. Coupled with an expansion toward middle latitudes of the region influenced by high latitudes, it is thus possible that both penetration and disturbance dynamo fields are simultaneously present. As noted earlier, the temporal evolution of the ion drift and the interplanetary electric field suggests that these two processes may not be clearly separated in time.

To appreciate the roles played by these mechanisms, it is important to locate the observation with respect to the auroral zone itself. This becomes particularly important during storm times when the normal two-cell convection pattern applied to the ionosphere at magnetic latitudes greater than 60° reaches to magnetic latitudes as low as 40°. Plate 1 shows the temporal evolution of the convection reversal boundary and the equatorward edge of the diffuse electron aurora observed from the DMSP F13 satellite during the large storm event of 20 November 2003. The top panel describes the north–south component of the IMF and the *Dst* index, allowing the storm period to be easily identified. The bottom panel indicates regions of sunward and antisunward convection with magnitudes represented by the color scale to the right. A solid green curve that designates the convection reversal boundary divides these regions. Another curve shown in black at lower latitudes defines the equatorward edge of the diffuse auroral electron precipitation. It is clear that magnetic latitudes near 50°, which might normally be associated with middle-latitude influences, are now fully engulfed in the auroral zone during the storm interval. Furthermore, near dusk the sunward (westward) flows normally associated with the auroral zone now exist at latitudes as low as 30°. Under such circumstances, the timescale for modification of the winds at latitudes near 40° is very short and even the dynamo wind field at the equator can be modified in periods less than 1 h after storm onset [*Maruyama et al.*, 2005].

Both upward and downward drift perturbations at the equator have a significant effect on the distribution of plasma both at the equator and at middle latitudes [*Heelis and Coley*, 2007]. The dramatic redistribution of plasma at low and middle latitudes, which is apparently associated with vertical drift perturbations, has been reported by several authors [*Tsurutani et al.*, 2004; *Mannucci et al.*, 2005] and modeled using sophisticated coupled models of the ionosphere and thermosphere [*Lin et al.*, 2005]. Plate 2, taken from the work of *Mannucci et al.* [2005], illustrates the total electron content (TEC) above the altitude of about 400 km observed as a function of latitude during the course of the large magnetic storm of 29 October 2003. As shown in the insert, each profile is taken at different longitudes over the United States near 1300 LT. Near 1900 UT, the latitude profile shows the normally seen TEC distribution with values ranging from 20 to 80 TEC units and the appearance of equatorial anomaly peaks near 15° magnetic latitude. During the storm, the TEC near the equator is reduced only slightly, whereas at latitudes beyond 20° the TEC is increased by up to a factor of 10. A peak in the TEC, which we may attribute to a signature of the equatorial anomaly, moves to magnetic latitudes between 20° and 30°, and a dramatic decrease in the TEC to values usually seen at auroral latitudes is marked by a shoulder of elevated TEC just poleward of 40° magnetic latitude. The possible role of electric fields and E×B drifts in producing these signatures will be discussed later.

Figure 2, taken from the work of *Heelis and Coley* [2007], shows a similar redistribution of the total ion number density in the topside ionosphere during the storm of 20 November 2003. Shown here is a sequence of latitude profiles obtained by the polar orbiting DMSP satellite as it crosses the equator near 1800 LT during the course of the storm. In this figure, the solid traces show the storm-time profile, and to the right the vertical drift perturbation observed at the magnetic equator during that pass is presented. The dotted profiles show the observation taken during the quiet-time passes on the days before and after the storm. A clear association between the plasma density variation near the equator and the upward drift perturbation at the equator is apparent. However, it is important to recognize that the latitude profiles of the ion density shown here and similarly the TEC profiles shown in Plate 2 are dependent on the time history of the ion drift and not simply the simultaneously observed value.

Further inspection of the data in Figure 2 shows, for example, that shortly after storm onset (1100 UT) the topside ion density is enhanced at middle latitudes prior to the enhancements observed at the equator and in the presence of downward drift perturbations at the equator. During the storm development, the equatorial anomaly signature produced by a large upward drift perturbation is later retained even in the presence of a downward drift perturbation. Finally, storm-time peaks in the topside ion density at middle and low latitudes become clearly separate features. We note that the peak in the topside density seen at middle latitudes marks a rapid decrease in density with increasing latitude. This is similar to the feature seen in the TEC profile in the form of a mid-latitude shoulder that marks a rapid decrease in TEC with increasing latitude near 40° magnetic.

Figure 3 provides a direct comparison of the latitude profile of topside ion density obtained from the DMSP satellite and TEC obtained from JPL using the global ground network of GPS receivers, during the storm of 6 November 2001. In this figure, the dashed curves represent the previous quiet-day observation and the solid curve represents the storm-time profile near 1800 LT. The latitude features identified earlier as the equatorial anomaly and the middle-latitude density enhancement or TEC shoulder are present in both parameters. However, the topside density shows hemispheric asymmetries associated with changes in the F peak height that do not appear in the TEC profile.

3. E×B DRIFTS, NEUTRAL WINDS, AND COROTATION

In seeking explanations for the storm-time behavior of the ionosphere, it must be appreciated that both E×B drifts and neutral winds (directly and through dynamo electric fields) can contribute to dramatic redistributions of the plasma. Furthermore, the effects of drifts and winds will be different in daylight and darkness, and different at low and middle latitudes. However, the ion number density distribution in the ionosphere reflects the time integrated contribution of both these drivers, and thus their effects may not be easily distinguishable in a single snapshot of an event provided by a satellite pass or a radar scan. The upward motion of the plasma during the daytime will move plasma to regions of lower recombination, whereas photoproduction will continue to replenish the ion number density just below the F peak. Thus, the F layer will increase in effective thickness with corresponding increases in the total electron content [*Balan et al.*, 1997]. Similarly, the downward motion of the plasma will increase the recombination rate in the presence of photoproduction leading to lower-than-expected ion number densities above and below the F peak.

At equatorial latitudes, E×B drifts are very effective in moving the plasma vertically. But in so doing, the magnetized plasma moves to regions of higher flux tube volume and field-aligned transport away from the equator is induced by the resulting plasma pressure gradients. *Hanson and Moffett* [1966] first presented models of the flux of plasma induced by upward E×B drifts during the daytime. They indicate that the combined effects of E×B drifts and field-aligned motion produce F region plasma fluxes that increase with increasing latitude and are directed almost horizontally near 10° magnetic latitude with downward fluxes present beyond 15° magnetic latitude. This pattern has been

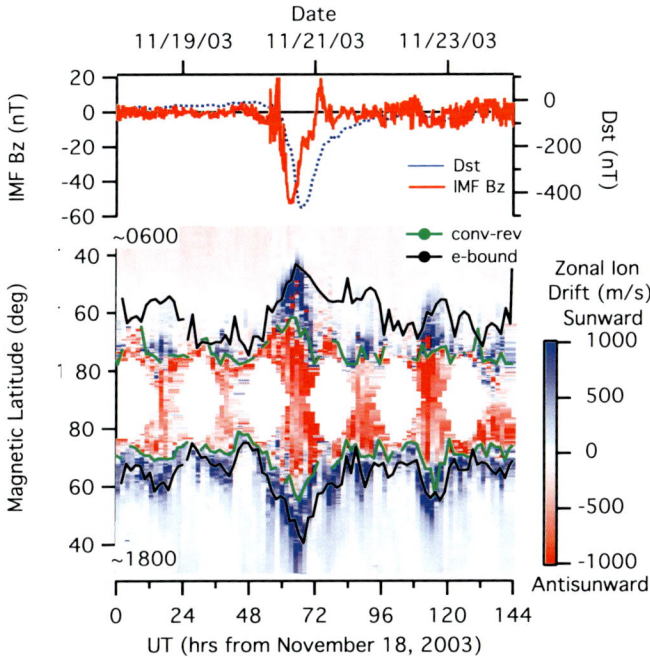

Plate 1. Storm-time evolution of the high-latitude convection pattern observed by DMSP F13 on 20 November 2003 shows the expansion of the high latitude region to middle latitudes and the penetration of zonal drifts to the equator.

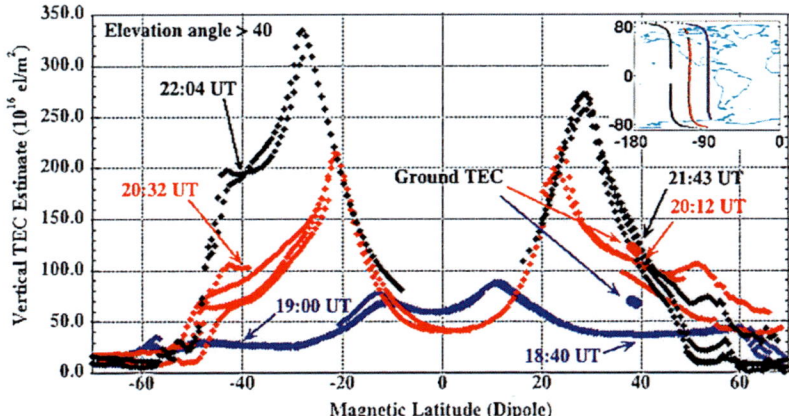

Plate 2. Storm-time latitude profiles of TEC obtained during the storm of October 2003 show the expansion of the equatorial anomaly and a middle-latitude enhancement that appears as a shoulder at the equatorward edge of the auroral region [after *Mannucci et al.*, 2005].

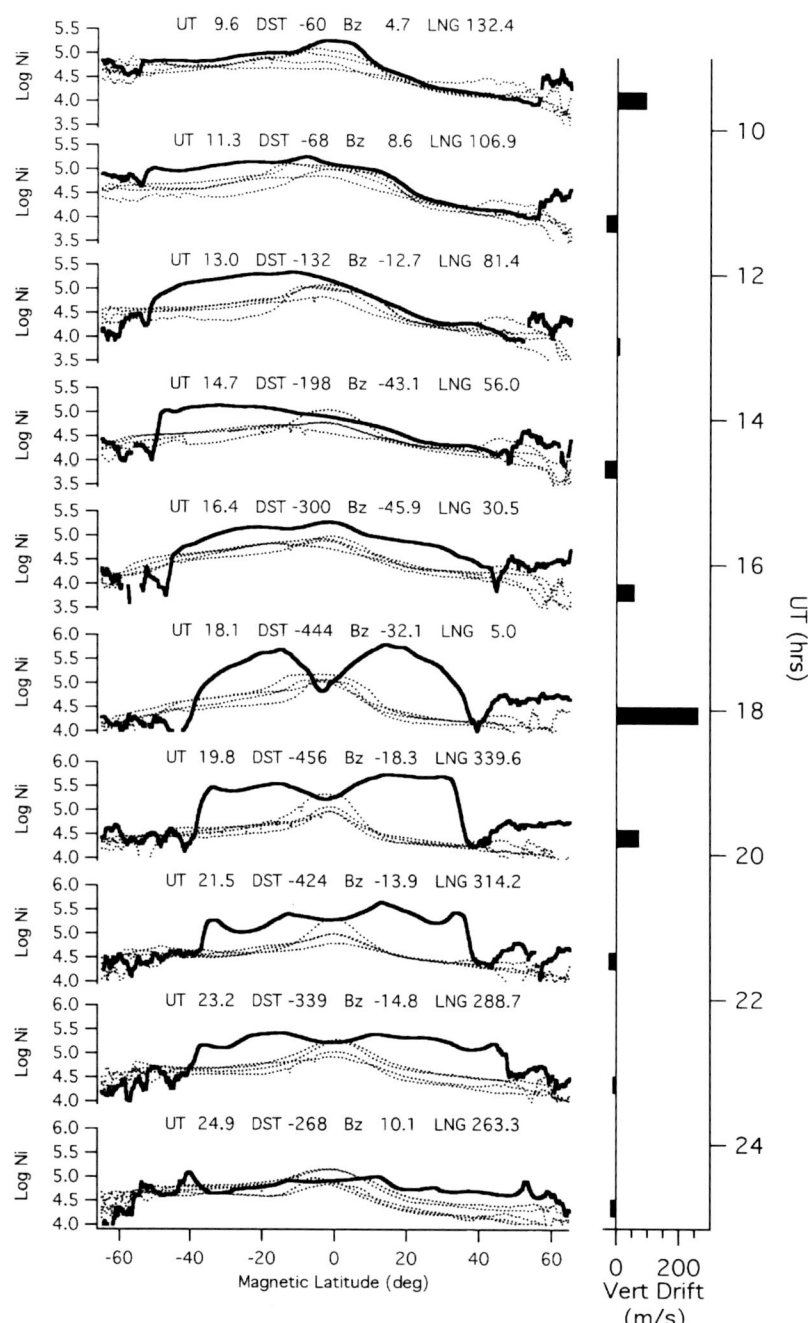

Figure 2. Sequence of ion number density profiles observed by DMSP F13 near 1800 LT for the storm of 20 November 2003. The solid traces correspond to the UT given above. Lighter traces show the profiles recorded 2 days before and after the event [after *Heelis and Coley*, 2007].

named the fountain effect and illustrates that the equatorial anomaly is produced by the horizontal and downward transport of plasma produced in the flux tube where the recombination rate is low.

It is important to recognize that near the equator chemistry and field-aligned transport act to rapidly reduce the vertical flux of plasma with altitude. Thus, substantial increases in the E×B drift will widen in latitude the region of enhanced

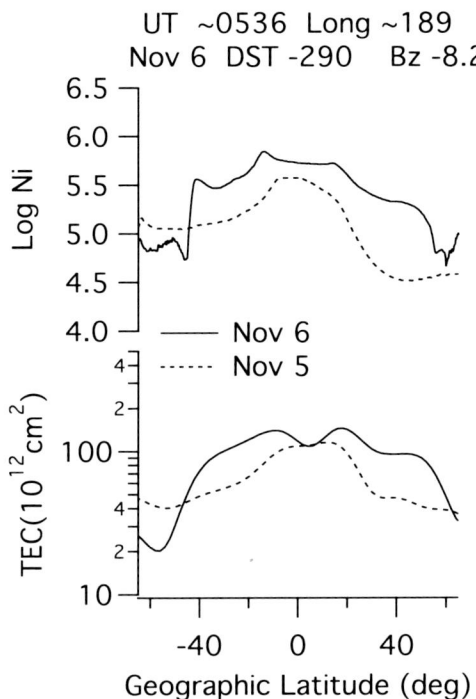

Figure 3. Simultaneous observation of the topside ion density and the total electron content for approximately the same geographic locations on 5 November 2001 (before a superstorm) and 6 November 2001 (during a superstorm) show comparable features corresponding to density enhancements in the equatorial anomaly and at middle latitudes at the equatorward edge of the auroral zone.

fluxes rather than increase in altitude the region of large vertical fluxes [*Balan and Bailey*, 1995]. Figure 4 offers a schematic representation of a so-called superfountain that will displace the anomaly peaks to higher latitudes. It emphasizes that plasma enhancements in the anomaly region are produced by the small upward, but largely horizontal, motion of plasma on neighboring flux tubes and not by the transport of plasma upward at the equator and along the magnetic flux tubes that thread the anomaly at higher latitudes. Neutral winds move the plasma primarily along the magnetic field and produce hemispheric asymmetries in the anomaly crests.

In the absence of photoproduction, during the evening and nighttime, the plasma is lost by recombination below the F peak. As the F peak density falls, the plasma in the topside diffuses rapidly downward along the magnetic field to replace it. Ion-neutral collisions ensure that the topside density preserves a diffusive equilibrium profile as the F peak density decays. At middle latitudes, E×B drifts will be accompanied by field-aligned diffusion such that the net transport of plasma is primarily horizontal [*Behnke et al.*, 1985; *Fejer*, 1993]. Thus, the F peak height is not affected and the plasma density is not changed significantly. However, meridional neutral winds will move the entire ionization profile vertically in the direction along the magnetic field and can lower or raise the F peak, and change the recombination rate. Thus, nighttime plasma density enhancements and depletions can be produced by meridional neutral winds at middle latitudes. In addition, the "shape preserving" attribute of the topside density variation can retain density enhancements produced in the daytime through much of the nighttime. For example, a localized upward drift perturbation during the daytime will produce an elevated F peak height and density. During the night, all plasma densities will decline but the relatively elevated height and density produced by the perturbation will be retained in a shape-preserving manner as they corotate into the evening sector in the absence of the upward drift perturbation.

Finally we point out that it is not necessary to invoke the same mechanism at low and middle latitudes for the production of plasma density and TEC enhancements. Indeed, the sometimes-independent behavior of these features would suggest that the formation of the density enhancements at middle and low latitudes should be treated separately. At low latitudes, neutral wind transport along the magnetic field is not effective in moving the plasma vertically. Thus, enhancements in the vertical E×B drift are a straightforward and verifiable candidate for the production of density enhancements. At middle latitudes, E×B drifts and neutral winds can combine to maximize the upward motion of the plasma or they may oppose to minimize the effect. During storm times, the neutral wind is expected to blow equatorward in response to energy inputs in the auroral zone. E×B drifts across the dayside are expected to be radially outward because of the expansion of the high-latitude convection

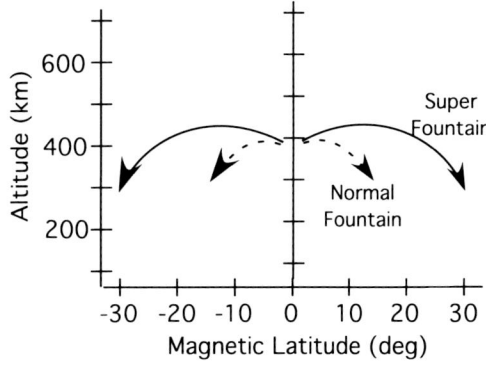

Figure 4. Schematic representation of the plasma flux induced by the so-called superfountain. Note that the plasma transport is largely horizontal.

pattern. Thus, across the afternoon sector both neutral winds and E×B drifts could contribute to enhancements in the ion density and TEC.

The simultaneous occurrence of storm-time perturbations in the ion density and the ion drifts at low and middle latitudes has led to many interesting hypotheses concerning the role that they play in phenomena observed at higher latitudes. Near the equatorward edge of the auroral zone, the so-called storm enhanced densities (SED) have been identified as latitudinally confined features that convect sunward across the afternoon sector and poleward toward the cusp [*Foster*, 1993]. These features have been identified with plasmaspheric erosion plumes of enhanced density that convect toward the magnetopause during major magnetic storms [*Foster et al.*, 2002]. The observation of westward flowing subauroral polarization streams (SAPS) and density enhancements in the evening sector have also led to speculation that these drifts may play a role in drawing plasma from the plasma density enhancement at middle latitudes to locations at higher latitudes and earlier local times [*Foster et al.*, 2005, 2007]. The original definition of SEDs attached a specific flow direction to the density for the enhancement to be termed a SED. More recently, the connection between SEDs and their convective motion has been loosened [*Coster et al.*, 2007], and detailed inspection of the data taken during storm times suggests that it is indeed prudent to consider separately the formation mechanisms for storm enhanced density features and the convective motion of the plasma.

Figure 5 shows a latitude profile of ion number density and zonal ion drift observed from the DMSP F13 satellite crossing the equator near 1800 LT during the superstorm of 20 November 2003. Vertical lines locate approximately the equatorward edge of the diffuse auroral electron precipitation that again emphasizes its location at much lower latitudes than might normally be expected. The solid curve represents the zonal ion drift in an inertial frame fixed with respect to the Sun, whereas the dashed curve shows the corotation velocity in this same frame. In this format, the westward flows associated with the auroral zone poleward of the electron precipitation can be easily identified. A SAPS can also be seen equatorward of the electron precipitation boundary.

The topside ion density profile clearly identifies large ion densities near the equator associated with an equatorial anomaly. The anomaly peaks are well separated from density enhancements at middle latitudes just equatorward of the auroral zone. In a corotating frame, the enhancements at middle latitudes are moving slowly to the west under the influence of what we will call a penetration field. They would thus qualify as storm-enhanced density. However, we point out that the westward drift does not exceed the corotation velocity at their location and thus near dusk these features are flowing away from the Sun and away from local noon. Thus, at the instant of this observation near dusk, subauroral flows are not associated with transport of enhanced density toward local noon. It should be recognized that enhanced F region densities, once produced, could persist for many hours depending on the production and loss processes at work. Thus, an instantaneous correspondence between spatial gradients in the convective flow and spatial gradients in the ion density should not be expected unless the configuration remains stable for several hours. During storm times, this is clearly not the case. It thus seems prudent to consider the possibility that in the evening sector subauroral flows identified as SAPS and subauroral density enhancements do not evolve simultaneously in time.

Although it is well established that enhanced densities associated with magnetic storm activity occur throughout the afternoon and evening sectors, good evidence exists from sequential TEC maps [*Foster et al.*, 2005] that only in the afternoon sector does the poleward extension of enhanced densities imply a poleward transport from the density enhancement reservoir at subauroral latitudes. In the afternoon sector, we point out that sunward auroral flows have westward and poleward components, and that the westward component is in opposition to the corotation velocity. Thus, the westward auroral flow and the eastward corotation can act to isolate a poleward (and upward) flow of plasma in sunlight, a condition that we have previously pointed out is most conducive to the creation of enhanced plasma density.

As an added note, we point out that not all spatial gradients in the ion number density at low and middle latitudes can be attributed to plasma transport by neutral winds and/or

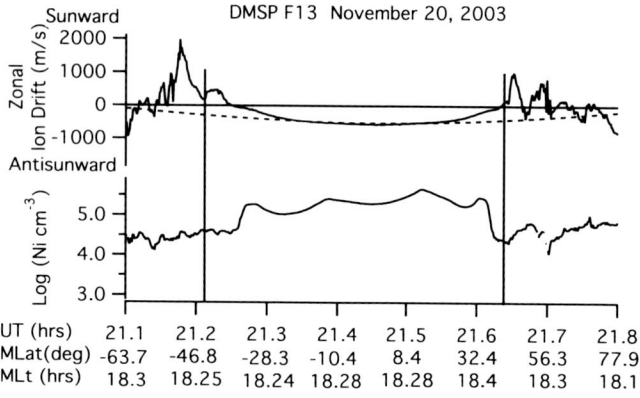

Figure 5. The zonal drift and topside density profile observed during a pole-to-pole pass of DMSP F13 during the storm of 20 November 2003. The zonal drift is shown in an inertial reference frame and the dashed line denotes the corotation velocity. Vertical lines locate the equatorial edge of diffuse auroral electron precipitation.

E×B drifts. Dramatic increases in the energy inputs to the thermosphere at high latitudes can produce large changes in the relative abundance of molecular nitrogen at a given altitude [*Immel et al.*, 2001; *Burns et al.*, 2006]. These changes are reflected in the ionospheric photochemistry leading to changes in the ionospheric density.

4. SUMMARY AND CONCLUSIONS

During large magnetic storms, the low- and middle-latitude ionosphere and thermosphere evolve over timescales of many (5–10) hours. During that time the dynamics of the ionosphere and thermosphere are significantly perturbed from the usual quasi-steady diurnal behavior upon which we base most of our understanding. It is important to recognize that even in the quasi-steady state the variability in the ion drifts near the equator is of the order of the average value of about 30 m/s. Thus, most attention is given to so-called superstorms when the drift perturbations significantly exceed this value. An increase in the electric potential across the magnetosphere effectively projects an image of a two-cell convection pattern into the plasmasphere. For example, for southward IMF conditions, the equatorial drift perturbations are up and down at noon and midnight, respectively, and east and west at dawn and dusk, respectively. Significantly large upward drifts are seen near the equator at dusk that are attributable to the influence of the ionospheric conductivity on closing the current from the inner magnetosphere [*Maruyama et al.*, 2005].

This projection of the high-latitude convection electric field, which produces large ion drift perturbations at the equator, is generally short-lived. That is, the convection pattern rapidly retreats from equatorial latitudes. However, over more extended periods, the auroral particle precipitation and the associated sunward auroral flows may be located in regions below magnetic latitudes of 60° that we would normally regard as rather benign middle latitudes. Dramatic enhancements in the ionospheric density and the total electron content lie just equatorward of the boundary of the newly enlarged auroral region, which is now located at middle latitudes. These enhancements exist across most of the dayside ionosphere and extend across the dusk terminator into the evening sector.

The middle-latitude density enhancements are frequently accompanied by enhancements in the density at lower latitudes produced by the short-lived upward drifts at the equator. Although an expanded equatorial fountain can displace the equatorial anomaly peaks to high latitudes, these peaks are frequently separated from the peaks associated with middle-latitude density enhancements immediately equatorward of the auroral zone. Thus, it appears prudent to consider the formation and evolution of these enhancements at middle and low latitudes as separate features even if they may at times be collocated.

Across the dayside, E×B drifts with an upward component appear to be the only viable mechanism to produce dramatic density enhancements at the equator. The penetration of the electric field produced by high-latitude magnetospheric potentials also appears to be the source of the drift perturbations. At middle latitudes, both equatorward neutral winds and poleward E×B drifts produce upward plasma motions that could account for density and TEC enhancements. Equatorward winds can be induced by storm-time energy inputs to the auroral region that heat the thermosphere, whereas poleward E×B drifts arise naturally across the dayside by virtue of an expanded high-latitude convection pattern even if the expansion does not involve a penetration field in the equatorial region.

Concerning the appearance of enhanced plasma densities in the evening side, it seems likely that these structures have corotated into the region from the dayside. Significant evidence exists for the stimulation of intense plasma irregularities in the equatorial region as the result of storm-time upward drift perturbations near dusk [*Basu et al.*, 2007]. Further investigations are necessary to reveal the extent to which the decay rate of enhanced densities at middle latitudes is modified by E×B drifts and winds.

Also present across the evening sector during the storm recovery phase are so-called SAPS [*Foster and Burke*, 2002]. These electric field structures are known to exist equatorward of the electron precipitation region in the evening [*Anderson*, 2004] and when combined with the corotation, the plasma in the SAPS region may be moving toward or away from the Sun. Examination of TEC maps covering the American sector appears to suggest that plasma from the middle-latitude density enhancements is transported poleward in the afternoon sector. For this plasma to first convect toward the Sun from much later local times before turning poleward would require the convective configuration to remain stable over a period of several hours during the storm. A more natural explanation for middle-latitude density enhancements that maximize in the afternoon sector and move poleward into the cusp region is to note that in this local time region the combined effects of corotation and the magnetospheric electric field will result in a net poleward (and upward) E×B drift in the presence of sunlight. These conditions both enhance the plasma density and provide the required motion toward the cusp. Further investigation of the poleward ion drifts associated with density variations across the afternoon and evening sectors is required to establish the effectiveness of this mechanism. Although it may not be necessary to invoke a subauroral polarization field to transport middle-latitude ion density enhancements toward the dayside, further study is required to determine the net sunward plasma flux that is supported by these flows. With these future observations and the incorporation of the results into numerical simulations, we may look forward to providing a clearer picture of the storm-time evolution of the low- and middle-latitude ionosphere.

Acknowledgments. This work is supported by NSF grant ATM-0436494 and NASA grant NNX07AF36G to the University of Texas at Dallas. The author thanks R.A. Wolf for useful insights into the properties of penetration fields, and A. Mannucci and Xiaoqing Pi at JPL for providing the TEC data for November 2001.

REFERENCES

Anderson, P. C. (2004), Subauroral electric fields and magnetospheric convection during the April 2002 geomagnetic storms, *Geophys. Res. Lett.*, *31*, L11801, doi:10.1029/2004GL019588.

Balan, N., and G. J. Bailey (1995), Equatorial plasma fountain and its effects: Possibility of an additional layer, *J. Geophys. Res.*, *100*(A11), 21,421–21,432.

Balan, N., G. J. Bailey, M. A. Abdu, K. I. Oyama, P. G. Richards, J. MacDougall, and I. S. Batista (1997), Equatorial plasma fountain and its effects over three locations: Evidence for an additional layer, the F_3 layer, *J. Geophys. Res.*, *102*(A2), 2047–2056.

Basu, S., Su. Basu, K. M. Groves, E. MacKenzie, M. J. Keskinen, and F. J. Rich (2005), Near-simultaneous plasma structuring in the midlatitude and equatorial ionosphere during magnetic superstorms, *Geophys. Res. Lett.*, *32*, L12S05, doi:10.1029/2004GL021678.

Basu, S., Su. Basu, F. J. Rich, K. M. Groves, E. MacKenzie, C. Coker, Y. Sahai, P. R. Fagundes, and F. Becker-Guedes (2007), Response of the equatorial ionosphere at dusk to penetration electric fields during intense magnetic storms, *J. Geophys. Res.*, *112*, A08308, doi:10.1029/2006JA012192.

Behnke, R., M. Kelley, C. Gonzalez, and M Larsen (1985), Dynamics of the Arecibo ionospheres—A case study approach, *J. Geophys. Res.*, *90*, 4448–4452.

Burns, A. G., W. Wang, T. L. Killeen, S. C. Solomon, and M. Wiltberger (2006), Vertical variations in the N_2 mass mixing ratio during a thermospheric storm that have been simulated using a coupled magnetosphere–ionosphere–thermosphere model, *J. Geophys. Res.*, *111*, A11309, doi:10.1029/2006JA011746.

Coster, A. J., M. J. Colerico, J. C. Foster, W. Rideout, and F. Rich (2007), Longitude sector comparisons of storm enhanced density, *Geophys. Res. Lett.*, *34*, L18105, doi:10.1029/2007GL030682.

Eccles, V. J. (1998), Modeling investigation of the evening prereversal enhancement of the zonal electric field in the equatorial ionosphere, *J. Geophys. Res.*, *103*(A11), 26,709–26,720.

Fejer, B. G. (1993), F region plasma drifts over Arecibo: Solar cycle, seasonal, and magnetic activity effects, *J. Geophys. Res.*, *98*(A8), 13,645–13,652.

Fejer, B. G., and L. Scherliess (1997), Empirical models of storm time equatorial zonal electric fields, *J. Geophys. Res.*, *102*(A11), 24,047–24,056.

Fejer, B. G., D. T. Farley, C. A. Gonzales, R. F. Woodman, and C. Calderon (1981), F region east–west drifts at Jicamarca, *J. Geophys. Res.*, *86*(A1), 215–218.

Fejer, B. G., S. A. Gonzalez, E. R. dePaula, and R. F. Woodman (1991) Average vertical and zonal F region plasma drifts over Jicamarca, *J. Geophys. Res.*, *96*, 13,901–13,906.

Fejer, B. G., J. W. Jensen, T. Kikuchi, M. A. Abdu, and J. L. Chau (2007), Equatorial ionospheric electric fields during the November 2004 magnetic storm, *J. Geophys. Res.*, *112*, A10304, doi:10.1029/2007JA012376.

Fesen, C. G., A. D. Richmond, and R. G. Roble (1993), Theoretical effects of geomagnetic activity on thermospheric tides, *J. Geophys. Res.*, *98*(A9), 15,599–15,612.

Foster, J. C. (1993), Storm-time plasma transport at middle and high latitudes, *J. Geophys. Res.*, *98*, 1675

Foster, J. C., and W. J. Burke (2002), SAPS: A new categorization for sub-auroral electric fields, *Eos Trans. AGU*, *83*(36), 393.

Foster, J. C., A. J. Coster, P. J. Erickson, J. Goldstein, and F. J. Rich (2002), Ionospheric signatures of plasmaspheric tails, *Geophys. Res. Lett.*, *29*(13), 1623, doi:10.1029/2002GL015067.

Foster, J. C., et al. (2005), Multiradar observations of the polar tongue of ionization, *J. Geophys. Res.*, *110*, A09S31, doi:10.1029/2004JA010928.

Foster, J. C., W. Rideout, B. Sandel, W. T. Forrester, and F. J. Rich (2007), On the relationship of SAPS to storm-enhanced density, *J. Atmos. Sol. Terr. Phys.*, *69*(3), 303, doi:10,101/j.jastsp.2006.07.021.

Hairston, M. R., and R. A. Heelis (1995), Response time of the polar ionospheric convection pattern to changes in the north–south direction of the IMF, *Geophys. Res. Lett.*, *22*(5), 631–634.

Hairston, M. R., K. A. Drake, and R. Skoug (2005), Saturation of the ionospheric polar cap potential during the October–November 2003 superstorms, *J. Geophys. Res.*, *110*, A09S26, doi:10.1029/2004JA010864.

Hanson, W. B., and R. J. Moffett, (1966) Ionization transport effects in the equatorial F region, *J. Geophys. Res.*, *71*, 5559–5572.

Hartman, W. A., and R. A. Heelis (2007), Longitudinal variations in the equatorial vertical drift in the topside ionosphere, *J. Geophys. Res.*, *112*, A03305, doi:10.1029/2006JA011773.

Heelis, R. A., and W. R. Coley (2007), Variations in the low- and middle-latitude topside ion concentration observed by DMSP during superstorm events, *J. Geophys. Res.*, *112*, A08310, doi:10.1029/2007JA012326.

Huang, C.-S., J. C. Foster, and M. C. Kelley (2005), Long-duration penetration of the interplanetary electric field to the low-latitude ionosphere during the main phase of magnetic storms, *J. Geophys. Res.*, *110*, A11309, doi:10.1029/2005JA011202.

Huba, J. D., G. Joyce, S. Sazykin, R. Wolf, and R. Spiro (2005), Simulation study of penetration electric field effects on the low- to mid-latitude ionosphere, *Geophys. Res. Lett.*, *32*, L23101, doi:10.1029/2005GL024162.

Immel, T. J., G. Crowley, J. D. Craven, and R. G. Roble (2001), Dayside enhancements of thermospheric O/N-2 following magnetic storm onset, *J. Geophys. Res.*, *106*, 15,471–15,488.

Kelley, M. C., J. J. Makela, J. L. Chau, and M. J. Nicolls (2003) Penetration of the solar wind electric field into the magnetosphere/ionosphere system, *Geophys. Res. Lett.*, *30*(4), 1158, doi:10.1029/2002GL016321.

Kil, H., S.-J. Oh, M. C. Kelley, L. J. Paxton, S. L. England, E. Talaat, K.-W. Min, and S.-Y. Su (2007), Longitudinal structure of the vertical $E \times B$ drift and ion density seen from ROCSAT-1, *Geophys. Res. Lett.*, *34*, L14110, doi:10.1029/2007GL030018.

Lin, C. H., A. D. Richmond, R. A. Heelis, G. J. Bailey, G. Lu, J. Y. Liu, H. C. Yeh, and S.-Y. Su (2005), Theoretical study of the low- and midlatitude ionospheric electron density enhancement during the October 2003 superstorm: Relative importance of the neutral wind and the electric field, *J. Geophys. Res.*, *110*, A12312, doi:10.1029/2005JA011304.

Mannucci, A. J., B. T. Tsurutani, B. A. Iijima, A. Komjathy, A. Saito, W. D. Gonzalez, F. L. Guarnieri, J. U. Kozyra, and R. Skoug (2005), Dayside global ionospheric response to the major interplanetary events of October 29–30, 2003 "Halloween Storms," *Geophys. Res. Lett.*, *32*, L12S02, doi:10.1029/2004GL021467.

Maruyama, N., A. D. Richmond, T. J. Fuller-Rowell, M. V. Codrescu, S. Sazykin, F. R. Toffoletto, R. W. Spiro, and G. H. Millward (2005), Interaction between direct penetration and disturbance dynamo electric fields in the storm-time equatorial ionosphere, *Geophys. Res. Lett.*, *32*, L17105, doi:10.1029/2005GL023763.

Roble, R. G., and E. C. Ridley (1994), A thermosphere–ionosphere–mesosphere–electrodynamics general circulation model (time-GCM): Equinox solar cycle minimum simulations (30–500 km), *Geophys. Res. Lett.*, *21*(6), 417–420.

Spiro, R. W., R. A. Wolf, and B. J. Fejer (1988) Penetrating of high-latitude-electric-field effects to low latitudes during SUNDIAL 1984, *Ann. Geophys.*, *6*, 39–49.

Sultan, P. J. (1996), Linear theory and modeling of the Rayleigh–Taylor instability leading to the occurrence of equatorial spread F, *J. Geophys. Res.*, *101*(A12), 26,875–26,891.

Toffoletto, F., S. Sazykin, R. W. Spiro, and R. A. Wolf (2003), Inner magnetospheric modeling with the Rice Convection Model, *Space Sci. Rev.* *107*(1), 175–196.

Tsurutani, B., et al. (2004), Global dayside ionospheric uplift and enhancement associated with interplanetary electric fields, *J. Geophys. Res.*, *109*, A08302, doi:10.1029/2003JA010342.

Wolf, R. A. (1970), Effects of ionospheric conductivity on convective flow of plasma in the magnetosphere, *J. Geophys. Res.*, *75*, 4677–4698.

R. A. Heelis, Hanson Center for Space Sciences, University of Texas at Dallas, 2601 North Floyd Road, Dallas, TX 75083-0688, USA. (heelis@utdallas.edu)

A Data-Model Comparative Study of Ionospheric Positive Storm Phase in the Midlatitude F Region

G. Lu,[1] L. P. Goncharenko,[2] A. J. Coster,[2] A. D. Richmond,[1] R. G. Roble,[1] N. Aponte,[3] and L. J. Paxton[4]

A strong positive storm phase was observed by both the Millstone Hill and Arecibo incoherent scatter radars during a moderate geomagnetic storm on 10 September 2005. The positive storm phase featured an interesting UT–altitude profile of the F region electron density enhancement that closely resembles the Greek letter Λ. The radar measurements showed that the uplift of the electron density peak height corresponded to a strong upward ion drift, whereas the subsequent falling of the peak height coincided with a downward ion drift. Using realistic, time-dependent ionospheric convection and auroral precipitation as input, the thermosphere–ionosphere electrodynamics general circulation model (TIEGCM) is able to reproduce the same Λ-like structure in the electron density profile, along with many large-scale features in electron temperature and vertical ion drift as observed by the radars. Over the 3-day period of 8–10 September, our simulation results show an error of 1%–4% for h_mF_2, electron, and ion temperatures at both radar locations. The estimated error for N_mF_2 is about 9% at Millstone Hill and 19% at Arecibo. However, the simulated vertical ion drifts are less accurate, with the normalized root-mean-square errors of 72% at Millstone Hill and 52% at Arecibo, due largely to model's inability to capture the large temporal fluctuations measured by the radars. However, it reproduces reasonably well the overall large-scale variations during the 3-day period, including the storm-time-enhanced upward ion drift that is responsible for the interesting F region density profile. The model is also able to reproduce the temporal and spatial total electron content variations as shown in the global GPS maps. The comparison with the GUVI O/N_2 is less satisfactory, although there is a general agreement in terms of relative O/N_2 changes during the storm in the longitudinal sector between 60°W and 80°W where the radars are located. The detailed data–model comparison carried out in this study is helpful not only to validate the model but also to interpret the complex observations. The

---

[1]High Altitude Observatory, National Center for Atmospheric Research, Boulder, Colorado, USA.
[2]Haystack Observatory, Massachusetts Institute of Technology, Westford, Massachusetts, USA.
[3]Arecibo Observatory, Arecibo, Puerto Rico, USA.
[4]Applied Physics Laboratory, Johns Hopkins University, Laurel, Maryland, USA.

TIEGCM simulations reveal that it is the enhanced meridional neutral wind, not the penetration electric field, that is the primary cause of the Λ structure of the *F* region electron density profile.

1. INTRODUCTION

Ionospheric disturbances are often categorized as a positive or negative storm phase if there is an increase or decrease of the *F* region peak electron densities with respect to their quiet time values. It is well-known that ionospheric storm effects are determined by a combination of chemical, dynamic, and electrodynamic processes. Increased O_2 and N_2 densities result in increased conversion of O^+ to O_2^+ and NO^+, which then rapidly recombine with electrons, resulting in a rapid decrease in electron density [*Rishbeth*, 1989; *Burns et al.*, 1995]. Although neutral composition changes are often attributed to the formation of a negative storm phase [*Prölss*, 1993], several studies have shown a positive storm phase as a result of a large increase in the O/N_2 ratio [e.g., *Burns et al.*, 1995; *Field et al.*, 1998]. Magnetospheric electric fields associated with a strongly southward interplanetary magnetic field (IMF) are most effective in producing large geomagnetic storms. They drive strong ion convection in the high-latitude polar regions, and a fraction of magnetospheric electric fields can penetrate to middle and low latitudes, prompting nearly simultaneous ionospheric disturbances there. The fast-moving ions driven by magnetospheric electric fields at high latitudes collide with neutrals to produce frictional heating or joule heating. The excessive joule heating dissipation in the high-latitude polar regions produces large pressure gradients that drive neutral winds equatorward toward middle and low latitudes, even into the opposite hemisphere. An enhanced equatorward (poleward) meridional wind pushes plasma up (down) along magnetic field lines due to their inclination with respect to Earth's surface, raising or lowering the *F* region electron density peak height accordingly. In addition, a polarization electric field can be created in the midlatitude regions by the storm-enhanced neutral winds to form the "disturbance dynamo" effect [*Blanc and Richmond*, 1980]. In a real storm event, several different processes may work in concert. Also, because of the complex interaction among the different processes, it is often very difficult to distinguish the effects of one process from another.

While the effects of neutral wind dynamics on ionospheric storms are fully recognized (see reviews by *Prölss* [1995], *Buonsanto* [1999], and *Mendillo* [2006]), in recent years, more attention has been paid to the prompt penetration electric fields, partly due to the fact that major geomagnetic storms tend to be associated with a strong southward IMF. *Huang et al.* [2005b] showed several cases in which the penetration electric field can last several hours after the IMF turns southward and geomagnetic activity remains high. The large penetration electric fields produce the so-called superfountain effect, which significantly intensifies the total electron content (TEC) [e.g., *Tsurutani et al.*, 2004; *Mannucci et al.*, 2005] and are found to be a primary cause of the dayside positive storm phase [e.g., *Huang et al.*, 2005a]. Therefore, the storm effects produced by such long-lasting penetration electric fields may have overshadowed the relatively weaker effects of disturbance neutral winds during major storms.

In this paper, we expand the work by *Lu et al.* [in press] on a moderate geomagnetic storm taking place on 10 September 2005. The event was depicted by a very interesting feature in the *F* region electron density profile that closely resembles the Greek letter Λ. The observational aspect of the event has been discussed in detail by *Goncharenko et al.* [2007]. Here we focus our attention on comparison of the simulation results from the Thermosphere–Ionosphere Electrodynamics General Circulation Model (TIEGCM) [*Richmond et al.*, 1992] with measurements obtained from two incoherent scatter radars located in the lower and higher midlatitudes, with Arecibo in Puerto Rico and Millstone Hill in Massachusetts, as well as with those from the global GPS receivers and the TIMED/GUVI instrument. Through such a comprehensive data–model comparison, we will not only validate our model's performance but also shed some new light on the underlying physical processes that are responsible for the Λ-shaped *F* region electron density variation.

It should be pointed out that a very similar Λ structure in the electron density profile was reported by *Roble et al.* [1978] during the September 1974 geomagnetic storm. Using a simplified two-dimensional (e.g., latitude versus altitude) thermospheric and ionospheric model, they concluded that the positive storm phase was produced by gravity waves generated by impulsive heating over the polar cap region. However, in that study, the heat source (assumed to be primarily joule heating) distribution as well as its temporal variation were prescribed to match the observed ionospheric property. In this study, we will use the more advanced three-dimensional model of the TIEGCM, along with more realistic magnetospheric energy inputs derived from the assimilative mapping of ionospheric electrodynamics (AMIE) procedure [*Richmond and Kamide*, 1988], to reexamine the conclusions made by *Roble et al.* [1978].

2. OBSERVATIONS AND MODEL COMPARISON

2.1. Geophysical Conditions and Model Inputs

The solar wind and geophysical conditions for the 8–10 September 2005 period are shown in Figure 1. The solar wind parameters measured by the ACE satellite have been time-shifted by 36 min to account for the solar wind propagation time required from the upstream location to the dayside magnetopause. From 0400 UT on 8 September until ~1400 UT on 9 September, the magnitude of IMF B_z was less than 5 nT. As a result, the geophysical condition of the ionosphere and magnetosphere was quiet, as indicated by the small values of Dst, AE, the polar cap potential drop and the power inputs of joule heating and auroral precipitation. Around 1400 UT on 9 September, an interplanetary shock arrived, along with a rapid increase in solar wind dynamic pressure. The pressure impulse prompted a storm sudden commencement as shown by the positive excursion in Dst, together with increases in AE, the polar cap potential drop and joule heating dissipation. The geophysical disturbances associated with the solar wind dynamic pressure impulse are interesting but not the focus of this study. Here we concentrate on the disturbance during the second half day on 10 September when a moderate geomagnetic storm took place following a southward turning of the IMF near 1600 UT. The storm had a minimum Dst value of about −70 nT and a maximum AE value of ~2500 nT. The cross polar cap potential drop and the hemispheric integrated joule heating both increased after the IMF turned southward. The magnitude of the hemispheric integrated joule heating rate was much larger in the northern hemisphere than in the southern hemisphere. As discussed by *Lu et al.* [in press], this hemispheric difference is due in part to the relatively sparse data coverage in the southern hemisphere, which resulted in smaller electric potential drops because of the interpolation of the limited data points when applying the AMIE procedure [*Lu et al.*, 1996]. The increase in auroral electron energy flux was rather subtle, and the hemispheric integrated auroral power was substantially smaller than joule heating dissipation during the event.

To simulate the ionospheric and thermospheric response to this moderate geomagnetic storm, we have used the realistic, time-dependent high-latitude ionospheric convection and auroral precipitation patterns derived from AMIE. The data input to AMIE was obtained from various space- and ground-based observations, including those from the

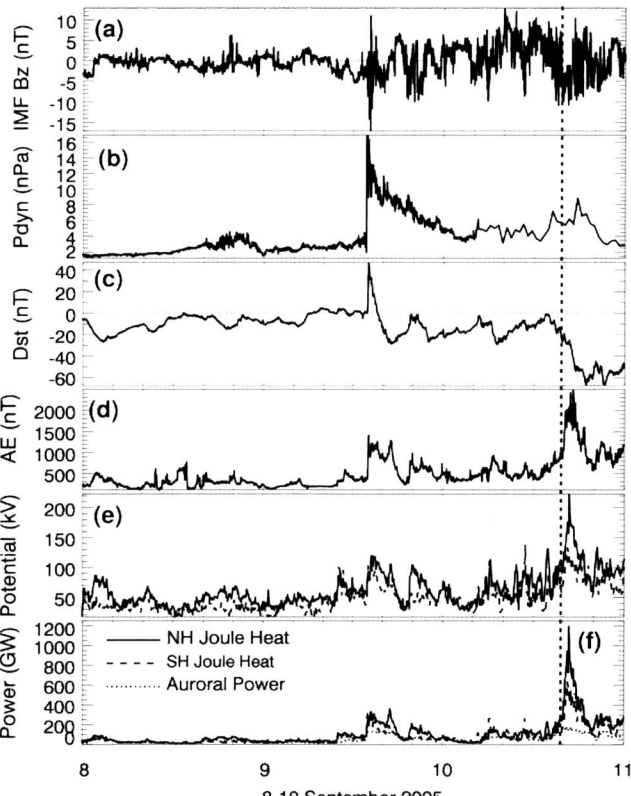

Figure 1. Distributions of the (a) IMF B_z component, (b) solar wind dynamic pressure, (c) Dst index, (d) AE index, (e) cross polar cap electric potential drop in the northern (solid line) and southern (dashed line) hemisphere, and (f) hemispheric integrated joule heating over the northern hemisphere (solid line) and southern (dashed line) and the northern hemispheric integrated auroral power (dotted line). The vertical dashed line marks the onset of the geomagnetic storm at about 1600 UT on 10 September.

Defense Meteorological Satellite Program (F13, 15, and 16) and NOAA (NOAA 15, 16, and 17) satellites, 12 Super Dual Auroral Radar Network radars (10 in the northern hemisphere and 2 in the southern hemisphere), and 178 ground magnetometers. Patterns of ionospheric convection, auroral electron energy flux, and characteristic energy, along with many other ionospheric electrodynamic fields, were derived in a 5-min cadence over both northern and southern hemispheres. The AMIE patterns were timely interpolated to drive the TIEGCM, which ran in a 2-min time step. Solar EUV and UV fluxes (which have a 1-min time resolution) were obtained from the empirical flare irradiance spectral model (FISM) [*Chamberlin et al.*, 2007] to replace the traditional F10.7 proxy (which is a daily average) for this particular event. At the lower boundary, the model incorporates the amplitudes of diurnal and semidiurnal tides at the model's lower boundary based on the global scale wave model [Hagan and Forbes, 2002]. For this study, we used the coarse-grid version of the TIEGCM, which has an effective 5° × 5° latitude–longitude grid with 29 constant pressure levels, extending from about 97 km up to 500–800 km, depending on solar cycle conditions. This study mainly concerns the altitude range between 100 and 500 km where the radar measurements were obtained.

2.2. Comparison With Radar Measurements

Plate 1 shows the measured and simulated electron density Ne, electron temperature Te, and vertical ion drift Wi over Millstone Hill from 1000 to 2400 UT on the quiet day of 8 September (left) and on the storm day of 10 September (right), respectively. The Millstone Hill radar is located at 42.6°N and 71.5°W, and the local time (LT) corresponds roughly to UT − 5. In comparison with the quiet day Ne distribution, there was a significant increase of Ne in the F region between ~1600 and 2300 UT on the storm day of 10 September, along with the increase of the F region peak height h_mF_2. The most striking feature is the UT–altitude profile of the Ne enhancement, which closely resembles the Greek letter Λ. Accompanied by the increase in Ne was a decrease in Te. The anticorrelation between Te and Ne is fully anticipated since the electron cooling rate is proportional to Ne [Schunk and Nagy, 2000]. The ion temperature (not shown), on the other hand, increased slightly during the storm [Goncharenko et al., 2007]. The measured vertical ion drift showed a large upward motion starting at ~1640 UT, which coincided with the initial uplift of h_mF_2. The ion drift then became downward at 1900 UT, about the same time when h_mF_2 started to fall.

At first glance, the TIEGCM simulations appear to be in a good quantitative agreement with the Millstone Hill radar measurements on both quiet and storm days. A similar Λ-like structure in the UT–altitude distribution of Ne is very well reproduced by the model, as is the anticorrelation between the simulated Ne and Te. There are, however, some qualitative differences between the observed and simulated Ne. For example, the simulated F region Te on 8 September starts to decrease too fast compared with the radar-measured Te shown in Plate 1c. However, during the storm, the simulations show a generally good agreement with the measurements, except for a short period around 2200 UT when the simulated Te becomes cooler by a few hundreds of degrees than the measured Te.

The simulated vertical ion drift also shows many large-scale features consistent with the radar observations. On the quiet day as well as prior to the storm onset, the Millstone Hill radar observed upward ion drift above ~350 km. Similar upward ion drift is seen in the simulations but at higher altitudes, mostly above 450 km. The prestorm upward ion drift, as found in the simulations, is associated with ion diffusion derived by the imbalance between the upward plasma pressure gradient force and the downward gravitational force on the plasma. At 1640 UT on 10 September, the time when the Millstone Hill radar started to observe strong upward ion drift, the simulation also shows a temporarily enhanced upward drift above 300 km followed by a more pronounced upward ion drift about 10 min later. In the region below 250 km, however, the simulated strong upward ion drift (> 25 m/s) lags behind the radar measurements by nearly 40 min.

Plate 2 shows the similar data–model comparison but over Arecibo. The Arecibo radar is at 18.3°N and 66.7°W and LT = UT − 4.4. It observed a similar Λ-like structure in the Ne profile as the Millstone Hill radar did. The rise of h_mF_2 at Arecibo was delayed until ~1750 UT, again coincident with the enhanced upward ion drift. The time delay in the initial uplift of h_mF_2 between Millstone Hill and Arecibo implies a propagation speed of 680 m/s for the traveling ionospheric disturbance (TID). The anticorrelation between Ne and Te was more pronounced in the Arecibo measurements.

The simulated Ne is in reasonably good agreement with the radar measurements. However, the simulations do not show the measured temporal density drop around 2100 UT on 8 September, and the simulated Ne is too large compared with the measured Ne in the upward lag of the Λ structure. The simulated Te is also in a good agreement with the measured Te on both the quiet day and the storm day. The simulated vertical ion drift agrees well with the measurements on 8 September. On 10 September, the prestorm F region vertical ion drift consisted of a strong downward flow, followed by a strong upward flow. The simulations display a similar downward-to-upward change in vertical ion drift at ~1030 UT, about 1 h too early with respect to the radar observations. A better agreement between the measured and simulated drifts is found after 1800 UT over Arecibo, including the hockey-stick-like structure of strong upward ion drift followed by downward ion drift.

To assess the model performance in a more quantitative fashion, we follow the same approach proposed by Pawlowski et al. [2008] who carried out a detailed data–model comparison for the entire month of September 2005, including the storm interval that we are analyzing here. Figure 2 shows the comparison of the observed and simulated parameters at the F region peak height, since h_mF_2 is commonly considered as a good representative of the F region variations during storms. The normalized root-mean-square (RMS) error and the cross-correlation coefficient (corr) of the model outputs are shown in each panel. According to

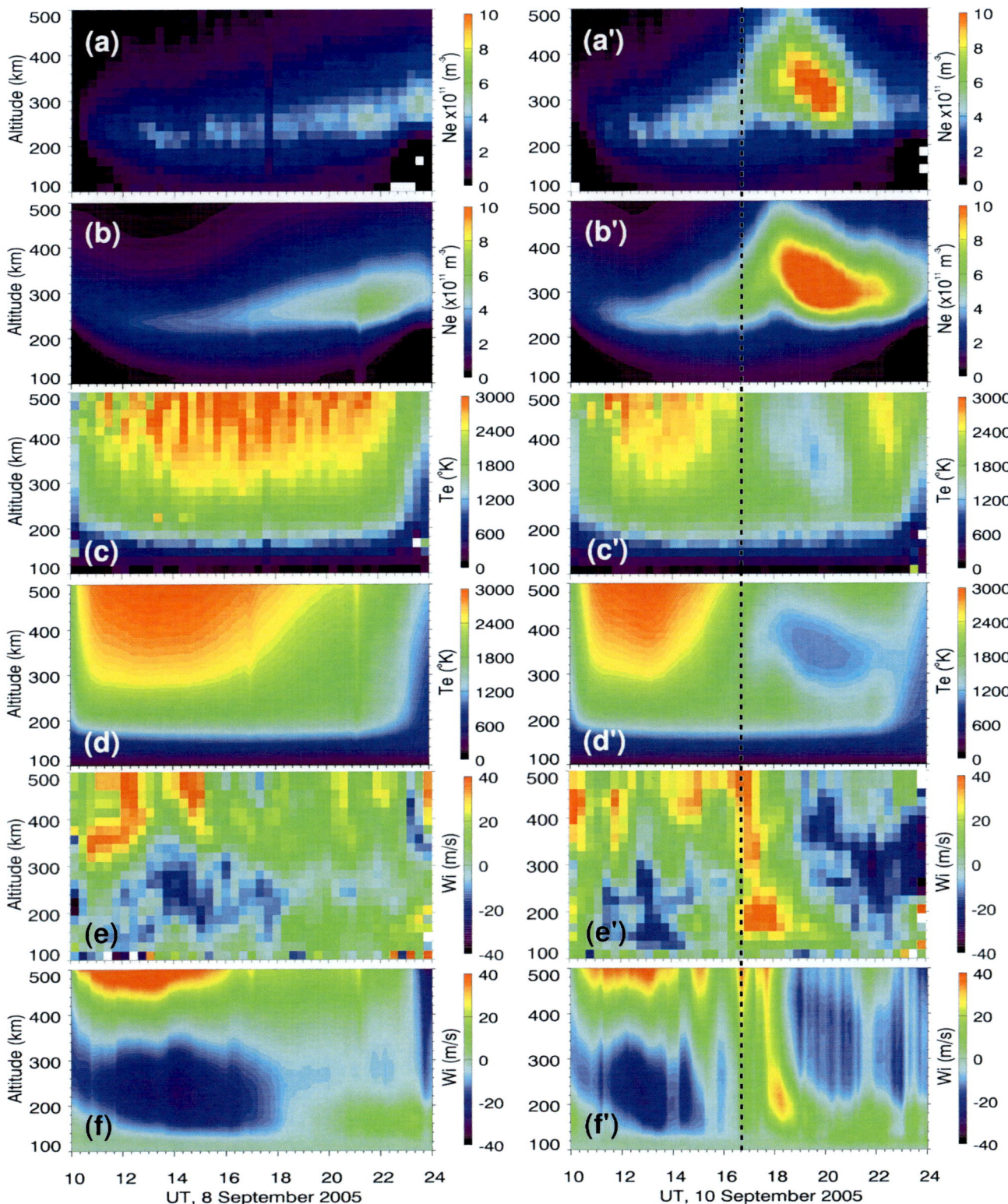

Plate 1. UT–altitude profiles of (a) measured electron density, (b) simulated electron density, (c) measured electron temperature, (d) simulated electron temperature, (e) measured vertical ion drift, and (f) simulated vertical ion drift for the (left) quiet day on 8 September and (right) storm day on 10 September over Millstone Hill. The vertical dashed line denotes the onset of upward lift of the electron density peak height.

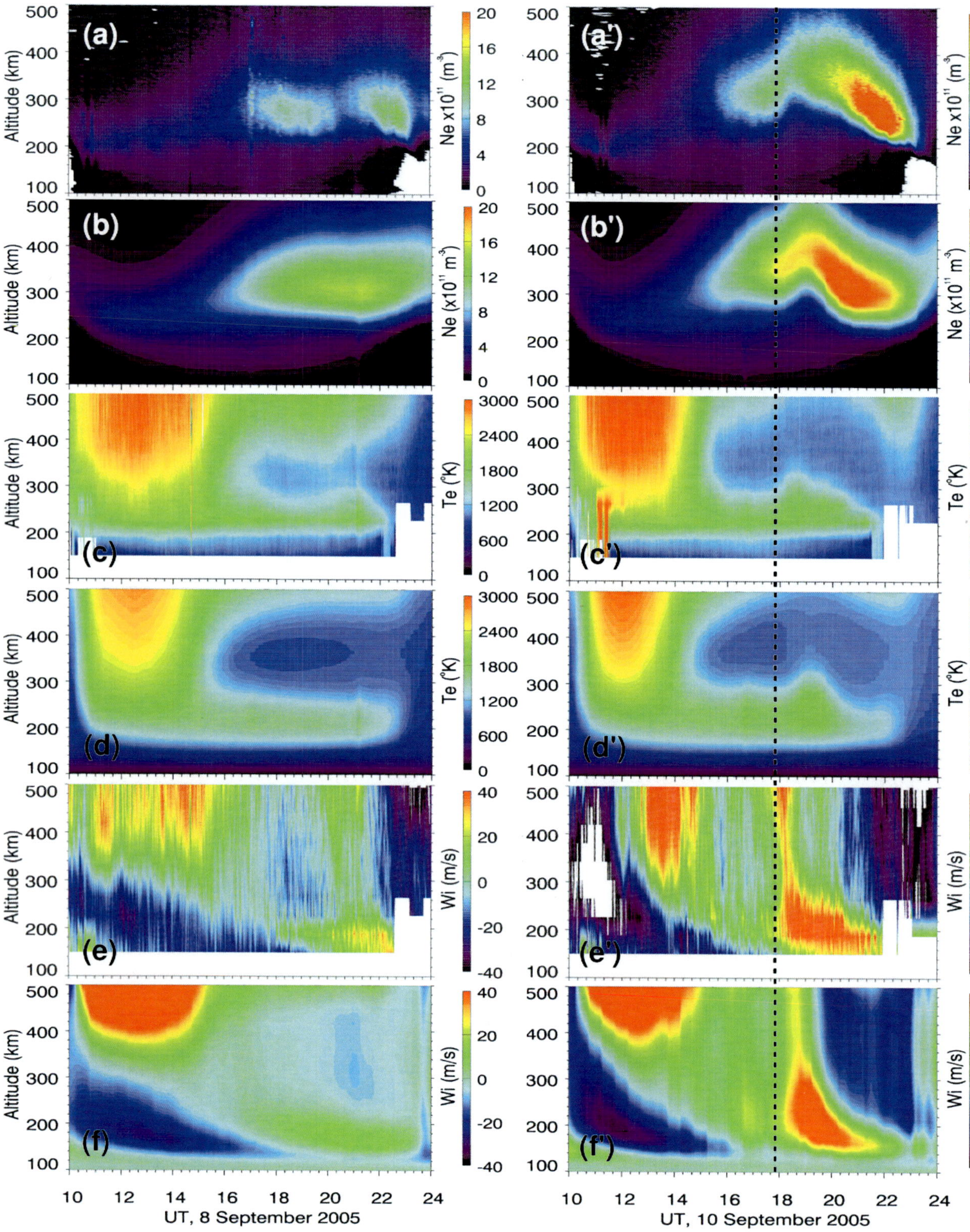

Plate 2. Similar to Plate 1, but over Arecibo.

Pawlowski et al. [2008], the normalized RMS error is defined as $\sqrt{\langle(F_{model} - F_{data})^2\rangle} / \sqrt{\langle F_{data}^2\rangle}$, where $\langle\ \rangle$ symbolizes taking a mean and F denotes a given parameter (e. g., the electron density, temperature, vertical ion drift). At Millstone Hill, the normalized RMS errors are about 1%–3%, except for the vertical ion drift, for which the error is substantially large at 72%. Our normalized RMS errors for the density and temperature are about an order of magnitude smaller than what *Pawlowski et al.* [2008] showed for their model validation. At Arecibo, our normalized RMS errors are 2% for h_mF_2, 19% for N_mF_2, 4% for Te, and 1% for Ti, but 52% for Wi. The large error in the simulated Wi at Millstone Hill is mainly associated with the large fluctuations in the observed Wi that the model is unable to capture partly due to the coarse grid size of our global model. On the other hand, the retrieval of the F region vertical ion drifts by IS radars may also be subject to some uncertainties [e.g., *Aponte et al.*, 2005]. Despite that, the model seems to do a reasonably good job in reproducing the general behavior of the vertical ion drift for the 3-day period, including the large upward drift during the storm on late September 10. The cross-correlation coefficients are around 0.70–0.90 for the density and temperature, which again are better than the values shown by *Pawlowski et al.* [2008]. However, the cross-correlation coefficient is somewhat lower (0.41 and 0.66 at the two radar locations, respectively) for Wi.

2.3. Comparison With GPS TEC Measurements

Plate 3 shows the comparison between the GPS TEC measurements and the simulated TEC from the TIEGCM. These TEC maps are plotted in a fixed local time range between 0400 and 2000 LT, as there was not much activity on the nightside. Although the GPS TEC maps suffer from the lack of data over the vast Pacific and Atlantic oceans, some storm-related TEC changes were discernible, particularly over the Central and North American sector (highlighted by the square box) and in the South Pacific region (highlighted by the oval-shaped area). There was a gradual increase in TEC in both these regions, with a maximum TEC enhancement reached around 2100 UT on 10 September.

There are some similarities between the simulated and observed TEC features, such as the TEC enhancements in the American and South Pacific regions. In the GPS maps, the

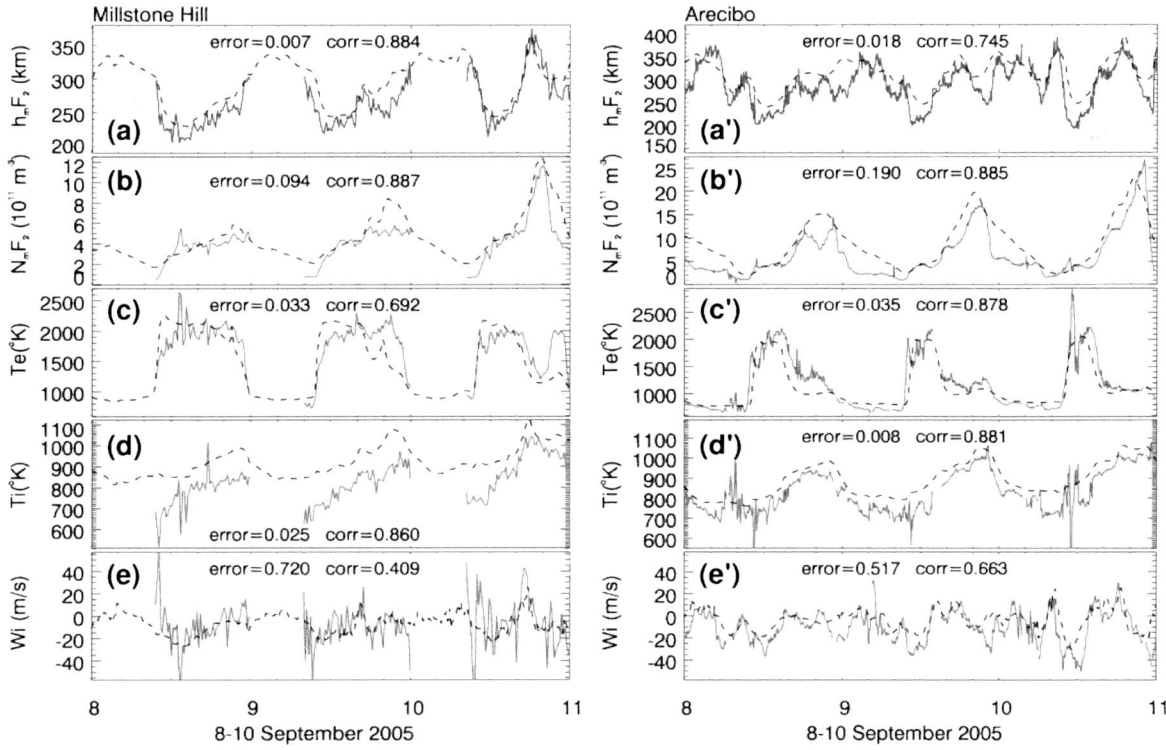

Figure 2. Comparison of observed (solid lines) and simulated (dashed lines) (a) and (a′) h_mF_2, (b) and (b′) N_mF_2, (c) and (c′) electron temperature Te at h_mF_2, (d) and (d′) ion temperature Ti at h_mF_2, and (e) and (e′) vertical ion drift Wi at h_mF_2 over (left) Millstone Hill and (right) Arecibo. The values of the normalized RMS error (error) and the cross-correlation coefficient (corr) are listed in each panel.

TEC enhancement over South Pacific appeared as a locally confined structure; in the model, it appears as a westward, slow-moving structure originated near the western edge of South America. From 1700 UT to 2100 UT, near Central America, the observed TEC increased from ~30 to over 70 TECU. In the simulation, TEC increases from ~25 to ~50 TECU. An important difference between the GPS TEC maps and the simulated TEC maps is that the simulated TEC values in the middle- and low-latitude regions are about 15%–35% smaller than the observations. This underestimation is partly due to the altitude limit of the TIEGCM (its upper boundary altitude is about 570 km for this storm, which took place near solar-minimum conditions).

2.4. Comparison With GUVI O/N_2 Measurements

Plate 4 shows the percent difference distributions of the ratio between the height integrated O and N_2 density, with the GUVI measurements shown on the top, and the simulation results that have been spatially and temporally extracted along the satellite track shown on the bottom. The comparison represents the percent change of O/N_2 on 10 September with respect to the quiet day of 8 September. There is some degree of agreement between the measurement and simulation in terms of the general morphology, for example, a depletion of O/N_2 over most parts of the northern and southern polar regions and a minor increase (~10%) for most of the middle-to-low-latitude region. An increase of O/N_2 is seen by both observations and simulations at the northern edge of Canada as well as between the longitude range of 120°W and 160°W, although the model results appear as much larger-scale structures. In the GUVI plot, there was a ~20% increase just off the U.S. east coast and ~20° eastward. A similar O/N_2 increase can be found in the simulations as well, but about 10° further eastward. In general, the agreement is reasonable in the longitude sector between 60°W and 80°W where the Millstone Hill and Arecibo radars are located. There are, of course, many qualitative differences between the measurements and simulations, particularly a pair of positive and negative O/N_2 changes over North America, in the simulation that cannot be found in the GUVI data.

3. DISCUSSION

The detailed data and model comparison shown in the previous section assure us that the coupled AMIE–TIEGCM has captured reasonably well some large-scale storm features in the *F* region. In this section, we attempt to explore the underlying physical mechanisms responsible for the observed ionospheric storm effects with the help of the numerical simulations.

As pointed out in the Introduction, neutral composition changes are known to play an important role in producing ionospheric disturbances during geomagnetic storms. To verify whether the composition change is a controlling factor in this case, *Lu et al.* [in press] compared the percent changes in *Ne* with the percent changes in O/N_2 [see *Lu et al.*, in press, Figure 4] and found no direct correlation between the electron density increase and the O/N_2 enhancement, implying that the composition changes are not the main cause of this particular positive storm phase. This conclusion is also consistent with the finding of *Goncharenko et al.* [2007] that the O/N_2 change played only a minor role in producing the observed positive storm phase. The weak compositional effect shown in this case is not surprising. As described by *Prölss* [1993], the equatorward transport of composition bulge is more pronounced in the postmidnight sector, where the storm-generated neutral winds are predominantly equatorward due to less interference by the poleward background winds present during daytime.

We now turn our attention to the possible effect that dynamic and electrodynamic processes may have on the observed ionospheric disturbances. The simulated vertical ion drift shown in Plates 1f and 2f is the sum of all contributions from the electric fields, meridional neutral wind, and ion diffusion [i.e., *Schunk and Nagy*, 2000]. To assess the relative contribution of neutral wind and the electric fields to vertical ion drift, Plates 5a to 5c shows the meridional wind, the vertical ion drift component due to meridional wind, and the vertical ion drift due to the electric field, respectively. As expected, the wind-driven vertical ion drift and the meridional wind are anticorrelated in the midlatitude region. Compared with the total vertical ion drift shown in Plates 1f and 2f, it is evident that storm-time vertical ion drift is primarily driven by the meridional wind surges. There are no significant changes in the electric field driven ion drift except for a very brief period around 1640 UT in both Millstone Hill and Arecibo. This temporal increase in upward ion drift is a result of the magnetospheric electric field penetration to midlatitudes. However, this leakage/penetration of magnetospheric electric field is a numerical rather than a well-simulated physical effect since the model was not coupled with an inner magnetospheric model such as in the study by *Maruyama et al.* [2005]. This weak penetration electric field, however, did produce a simultaneous increase in TEC seen by the ground GPS receivers across several latitudes at 1630 UT [*Goncharenko et al.*, 2007]. The penetration electric field in our simulation is thus delayed about 10 min compared with the observations, which can be partly attributed to the fact that the TIEGCM outputs were saved in a 10-min cadence although the model itself was running in a 2-min time step.

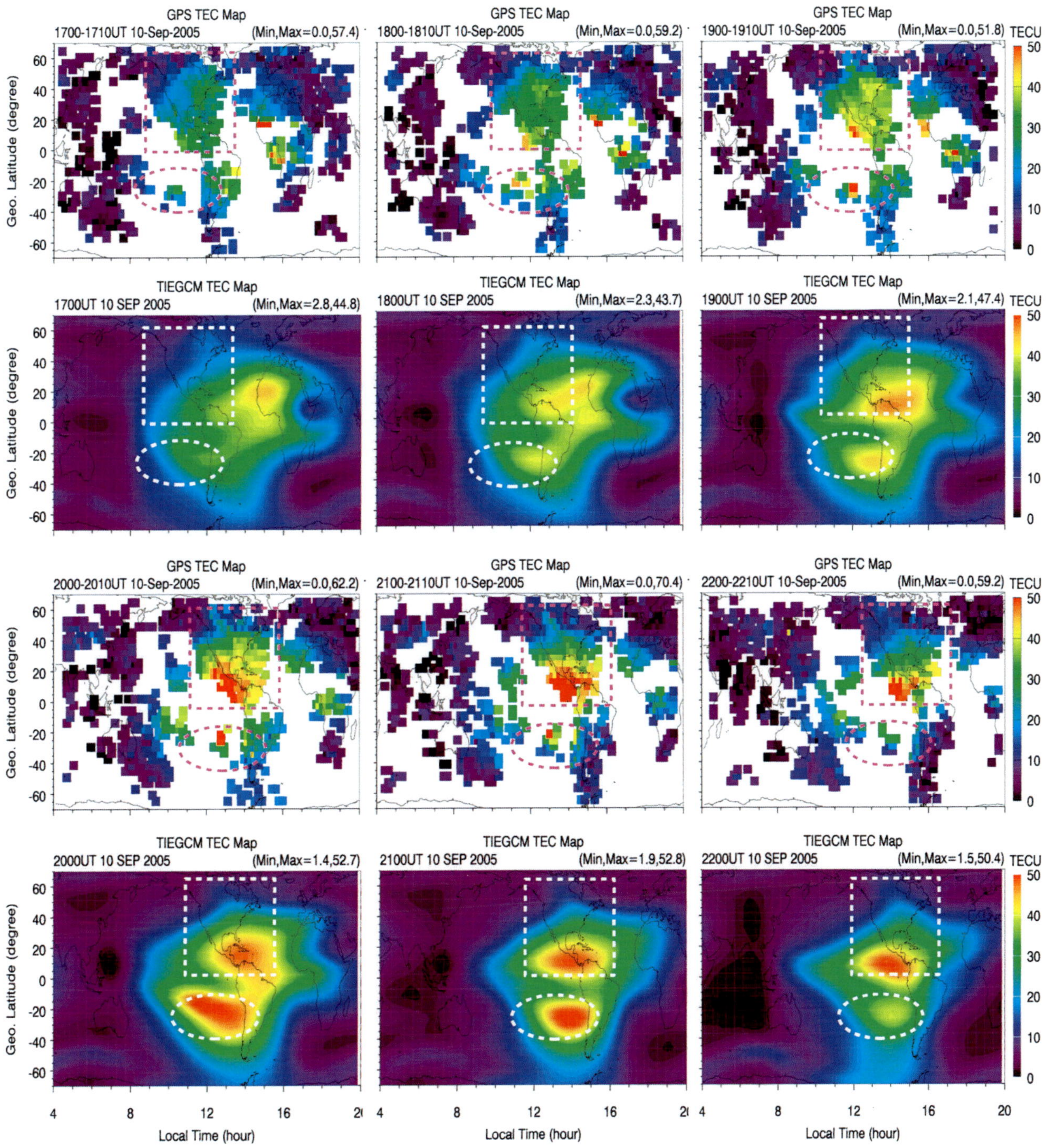

Plate 3. Comparison of measured GPS TEC and simulated TEC distributions.

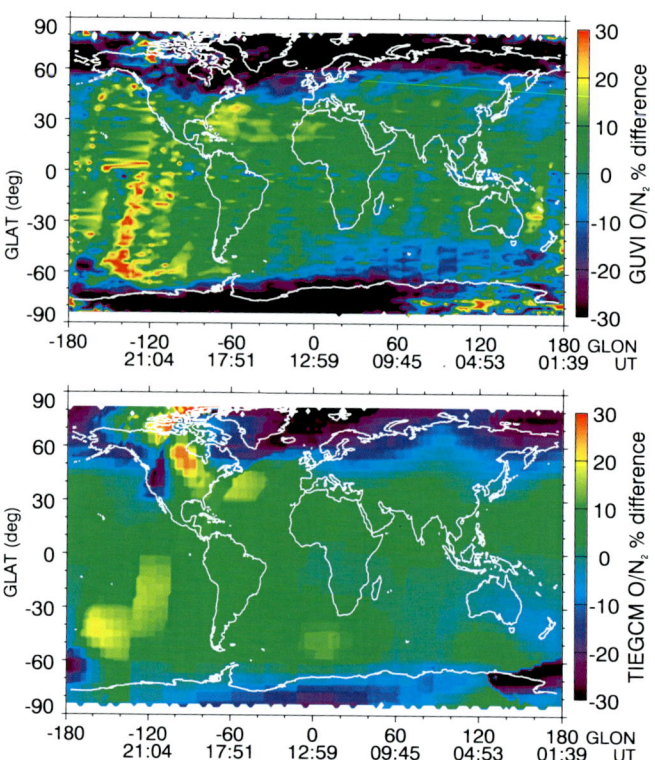

Plate 4. Comparison of the O/N$_2$ percent difference from the (top) GUVI measurements and (bottom) TIEGCM simulations.

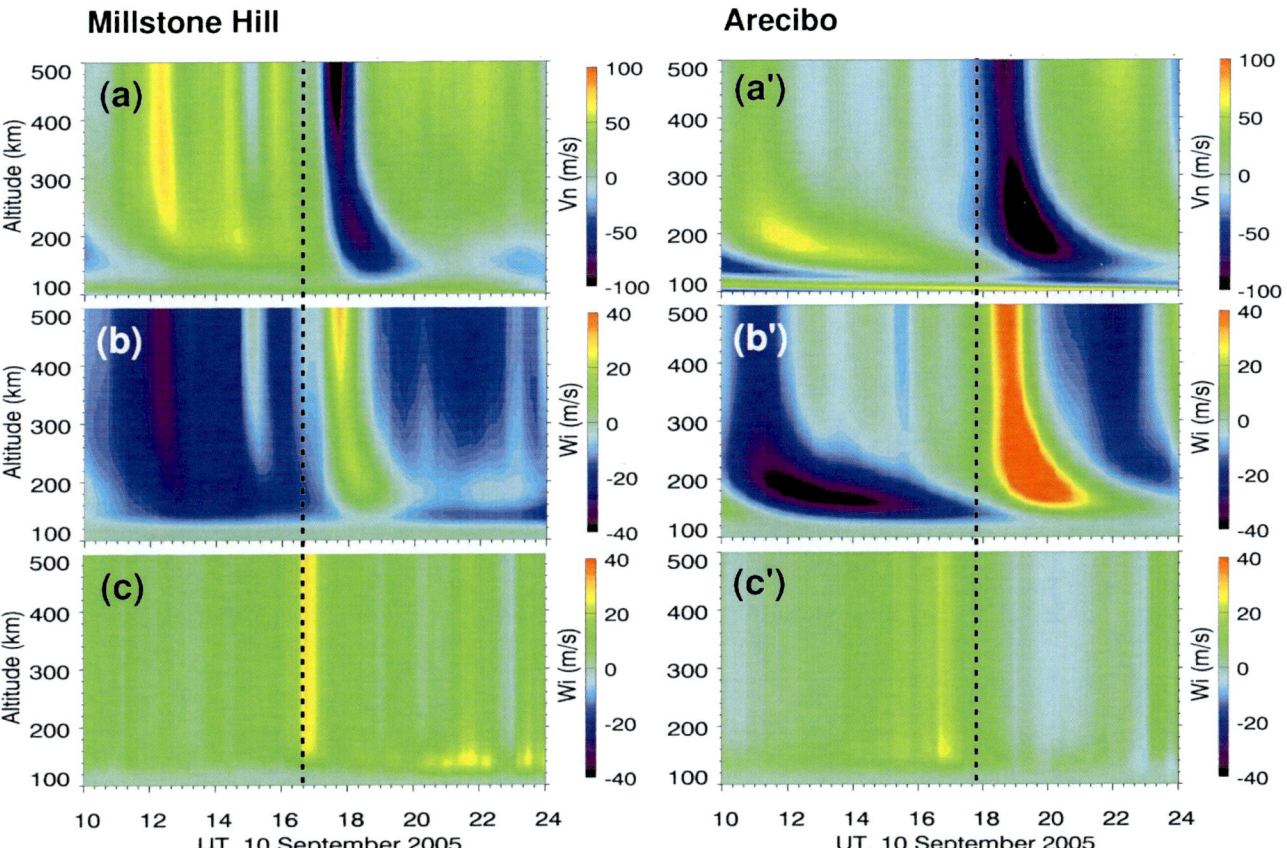

Plate 5. (a) and (a′) Simulated UT–altitude profiles of meridional wind, (b) and (b′) vertical ion drift due to neutral wind, and (c) and (c′) vertical ion drift due to electric field (bottom row). The vertical dashed lines mark the onset of the uplift of the electron density peak height over (left) Millstone Hill and (right) Arecibo. Positive value corresponds to northward meridional wind or upward ion drift.

As shown by *Lu et al.* [in press, Figure 3], there is a clear signature of traveling atmospheric disturbance in meridional wind, which propagates at a phase speed of ~700 m/s. The equatorward wind surge pushes ions upward in the northern hemisphere and downward in the southern hemisphere, with the demarcation near the local magnetic equator. However, there was very little change in zonal ion drift during the storm. They attributed the ineffectiveness for neutral winds to generate a disturbance dynamo field to the fact that the relatively large daytime E region conductivity may effectively short out the F region dynamo-driven currents, a mechanism proposed by *Rishbeth* [1997].

4. CONCLUSION

The comprehensive storm-time observations by the Millstone Hill and Arecibo radars provided an excellent opportunity to put the TIEGCM to a test. Although it was a moderate storm with a minimum Dst value of −70 nT and a maximum AE value of 2500 nT, the ionosphere exhibited some very interesting characteristics, such as the Λ-shaped profile of the F region electron density. Through the detailed data–model comparison, we have shown that the coupled AMIE–TIEGCM is able to reproduce many observed storm features of the ionosphere, including the Λ structure in the F region electron density profile. The quantitative error assessment indicates that our simulation results have an error of 1%–4% for h_mF_2, electron and ion temperatures at both radar locations during the 3-day period of 8–10 September 2005. The estimated error for N_mF_2 is about 9% at Millstone Hill and 19% at Arecibo. However, the simulated vertical ion drifts are less accurate, with the normalized RMS errors at 72% at Millstone Hill and 52% at Arecibo. The model is unable to capture the large temporal fluctuations shown in the radar observations, but it does reproduce reasonably well the overall large-scale variations during the 3-day period, including the storm-time-enhanced upward ion drift.

The TIEGCM simulations reveal that the primary cause of this dayside positive storm phase is the storm-enhanced meridional neutral wind. The enhanced joule heating in the high-latitude auroral zone produces a strongly equatorward/southward meridional wind surge that pushes plasma upward at midlatitudes. The subsequent northward wind associated with the rarefaction waves then pushes plasma downward, causing the F region peak height to drop. The neutral wind surges propagate in the form of gravity waves at a speed of ~700 m/s, consistent with the time delay observed between the two radars. The detailed component analysis of the simulated ion drift shows no significant dynamo electric field produced by the wind surges in this case study. Our simulations also confirm that for this particular storm event, both composition changes and the penetration magnetospheric electric fields have played a very minor role in producing the observed positive storm phase. Our study presented here is fully consistent with the previous findings by *Roble et al.* [1978], and it reiterates the importance of neutral wind effects on ionospheric disturbances.

Acknowledgments. We are grateful to many colleagues who provided various satellite- and ground-based data that have used in AMIE for this study. The ACE data were obtained from the NASA CDAW website. The work at HAO was supported under NASA's Sun–Earth Connection Guest Investigators and Living With a Star programs. Work at the Haystack Observatory was supported in part by the NASA grant NAG5-13602 and by NSF grant 0455831. NCAR is sponsored by the NSF. Millstone Hill radar observations and analysis are supported by an NSF cooperative agreement with MIT. The Arecibo Observatory is operated by Cornell University with support from a cooperative agreement with the NSF.

REFERENCES

Aponte, N., M. J. Nicolls, S. A. González, M. P. Sulzer, M. C. Kelley, E. Robles, and C. A. Tepley (2005), Instantaneous electric field measurements and derived neutral winds at Arecibo, *Geophys. Res. Lett.*, 32, L12107, doi:10.1029/2005GL022609.

Blanc, M., and A. D. Richmond (1980), The ionospheric disturbance dynamo, *J. Geophys. Res.*, 85, 1669–1686.

Buonsanto, M. J. (1999), Ionospheric storms—A review, *Space Sci. Rev.*, 88, 563–601.

Burns, A. G., T. L. Killeen, G. R. Carignan, and R. G. Roble (1995), Large enhancements in the O/N_2 ratio in the evening sector of the winter hemisphere during geomagnetic storms, *J. Geophys. Res.*, 100, 14,661–14,671.

Chamberlin, P. C., T. N. Woods, and F. G. Eparvier (2007), Flare Irradiance Spectral Model (FISM): Daily component algorithms and results, *Space Weather*, 5, S07005, doi:10.1029/2007SW000316.

Field, P. R., H. Rishbeth, R. J. Moffett, D. W. Idenden, T. J. Fuller-Rowell, G. H. Millward, and A. D. Aylward (1998), Modeling composition changes in F-layer storm, *J. Atmos. Sol. Terr. Phys.*, 60, 523–543.

Goncharenko, L. P., J. C. Foster, A. J. Coster, C. Huang, N. Aponte, and L. J. Paxton (2007), Observations of a positive storm phase on September 10, 2005, *J. Atmos. Sol. Terr. Phys.*, 69, 1253–1272.

Hagan, M. E., and J. M. Forbes (2002), Migrating and nonmigrating diurnal tides in the middle and upper atmosphere excited by tropospheric latent heat release, *J. Geophys. Res.*, 107(D24), 4754, doi:10.1029/2001JD001236.

Huang, C.-S., J. C. Foster, L. P. Goncharenko, P. J. Erickson, W. Rideout, and A. J. Coster (2005a), A strong positive phase of ionospheric storms observed by the Millstone Hill incoherent scatter radar and global GPS network, *J. Geophys. Res.*, 110, A06303, doi:10.1029/2004JA010865.

Huang, C.-S., J. C. Foster, and M. C. Kelley (2005b), Long-duration penetration of the interplanetary electric field to the low-latitude ionosphere during the main phase of magnetic storms, *J. Geophys. Res.*, *110*, A11309, doi:10.1029/2005JA011202.

Lu, G., et al. (1996), High-latitude ionospheric electrodynamics as determined by the assimilative mapping of ionospheric electrodynamics procedure for the conjunctive SUNDIAL/ATLAS 1/GEM period of March 28–29, 1992, *J. Geophys. Res.*, *101*, 26,697–26,718.

Lu, G., L. P. Goncharenko, A. D. Richmond, R. G. Roble, and N. Aponte (2008), A dayside ionospheric positive storm phase driven by neutral winds, *J. Geophys. Res.*, *113*, A08304, doi:10.1029/2007JA01289.

Mannucci, A. J., B. T. Tsurutani, B. A. Iijima, A. Komjathy, A. Saito, W. D. Gonzales, F. L. Guarnieri, J. U. Kozyra, and R. Skoug (2005), Dayside global ionospheric response to the major interplanetary events of October 29–30, 2003 "Halloween Storms", *Geophys. Res. Lett.*, *32*, L12S02, doi:10.1029/2004GL021467.

Maruyama, N., A. D. Richmond, T. J. Fuller-Rowell, M. V. Codrescu, S. Sazykin, F. R. Toffoletto, R. W. Spiro, and G. H. Millward (2005), Interaction between direct penetration and disturbance dynamo electric fields in the storm-time equatorial ionosphere, *Geophys. Res. Lett.*, *32*, L17105, doi:10.1029/2005GL023763.

Mendillo, M. (2006), Storms in the ionosphere: Patterns and processes for total electron content, *Rev. Geophys.*, *44*, RG4001, doi:10.1029/2005RG000193.

Pawlowski, D. J., A. J. Ridley, I. Kim, and D. S. Bernstein (2008), Global model comparison with Millstone Hill during September 2005, *J. Geophys. Res.*, *113*, A01312, doi:10.1029/2007JA012390.

Prölss, G. W. (1993), On explaining the local time variation of ionospheric storm effects, *Ann. Geophys.* 1–9.

Prölss, G. W. (1995), Ionospheric F-region storms, in *Handbook of Atmospheric Electrodynamics*, edited by H. Volland, pp. 195–235, CRC Press, Boca Raton, Fla.

Richmond, A. D., and Y. Kamide (1988), Mapping electrodynamic features of the high-latitude ionosphere from localized observations: Technique, *J. Geophys. Res.*, *93*, 5741–5759.

Richmond, A. D., E. C. Ridley, and R. G. Roble (1992), A thermosphere/ionosphere general circulation model with coupled electrodynamics, *Geophys. Res. Lett.*, *19*, 601–604.

Rishbeth, H. (1989), F-region storms and thermospheric circulation, in *Electromagnetic Coupling in the Polar Clefts and Caps*, edited by P. E. Sandholt, and A. Egeland, pp. 393–406, Kluwer Acad., Norwell, Mass.

Rishbeth, H. (1997), The ionospheric E-layer and F-layer dynamo—A tutorial review, *J. Atmos. Sol. Terr. Phys.*, *99*, 1873–1880.

Roble, R. G., A. D. Richmond, W. L. Oliver, and R. M. Harper (1978), Ionospheric effects of the gravity wave launched by the September 18, 1974, sudden commencement, *J. Geophys. Res.*, *83*, 999–1009.

Schunk, R. W., and A. F. Nagy (2000), *Ionospheres—Physics, Plasma Physics, and Chemistry*, Cambridge Univ. Press, New York.

Tsurutani, B., et al. (2004), Global dayside ionospheric uplift and enhancement associated with interplanetary electric fields, *J. Geophys. Res.*, *109*, A08302, doi:10.1029/2003JA010342.

N. Aponte, Arecibo Observatory, HC03 Box 53995, Arecibo, P. R. 00612, USA.

A. J. Coster and L. P. Goncharenko, Haystack Observatory, Massachusetts Institute of Technology, Off Route 40, Westford, MA 01886-1299, USA.

G. Lu, A. D. Richmond, and R. G. Roble, High Altitude Observatory, National Center for Atmospheric Research, 1850 Table Mesa Drive, Boulder, CO 80305, USA. (ganglu@ucar.edu)

L. J. Paxton, Applied Physics Laboratory, Johns Hopkins University, 11100 Johns Hopkins Road, Laurel, MD 20723-6099, USA.

High-Resolution Observations of Subauroral Polarization Stream-Related Field Structures During a Geomagnetic Storm Using Passive Radar

Melissa G. Meyer

Department of Electrical and Computer Engineering, Michigan Technological University, Houghton, Michigan, USA

In observations made with the Manastash Ridge Radar (MRR, the University of Washington's passive, VHF coherent radar) during July 2004, we detected a large-scale wavelike structure (in latitude and/or time) that propagated through the field of view. The amplitude and period of these equatorward-moving oscillations in backscatter intensity closely resemble those reported by the Millstone Hill Atmospheric Science Group [*Foster et al.*, 2004] in their studies of spatial and temporal variations in the subauroral polarization stream (SAPS) electric field. We present supporting evidence from Defense Meteorological Satellite Program (DMSP) satellite measurements to show that a SAPS channel was present in the MRR field of view during the time in question. By treating individual backscattering irregularities (present due to ionospheric two-stream instabilities) as tracers for electric field structure within the larger and longer-lived SAPS channel, we find quasi-periodic oscillations in the field structure within the channel as well as an equatorward motion of the entire SAPS structure.

1. INTRODUCTION

The subauroral polarization stream (SAPS) is a midlatitude phenomenon of interest in the study of magnetosphere–ionosphere coupling. We use ground-based coherent scatter radar to observe the "footprint" of the SAPS in the ionospheric E layer. By treating individual backscattering plasma irregularities (present due to ionospheric two-stream instabilities) as tracers for electric field structure within the SAPS channel, we can monitor the temporal evolution and spatial structures associated with the SAPS. During an event on 17 July 2004, we find quasi-periodic oscillations in the field structure within the SAPS channel as well as an apparent equatorward motion of the entire structure.

1.1. The Manastash Ridge Radar

The University of Washington passive radar, known as the Manastash Ridge Radar (MRR), detects scatter from targets of interest by utilizing illumination by transmitters of opportunity [*Sahr and Lind*, 1997]. Figure 1 illustrates the instrument's operation. Using commercial FM radio broadcasts (at VHF), we can obtain a range resolution of 1.5 km. We also perform interferometry with the data [*Meyer and Sahr*, 2004], allowing us to resolve scattering volumes in the cross-beam dimension and facilitating the localization of echoes within the radar field of view. The receivers are situated in Washington State, yielding an effective field of view in the subauroral region over southwestern Canada.

We are able to use the MRR for ionospheric science applications because the radar detects Bragg scatter from unstable ion acoustic waves excited by the modified two-stream instability [*Farley*, 1963]. Any electric field with above-threshold magnitude can cause plasma density irregularities to erupt in its path. Coherent scatter radars are sensitive to

Midlatitude Ionospheric Dynamics and Disturbances
Geophysical Monograph Series 181
Copyright 2008 by the American Geophysical Union.
10.1029/181GM08

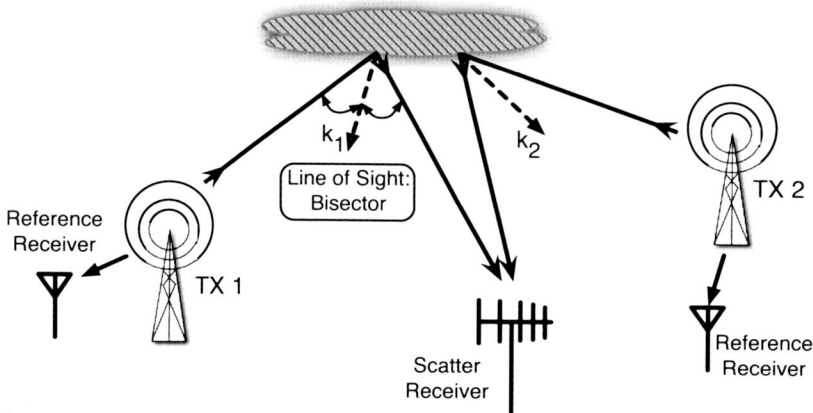

Figure 1. An illustration of the MRR system. The reference receivers (located in Seattle and Spokane, Wash.) record copies of broadcasts from area transmitters, while the scatter receiver (located on Manastash Ridge in central Washington) records the signals scattered from targets. The radar lines of sight bisect the bistatic angles.

these irregularities; thus, we can use them as tracers for observing electric field structures.

1.2. SAPS as Evidence of M–I Coupling

The SAPS is an inclusive term for the subauroral disturbance electric fields observed during geomagnetic storms [*Foster and Burke*, 2002]. SAPS is characterized by strong sunward plasma drift over a latitudinally broad region a few degrees below the auroral oval, occurring on the nightside and lasting up to several hours [*Coster and Foster*, 2007].

A positive feedback process associated with SAPS resulting in strong electric field structures at midlatitude may explain many of the plasma irregularity observations that the MRR (with a subauroral field of view in all but the most disturbed periods) has made. This concept begins with field-aligned current closing through a midlatitude ionospheric region of low density (and therefore low conductivity). The electric field must intensify to maintain the current despite the low conductivity of the region. This heats the existing plasma, encouraging recombination and further reducing the density and conductivity in the region. In turn, the electric field intensifies, feeding the instability mechanism.

The circumstances necessary to begin the feedback instability, to cause the SAPS field and modulate it with waves, are still under investigation. For example, *Mishin et al.* [2003] discuss several wave generation mechanisms as possible explanations for irregularities within SAPS structures. *Streltsov and Foster* [2004] attribute the initial seeding mechanism to a storm-enhanced density gradient at the plasmapause.

The resulting observable effects associated with SAPS are an enhanced, northward-pointing electric field and corresponding enhanced plasma drift in the sunward direction.

The SAPS electric fields occur at F region heights, but can map down along field lines to the E region. Coherent scatter radars such as the MRR can detect the plasma density irregularities that occur in regions with elevated electric

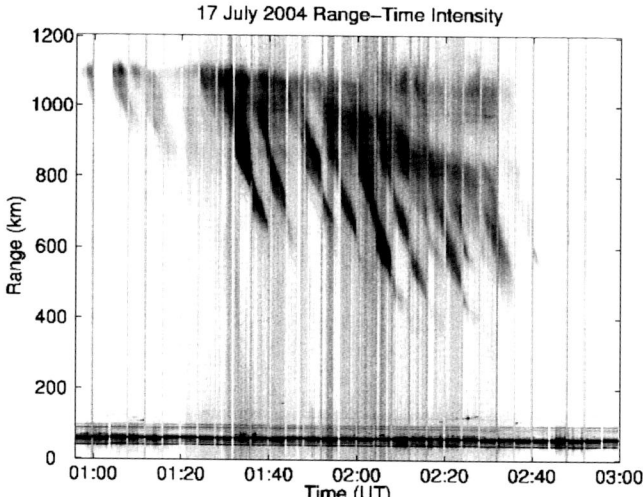

Figure 2. A range–time intensity diagram from the MRR observations on 17 July 2004. Ground clutter can be seen along the bottom edge at approximately 70 km range; above, electric field structures possibly due to SAPS convect equatorward through the radar field of view.

field strength. The MRR, situated between middle and high latitudes (47°N), is in a very useful position to contribute to the study of the SAPS.

2. THE 17 JULY 2004 GEOMAGNETIC EVENT

On 17 July 2004, the *Dst* index took a sharp dip down from approximately 0 to −80. The magnetic storm continued over 10 more days, with *Dst* reaching −197 on 27 July. We observed significant coherent backscatter with the MRR during this period. Due to the availability of corroborating satellite data, we will focus here on the observations made during the hours of 0000 and 0300 UT on 17 July. *Kp* was only moderately high on this date, reaching 6.0 during the abovementioned hours and falling off afterward (however, *Kp* reached 9.0 on 27 July).

Figure 2 shows MRR backscattered power versus range and time over the relevant period on 17 July 2004. Along the bottom of the figure, the ever-present ground clutter signature can be seen at a range of approximately 70 km. At 1150-km range, the backscattering region (the *E* region ionosphere subject to Farley–Buneman turbulence) falls below the horizon, and this range cutoff is evident in Figure 2. Between the ranges of 400 and 1150 km, ionospheric coherent scatter from plasma density irregularities can be seen. The MRR detects coherent backscatter arising from various mid- to high-latitude phenomena, including SAPS-related fields as well as the auroral electrojet and polarization fields on the edges of charged particle precipitation regions.

3. DMSP OBSERVATIONS

We are fairly confident that a SAPS electric field, rather than another phenomenon, caused the 17 July 2004 irregularities. Two overflights of the Defense Meteorological Satellite Program (DMSP) satellite F13 bracket the MRR field of view during the period when the MRR observed irregularity echoes. The satellite data show evidence of a SAPS channel within the radar field of view, with the auroral oval directly poleward.

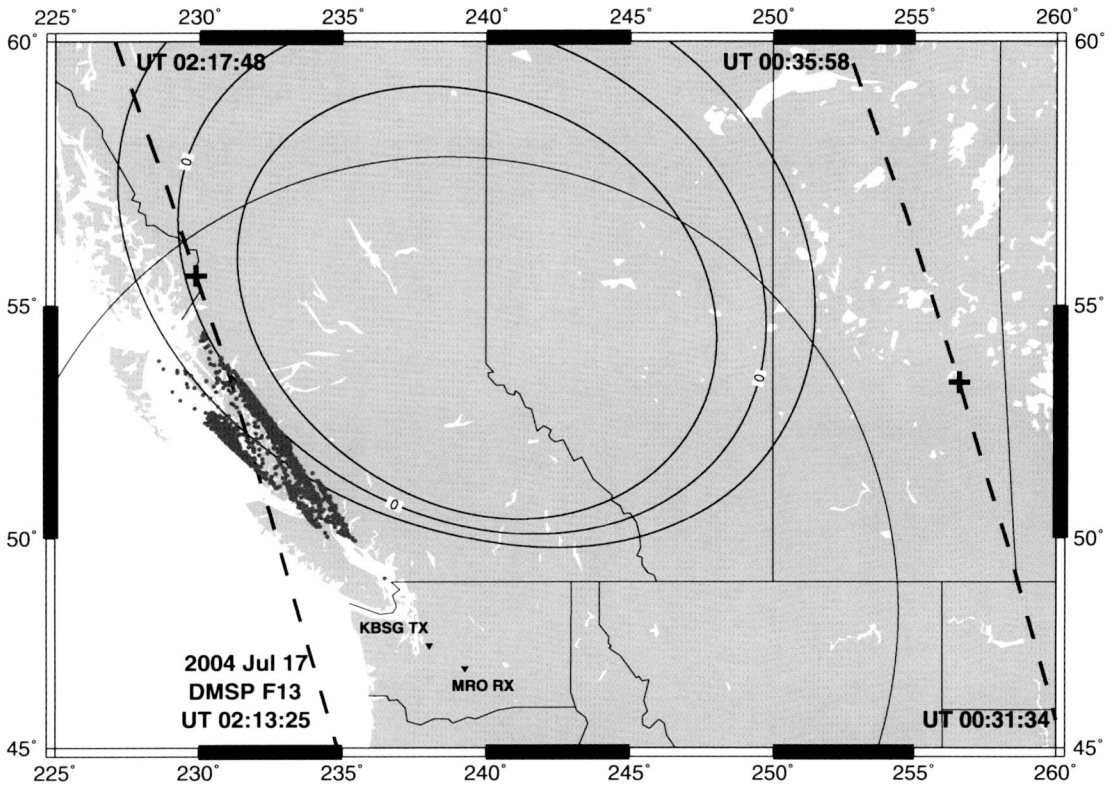

Figure 3. DMSP F13 passes (heavy dashed lines) over the MRR field of view on 17 July 2004. In both passes (beginning 0031:34 and 0213:25 UT), the satellite is traveling northward. The pluses ("+") mark the equatorward boundary of energetic electron precipitation for each pass. Irregularities observed by the MRR during the overpass periods are overlaid (gray circles). Latitude (°N) and longitude (°E) are provided, along with transmitter and receiver locations (inverted triangles), radar maximum range, and contours of constant magnetic aspect angle for 100-km altitude (0° and ±1°).

We establish the equatorward boundary of the auroral oval using energetic electron precipitation measurements made by DMSP satellite F13. Figure 3 shows the flight path over the MRR field of view for two consecutive passes at approximately 0030 UT and 0210 UT. MRR irregularity observations that occurred during the overflights are overlaid on the map (dark gray circles). These mappings were made by using radar range and interferometric data along with the constraint that the backscatter occurred as near as possible to the perpendicular magnetic aspect angle contour, plotted along with ±1° deviations for the assumed altitude of 100 km in Figure 3 [*Meyer*, 2006, p. 184]. Significant deviations from perpendicular aspect angle apparent in Figure 3 are likely due to variation in the altitude of the irregularities. For the MRR view geometry at 1000-km range, magnetic aspect angles vary by roughly 0.5° per 10-km altitude [*Meyer*, 2006, p. 171].

The auroral oval boundary, marked on Figure 3 with "+," is within the MRR field of view (shown as a contour of constant range at 1200 km), but significantly poleward of the irregularities detected by the MRR. From this, we infer that the irregularities are not due to auroral electrojet-related disturbances.

Inspecting the plasma density and velocity measured by DMSP F13, we find evidence of a SAPS channel: a density depletion in combination with enhanced sunward plasma drift [*Foster and Vo*, 2002]. Figure 4 shows plasma horizontal drift velocity and density measured along the satellite path on 17 July 2004, 0210–0219 UT (western F13 pass in Figure 3). Sunward plasma drift has positive velocity. Geodetic latitude increases in the same direction as time; the dashed lines mark the auroral precipitation region boundary.

4. PERIODIC MODULATION OF SAPS CHANNEL ELECTRIC FIELD

Using the evidence from the DMSP satellite data, we conclude that the MRR coherent scatter on 17 July 2004 likely arose from electric fields associated with the SAPS channel. We can now use the MRR observations to study characteristics of the SAPS, adding to existing observations [*Erickson et al.*, 2002]. Simultaneous observations of E region coherent backscatter and electric field magnitude on the same L shell using the Millstone Hill radar demonstrated that electric field magnitude varies linearly with decibels of backscattered power near 440 MHz [*Foster and Erickson*, 2000]. Whether this linearity result may safely be extended to frequencies near 100 MHz is an open research question. However, we may still interpret the irregularities of Figure 2 as tracers of electric field amplitude structures during a SAPS event. We do not expect plasma density gradients to be a significant irregularity generation mechanism, since the waves we observe at 100 MHz have a 1.5-m scale size.

Figure 2 shows a fine-scale wavelike modulation propagating through the SAPS electric field, which is itself apparently drifting (more slowly) equatorward. The period of the modulation is approximately 5 min (we count 12 "fingers" in Figure 2 between 0130 and 0230 UT). The electric field substructures (referred to as subauroral ionization drifts, or SAIDs, by some authors) seem, via successively appearing irregularities, to be propagating equatorward at an average phase velocity of 415 m/s, while the entire channel drifts equatorward at approximately 140 m/s. We used range–rate methods to determine these speeds.

We can discount the possibility of a stationary structure under which the radar moves, since E region plasma corotates with the Earth, and furthermore, from 50° to 55° latitude, apparent speed due to Earth's rotation is between 260 and 300 m/s—too slow to account for the SAIDs observations and too fast to account for the SAPS channel motion.

The Millstone Hill Atmospheric Science Group published a very similar radar observation [*Foster et al.*, 2004] made with the UHF incoherent scatter radar, one that like MRR is located at a midlatitude/subauroral site (42.6°N, 288.5°E). Their observation, reproduced in Figure 5, shows an overall channel motion of approximately 150 m/s, subfeature motion up to 800 m/s, and a modulation period varying between 3 and 5 min (3–5 mHz). The observed behavior of the SAPS channel is roughly consistent in these two instances, indicat-

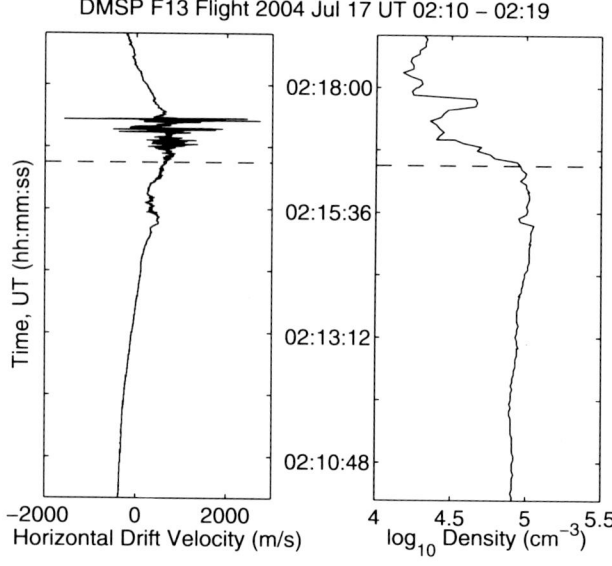

Figure 4. Plasma velocity and density during DMSP F13 pass on 17 July 2004, 0210–0219 UT (western track in Figure 3). The dashed lines mark the auroral precipitation region boundary.

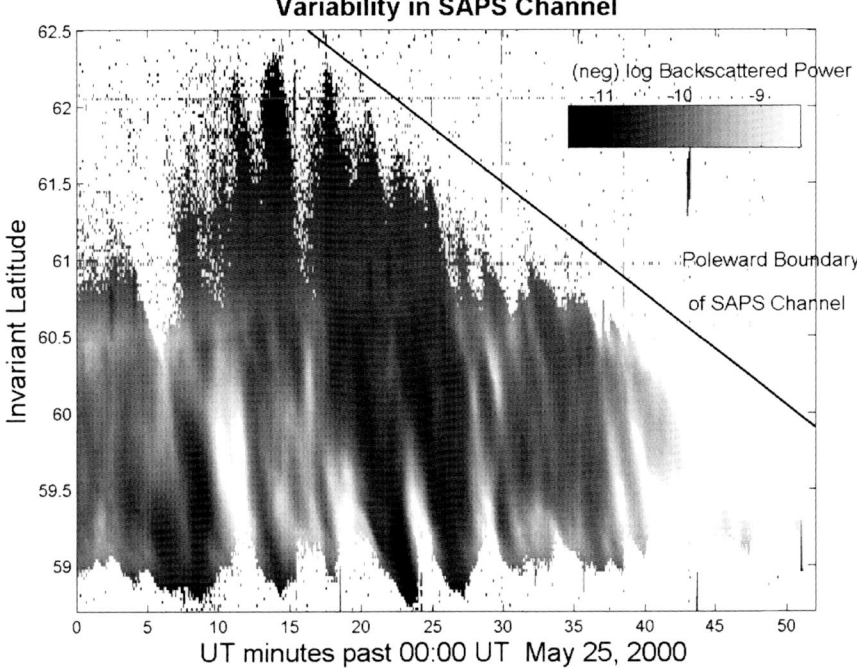

Figure 5. Electric field strength versus time and invariant latitude, inferred from coherently backscattered power measured by the Millstone Hill UHF incoherent scatter radar [*Foster et al.*, 2004].

ing that the phenomenon is caused by a recurrent magnetosphere–ionosphere coupling condition and that the SAPS channel has consistent characteristics over the span of U.S. longitudes.

5. SUMMARY

We have used a new remote sensing technology, passive radar, to assist in characterizing midlatitude evidence of magnetosphere–ionosphere coupling. The MRR offers resolution unparalleled among auroral radars, as well as the possibility of remaining in a continuous mode of operation indefinitely (since only receive equipment must be maintained). However, we require independent information (e.g., DMSP satellite data) to determine whether a SAPS field is likely causing the MRR backscatter.

For the July 2004 geomagnetic event, the required satellite data were available, and we analyzed the MRR backscatter in terms of the SAPS. We detected equatorward motion of both the entire SAPS channel (100–150 m/s) as well as waves within the channel (much faster). The wavelike periodic modulation of the SAPS had a period on the order of a few minutes. These observations are consistent with data from the Millstone Hill radar, which observes the SAPS channel from the eastern side of the United States.

Provided we can rule out the auroral electrojet as a source of plasma irregularity excitation, we may now utilize the growing network of passive radar receivers along the United States–Canada border to monitor and characterize the subauroral footprints of magnetosphere–ionosphere coupling.

Acknowledgments. The author is grateful to F. J. Rich (AFRL, Hanscom AFB) for providing access to DMSP F13 satellite data and to John Sahr for the MRR data. She also thanks John Foster and Frank Lind for helpful conversations.

REFERENCES

Coster, A., and J. Foster (2007), Space-weather impacts of the subauroral polarization stream, *Radio Sci. Bull.*, *321*, 28.

Erickson, P. J., J. C. Foster, and J. M. Holt (2002), Inferred electric field variability in the polarization jet from Millstone Hill E region coherent scatter observations, *Radio Sci.*, *37*, doi.10.1029/2000RS002531.

Farley, D. T. (1963), A plasma instability resulting in field-aligned irregularities in the ionosphere, *J. Geophys. Res.*, *68*, 6083–6097.

Foster, J. C., and H. B. Vo (2002), Average characteristics and activity dependence of the subauroral polarization stream, *J. Geophys. Res.*, *107*(A12), 1475, doi:10.1029/2002JA009409.

Foster, J. C., and P. J. Erickson (2000), Simultaneous Observations of E-Region Coherent Backscatter and Electric Field Amplitude at F-Region Heights with the Millstone Hill UHF Radar, *Geophys. Res. Lett.*, *27*(19), 3177–3180.

Foster, J. C., and W. J. Burke (2002), SAPS: A new characterization for sub-auroral electric fields, *Eos Trans. AGU*, *83*, 393.

Foster, J. C., P. J. Erickson, F. D. Lind, and W. Rideout (2004), Millstone Hill coherent-scatter radar observations of electric field variability in the sub-auroral polarization stream, *Geophys. Res. Lett.*, *31*, L21803, doi:10.1029/2004GL021271.

Meyer, M. G. (2006), Remote sensing of localized ion acoustic waves with multistatic passive radar, Ph.D. thesis, Univ. of Wash., Seattle.

Meyer, M. G., and J. D. Sahr (2004), Passive coherent scatter radar interferometer implementation, observations, and analysis, *Radio Sci.*, *39*, RS3008, doi:10.1029/2003RS002985.

Mishin, E. V., W. J. Burke, C. Y. Huang, and F. J. Rich (2003), Electromagnetic wave structures within subauroral polarization streams, *J. Geophys. Res.*, *108*(A8), 1309, doi:10.1029/2002JA009793.

Sahr, J. D., and F. D. Lind (1997), The Manastash Ridge Radar: A passive bistatic radar for upper atmospheric radio science, *Radio Sci.*, *32*, 2345–2358.

Streltsov, A. V., and J. C. Foster (2004), Electrodynamics of the magnetosphere–ionosphere coupling in the nightside subauroral zone, *Phys. Plasmas*, *11*, 1260, doi:10.1063/1.1647,139.

M. G. Meyer, Department of Electrical and Computer Engineering, Michigan Technological University, 1400 Townsend Drive, Houghton, MI 49931, USA. (mgmeyer@mtu.edu)

Ionization Dynamics During Storms of the Recent Solar Maximum

C. N. Mitchell, P. Yin, P. S. J. Spencer, and D. Pokhotelov

Department of Electronic and Electrical Engineering, University of Bath, Bath, UK

Storms from the recent solar maximum have been studied to reveal the dynamics of ionospheric electron density over the northern hemisphere. The observation technique is GPS imaging, where line integral total electron content observations are inverted into three-dimensional time-dependent maps of electron density. Three different storms have been studied: July 2000, October 2003, and November 2003. The October 2003 storm enhancements in the TEC over North America start to occur before any obvious changes in the convection pattern. For each storm, the main phase involves a sudden uplift in the F layer altitude, extending across the entire mid-latitude region and propagating westward (from Europe to the United States). For the November 2003 storm, there is also a distinct north–south delay in the uplift. Future modeling techniques that will use the electron density images to determine the role of electric field, composition, and neutral wind in these phenomena are described.

1. INTRODUCTION

It became apparent during the last solar maximum that networks of dual-frequency GPS receivers, deployed primarily for geodetic purposes, could provide information about the ionosphere across a wide area. Although total electron content (TEC) changes during storms had already been studied extensively (see a review by *Mendillo* [2006]), there had never before been such a quantity of TEC data recorded simultaneously across a wide area. Consequently, it was the first time that the progress of an ionospheric storm could be viewed continuously across many different sectors [*Coster et al.*, 2001], and large-scale ionospheric structures could be related to magnetospheric events. Currently, there are several hundred dual-frequency GPS receivers in the northern hemisphere, and due to the dispersive nature of the ionosphere these receivers provide networks of TEC data. The two carrier frequencies can be differenced both in terms of phase advance and group delay, and the differencing enables most of the error terms, such as those associated with satellite and receiver clocks and ephemeris, to be minimized. The most important error terms left are the satellite and receiver interfrequency biases (IFBs), and to use the group delay measurement these must be removed. A comprehensive description of the method to obtain slant TEC from GPS measurements is given by *Manucci et al.* [1995].

Maps of the large-scale distribution of electron density can provide important information about ionospheric storm activity and progression. One approach that is useful for ionospheric investigation using GPS TEC is to map calibrated slant TEC values into the equivalent vertical TEC using a geometric mapping function associated with a shell height representative of the mean altitude for the ionospheric electron density [e.g., *Leitinger et al.*, 1975]. An alternative concept for ionospheric studies using GPS data has its roots in ionospheric tomography, as proposed by K.C. Yeh and his team at the Beacon Satellite Symposium in 1986 [*Austen et al.*, 1986]. In this approach, slant TEC values are used in a linear inversion to reveal the spatial distribution of the electron density. The inversion is not straightforward because the data coverage is uneven and limited in the angular distribution. It falls into a class of problems known as underdetermined inversions that can be solved using appropriate mathematical constraints. Reviews of ionospheric tomography and imaging can be found in *Mitchell* [2002] and *Bust and Mitchell* [2008].

There are two different ways to use GPS data in ionospheric imaging. The first involves a precalibration of the TEC whereby the differential delays are used to calibrate the differential phase values and appropriate adjustments are made for the IFBs. This results in an absolute value of slant TEC at a given time and can be used in a three-dimensional (3-D) inversion algorithm. The inversion then reveals the spatial electron density field and consecutive images then show changes in the electron density. The second approach involves differencing adjacent differential-phase values within data arcs collected between a single satellite and receiver, thus removing the need to calculate IFBs. Since the ionosphere changes during the time between these measurements (GPS satellites move relatively slowly), the data are used in a 4-D inversion algorithm to simultaneously reveal both electron density and the change in the electron density.

2. METHOD

The multi-instrument data analysis system (MIDAS) algorithm [*Mitchell and Spencer*, 2003] typically uses the 4-D approach. An alternative 3-D time-dependent Kalman filter implementation can be used for regions of sparse data and/or fast-moving plasma [*Spencer and Mitchell*, 2008], and is therefore particularly relevant to storm studies. The mathematical details of the inversion approaches are not repeated here. Instead, the concepts are highlighted. Plate 1 shows diagrammatically the implementation of the inversion approach. Plate 1a shows the ray path geometry for a typical image. Plate 1b shows the electric field map produced from the Weimer model and Plate 1c magnetic field vectors. Plate 1d shows the resulting convection map that is applied in the Kalman filter as described below.

The standard Kalman filter implementation for ionospheric imaging uses the output from the previous step as an initialization of the next image. The problem with this approach in the polar region is that for an efficient algorithm with image steps of typically 10 min, the polar cap convection moves the plasma a significant distance in that time. For example, during the October 2003 storm the $E \times B$ convection in the polar cap could result in a movement of several hundred kilometers in 10 min. Thus a prior estimate for the filter could be inappropriate. The prior image was therefore convected forward in accordance with the data-driven Weimer model [*Weimer*, 1995] for the electric field and the IGRF model for the magnetic field, thus providing a more appropriate input to the Kalman filter. This is necessary for imaging of fast-moving polar cap structures such as those that are found under certain storm conditions where the interplanetary magnetic field is southward. This approach avoids the need of a prior electron density model for the polar cap since the "model" is constructed from a projection of the prior solution. It is recognized that any convection model or measurement could be used in this algorithm, for example, Doppler velocities produced from the Super Dual Auroral Radar Network (SuperDARN) radars could be incorporated into the solution.

3. RESULTS AND SUMMARY

Example results from the 30 October 2003 storm are shown. Plate 2 shows the cross-polar plasma transport in the northern hemisphere between 1900 and 2200 UT. Electron density variations at 1900, 2000, 2100, and 2200 UT are presented in Plate 2. Attributed partly to solar radiation, large amounts of plasma are produced over North America at about 1900 UT, which can be clearly seen in Plate 2a. One hour later, electron densities in the dayside ionosphere were significantly enhanced at higher altitudes, whereas some of the plasma was transported toward the polar region under the influence of convection flow in the high latitudes. The plasma drifted across the polar region to the nightside in northern Europe, as shown in Plate 2c. As ionization became largely enhanced toward the west of North America at about 2200 UT, more cross-polar plasma was convected antisunward, even toward mid latitudes in Europe.

New observations of storm-time dynamics from MIDAS imaging include:

- Imaging of dramatic elevation of the *F* layer of the ionosphere over Europe and the United States [*Spencer and Mitchell*, 2001; *Yin et al.*, 2004].
- Convection of the plasma from the dayside ionosphere over the United States toward the nightside in Europe within polar cap patches associated with GPS phase and amplitude scintillation at mid and high latitudes in Europe [*Mitchell et al.*, 2005].
- The *F* layer elevation propagates from high latitudes to lower latitudes for the November 2003 storm, but for the October 2003 and July 2000 storms the elevation is simultaneous across all latitudes [*Yin et al.*, 2006].
- All three storms (November 2003, October 2003, and July 2000) show an east–west time delay in the peak height elevation, that is, first, in the European sector, then the east coast of the United States, then later occurring in the west coast of the United States [*Yin et al.*, 2006].

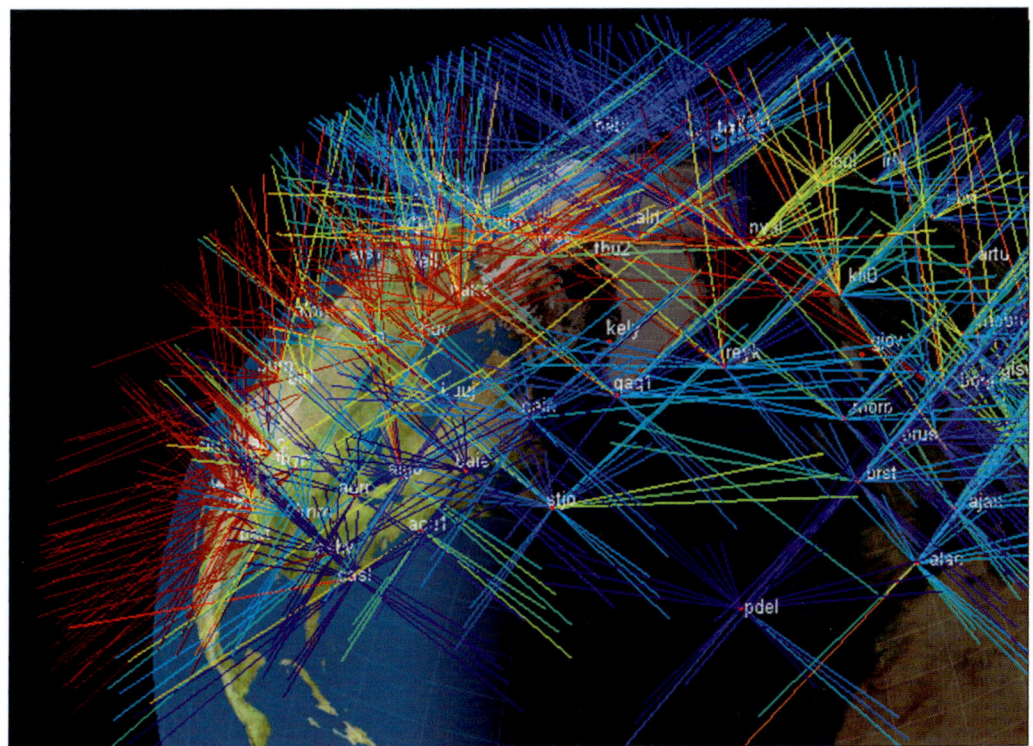

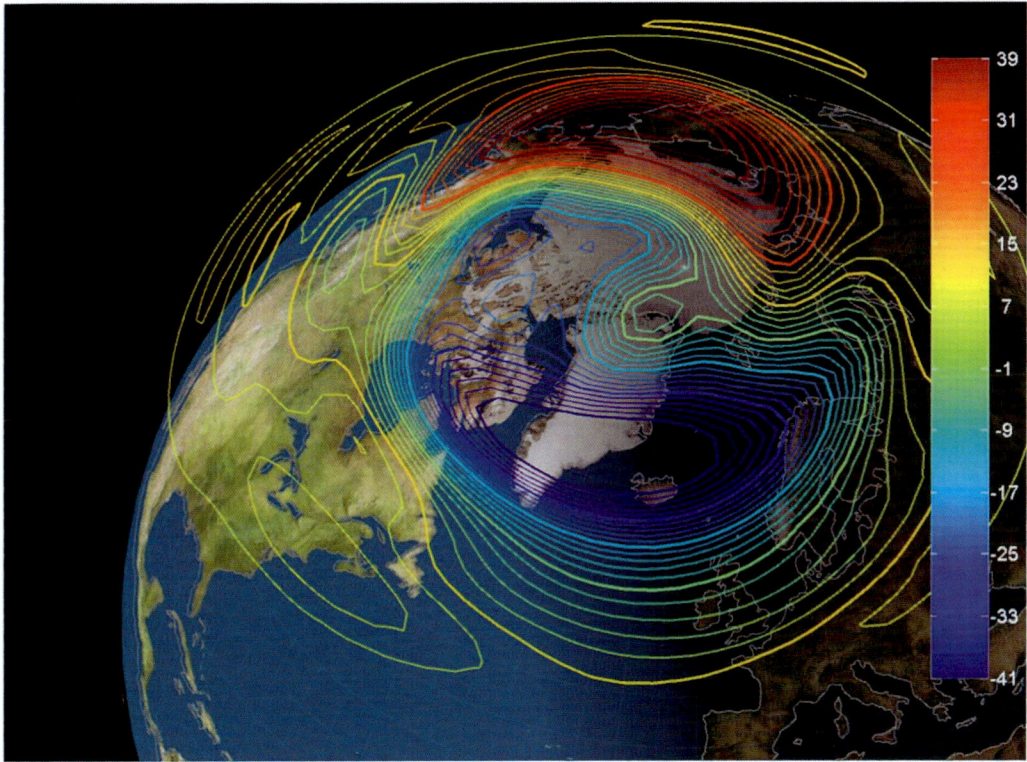

Plate 1. Diagrams showing the inputs to the algorithm for 30 October 2003 at 2100 UT: (a) GPS satellite to receiver ray paths up to 1000 km altitude; (b) electric field potential contours (kV); (c) magnetic field vectors from the IGRF model; and (d) E × B drift.

86 IONIZATION DYNAMICS DURING STORMS OF THE RECENT SOLAR MAXIMUM

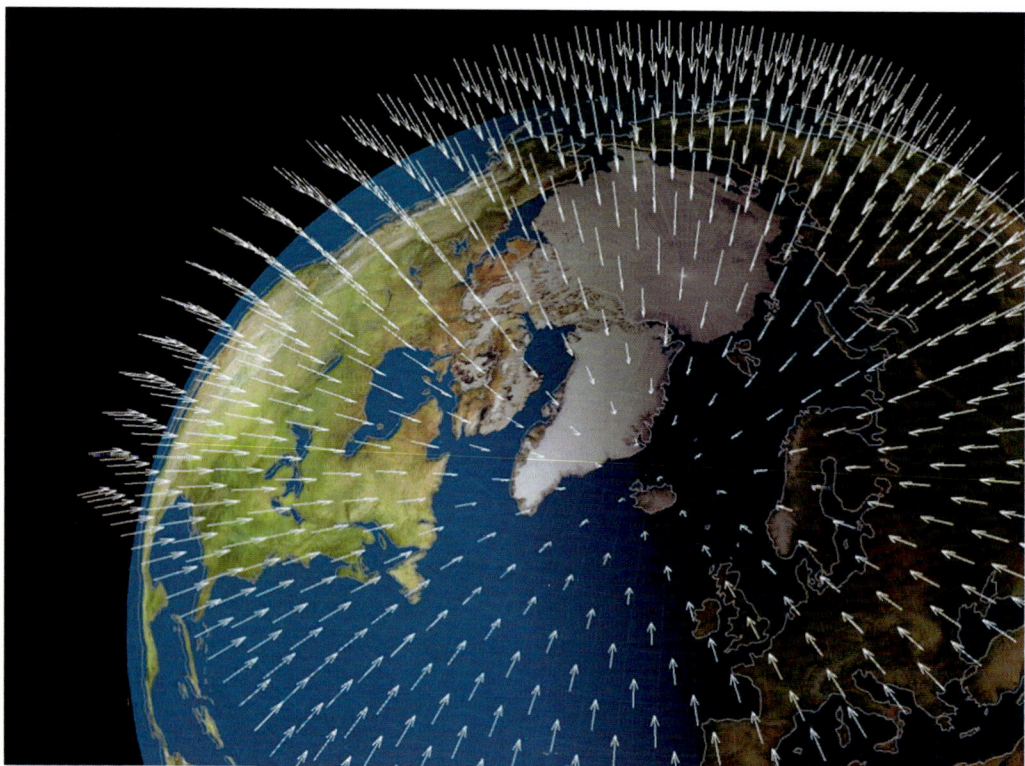

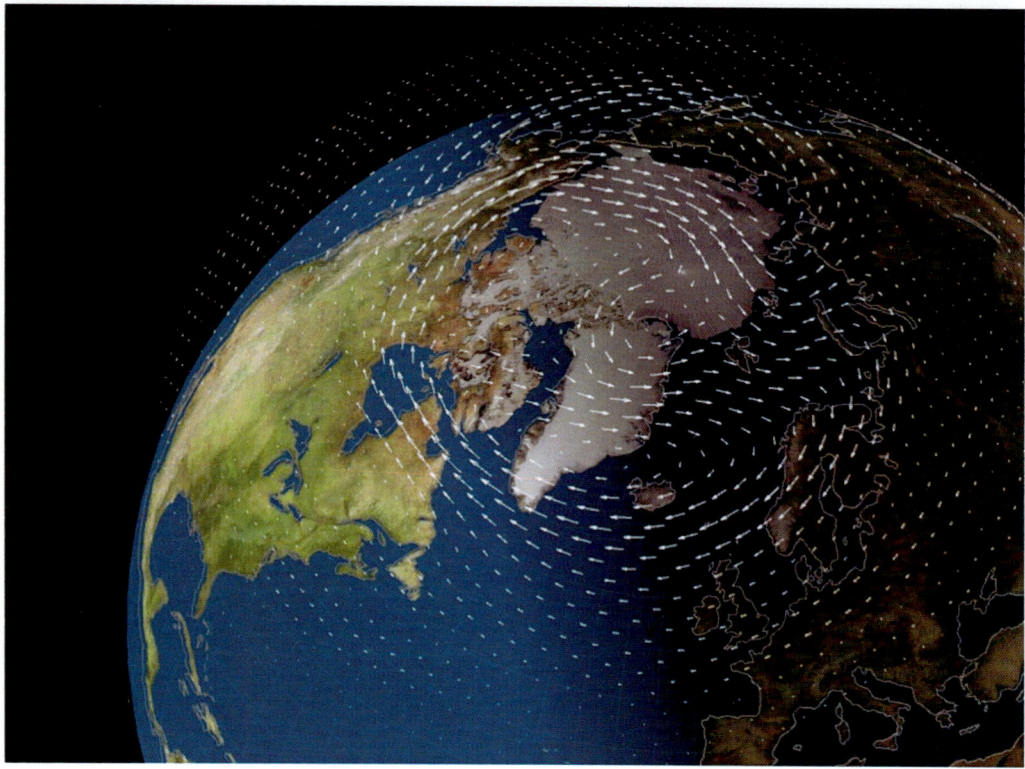

Plate 1. (continued)

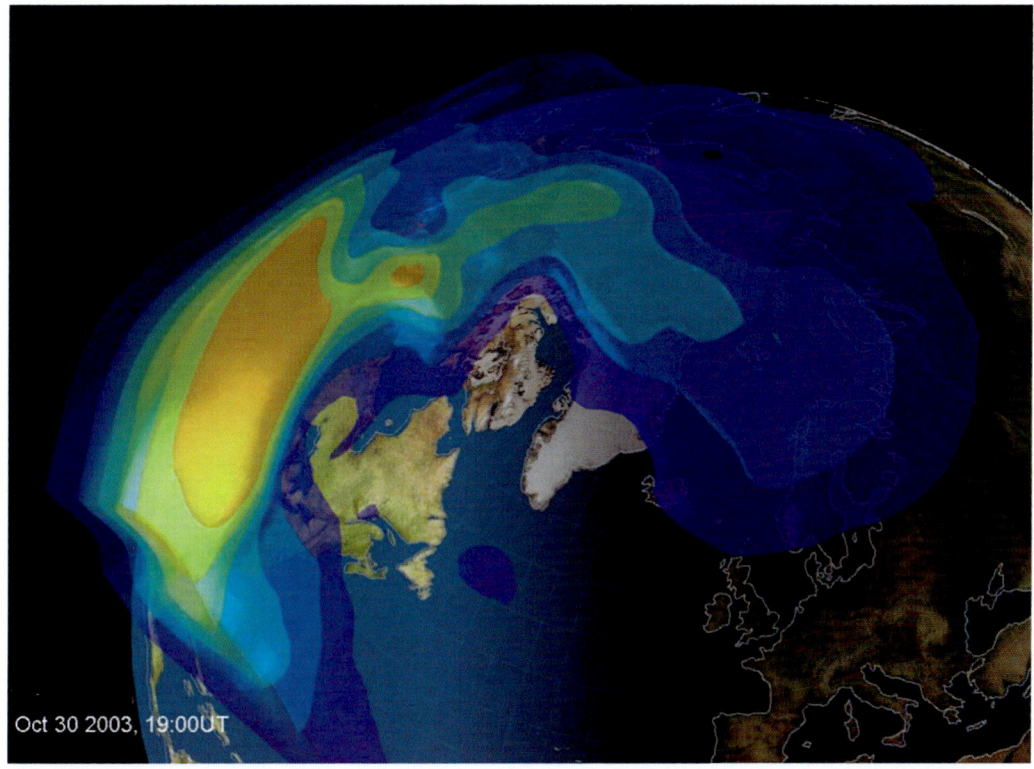
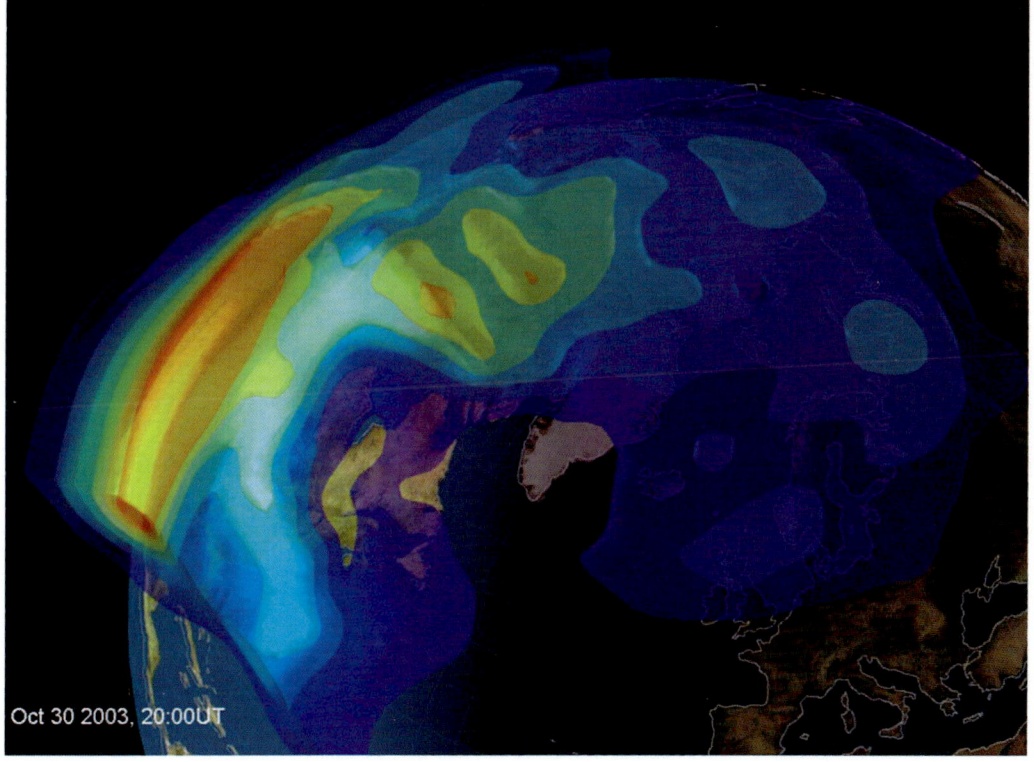

Plate 2. Isocontours of electron density (in steps of 2×10^{11} m^{-3}) for 30 October 2003 at: (a) 1900 UT, (b) 2000 UT, (c) 2100 UT, and (d) 2200 UT.

88 IONIZATION DYNAMICS DURING STORMS OF THE RECENT SOLAR MAXIMUM

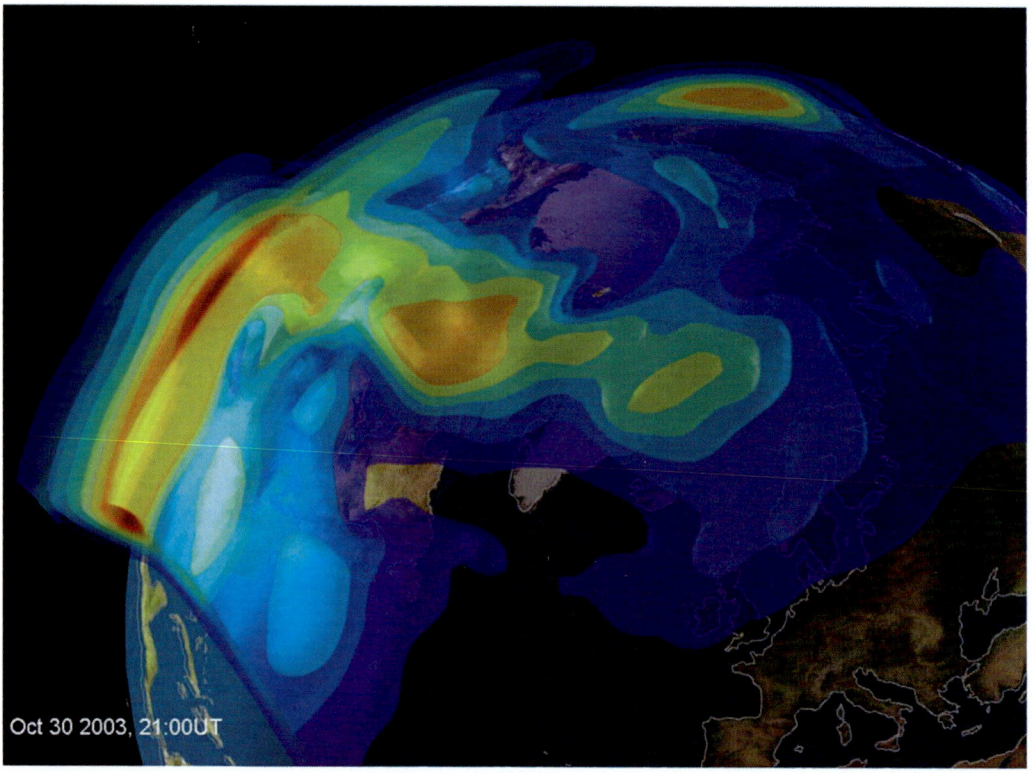

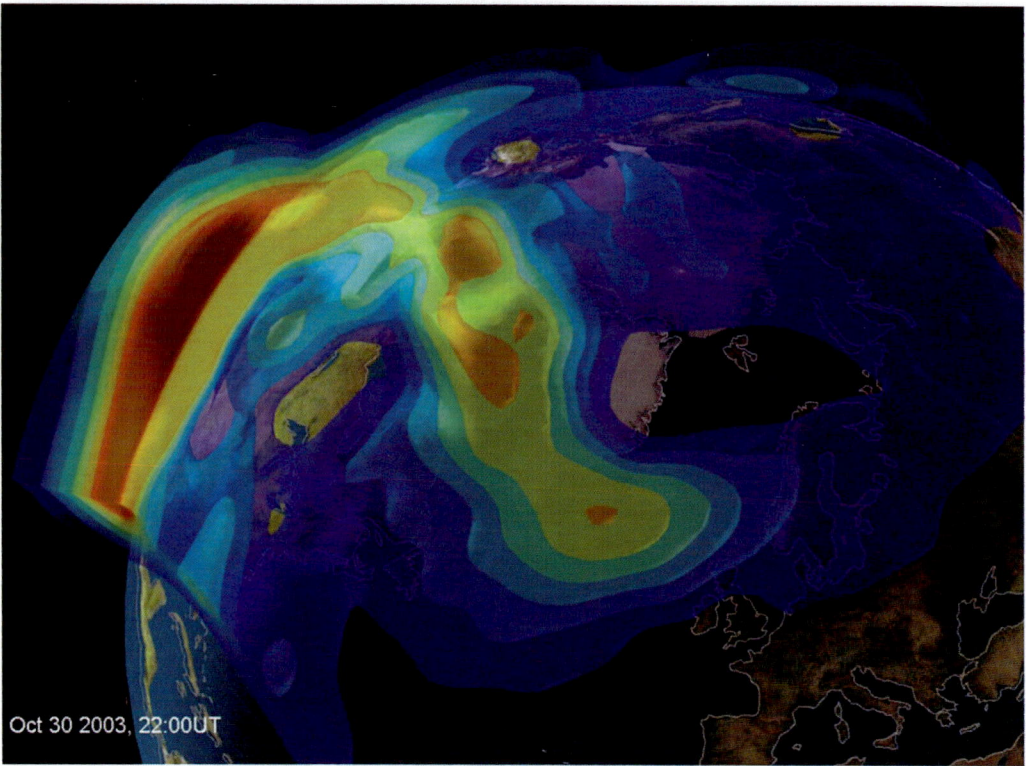

Plate 2. (continued)

- Uplifts in the U.S. sector are always accompanied by TEC/electron density enhancements, but those in the European sector are accompanied by decreasing electron densities/TEC [*Yin et al.*, 2006].
- Enhancement in TEC on 30 October 2003 starts just after sunrise (1300 UT) and grows into a longitudinally limited feature over North America

4. FUTURE WORK

Now that imaging has reached maturity, it is appropriate to ask the question "what physics can we discover from the images?" If a change in electron density is observed, then how can we interpret where that change came from? Even in the simplest case a physical model or physical assumption must be invoked. For example, if a high-density region of plasma changes location in two consecutive images there are at least three different explanations, which become more or less likely depending on the region of the Earth. If the high-density region is at mid latitude, then it is likely to move due to the location of the maximum solar radiation as the Earth rotates. If it is in the auroral zone in the E region, then it could be changing regions of precipitation and the effect of rapid recombination. In the polar cap, convection is very likely to dominate during active times.

To take a more systematic approach in interpreting the images, efforts are currently underway to use physical models to aid in image interpretation. The ability of a physical model to replicate the electron density changes observed with MIDAS therefore puts a constraint on the models that can potentially probe the underlying physics. This provides a framework for image interpretation that can be used in a systematic manner. It does, however, bring forward further questions of uniqueness in model drivers. To remove ambiguities in defining the dominant physical process responsible for certain changes in the ionospheric dynamics, it is necessary to use a number of datasets from various experiments. For instance, global changes in the O/N_2 ratio can be deduced from TIMED (Thermosphere, Ionosphere, Mesosphere Energetics and Dynamics) GUVI (Global Ultraviolet Imager) data [*Paxton et al.*, 2003] and the neutral wind dynamics can be measured by Fabry-Perot interferometers [*Aruliah et al.*, 2005]. One has to remember, however, that during major magnetic storms some ionospheric measurement techniques become unusable or provide very limited information. Extensive auroral networks of ground magnetometers and SuperDARN radars cannot provide a correct picture of the storm-time ionospheric convection due to the expansion of the convection flow to lower latitudes. This emphasizes the importance of in situ spacecraft observations of the storm-time ionospheric convection. Examples include the DMSP thermal plasma drift experiment [*Rich and Hairston*, 1994] and the expansion of ground-based networks to lower latitudes (e.g., the StormDARN project; *Grocott et al.* [2006]). It is our expectation that model optimization approaches will lead to time-dependent bounds on the possible physical drivers for the major storms that can further advance our understanding of the storm processes.

Acknowledgments. The authors are grateful for the use of GPS data from the International GNSS Service and for the use of the Weimer model. They acknowledge support from the UK research councils: STFC and EPSRC.

REFERENCES

Aruliah, A. L., E. M. Griffin, A. D. Aylward, E. A. K. Ford, M. J. Kosch, C. J. Davis, V. S. C. Howells, S. E. Pryse, H. R. Middleton, and J. Jussila (2005), First direct evidence of meso-scale variability on ion-neutral dynamics using co-located tristatic FPIs and EISCAT radar in Northern Scandinavia, *Ann. Geophys.*, *23*(1), 147–162.

Austen, J. R., S. J. Franke, C. H. Liu, and K. C. Yeh (1986), Application of computerized tomography techniques to ionospheric research, *Proc. Int. Beacon Satell. Symp.*, *25*, Oulu, Finland.

Bust, G. S., and C. N. Mitchell (2008), History, current state and future directions of ionospheric imaging, *Rev. Geophys.*, *46*, RG1003, doi:10.1029/2006RG000212.

Coster, A. J., J. C. Foster, P. J. Erickson, and F. J. Rich (2000), Regional GPS mapping of storm enhanced density during the 15–16 July 2000 geomagnetic storm, *Proceedings of the 14th International Technical Meeting of the Satellite Division of the Institute of Navigation ION GPS*.

Grocott, A., M. Lester, M. L. Parkinson, T. K. Yeoman, P. L. Dyson, J. C. Devlin, and H. U. Frey (2006), Toward a synthesis of substorm electrodynamics: HF radar and auroral observations, *Ann. Geophys.*, *24*(12), 3365–3381.

Leitinger, R., G. Schmidt, and A. Tauriainen (1975), An evaluation method combining the differential Doppler measurements from two stations that enables the calculation of the electron content of the ionosphere, *J. Geophys. Res.*, *41*, 201–213.

Manucci, A. J., B. D. Wilson, and C. D. Edwards (1993), A new method for monitoring the Earth's ionospheric total electron content using the GPS global network, *Proc. of the Institute of Navigation GPS-93*, Salt Lake City, Utah, Sept. 22–24.

Mendillo, M. (2006), Storms in the ionosphere: Patterns and processes for total electron content, *Rev. Geophys.*, *44*, RG4001, doi:10.1029/2005RG000193.

Mitchell, C. N. (2002), Imaging of near-Earth space plasma, *Philos. Trans. R. Soc. London, Ser. A*, *360*(1801), 2805–2818.

Mitchell, C. N., and P. Spencer (2003), A three-dimensional time-dependent algorithm for ionospheric imaging using GPS, *Ann. Geophys.*, *46*(4), 687.

Mitchell, C. N., L. Alfonsi, G. De Franceschi, M. Lester, V. Romano, and A. W. Wernik (2005), GPS TEC and scintillation measurements from the polar ionosphere during the October 2003 storm, *Geophys. Res. Lett.*, *32*, L12S03, doi:10.1029/2004GL021644.

Paxton, L. J., D. Morrison, D. J. Strickland, M. G. McHarg, Y. Zhang, B. Wolven, H. Kil, G. Crowley, A. B. Christensen, and C.-I. Meng (2003), The use of far ultraviolet remote sensing to monitor space weather, *Adv. Space Res.*, *31*(4), doi:10.1016/S0273-1177(02)00886-4.

Rich, F. J., and M. Hairston (1994), Large-scale convection patterns observed by DMSP, *J. Geophys. Res.*, *99*(A3), 3827–3844.

Spencer, P., and C. N. Mitchell (2008), Imaging of fast-moving structures in the polar cap, *Ann. Geophys.*, in press.

Spencer, P. S. J., and C. N. Mitchell (2001), Multi-instrument inversion technique for ionospheric imaging, *Proceedings of International Beacon Satellite Meeting*, Boston.

Weimer, D. R. (1995), Models of high-latitude electric potentials derived with a least error fit of spherical harmonic coefficients, *J. Geophys. Res.*, *100*, 19,595–19,607.

Yin, P., C. N. Mitchell, P. S. J. Spencer, and J. C. Foster (2004), Ionospheric electron concentration imaging using GPS over the USA during the storm of July 2000, *Geophys. Res. Lett.*, *31*, L12806, doi:10.1029/2004GL019899.

Yin, P., C. N. Mitchell, and G. Bust (2006), Observations of the *F* region height redistribution in the storm-time ionosphere over Europe and the USA using GPS imaging, *Geophys. Res. Lett.*, *33*, L18803, doi:10.1029/2006GL027125.

C. N. Mitchell, D. Pokhotelov, P. S. J. Spencer, and P. Yin, Department of Electronic and Electrical Engineering, University of Bath, Bath BA2 7AY, UK. (c.n.mitchell@bath.ac.uk)

Mapping the Time-Varying Distribution of High-Altitude Plasma During Storms

G. S. Bust and Geoff Crowley

Atmospheric & Space Technology Research Associates, San Antonio, Texas, USA

1. INTRODUCTION

Space weather refers to variability that occurs in the near-Earth environment between the Sun and Earth that has an effect on human systems. While there is daily variability, much like traditional "weather," the most severe effects on human systems occur during space weather storms. The ionosphere, one part of the coupled Sun–Earth system, has a particularly dramatic effect on RF systems during storms. Ionospheric storms are the final phase in a chain of physical processes that start at the Sun and involve complicated interactions and coupling of the solar wind, the interplanetary magnetic field, the Earth's magnetosphere, ionosphere, and thermosphere. During the most disturbed conditions, storms have a global impact on the distribution and levels of ionization. The expansion of the auroral oval and the equatorward movement of the main trough can result in severe gradients in total electron content (TEC) in both space and time. While the main phase of a storm may last less than a day, the entire recovery time can last for many days. Reviews of ionospheric storms can be found by *Prölss* [1995] and *Buonsanto* [1999].

Recently, two-dimensional TEC maps obtained from global ground GPS receivers have shed new light on the space–time properties of midlatitude storm enhanced density (SED) and its relationship to plasmaspheric processes [*Foster et al.*, 2002; *Coster et al.*, 2003; *Foster and Rideout*, 2005]. Four-dimensional (4-D) imaging [*Mitchell*, 2002; *Mitchell and Spenser*, 2003; *Bust et al.*, 2004], by providing time-evolving maps of electron density and its gradients over wide spatial scales, provides a unique method of investigating the physical dynamics of ionospheric storms [*Bust et al.*, 1997, 2007; *Mitchell et al.*, 2005; *Yin et al.*, 2004, 2006]. For the case study of 30 October and 20 November 2003 storms, analysis of a coordinated set of observations obtained from IDA3D that extend from the equatorial region in the American sector through midlatitudes all the way into the polar cap is presented. The analysis made use of results from IDA3D 4-D imaging algorithms. It is only through such 4-D imaging methods that such a large-scale global analysis of the ionospheric response to a major geomagnetic storm can be obtained. In particular, we investigate whether enhanced midlatitude plasma is due to high-altitude transport of equatorial plasma, and if so, what are the similarities and differences between the 30 October and 20 November storms.

2. APPROACH

2.1. Introduction to IDA3D

IDA3D [*Bust et al.*, 2000, 2004] organizes the data into spatial maps of electron density via an objective analysis technique. These spatial maps are projected forward in time through a Gauss–Markov Kalman filter [*Gelb*, 1974] where they are used to initialize the next analysis. However, while the background model can be a first principles ionospheric model, there is no feedback of the spatial maps into the first principles model and no correction or update to the model is made. IDA3D is an objective analysis algorithm (step 2 in the data assimilation cycle above), based upon three-dimensional variational (3DVAR) data assimilation [*Daley and Barker*, 2001; *Daley*, 1991]. The basic idea is to create a procedure for combining a model output with actual measurement data. Obviously, sparsely measured quantities could create discontinuities if they were used by directly placing into a model output. In addition, instruments will measure a quantity over a given region of space, and this measurement in itself will have a region over which it can be considered. 3DVAR is a statistical minimization procedure that deals with these issues. It takes into account data, data error covariances, a background model, and the background model error covariances.

Currently, IDA3D uses a nonlinear iterative approach that solves for the log of the density, which guarantees the density is positive definite. The nonlinear approach also makes it straightforward to add data sources nonlinearly related to density such as ionosonde virtual height versus frequency, EUV radiances, and oblique HF data.

2.2. Description of Data Sets Used in This Study

For this study, IDA3D ingested ground-based GPS TEC data from more than 1100 stations globally, ground-based TEC data from nine "beacon 150/400 MHz" stations, satellite in situ measurements of electron density from DMSP and Challenging Minisatellite Payload for Geophysical Research and Application (CHAMP), GPS occultation data from CHAMP and PicoSAT, and over satellite electron content (OSEC) from CHAMP, Gravity Recovery and Climate Experiment (GRACE), and SACC.

The OSEC data are of particular interest for this study. CHAMP and GRACE are at altitudes of ~400 km, near the peak of the F layer. The dual-frequency GPS navigation receiver on board the satellites looks up to the GPS constellation, typically collecting data every 10 s from 8–12 satellites at a time. As such, the OSEC data provide a unique capability to image the topside ionosphere and plasmasphere and can be used to interpret the redistribution of high-altitude plasma during magnetic storms [*Heise et al.*, 2002; *Jakowski et al.*, 2003; *Stankov et al.*, 2003; *Mannucci et al.*, 2005].

3. RESULTS

The two super storms of 30 October and 20 November 2003 both show the formation, evolution, and decay of SED over the U.S. sector (here, an SED is considered to be defined by both the midlatitude enhanced bulge of plasma and the narrow plume of ionization that extends over the pole). However, while both storms have similar characteristics, and in both cases SED forms in the postnoon period in the United States, the details of the SED structures between the two storms are quite different. Plate 1 shows fully formed SED for the two storms side by side. The most striking feature of contrast between the two storms is for 30 October, the high-density plasma bulge forms off the west coast of the United States and Mexico, with the orientation aligned with the California/Mexico coastline, while the narrow plume of plasma or tongue of ionization (TOI) stretches back in a southwest to northeast direction up over the pole. For the 20 November case, the bulge forms off the southeastern coast of the United States in the Caribbean, with the TOI stretching from southeast to northwest up over the pole. The second feature of interest is that the 20 November SED forms ~1.5 h earlier in time than the 30 October case. The third interesting feature is the high-density bulge extends too much higher latitudes on 30 October than on 20 November. For the 30 October, it extends to ~45° geographic latitude while 20 November extends to ~35°.

When contrasting the similarities and differences among the formation, evolution, and decay of midlatitude SED for 30 October and 20 November 2003, it is useful to start with the time history of the equatorial vertical drifts over Jicamarca for the two storms. The daytime drifts are obtained from the magnetometer method developed by *Anderson et al.* [2002, 2004, 2006] and given in Plate 2. In many ways, the equatorial drifts are similar for the two storms. Both of the initial prompt penetration electric (PPE) field effects occur near local noon. On 20 November, the vertical drift rises up to 80 m/s at about 1130 LT, while for 30 October, the increase to 50 m/s at about 1300 LT. In both cases, the enhancement in vertical drift above climate persists until about 1700 LT, slowly decreasing from the peak values at 1400 LT back to climatalogical values. Finally, in both cases the drifts are lower than climate earlier in the day (~0800 LT). However, there are also significant differences. The most obvious difference is the 20 November maximum drift is 80 m/s while for 30 October it is 50 m/s. There are other significant differences however. On 20 November, the 80 m/s upward drift persists approximately constant for ~2.5 h while for 30 October, the drift rises to 50 m/s and then back to climate in a period of a little less than one hour. Then, approximately an hour later at 1400 LT, it increases again to ~50 m/s. So while there is a single strong upward vertical drift for 20 November, for 30 October, the response is more of two impulsive increases in vertical drift, with the second increase decaying more slowly. Finally, the initial upward surge on 20 November begins ~1.5 h earlier than for 30 October.

The interesting question is how do these differences in equatorial vertical drifts impact the transport of high-altitude plasma to the midlatitude sector and the formation of the SED? To address this question, we examine IDA3D images of electron density in the American sector. In particular, we look at images during time periods that the CHAMP or GRACE satellites were in the American sector. As discussed above, both of these satellites have navigation GPS receivers on board that provide upward looking, so called, over satellite electron content. These data sets are sensitive to ionospheric and plasmaspheric plasma between ~400 km altitude and 20,000 km altitude. As such, they provide a unique

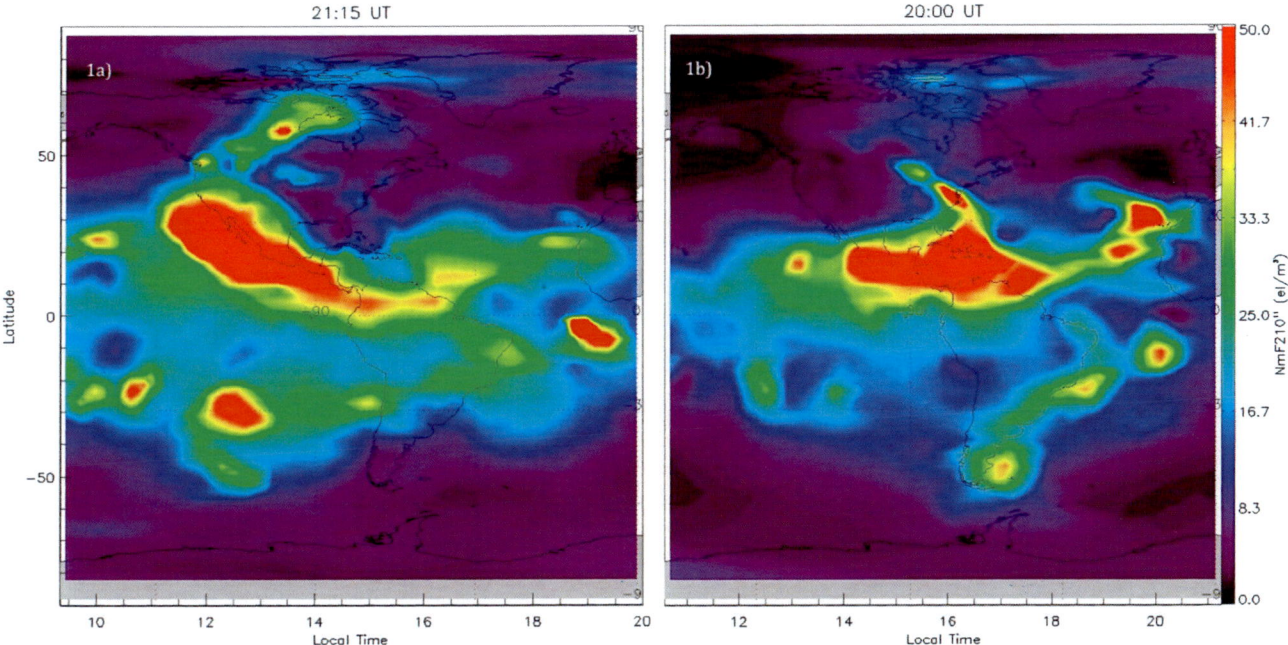

Plate 1. A comparison between the fully formed SED on 30 October and 20 November 2003. Presented are IDA3D horizontal images of peak electron density for (a) 30 October and (b) 20 November. Note that the horizontal axis shows the local time for each longitude sector.

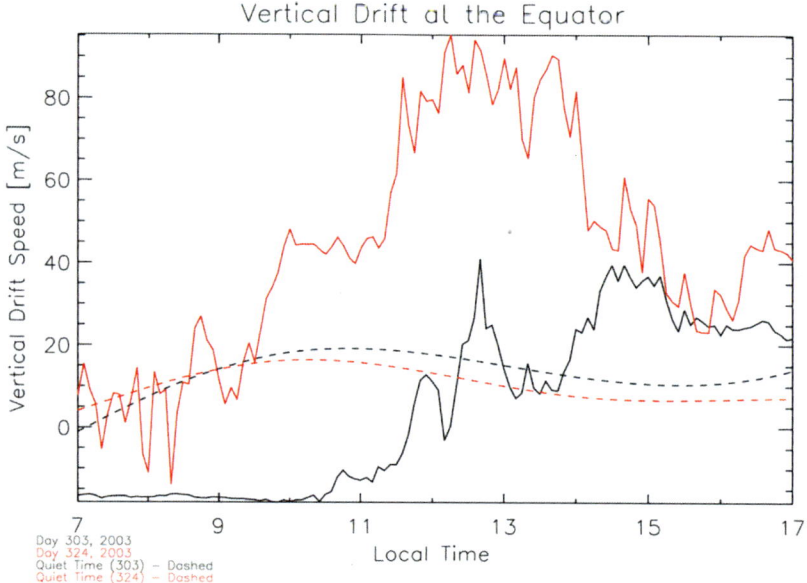

Plate 2. The equatorial vertical drifts for 30 October (solid black curve) and 20 November 2003 (solid red curve). The equatorial daytime drifts are estimated by the method developed by *Anderson et al.* [2004, 2006]. The dashed black and red curves represent the quiet time empirical values.

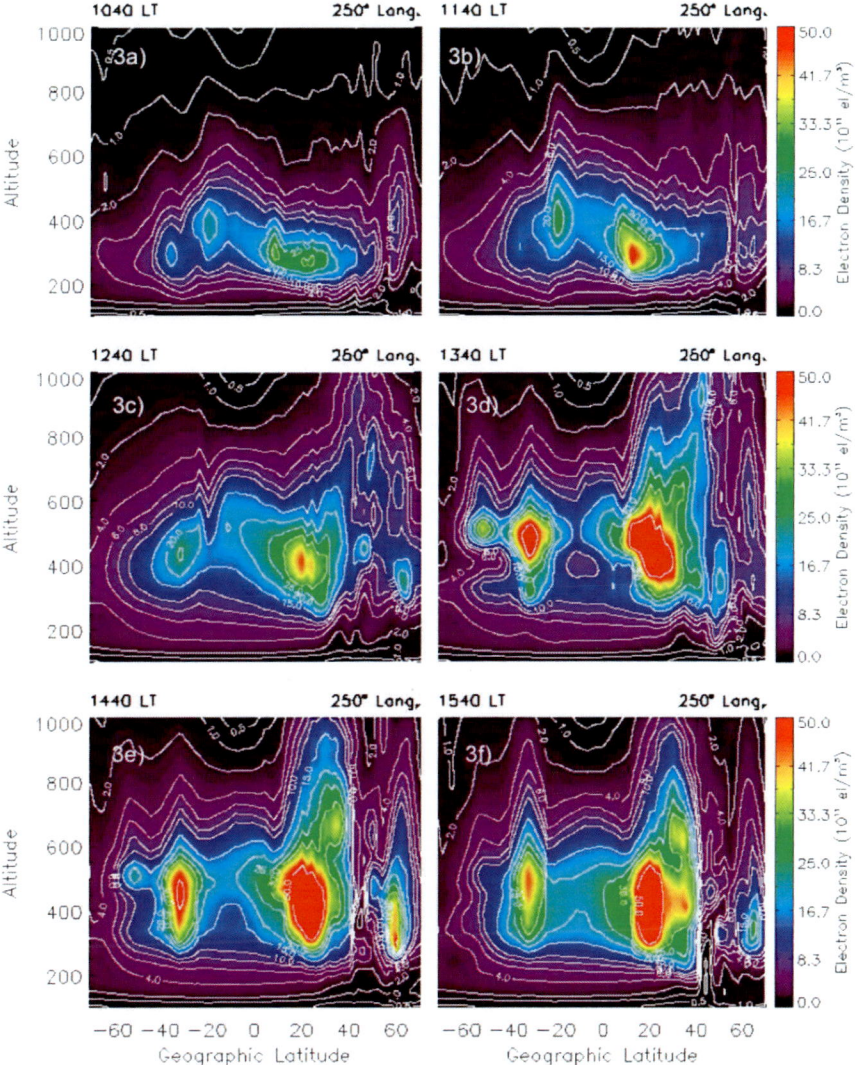

Plate 3. A sequence of six IDA3D slices of electron density (latitude versus altitude) for 30 October 2003. The slices are taken through a constant geographic longitude of 250°. Shown are the local times for each slice.

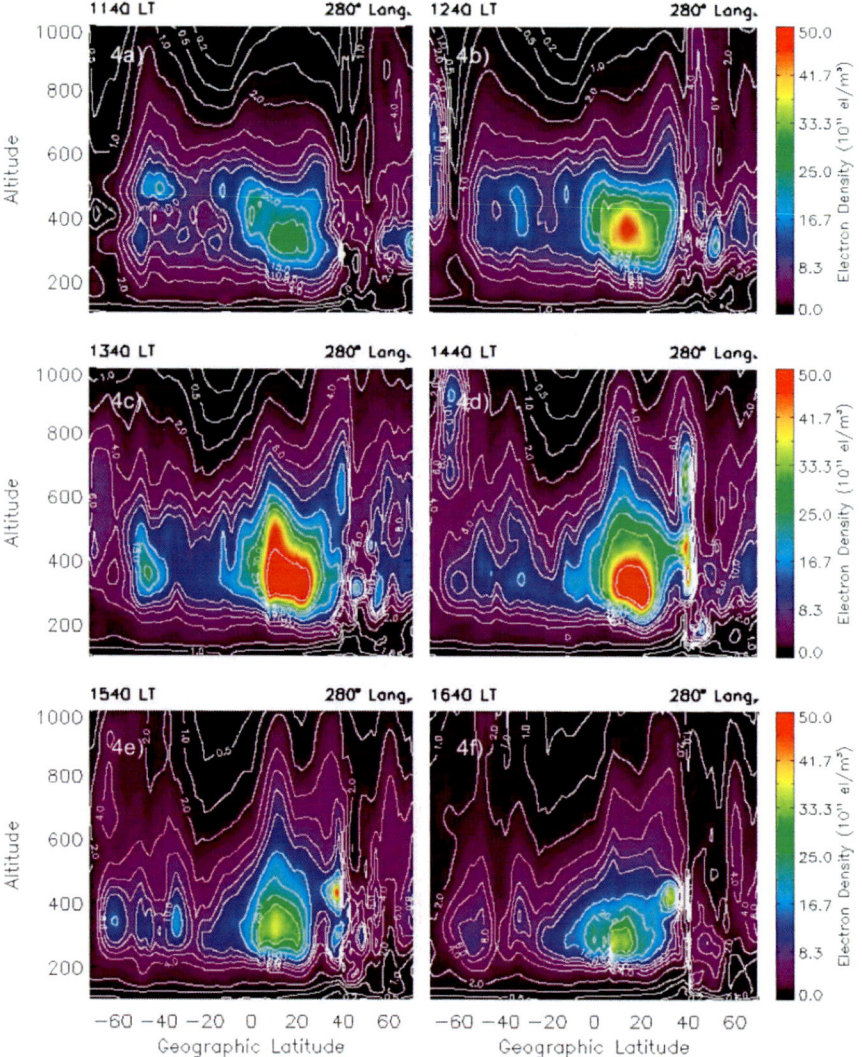

Plate 4. A sequence of six IDA3D slices of electron density (latitude versus altitude) for 20 November 2003. The form is similar to Plate 3, except the slices are taken through a constant 280° geographic longitude. As in Plate 3, local times are shown for each slice.

insight into the distribution of high-altitude plasma during storms and can help us understand how plasma equatorial plasma dynamics is coupled to midlatitude SED formation.

Further details are revealed by looking at a sequence of slices of IDA electron density images along a constant longitude. Plate 3 shows a sequence of six images of electron density for 30 October. The horizontal axis is along a constant longitude of 250°, and the vertical axis is altitude. The local times in the figure range from just before the start of the increase in vertical drift in Plate 2 (solid black curve) through 1540 LT. The six images show (1) the increase in altitude of the equatorial plasma (Plates 3a and 3b), (2) the separation of the anomaly peaks (Plates 3c and 3d), (3) the sharpening of the horizontal gradient of plasma at the poleward edge of the anomaly, along with the upward distribution of plasma along the poleward edge of the anomaly (Plates 3c–3f), and (4) the separation and growth of the SED plume (Plates 3c–3f). Plate 4 presents a similar set of six images along a longitude of 280° for 20 November. The images begin shortly after the rise in vertical drift in figure (solid red curve) (1140 LT) and continue through 1640 LT. When comparing with 30 October, the first striking feature is that at 1140–1240 LT (Plates 4a and 4b), shortly after the PPE produces the strong vertical drift, the anomalies are already well separated (~20° geographic latitude) with the northern peak much more intense than the southern peak. This maybe due to the previous storm history for this day, and this preconditioning of the anomaly peaks may influence the subsequent development of the SED. In addition, there is already evidence of an uplifting of the plasma to high altitudes at ~47.5° latitude in Plates 4a and 4b that was not present in the first two hours of the 30 October case (Plates 3a and 3b). The other interesting difference is that the main equatorial anomaly region on 20 November does not really rise with altitude. In fact, the entire equatorial region does not really rise to higher altitudes, despite the strong equatorial vertical drifts. It maybe that the off-equatorial conductance distribution is different for 20 November, thus restricting the electric field to the equatorial region, or perhaps winds play an important role. Regardless, the differences in equatorial uplift between the two storms are striking. Plates 4c–4f present results that are similar to Plates 3c–3f, although in different longitude sectors and at different local times. Both sets of figures show a continued separation of the anomaly region with latitude as time progresses. The latitudinal gradient on the poleward edge of the midlatitude bulge (~45°–47°) becomes more pronounced, and the maximum plasma density in the bulge increases as time progresses. Finally, both sets of figures show high-density plasma extending up to ~1000 km altitude at the poleward edge of the enhancement region. However, for 20 November, the anomaly peaks and the midlatitude plasma bulge are not raised in altitude as compared with the 30 October case. Despite the higher vertical drifts (~80 m/s) that last for an extended period, the off-equatorial plasma is not uplifted to high altitudes.

4. SUMMARY

Comparisons have been made between the two geomagnetic superstorms that occurred on 30 October and 20 November 2003. The comparison has focused on the development of enhanced plasma SED in the noon and afternoon in the midlatitude American sector. In particular, the relationship among the onset of prompt penetration fields at the equator, growth and separation of the equatorial anomalies, and the formation and evolution of the SED has been discussed.

While the two storms both have an SED bulge form over the southern United States that then leads to plumes of TOI stretching across the magnetic pole, the details of when and how the plasma bulges form are quite different. The 30 October event starts ~1 h later in universal time than 20 November but earlier in local time, and although the equatorial vertical drift is smaller and lasts for a shorter period than 20 November, the equatorial plasma is lifted to altitudes >500 km, while for 20 November, there is little to no uplift. In addition, there appears to be preconditioning of the equatorial region on 20 November, perhaps from storm effects that occurred earlier in the day. Thus, while there appears to be a strong linkage among the appearance of PPE, growth and separation of the equatorial anomalies, development of an SED midlatitude bulge, and subsequent formation of TOI (see Plates 3 and 4), the details of that linkage and how it varies for different storms still require elucidation.

One of the outstanding research questions concerns the details of how the SED bulge forms over the southern United States. It appears from the analysis that it is an extension of the equatorial anomaly region. However, the peak plasma density can double (from 2.5–5.0 E12 el/m^3) in 30 minutes. The relative roles of electric fields, winds, and composition remain to be understood in the context of the detailed formation of the SED. A second question relates to how the midlatitude bulge, which is not moving at a high speed, transforms into a TOI that is moving poleward at large plasma velocities. From Plates 3 and 4, it appears that the plasma is lifted to high altitudes (~700 km or higher) near 45°–47° latitude and then streams into the polar cap. If this is the case, is the uplift due to local electric fields, winds, or both?

One approach to investigating these questions is to use the results of the ionospheric imaging, which consists of a 4-D scalar field of electron density plus estimate of error to estimate the electric fields and winds through analysis of the electron density continuity equation. Such a "reverse engineering" approach to the data analysis for cases where there are only one or two dominant physical drivers that determine the plasma distribution can, in the best case, determine the relative importance of the winds, fields, and composition and, in the worst case, at least constrain the possible values the drivers can take, thus providing additional information to first principles models. Such work is ongoing, and progress in addressing the outstanding research questions is expected in the near future.

Acknowledgments. Development of IDA3D was supported by the Office of Naval Research under grant N00014-97-1-0236. IDA3D investigations of the equatorial ionosphere and polar cap were supported through the National Science Foundation under grants ATM-0513826 and ATM-0228467, respectively.

REFERENCES

Anderson, D., A. Anghel, K. Yumoto, M. Ishitsuka, and E. Kudeki (2002), Estimating daytime vertical ExB drift velocities in the equatorial F-region using ground-based magnetometer observations, *Geophys. Res. Lett.*, 29(12), 1596, doi:10.1029/2001GL014562.

Anderson, D., A. Anghel, J. Chau, and O. Veliz (2004), Daytime vertical ExB drift velocities inferred from ground-based magnetometer observations at low latitudes, *Space Weather*, 2, S11001, doi:10.1029/2004SW000095.

Anderson, D., A. Anghel, J. L. Chau, and K. Yumoto (2006), Global, low-latitude, vertical ExB drift velocities inferred from daytime magnetometer observations, *Space Weather*, 4, S08003, doi:10.1029/2005SW000193.

Buonsanto, M. J. (1999), Ionospheric storms—A review, *Space Sci. Rev.*, 88, 563–601.

Bust, G. S., T. L. Gaussiran, II, and D. S. Coco (1997), Ionospheric observations of the November 1993 storm, *J. Geophys. Res.*, 102(A7), 14,293–14,304, doi:10.1029/96JA03994.

Bust, G. S., D. Coco, and J. Makela (2000), Combined Ionospheric Campaign 1: Ionospheric tomography and GPS total electron count (TEC) depletions, *Geophys. Res. Lett.*, 27(18), 2849–2852, doi:10.1029/2000GL000053.

Bust, G. S., T. W. Garner, and T. L. Gaussiran, II (2004), Ionospheric Data Assimilation Three-Dimensional (IDA3D): A global, multisensor, electron density specification algorithm, *J. Geophys. Res.*, 109, A11312, doi:10.1029/2003JA010234.

Bust, G. S., G. Crowley, T. W. Garner, T. L. Gaussiran, II, R. W. Meggs, C. N. Mitchell, P. S. J. Spencer, P. Yin, and B. Zapfe (2007), Four dimensional GPS imaging of space-weather storms, *Space Weather*, 5, S02003, doi:10.1029/2006SW000237.

Coster, A. J., J. Foster, and P. Erickson (2003), Monitoring the ionosphere with GPS, Space Weather, *GPS World*, 14(5), 42–49.

Daley, R. (1991), *Atmospheric Data Analysis*, Cambridge University Press, Cambridge Atmospheric and Space Science Series, Cambridge University, Cambridge, UK.

Daley, R., and E. Barker (2000), *The NAVDIS Source Book, NRL/PU/7530-00-418*, Naval Research Laboratory, Monterey, Calif.

Foster, J. C., and W. Rideout (2005), Midlatitude TEC enhancements during the October 2003 superstorm, *Geophys. Res. Lett.*, 32, L12S04, doi:10.1029/2004GL021719.

Foster, J. C., A. J. Coster, P. J. Erickson, J. Goldstein, and F. J. Rich (2002), Ionospheric signatures of plasmaspheric tails, *Geophys. Res. Lett.*, 29(13), doi:10.1029/2002GL015067.

Gelb, A. (1974), *Applied Optimal Estimation*, MIT Press, Cambridge, Mass.

Heise, S., N. Jakowski, A. Wehrenpfennig, Ch. Reigber, and H. Lühr (2002), Sounding of the topside ionosphere/plasmasphere based on GPS measurements from CHAMP: Initial results, *Geophys. Res. Lett.*, 29(14), doi: 10.1029/2002GL014738.

Jakowski, N., S. Heise, K. Tsybulya, and A. Wehrenpfennig (2003), Sounding the ionosphere by GPS measurements on CHAMP, *Geophys. Res. Abstr.*, 5, Abstract 09663.

Mannucci, A. J., B. T. Tsurutani, B. A. Iijima, A. Komjathy, A. Saito, W. D. Gonzales, F. L. Guarnieri, J. U. Kozyra, and R. Skoug (2005), Dayside global ionospheric response to the major interplanetary events of October 29–30, 2003 "Halloween Storms," *Geophys. Res. Lett.*, 32, L12S02, doi:10.1029/2004GL021467.

Mitchell, C. N. (2002), Imaging of near-Earth space plasma, *Philos. Trans. R. Soc. London Ser. A*, 360(1801), 2805–2818.

Mitchell, C. N., and P. S. J. Spencer (2003), A three-dimensional time-dependent algorithm for ionospheric imaging using GPS, *Ann. Geophys.*, 46(4), 687–696.

Mitchell, C. N., L. Alfonsi, G. De Franceschi, M. Lester, V. Romano, and A. W. Wernik (2005), GPS TEC and scintillation measurements from the polar ionosphere during the October 2003 storm, *Geophys. Res. Lett.*, 32, L12S03, doi:10.1029/2004GL021644.

Prölss, G. W. (1995), Ionospheric F-region storms, in *Handbook of Atmospheric Electrodynamics*, vol. II, edited by H. Volland, pp. 195–248, CRC Press, Boca Raton, Fla.

Stankov, S. M., N. Jakowski, S. Heise, P. Muhtarov, I. Kutiev, and R. Warnant (2003), A new method for reconstruction of the vertical electron density distribution in the upper ionosphere and plasmasphere, *J. Geophys. Res.*, 108(A5), 1164, doi:10.1029/2002JA009570.

Yin P., C. N. Mitchell, P. S. J. Spencer, and J. C. Foster (2004), Ionospheric electron concentration imaging using GPS over the USA during the storm of July 2000, *Geophys. Res. Lett.*, 31, L12806, doi:10.1029/2004GL019899.

Yin, P., C. N. Mitchell, and G. S. Bust (2006), Observations of the F region height redistribution in the storm-time ionosphere over Europe and the USA using GPS imaging, *Geophys. Res. Lett.*, 33, L18803, doi:10.1029/2006GL027125.

G. S. Bust and G. Crowley, Atmospheric & Space Technology Research Associates, 12703 Spectrum Drive, Suite 101, San Antonio, TX 78249, USA. (gbust@astraspace.net)

Interplanetary Causes of Middle Latitude Ionospheric Disturbances

Bruce T. Tsurutani,[1] Ezequiel Echer,[2] Fernando L. Guarnieri,[3] and Olga P. Verkhoglyadova[1,4]

The solar and interplanetary causes of major middle latitude ionospheric disturbances are reviewed. Solar flare photons can cause abrupt (within ~5 min), 30% increases in ionospheric total electron content, a feature that can last for tens of minutes to hours, depending on the altitude of concern. Fast interplanetary coronal mass ejection sheath fields and magnetic clouds can cause intense magnetic storms if the field in either region is intensely southward for several hours or more. If the field conditions in both regions are southward, "double storms" will occur. Multiple interplanetary fast forward shocks "pump up" the sheath magnetic field, leading to conditions that can lead to superstorms. Magnetic storm auroral precipitation and Joule heating cause pressure waves that propagate from subauroral latitudes to middle and equatorial latitudes. Shocks can create middle latitude dayside auroras as well as trigger nightside subauroral supersubstorms. Solar wind ram pressure increases after fast shocks can lead to the formation of new radiation belts under proper conditions. Prompt penetration electric fields can cause a dayside ionospheric superfountain, leading to plasma transport from the equatorial region to middle latitudes. The large amplitude Alfvén waves present in solar wind high-speed streams cause sporadic magnetic reconnection, plasma injections, and electromagnetic chorus wave generation. Energetic electrons interacting with chorus (and PC5) waves are accelerated to hundreds of keV up to MeV energies.

1. INTRODUCTION

There are a number of possible routes that various types of signals can take going from interplanetary space to the Earth's ionosphere. Signals can pass through the dayside magnetosphere, they can first go to the tail/plasmasheet and then travel through the nightside magnetosphere, they can pass along magnetic field lines, or they can travel from the polar ionosphere to the equator through the Earth-ionospheric wave guide. For this review, we will not focus on how the signals travel from interplanetary space (and ultimately from the Sun) to the middle latitude ionosphere, but more on what types of interplanetary phenomena are important for this process and what are the middle latitude ionospheric responses. For our definition of what the middle latitude region is, we will take a liberal viewpoint, identifying ionospheric responses that occur equatorward of the normal location of auroral displays (the auroral zone) and poleward of the magnetic equator/equatorial electrojet. During strong geomagnetic activity, auroral displays move equatorward, and equatorial ionospheric phenomena move poleward, making the term "middle latitude" somewhat ambiguous. It is felt that it would be better to err on the side of overcompleteness for this particular review.

---

[1] Jet Propulsion Laboratory, California Institute of Technology, Pasadena, California, USA.

[2] Brazilian National Institute for Space Research (INPE), Sao Jose dos Campos, Brazil.

[3] University of Paraiba Valley (UNIVAP), Sao Jose dos Campos, Brazil.

[4] CSPAR, University of Alabama in Huntsville, Huntsville, Alabama, USA.

Midlatitude Ionospheric Dynamics and Disturbances
Geophysical Monograph Series 181
Copyright 2008 by the American Geophysical Union.
10.1029/181GM11

Solar photons create the Earth's ionosphere by photoionization of atmospheric neutrals, and the solar wind forms and shapes the Earth's magnetosphere. For this review, we will focus on solar/interplanetary features that cause significant changes in the baseline ionospheric and magnetospheric values. This is the topic known as "space weather." The three energy transfer mechanisms that will be discussed in detail are: (1) solar photoionization, (2) solar wind ram pressure pulses, and (3) magnetic reconnection between interplanetary magnetic fields (IMFs) and the Earth's magnetic fields.

2. RESULTS

2.1. Solar Cycle-Dependent Phenomena

The sun has an approximate 22-year cycle (the Hale cycle), within which the polar magnetic fields reverse polarity each half-cycle, or approximately 11 years [*Schwabe*, 1843; *Babcock*, 1961]. During each half-cycle, there are two well-defined phases: solar maximum, when there is the largest number of sunspots (defined after the fact); and solar minimum, when there is a minimum number of sunspots. The three main types of solar activity that have severe consequences for space weather can be generally defined in relation to these solar phases. The greatest number of solar flares occurs during and near solar maximum [*Svestka*, 1976]. Coronal mass ejections (CMEs) take place in association with the more intense solar flares [*Tsurutani et al.*, 1988; *Tang et al.*, 1989; *Webb and Howard*, 1994; *Gopalswamy*, 2006]. CMEs occur most frequently approximately 1 year before solar maximum to 2 or 3 years after it.

It should be noted that solar flares can occur without CMEs and CMEs can occur without flares. However, interplanetary CMEs (ICMEs) and/or their upstream sheaths are the cause of almost all major magnetic storms at Earth [*Echer et al.*, 2008]. Solar flares, in turn, are associated with almost all, if not all, of these CMEs. Thus in practical terms, for the big storm-producing events, CME release and solar flares are almost synonymous.

After solar maximum has been reached, polar coronal holes begin to form [*Harvey et al.*, 2000; *Harvey and Recely*, 2002]. During the declining phase of the solar cycle (from maximum to minimum), these holes grow increasingly larger and can extend to the solar magnetic equator or even cross it. High-speed solar wind streams emanate from these "holes" [*Krieger et al.*, 1973]. When the coronal holes extend to sufficiently low heliolatitudes (or there are isolated coronal holes at these latitudes), the high-speed streams will hit the Earth's magnetosphere and cause enhance geomagnetic activity.

It should be noted that solar flares and CMEs are not limited to the solar maximum interval or its proximity. Intense flaring and very high-speed CMEs can occur during solar minimum. But the occurrence frequency is greatly reduced during the latter solar phase. The ratio of the number of major flares at solar maximum to minimum is greater than 10:1. The ratio of the number of fast CMEs at solar maximum to minimum is ~10:1 [*Webb and Howard*, 1994; *St Cyr et al.*, 2000; *Schwenn*, 2006]. The ratio of number of major magnetic storms with $Dst < -100$ nT is also ~10:1 [*Tsurutani et al.*, 2006a; *Gonzalez et al.*, 2007].

2.2. Solar Effects on the Ionosphere

Solar flares are sudden bursts of solar irradiance that last for tens of minutes [*Carrington*, 1859]. They have an asymmetric intensity–time profile, with a fast rise time and a slow decay. The spectra of flares are much different than the regular solar spectrum. Flares have the greatest intensity increases at the shortest wavelengths, for example, in the X-ray, ultraviolet (UV), and extreme ultraviolet (EUV) wavelength range (we refer the reader to *Sturrock* [1980] for a review of the physical properties of solar flares). The amount of intensity increase in visible light is usually negligible. As an example, the first reported solar flare [*Carrington*, 1859] only had a factor of ~2 intensity increase in visible light. On the other hand, the intensity increase in the X-ray range for intense X-class flares can be 3 orders of magnitude or greater [*Tsurutani et al.*, 2005].

Solar flares are especially effective in creating short-duration dayside ionospheric disturbances, called sudden ionospheric disturbances [*Thome and Wagner*, 1971; *Mitra*, 1974; *Donnelly*, 1976; *Afraimovich*, 2000]. The penetration of the solar flare photons into the dayside atmosphere depends on the energy of the photons. The greater the energy, the deeper the penetration. Photons of X-ray energies penetrate to depths of ~90- to 110-km altitude. They lose their energy by photoionization of atmospheric atoms and molecules. Energetic secondary electrons are also produced by this process. The secondaries, in turn, lead to further cascading of the ionization process. Because of the high densities at these altitudes, recombination time scales are relatively short, approximately tens of minutes.

The EUV/UV part of the solar flare spectrum causes effects that are longer lasting. Because of the lower energy of the photons in this part of the spectrum, these photons are stopped relatively high in the atmosphere, hundreds of kilometers above the Earth. However, since the recombination time scale at these altitudes approximately last for hours, the effects (enhanced dayside ionization) last longer [*Tsurutani et al.*, 2005].

Plate 1 shows the total electron content (TEC) enhancement over NKLG, a station at Libreville, Gabon, located near the Sun's subsolar point (11.7 LT, 0.4° latitude) during the 28 October 2003 (Halloween) solar flare. This flare had a magnitude of X-17 as measured by the GOES satellite in a ~1- to 8-Å (X-ray) bandwidth. This flare was very intense in X-rays, but not as intense as the 4 November 2003 event (however, see discussion in Thomson et al. [2004]). On the other hand, as measured in 260- to 340-Å EUV wavelengths (SOHO SEM; Judge [1998]), the 28 October event is the largest on record. See Tsurutani et al. [2005] for further discussion. The NKLG TEC enhancement rose to an ~25 TECU increase (a TECU is 10^{16} el m^{-2} column density) in ~5 min. The background density was ~82 TECU at the time, so this flare lead to a ~30% TEC increase in ~5 min. The ionization effect was noted at all latitudes and longitudes in the dayside ionosphere.

2.3. Interplanetary CMEs, Shocks, and Sheaths

A CME evolves as it propagates through the solar corona and into interplanetary space. What is detected at 1 AU upstream of the Earth may be significantly different from that which was emitted at the sun [Hu and Sonnnerup, 2001; Riley and Crooker, 2004; Liu et al., 2006; Lepping et al., 2007]. For this reason, the plasma and magnetic field features of a CME in interplanetary space have been named an ICME [Tsurutani et al., 1997]. The bright outer loops of the CME and the trailing filaments noted in coronagraph images close to the sun have rarely been detected at 1 AU. However, the most geoeffective part of the CME, the "dark region" in coronagraph images, is believed to be a magnetic cloud (MC) and is indeed detected by in situ plasma and magnetic field measurements made at 1 AU. A characteristic used to identify MCs is an unusually low plasma beta, the ratio of plasma thermal pressure to magnetic pressure [Farrugia et al., 1997; Tsurutani and Gonzalez, 1997]. The plasma beta in MCs is typically ~0.1 and can be as low as ~0.001, whereas the normal solar wind has a beta of ~1.0. MCs are also characterized by a lack of waves and discontinuities [Tsurutani et al., 1988].

If the CME is ejected from the sun at a high speed, a shock will form along the antisolar front. In magnetohydrodynamic terminology, this is a fast (magnetosonic or fast) shock [Kennel et al., 1985]. Since it is propagating in the same direction as the ICME, it is called a forward shock. Hence, the complete nomenclature is fast forward shock. For brevity, some people simply call it a "shock." The strength (Mach number) of the shock depends on the relative speed of the ICME to the upstream solar wind. The higher the differential speed of the ICME, the higher the Mach number of the shock (it should be noted that the speed of the shock into the upstream medium is not simply the speed of the plasma behind the shock). To calculate the shock speed, the shock normal must first be calculated. The speed is determined by calculating the Rankine–Hugoniot conservation laws along that direction. Details can be found in Tsurutani and Lin [1985].

The solar wind plasma decelerates from a super-(magneto)sonic speed to a sub-(magneto-)sonic speed across a shock. Thus both the plasma density and the magnetic field intensity are greatly enhanced (compressed). At 1 AU, typical interplanetary shocks have magnetosonic Mach numbers of ~1 to 3. In some extreme cases, they have higher values. What is important for space weather effects is that the plasma densities increase across fast forward shocks. The downstream to upstream density ratios are about equal to the Mach numbers (up to a value of 4.0) [Kennel et al., 1985]. Thus when shocks impinge upon the Earth's magnetosphere, the instantaneous higher plasma densities (and higher speeds) lead to significant ram pressure increases. This will cause the magnetosphere to be compressed until new pressure equilibria are reached. Some specific ram pressure increase consequences will be discussed later in this paper.

The plasma and magnetic fields in the region from the shock up to the MC is called the sheath, in analogy to the Earth's magnetosheath. In the case of the Earth's bow shock, it is a fast reverse shock. It is called a reverse shock because it is propagating in the solar wind plasma in a direction counter to the solar wind flow. This sheath plasma and magnetic field are not part of the ICME, but are compressed and heated slow solar wind plasma. The Earth's magnetosheath is similar in nature. The plasma and magnetic fields within the magnetosheath flow around the magnetosphere, for the most part.

The magnetic fields within the sheath are substantially higher than the ambient upstream magnetic fields, by a factor of approximately the Mach number (roughly the same as the plasma density) up to a maximum of ~4.0. Therefore, if the sheath fields have southward orientations and they impinge upon the Earth, they will cause intense magnetic reconnection and thus intense magnetic storms [Tsurutani et al., 1988]. MC magnetic fields are intrinsically intense (the field magnitude is also dependent on the MC speed; Gonzalez et al. [1998]). If the field orientation is southward, a magnetic storm will occur. If they are northward, there will be geomagnetic quiet [Tsurutani and Gonzalez, 1995]. Past studies have indicated that about half of all intense magnetic storms are due to (southwardly oriented) sheath fields and the other half due to MC fields [Tsurutani et al., 1992; Echer and Gonzalez, 2004; Gonzalez et al., 1994, 2007; Echer, 2008]. If the field is southward in both places, a "double storm" will occur [Tsurutani et al., 1988, 1992; Kamide et al., 1998].

Plate 2 is an example of some of the features discussed above. The event, which occurred on 7–8 November 2004, was selected by the CAWSES program for detailed study. The interplanetary data are taken from the Advanced Composition Explorer (ACE) satellite [*Stone et al.*, 1998], located ~0.01 AU upstream of the Earth. The top nine panels are solar wind parameters. From top to bottom, they are: proton temperature, solar wind speed, density, four panels of the IMF, the interplanetary electric field y component, ram pressure, and plasma beta. The bottom panel is the *Dst* index. The *Dst* index has not been time-shifted for the delay of the solar wind propagation from ACE to the Earth because the time shift is different for different parts of the event/interval.

There are three vertical markers on the left side labeled "FS" for forward shock (fast forward shock). Each shock had a compression factor of about 2 times, that is, the density and field magnitude increase across it by about a factor of 2. Thus, the magnetic field magnitude increased by a total of ~8 times from the crossing of the three shocks. Positive sudden impulses (SI+) are noted in the *Dst* index for the three shock crossings. The high ram pressures after the shocks cause magnetospheric compressions and the SI+ events (see discussion about SI+, SI–, SC, and SSC in *Joselyn and Tsurutani* [1990]).

The IMF is thus "pumped up" across each of these shocks. After the third shock, the magnetic field reached a magnitude of ~44 nT. However, the direction of the magnetic field is northward and these intense, shocked fields do not cause magnetic storm main phases.

The storm main phase is intensified by the southward magnetic fields within the MC. The magnetic cloud is indicated in the plate by the horizontal bar. As previously noted, this is identified by a low plasma beta. This superstorm reached an intensity of $Dst = -373$ nT.

The vertical line at ~04 UT on 4 November is a reverse wave. The jump parameters indicate that it is not a reverse shock [*Tsurutani et al.*, 2008]. It occurs within the MC and leads to a decrease in the magnetic field magnitude. What is particularly significant here is that it causes a decrease in the southward field component and is the cause of the start of the recovery phase of the magnetic storm. More details concerning this event are given by *Tsurutani et al.* [2008].

2.4. Shocks Causing Dayside Auroras and Triggering Substorms

When interplanetary fast forward shocks and their trailing sheath plasmas impinge upon the magnetosphere, they compress the magnetosphere and the plasma contained therein. Adiabatic compression of the magnetospheric plasma in the direction perpendicular to the magnetic field leads to asymmetric plasma distributions. These distributions are unstable and in turn lead to plasma instabilities with concomitant plasma wave growth. The plasma waves will scatter the particles and cause their precipitation into the upper atmosphere/ionosphere [*Zhou and Tsurutani*, 1999; *Tsurutani et al.*, 2001; *Zhou et al.*, 2003a]. The ram compression will also create magnetospheric Alfvén waves, which will propagate to the ionosphere and heat auroral plasma through wave damping [*Haerendel*, 1994; *Tsurutani and Zhou*, 2003]. If the shock is particularly strong, the effects will occur at dayside auroral to middle latitudes.

Strong shocks will also compress the near-Earth magnetotail as they propagate past the Earth and down the tail. If the magnetosphere has been "primed" by southward magnetic fields before the arrival of the shocks, substorm triggering will occur [*Schieldge and Siscoe*, 1970; *Kawasaki et al.*, 1971; *Burch*, 1972; *Kokubun et al.*, 1977; *Zhou and Tsurutani*, 2001; *Tsurutani and Zhou*, 2003]. These substorms are particularly noteworthy in that they have unusually large intensities. Their effects can easily reach middle latitudes of the nightside ionosphere. These have been called supersubstorms.

An example of dayside aurora and a supersubstorm caused by an intense shock (and trailing sheath) on 24 September 1998 are shown in Figure 1 and Plate 3. Figure 1 shows the shock as detected by the ACE spacecraft. The magnetic field increase from ~15 to ~45 nT across the shock, a factor of ~3.0. The last two panels give the *AE* and *AL* indices. The time delay for the solar wind to propagate from ACE to the near-Earth magnetotail has not been removed. Most of the time delay is due to this propagation delay feature. The substorm triggered by the shock compression of the tail has an intensity of ~2000 nT. It is a supersubstorm.

Plate 3 shows the UV images of this event. These images were taken by the Polar UV imager [*Torr et al.*, 1995] in the ~1700-Å Lyman–Birdge–Hopfield (LBH) bandwidth. The cadence was ~1 min 14 s. Time increases from the top left to the right and then on the bottom left. The images were taken of the northern polar region, with the pole at the center of the image. Local noon is at the top and midnight at the bottom.

Before the shock arrival, there was some auroral activity in the midnight sector, from 2100 LT to 0200 LT. This can be noted in the first three images. A sudden brightening of aurora on the dayside is noted in the image taken at 2345:57 UT. This is due to shock compression of the magnetosphere and the mechanisms discussed earlier. This aurora was not present in the image taken ~1 min earlier. At 2347:11 UT, the nightside aurora becomes intensified, indicating that the supersubstorm had begun. By 2352:05 UT, the supersubstorm had become fully developed, extending from 18 LT through midnight to 04 LT, and from 73° to well below 60° MLAT. These supersubstorms have large latitudinal and longitudinal extents.

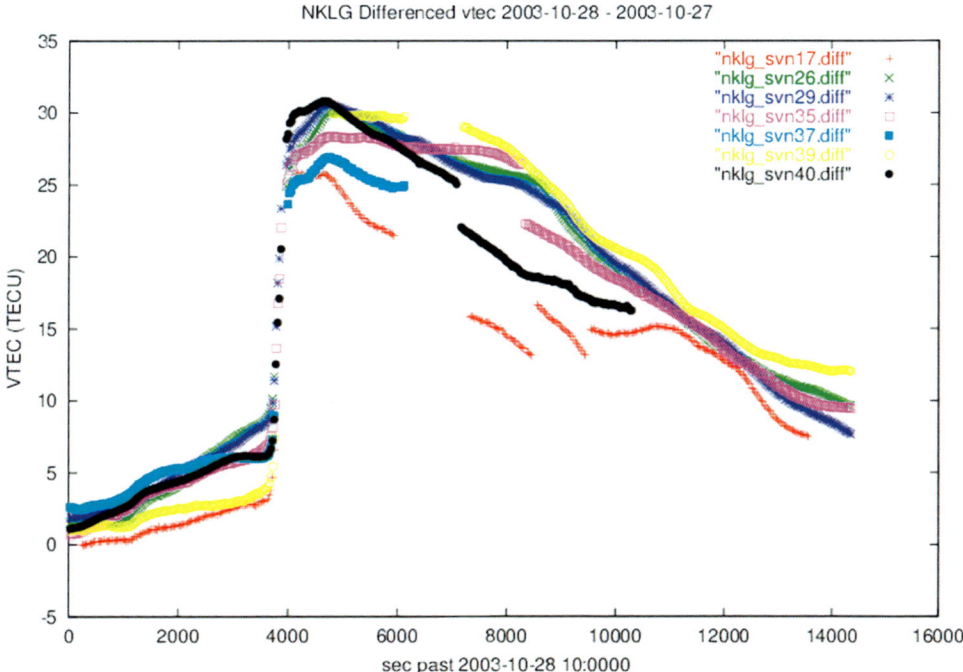

Plate 1. Ground-based GPS receiver data from NKLG, a Libreville Gabon station close to the subsolar point at the time of the 28 October 2003 solar flare. The TEC derived from multiple satellite tracks are shown. The TEC rises over 25 TECU in ~5 min.

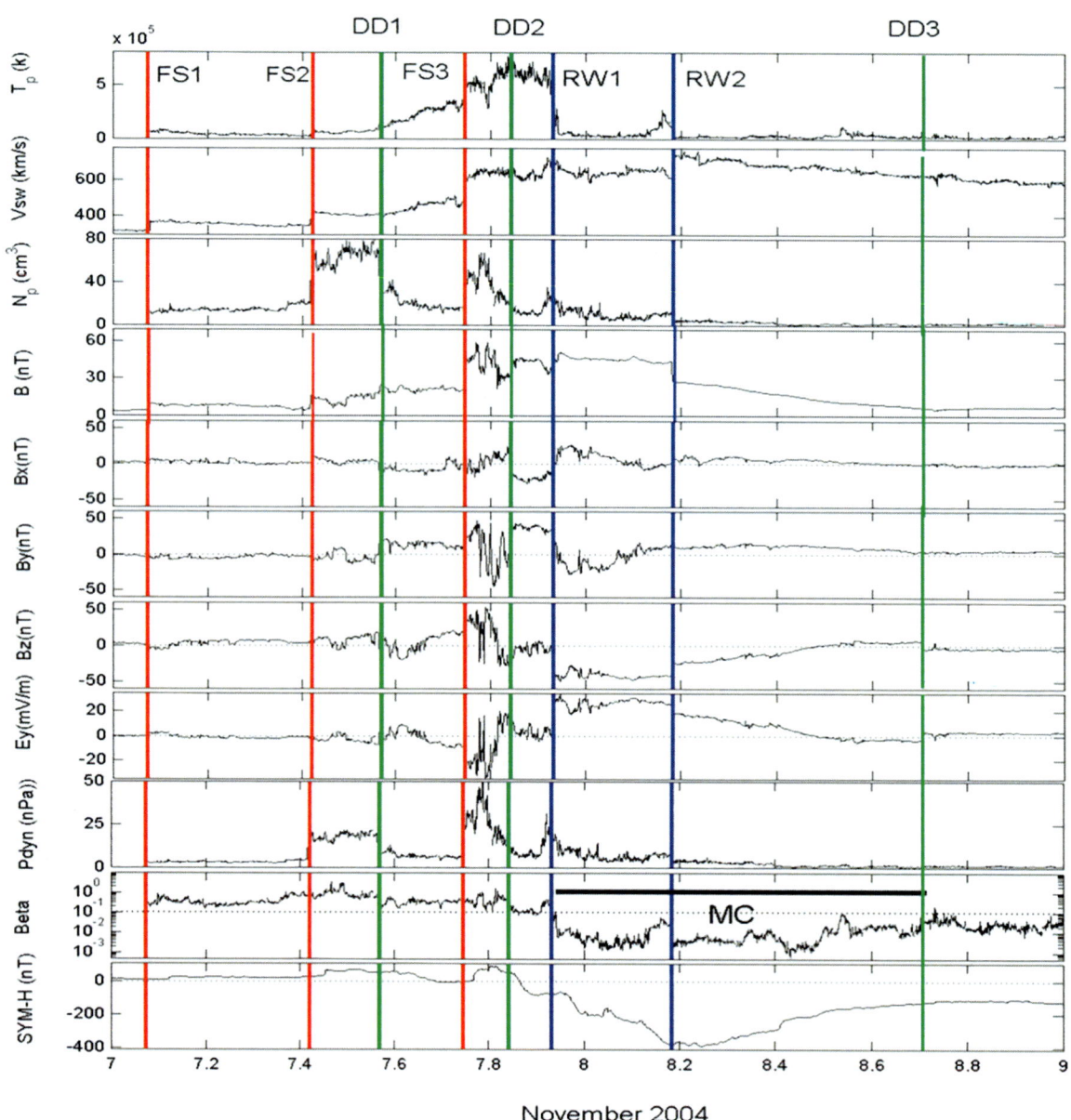

Plate 2. A complex interplanetary event on 7–8 November 2004. The event had three forward shocks, pumping the IMF to higher magnitudes. A magnetic cloud southward field causes a superstorm. A reverse wave is responsible for a reduction in magnetic reconnection rate, and the onset of the storm recovery phase.

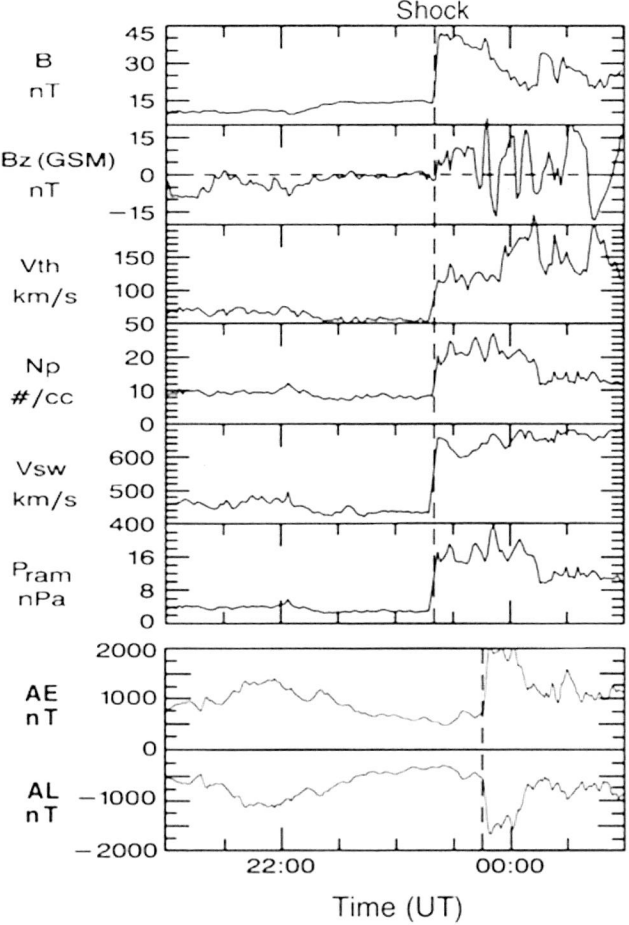

Figure 1. An interplanetary shock and geomagnetic response for an event on 24 September 1998. The shock compression triggers a supersubstorm.

2.5. Shock Creation of New Radiation Belts

Interplanetary shocks can cause the acceleration of energetic particles at and just upstream of their surfaces [*Tsurutani et al.*, 1982; *Tsurutani and Lin*, 1985; *Sanderson et al.*, 1985; *Kallenrode et al.*, 1992]. They accelerate particles continuously from the time of their formation at ~5 Rs [*Tsurutani et al.*, 2003] from the surface of the sun to 1 *AU* and beyond (*Kahler et al.* [1984]; see reviews in *Reames* [1999] and *Kallenrode* [2003]). The solar flare reconnection process is believed to also accelerate energetic particles at the flare/reconnection site. Some suggested mechanisms can be found in *Craig and Litvinenko* [2002], *Wu* [1996], and *Haerendel* [2001]. Both types of particles, those accelerated at shocks and those accelerated at the flare sites, are guided along the IMF lines, due to their low energy densities relative to those of the magnetic fields. If the magnetic field lines that contain the energetic particles lead to the vicinity of the Earth, the energetic particles may enter the magnetosphere.

It has been shown that the rapid compression of the Earth's magnetosphere due to the ram pressure of the shock/sheath energizes trapped and semitrapped particles within the magnetosphere by adiabatic compression. Since the time scale for compression is approximately minutes, the particles' third adiabatic invariants are broken and the semitrapped particles may become trapped. Several cases of this occurrence have been illustrated in the literature [*Mullen et al.*, 1991; *Lorentzen et al.*, 2002; *Mazur et al.*, 2005] as well as computer modeling [*Hudson et al.*, 1997, 2004; *Elkington et al.*, 2003, 2004], giving details of the trapping mechanism. When this happens, a new radiation belt is formed. Plate 4 shows computer simulation results of new radiation belt formation under a variety of solar wind conditions.

All of the above discussion concerning ram pressure effects had to do with changes on time scales of shock crossings, or approximately minutes. It has been shown that ram pressure changes that occur much more slowly than shock crossings will also have geomagnetic effects. There is empirical evidence that dayside auroras are associated with ram pressure changes occurring over time scales of hours [*Zhou et al.*, 2003b; *Zhou and Tsurutani*, 2004]. However, in these cases, clear understanding of the physical mechanisms is presently lacking.

2.6. Interplanetary Electric Fields and the Dayside Ionospheric Superfountain Effect

When the IMF is southwardly oriented, an interplanetary spacecraft would observe a motional electric field in the "dawn-to-dusk" direction. With magnetic reconnection between the interplanetary fields and the Earth's magnetopause fields [*Dungey*, 1961], the magnetosphere will experience an electric field in the same general direction, but diminished in amplitude. An approximate ratio is that the magnetospheric/polar ionospheric electric field is 5–10% of the intensity of the interplanetary electric field [*Gonzalez et al.*, 1989; *Kelley et al.*, 2003].

This magnetospheric/polar ionospheric electric field has been noted to propagate to the dayside and nightside equatorial ionosphere [*Obayashi*, 1967; *Nishida*, 1968; *Onwumechili et al.*, 1973; *Kelley et al.*, 1979] with very little delay, essentially the speed of light [*Kikuchi and Araki*, 1979]. For obvious reasons, these have been called prompt penetration electric fields (PPEFs). It had previously been thought that field-aligned shielding currents would be set up after ~30 min so that the effects of PPEFs on the equatorial ionosphere would be minimized [*Southwood*, 1977; *Southwood and Wolf*, 1978]. However, recent observations have indicated

106 INTERPLANETARY CAUSES OF MIDDLE LATITUDE IONOSPHERIC DISTURBANCES

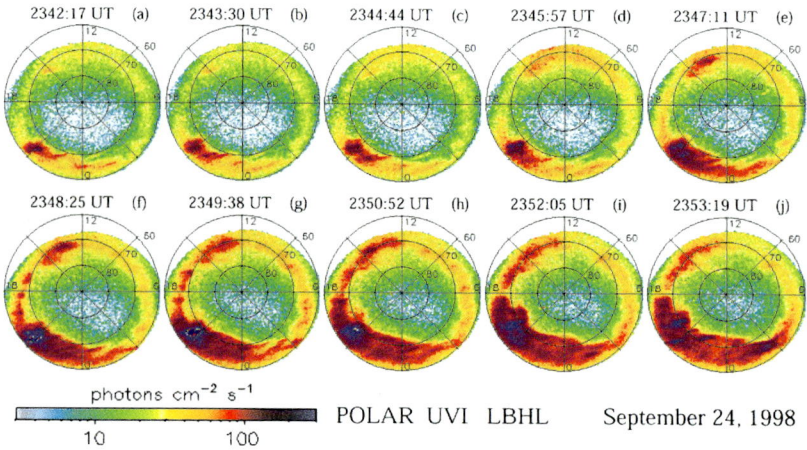

Plate 3. The auroral response for the event shown in Figure 1. Shock compression causes a dayside aurora and triggers a supersubstorm that reaches middle latitudes on the nightside.

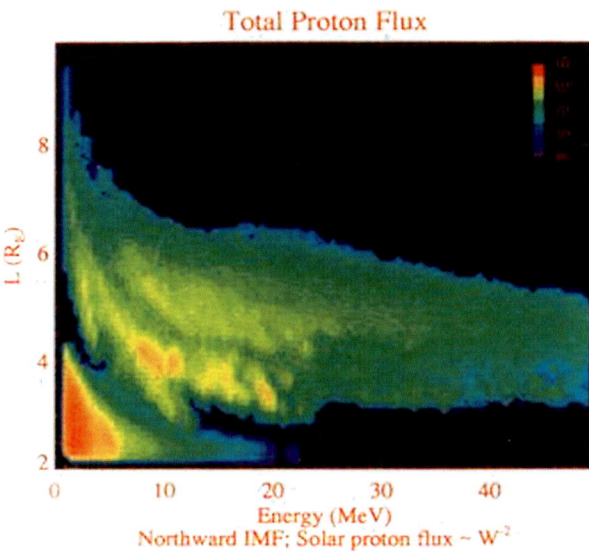
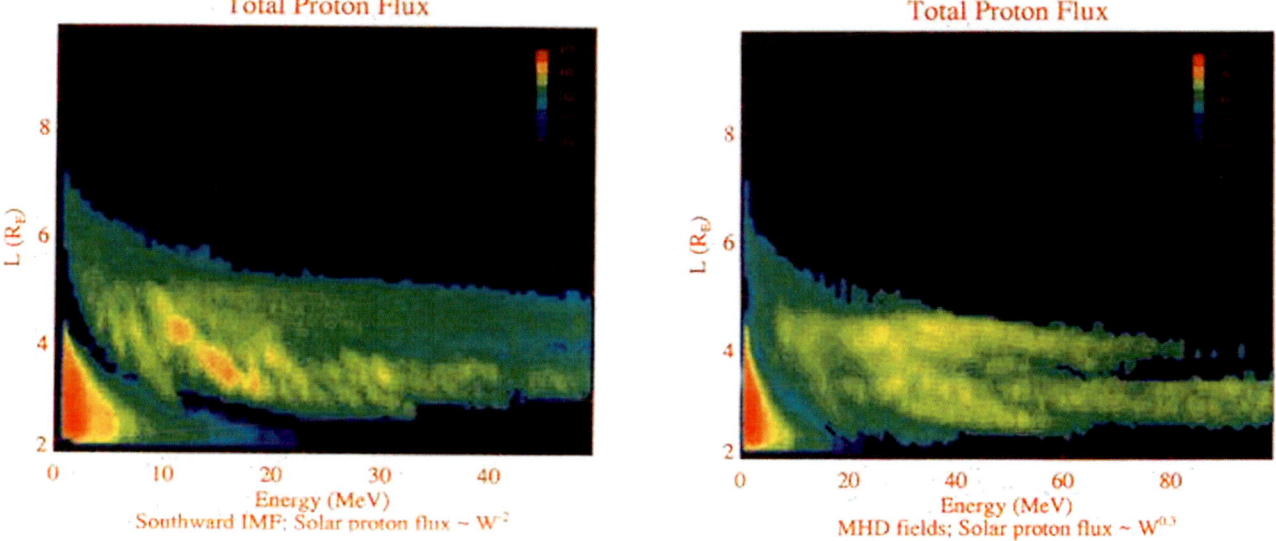

Plate 4. Modeling of shock-induced formation of a new radiation belt under a variety of interplanetary conditions.

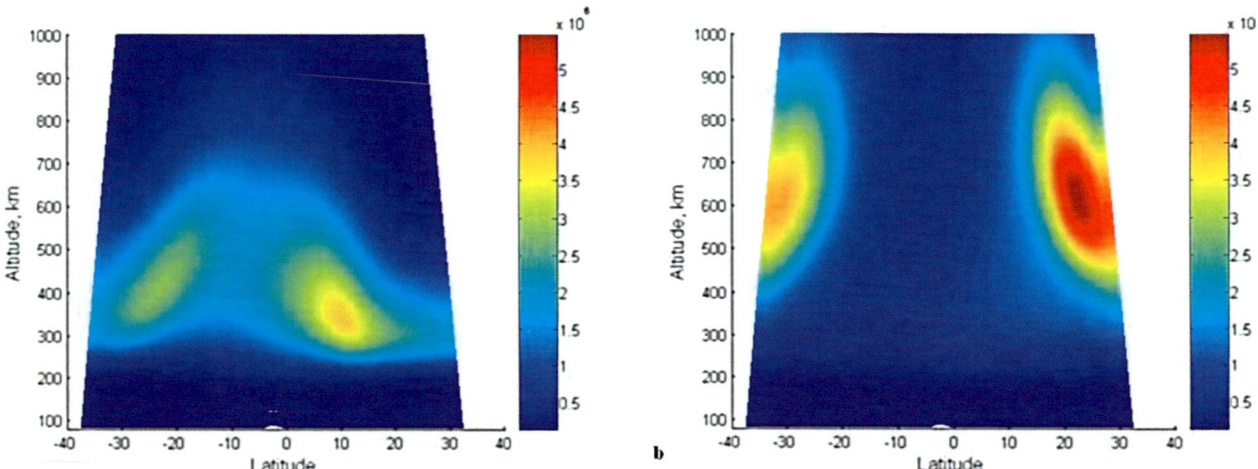

Plate 5. SAMI2* modeling of the dayside ionosphere at ~2:00 pm local time. Plate 5a is for quiet-time conditions. Plate 5b is for a PPEF of ~4 mV/m applied for ~2 h. The EIAs are lifted to higher altitudes and higher latitudes.

that the electric fields can last for several hours [*Tsurutani et al.*, 2004; *Maruyama et al.*, 2004; *Mannucci et al.*, 2005; *Sahai et al.*, 2005] and even up to ~5 h [*Huang et al.*, 2005] on the dayside without significant shielding effects taking place. This topic is one of intense research efforts and thus our understanding of this process is rapidly evolving.

What are the effects of the presence of these ionospheric electric fields? Near the magnetic equator where the Earth's magnetic field lines are relatively horizontal, the electric fields will cause uplift (on the dayside) and downdraft (on the nightside) due to $E \times B$ convection. On the dayside, the equatorial ionization anomalies (EIAs) [*Namba and Maeda*, 1939], previously called the Appleton anomalies, which are nominally located at ~±10° from the magnetic equator, are not only lifted upward but also to high magnetic latitudes due to the $E \times B$ convection [*Tsurutani et al.*, 2006b, 2008; *Verkhoglyadova et al.*, 2006, this volume]. This is one mechanism of plasma transport of equatorial plasma to middle latitudes. Once the dayside ionosphere has been uplifted, solar photoionization will create a new ionosphere at lower altitudes [*Tsurutani et al.*, 2004]. Thus, during dayside ionospheric superfountain events, the TEC of the ionosphere can be enhanced by a factor of 6 times at ~30° magnetic latitudes.

The $E \times B$ convection direction for a dawn-to-dusk electric field is downward in the nightside equatorial region. As plasma is convected to lower altitudes, rapid recombination takes place, lowering the TEC of the nightside ionosphere.

Both of these local time effects take place simultaneously during magnetic storms. The dayside increase in TEC is known as a positive ionosphere storm and the nightside decrease is a negative ionospheric storm.

2.7. Disturbance Dynamos

During magnetic storms, the vast energy input into the auroral zone ionosphere due to particle precipitation and Joule heating leads to the expansion of the neutral atmosphere. This expansion will propagate to both higher and lower latitudes and also to different longitudes. Locally, this pressure wave will convect the ionosphere downward and create a dynamo action [*Blanc and Richmond*, 1980; *Scherliess and Fejer*, 1997; *Richmond and Lu*, 2000].

The location of "auroral" precipitation and Joule heating moves to subauroral zone magnetic latitudes during magnetic storms. These neutral winds associated with the heating alter the mid-latitude ionosphere by moving plasma along field lines to regions of altered composition [*Proelss*, 1997; *Fuller-Rowell*, 1997; *Buonsanto*, 1999]. The flow from high- to low- latitudes, through continuity, causes upwelling of the neutral gas and an increase of nitrogen molecules relative to oxygen atoms. This, in turn, leads to increased ion recombination rates and ionospheric plasma density reductions. The heating is also not restricted to the nightside, but can occur on the dayside as well. Thus, accurate modeling of specific storms is very useful to understand middle latitude ionospheric effects.

There is a propagation delay time for the disturbance dynamo effects to reach the dayside equator. A number of models have estimated this delay to range from ~3 to 4 h [*Proelss*, 1995]. At this time, there have been very few studies done that include both PPEFs and disturbance dynamo effects. This is an interesting area of research.

Plate 5 shows a model of oxygen ion uplift during the 30 October 2003 magnetic superstorm [*Tsurutani et al.*, 2006c, 2007]. A modified NRL SAMI2 code [*Huba et al.*, 2000] was used for the simulations. Plate 5a shows the normal ionosphere at ~2:00 pm. The quiet-time EIAs are located at ~±10°. A ~ 4 mV/m electric field was calculated for the intensity of the PPEF. The electric field was assumed to exist for ~2 hours, from noon to ~2:00 pm. The results of simulations using a modified NRL SAMI2 code are shown in Plate 5b. The $E \times B$ convection of the EIAs moves the high-density regions to higher latitudes and higher altitudes.

Ionospheric electrons display a similar pattern. Because the low-altitude ionosphere is uplifted, solar photoionization begins to create a new ionosphere at lower altitudes, increasing the TEC of the ionosphere [*Verkhoglyadova et al.*, 2006, this volume]. This can also be noted in Plate 5b.

The rapid uplift of the EIAs leads to substantial ion-neutral drag and increases in the atmospheric neutral densities at ~400 to 600 km [*Tsurutani et al.*, 2006c, 2007]. This can lead to rapid low-altitude satellite deceleration during magnetic storms. This effect is expected to be particularly large during superstorms when PPEF magnitudes are largest. However, so far very few events have been studied. Further confirmation of neutral uplift is needed, both experimentally and by modeling.

2.8. Interplanetary High-Speed Streams and Corotating Interaction Regions

High-speed, ~750 to 800 km/s streams emanate from coronal holes [*Krieger et al.*, 1973; *Neupert and Pizzo*, 1974; *McComas et al.*, 2002]. At this time, the exact mechanism for the rapid acceleration from the solar "surface" is not known, but most solar scientists speculate that some form of wave-particle interaction is causing the plasma heating and acceleration.

As a fast stream propagates outward from the sun, it interacts with the slow solar wind (V_{sw} ~350–400 km/s), forming a region of compressed plasma and magnetic fields. This compressed region is called an interaction region. Since many coronal holes are long-lasting (more than one solar rotation), the compressed field and plasma region appears to corotate with the sun and thus has been called a corotating interaction region (CIR) [*Smith and Wolfe*, 1976; *Pizzo*, 1985; *Balogh et al.*, 1999].

CIRs have magnetic field intensities of ~20–30 nT (the nominal solar wind magnetic fields are typically only ~5–10 nT), but because the B_z component within CIRs are characteristically highly fluctuating, the geomagnetic activity created by the impingement of CIRs onto the magnetosphere is typically only low to moderate [*Tsurutani et al.*, 1995, 2006a, 2006b; *Alves et al.*, 2006; *Kozyra et al.*, 2006; *Richardson*, 2006; *Turner et al.*, 2006]. Magnetic storms of intensity Dst <–100 nT caused by CIRs are relatively rare.

At Earth orbit, the CIRs and high-speed streams are generally found following the high plasma density region surrounding the heliospheric current sheet [*Smith et al.*, 1978]. This high plasma density region or heliospheric plasmasheet [*Winterhalter et al.*, 1994] can reach values of up to 100 cm^{-3} or higher. Although this phenomenon is located in the slow solar wind, the ram pressure effects can be substantial [*Tsurutani et al.*, 1995], even more than at shocks in front of fast ICMEs. Because CIRs typically do not form forward shocks by 1 AU (CIR shocks generally form at ~1.5 to 2.0 AU from the sun), these ram pressure increases are more gradual than at ICME shocks.

2.9. High-Speed Streams and Embedded Alfvén Waves

A characteristic of high-speed solar wind streams is that they carry nonlinear Alfvén waves with them [*Tsurutani et al.*, 1994]. The local Alfvén wave phase speed is typically only ~50 to 70 km/s, so they are basically convected by the solar wind. The amplitude of the waves are exceedingly large, typically $\Delta B/|B| \sim 1-2$. When the Alfvén waves have southward components, they can cause reconnection at the magnetopause and lead to sporadic plasma injections into the nightside magnetosphere. The auroral consequences of these injections are called high-intensity, long-duration, continuous AE activity (HILDCAA) events [*Tsurutani and Gonzalez*, 1987].

An example of a HILDCAA event is shown in Figure 2. The spacecraft was IMP-8, an Earth-orbiting satellite that

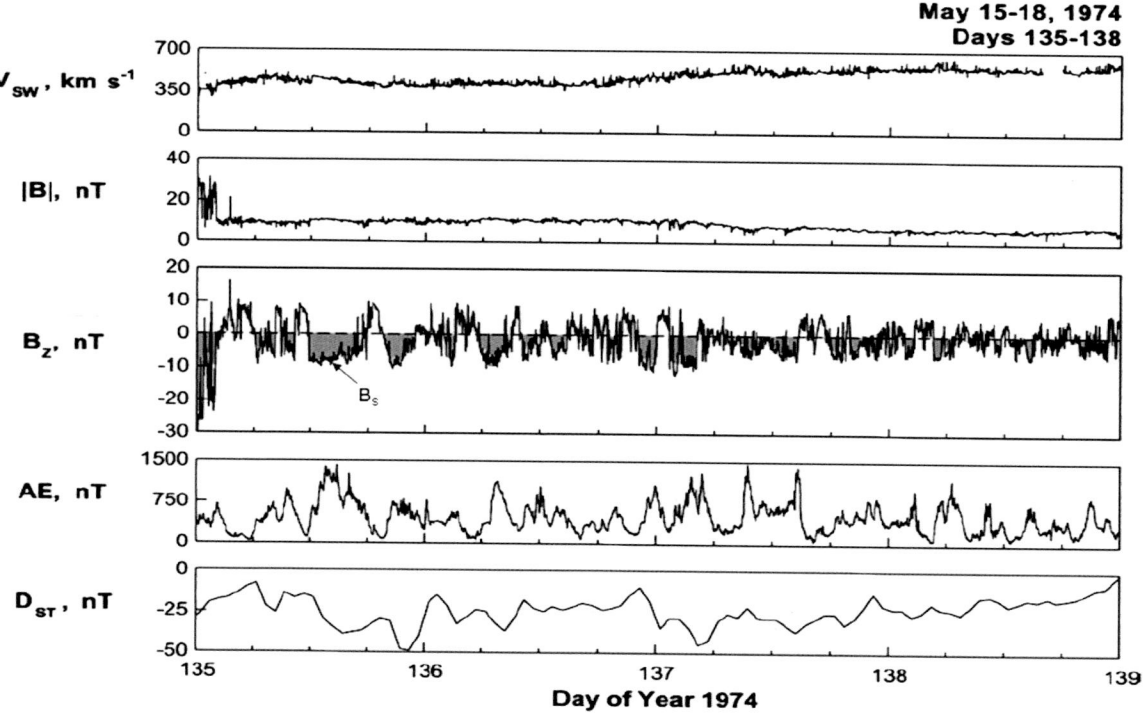

Figure 2. Interplanetary Alfvén waves and HILDCAAs. The southward excursions of the Alfvenic fluctuations are correlated with *AE* increases and *Dst* decreases.

was in the interplanetary medium at the time. The panels from top to bottom represent the solar wind speed, magnetic field magnitude, magnetic field B_z component, *AE* index, and *Dst* index.

The Alfvenic fluctuations can be clearly noted in the third panel, which contains B_z. For each large southward excursion of the IMF, there is an *AE* increase. Concurrent with this is a small *Dst* decrease. An interpretation of these features is that magnetic reconnection leads to nightside plasma injections [*Soraas et al.*, 2004, 2005]. Subsequent particle precipitation into the auroral zone causes ionospheric conductivity enhancements and auroral electrojet intensifications. Because much of the injected plasma is trapped within the magnetosphere (minilocalized ring current clouds?), there is a slight increase in ring current intensification, thus the *Dst* decrease.

If there are large polar coronal holes both in the northern and southern parts of the sun during the declining phase of the solar cycle, the Earth's magnetosphere can be bathed in the high-speed solar wind for the whole solar rotation period of ~27 days. The Alfvén wave fluctuations can continuously inject plasma into the magnetosphere, but not deeply, only to distances $L > 4.0$. These shallow injections supplant the loss of ring current particles and maintain *Dst* at a low level as noted in Figure 2.

When there are two high-speed streams impinging upon the Earth's magnetosphere for the whole solar rotation period, and the high-speed streams last for a large part of the year, and the energy input into the magnetosphere/ionosphere can be greater than during a solar maximum year [*Tsurutani et al.*, 1995, 2006a; *Kozyra et al.*, 2006; *Guarnieri*, 2006; *Turner et al.*, 2006]. Although the energy input into the ionosphere is primarily at auroral zone latitudes, this may have consequences for middle latitudes as well. One question that needs to be investigated is "what does the disturbance dynamo look like under these conditions?" "Can there be a steady-state disturbance of the middle latitudes or does it come in "waves"?

2.10. HILDCAAs and Relativistic Electrons

Because plasma is continuously injected into the nightside magnetosphere, the compressed energetic electrons will drift (due to magnetic field gradients and curvature) toward local dawn. The anisotropic ~10- to 100-keV electrons will generate electromagnetic chorus [*Tsurutani and Smith*, 1974, 1977; *Meridith et al.*, 2001], which will, in turn, pitch angle scatter the particles through cyclotron resonant interactions. In addition, it has been hypothesized that waves interacting with the upper end of the electron spectrum can accelerate the particles to hundreds of keV or even tens of MeV energies [*Horne et al.*, 1998, 2006; *Meredith et al.*, 2003; *Trakhtengerts et al.*, 2003]. Another mechanism is radial diffusion associated with the resonance of PC5 waves [*Hudson et al.*, 1999; *Li and Temerin*, 2001; *O'Brien et al.*, 2001; *Elkington et al.*, 2003; *Li*, 2006]. It has been noted that PC5 waves are enhanced during HILDCAA events/high-speed streams. These relativistic electrons [*Paulikas and Blake*, 1979; *Baker et al.*, 1986] will have access to the entire magnetosphere and ionosphere.

Figure 3 shows a high-speed stream (first panel) and the embedded Alfvén wave structures. The IMF B_z fluctuations are shown in the seventh panel from the top. The eighth panel, the *Dst* index, shows that there is very little ring current enhancement throughout this event. Because the *Dst* deviation is so small, this would typically not be considered as a magnetic storm. The *AE*, and *AL* and *AU* indices are shown in the ninth and tenth panels, respectively. It can be noted that there are intense, continuous auroral electrojet activity associated with the Alfvenic fluctuations. This is a HILDCAA event.

Plate 6 shows the same event, but with different information (from top to bottom): ~400-keV electron fluxes, chorus intensities, ground-based PC5 wave intensities, IMF B_z, solar wind ram pressure, and the *AE* and *Dst* indices. The fourth through seventh panels were replotted in higher time resolution for continuity between the figure and the plate. The relativistic electron data are taken from the Polar CAMICE instrument, the chorus emissions from the Akebono satellite, and the PC5 wave data from a Canadian magnetometer chain. The interplanetary data were taken from the ACE satellite.

Chorus emissions and PC5 waves are enhanced during the first 3 days of the high-speed stream, 23–25 July 1988. This is the same interval where the AE level is generally high (the HILDCAA event). The chorus can be easily explained by the continuous injection of energetic 10- to 100-keV electrons mentioned previously. The mechanism for PC5 wave generation is less well understood. It could be associated with the sporadic plasma injections or it could be related to a Kelvin–Helmholtz instability occurring at the flanks of the magnetopause. Further studies are needed to clarify the source of the latter waves.

The relative electron flux begins to increase on 24 July, the second day of the HILDCAA event. It maximizes near the end of 24 July and continues to exist through 29 July and beyond. The relativistic electrons are believed to be accelerated by two different processes: cyclotron resonant interactions with chorus [*Horne et al.*, 2006] and drift resonance with the PC5 waves [*Hudson et al.*, 1999; *Li*, 2006]. The first process directly energizes a small percentage of the particles by wave absorption and the second leads to the breaking of the third adiabatic invariant and the radial diffusion of the

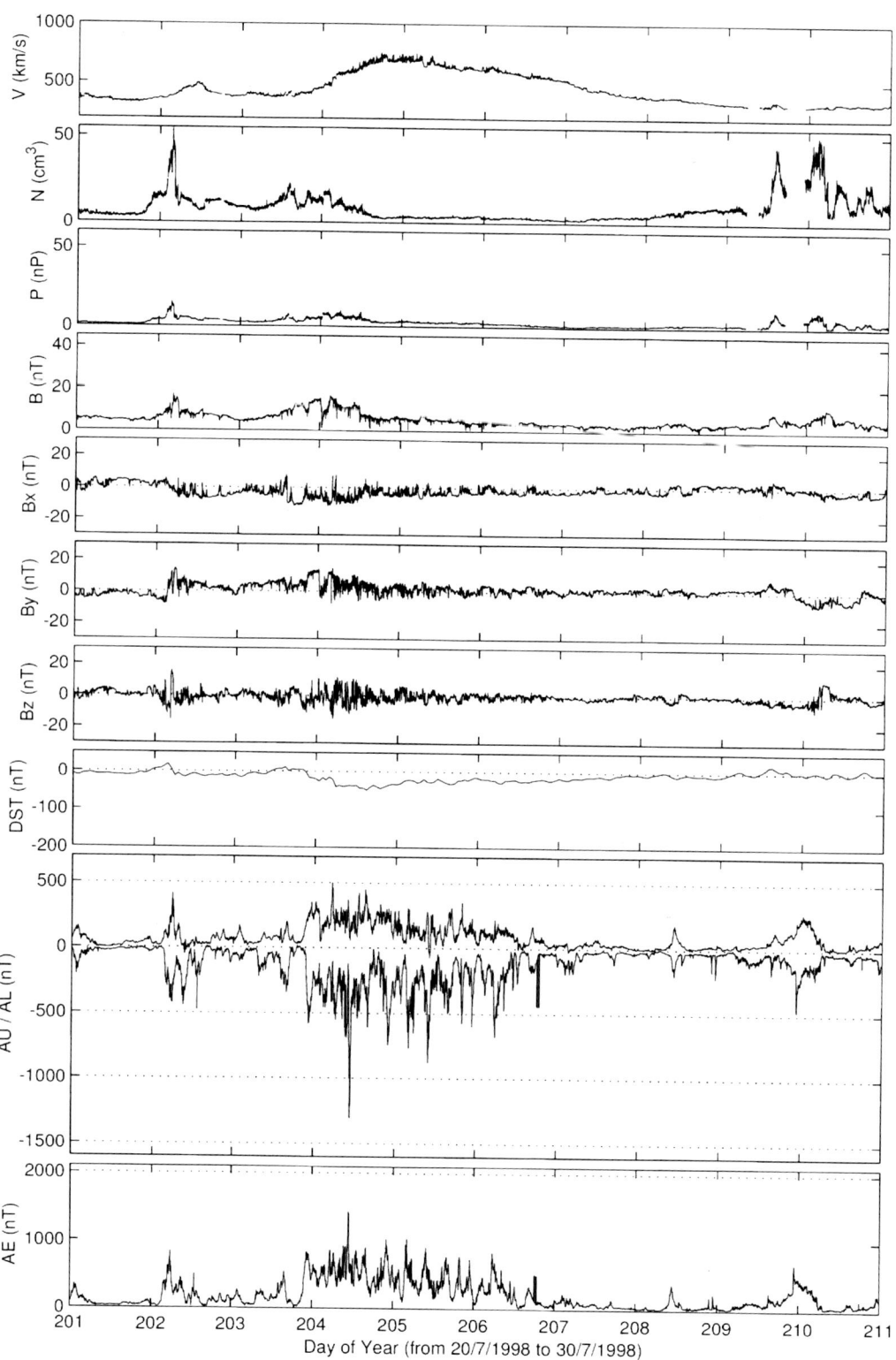

Figure 3. A high-speed stream, Alfvenic fluctuations and a HILDCAA from 22 to 25 July 1998.

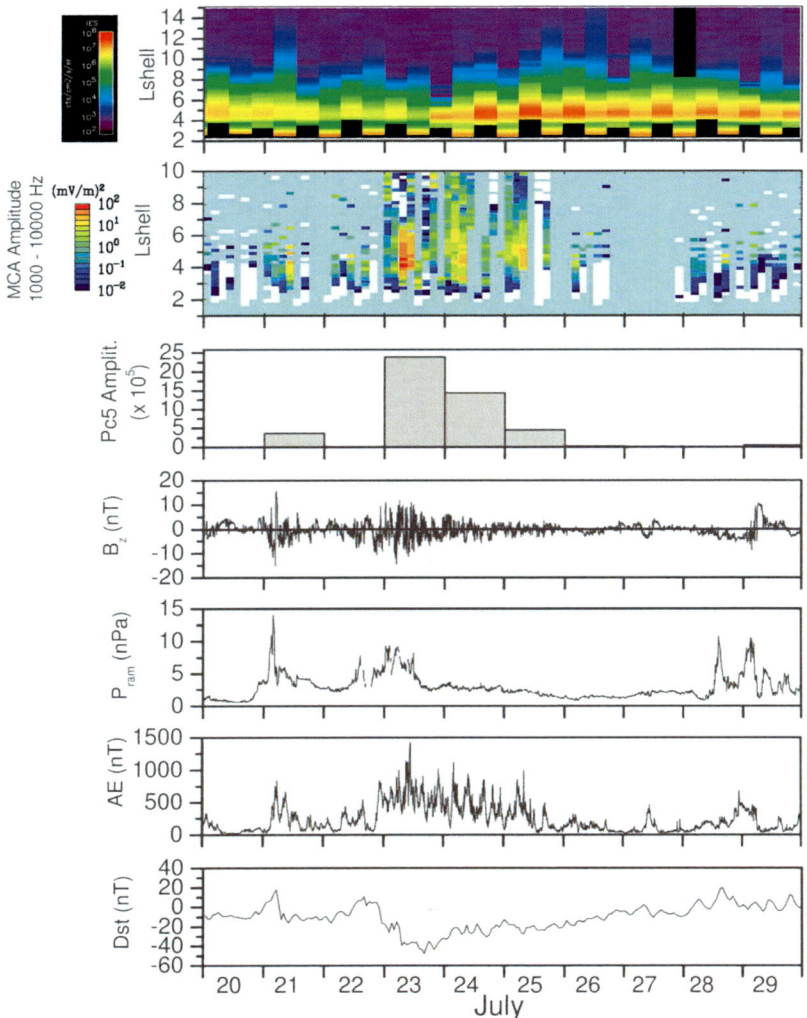

Plate 6. Same event as Figure 3, but showing enhanced chorus emissions, PC5 waves, and relativistic electrons during the HILDCAA event.

114 INTERPLANETARY CAUSES OF MIDDLE LATITUDE IONOSPHERIC DISTURBANCES

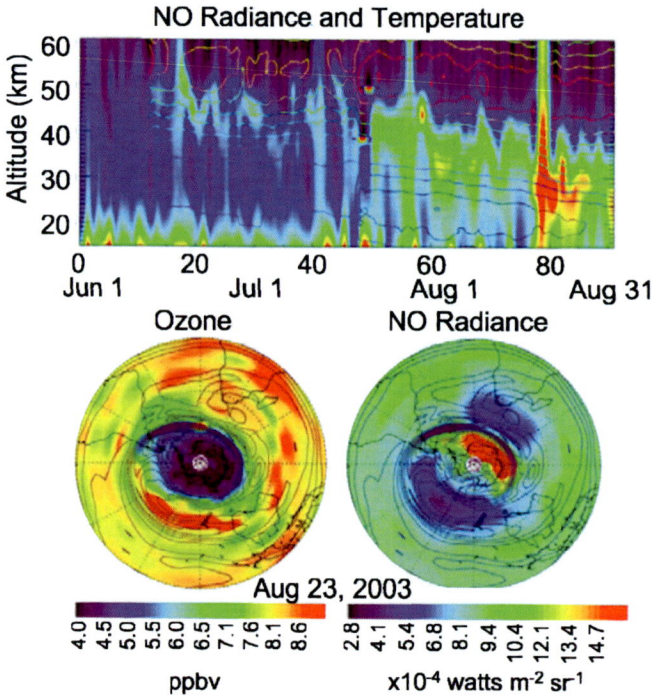

Plate 7. (a) NO radiation and temperature height dependence for an event from 1 June to 31 August 2003 during an interval of high-speed streams. (b) Southern polar ozone depletion on 31 August (left) and NO entrainment by the polar vortex (right).

electrons. In this latter process, a small percentage of resonant particles diffuse inward and become energized.

2.11. Energetic Electrons and NOy Production

When the relativistic electrons are scattered by waves into the upper atmosphere, one of the processes that occurs is a chain of chemical reactions that results in the production of NO_y [*Thorne*, 1980]. Subsequent transport of NO_y down into the stratosphere will lead to the catalytic reaction $NO + O_3 \rightarrow NO_2 + O_2$ and $NO_2 + O \rightarrow NO + O_2$.

Plate 7 shows an example where this NO_y production and downward transport has taken place [*Kozyra et al.*, 2006]. In this case, the NO_y is produced by long-duration precipitation of 1- to 10-keV electrons associated with HILDCAAs. The precipitation occurs high (~100 km) in the atmosphere, and in this case the southern polar vortex has entrained the NO_y to maintain high concentration levels. The downward descent of these molecules leads to the decrease of ozone as noted in the plate.

Relativistic electrons are fewer in number than the 1- to 10-keV electrons, but they have access to the middle latitude ionosphere. The chemical effects that they have on the ozone balance are not well understood at this time.

3. SUMMARY

We have attempted to give a brief review of the solar and interplanetary causes of middle latitude ionospheric phenomena. We have focused on "space weather" topics, that is, large changes from the nominal baseline values. Three basic "forcing phenomena" were examined: (1) high levels of photons associated with solar flares, (2) solar wind ram pressure increases associated with fast forward shocks antisunward of ICME structures and also associated with the heliospheric plasmasheet, and (3) magnetic reconnection. Magnetic reconnection is associated with southward interplanetary sheath fields, southward directed MC fields, and the southward components of Alfvén waves embedded in high-speed streams. One consequence of magnetic reconnection with sheath and MC fields are intense magnetic storms. Magnetic storms create well-known disturbance dynamos whose enhanced pressure disturbances propagate from subauroral latitudes equatorward. PPEFs, another product of magnetic reconnection, lead to plasma transport from the equatorial region to higher latitudes.

We hope that the reader will have gained a broad perspective of space weather as it pertains to the middle latitude ionosphere and will be able to identify fruitful areas of personal and joint research.

Acknowledgments. Portions of this work were conducted at the Jet Propulsion Laboratory, California Institute of Technology, Pasadena, CA, under contract with NASA. E.E. would like to thank the CNPq agency (PQ-300104/2005-7 and 470706/2006-6) for financial support.

REFERENCES

Afraimovich, E. L. (2000), GPS global detection of the ionospheric response to solar flares, *Radio Sci.*, *35*, 1417.

Alves, M. V., E. Echer, and W. D. Gonzalez (2006), Geoeffectiveness of corotating interaction regions as measured by *Dst* index, *J. Geophys. Res.*, *111*, A07S5, doi:10.1029/2005JA011379.

Babcock, H. W. (1961), The topology of the sun's magnetic field and the 22-year cycle, *Astrophys. J.*, *133*, 572.

Baker, D. N., J. B. Blake, R. W. Klebesadel, and P. R. Higbie (1986), Highly relativistic electrons in the Earth's outer magnetosphere, 1. Lifetimes and temporal history 1979–1984, *J. Geophys. Res.*, *91*, 4265–4276.

Balogh, A., J. T. Gosling, J. R. Jokipii, R. Kallenbach, and H. Kunow (Eds.) (1999), Corotating Interaction Regions, *Space Sci. Rev.*, 89.

Blanc, M., and A. D. Richmond (1980), The ionospheric disturbance dynamo, *J. Geophys. Res.*, *85*, 1669–1686.

Buonsanto, M. J. (1999), Ionospheric storms—a review, *Space Sci. Rev.*, *88*, 563.

Burch, J. L. (1972), Preconditions for the triggering of polar magnetic substorms by storm sudden commencements, *J. Geophys. Res.*, *77*, 5629–5632.

Carrington, R. C. (1859), Description of a singular appearance seen in the Sun on September 1, 1859, *Mon. Not. R. Astron. Soc.*, *XX*, 13.

Craig, I. J. D., and Y. E. Litvinenko (2002), Particle acceleration scalings based on exact analytic models for magnetic reconnection, *Astrophys. J.*, *570*(1), 387–394, Part 1.

Donnelly, R. F. (1976), Empirical models of solar flare x ray and EUV emissions for use in studying their *E* and *F* region effects, *J. Geophys. Res.*, *81*, 4745–4753.

Dungey, J. W. (1961), Interplanetary magnetic field and the auroral zones, *Phys. Rev. Lett.*, *6*, 47.

Echer, E., and W. D. Gonzalez (2004), Geoeffectiveness of interplanetary shocks, magnetic clouds, sector boundary crossings and their combined occurrence, *Geophys. Res. Lett.*, *31*, L09808, doi:10.1029/2003GL019199.

Echer, E, W. D. Gonzalez, B. T. Tsurutani, and A. L. C. Gonzalez (2008), Interplanetary conditions causing intense geomagnetic storms (Dst <−100 nT) during solar cycle 23 (1996–2006), *J. Geophys. Res.*, *113*, A05221, doi:10.1029/2007JA012744.

Elkington, S. R., M. K. Hudson, and A. A. Chan (2003), Resonant acceleration and diffusion of outer zone electrons in an asymmetric geomagnetic field, *J. Geophys. Res.*, *108*(A3), 1116, doi:10.129/2001JA009202.

Elkington, S. R., M. Wiltberger, A. A. Chan, and D. N. Baker (2004), Physical models of the geospace radiation environment, *J. Atmos. Sol. Terr. Phys.*, *66*, 1371.

Farrugia, C. J., L. F. Burlaga, and R. P. Lepping (1997), Magnetic clouds and the quiet-storm effect at Earth, in *Magnetic Storms, Geophys. Monogr. Ser.*, vol. 98, edited by B. T. Tsurutani et al., pp. 91–106, AGU, Washington, D. C.

Fuller-Rowell, T. M., M. V. Codrescu, R G. Roble, and A. D. Richmond (1997), How does the thermosphere and ionosphere react to a geomagnetic storm?, in *Magnetic Storm, Geophys. Monogr. Ser.*, vol. 98, edited by B. T. Tsurutani et al., pp. 203–225, AGU, Washington, D. C.

Gonzalez, W. D., B. T. Tsurutani, A. L. C. Gonzalez, E. J. Smith, F. Tang, and S.-I. Akasofu (1989), Solar wind-magnetosphere coupling during intense magnetic storms (1978-1979), *J. Geophys. Res.*, 94, 8835–8851.

Gonzalez, W. D., J. A. Joselyn, Y. Kamide, H. W. Kroehl, G. Rostoker, B. T. Tsurutani, and V. M. Vasyliunas (1994), What is a geomagnetic storm?, *J. Geophys. Res.*, 99, 5771–5792.

Gonzalez., W. D., A. L. C. de Gonzalez, A. Dal Lago, B. T. Tsurutani, J. K. Arballo, G. K. Lakhina, B. Buti, C. M. Ho, and S.-T. Wu (1998), Magnetic cloud field intensities and solar wind velocities, *Geophys. Res. Lett.*, 25, 963–966.

Gonzalez, W. D., E. Echer, A. L. Clua-Gonzalez, and B. T. Tsurutani (2007), Interplanetary origin of intense geomagnetic storms (Dst <−100 nT) during solar cycle 23, *Geophys. Res. Lett.*, 34, L06101, doi:10.1029/2006GL028879.

Gopalswamy, N. (2006), Coronal mass ejections of solar cycle 23, *J. Astrophys. Astron.*, 27, 243, Jun–Sept.

Guarnieri, F. L. (2006), The nature of auroras during high-intensity long-duration continuous AE activity (HILDCAA) events: 1998 to 2001, in *Recurrent Magnetic Storms: Corotating Solar Wind Streams, Geophys. Monogr. Ser.*, vol. 167, edited by B. T. Tsurutani et al., pp. 235–244, AGU, Washington D. C.

Haerendel, G. (1994), Acceleration from field-aligned potential drops, *Astrophys. J. Suppl. Ser.*, 90, 765.

Haerendel, G. (2001), Auroral acceleration in astrophysical plasmas, *Phys. Plasmas*, 8, 2365.

Harvey, K. L., and F. Recely (2002), Polar coronal holes during cycles 22 and 23, *Sol. Phys.*, 211, 31.

Harvey, K., S. Suess, M. Aschwanden, M. Guhathakurta, J. Harvey, D. Hathaway, B. LaBonte, N. Sheeley, and B. T. Tsurutani (2000), A NASA workshop on coronal holes near solar maximum and over the solar cycle, *NASA White Paper*, December.

Horne, R. B. (1998), The contribution of wave–particle interactions to electron loss and acceleration in the Earth's radiation belts during geomagnetic storms, in *Rev. Radio Sci. 1999–2002*, edited by W. R. Stone, p. 801, Wiley, New York.

Horne, R. B., N. P. Meredith, S. A. Glauert, A. Varotsou, D. Boscher, R. M. Thorne, Y. Y. Shprits, and R. R. Anderson (2006), Mechanisms for the acceleration of relativistic electrons, in *Recurrent Magnetic Storms: Corotating Solar Wind Streams, Geophys. Monogr. Ser.*, vol. 167, edited by B. T. Tsurutani et al., pp.151–174, AGU, Washington, D. C.

Hu, Q., and B. U. Ö. Sonnerup (2001), Reconstruction of magnetic flux ropes in the solar wind, *Geophys. Res. Lett.*, 28, 467–470.

Huang, C.-S., J. C. Foster, L. P. Goncharenko, P. J. Erickson, W. Rideout, and A. J. Coster (2005), A strong positive phase of ionospheric storms observed by the Millstone Hill incoherent scatter radar and global GPS network, *J. Geophys. Res.*, 110, A06303, doi:10.1029/2004JA010865.

Huba, J. D., G. Joyce, and J. A. Fedder (2000), Sami2 is another model of the ionosphere (SAMI2): A new low-latitude ionosphere model, *J. Geophys. Res.*, 105(A10), 23,035–23,053.

Hudson, M. K., S. R. Elkington, J. G. Lyon, V. A. Marchenko, I. Roth, M. Temerin, J. B. Blake, M. S. Gussenhoven, and J. R. Wygant (1997), Simulations of radiation belt formation during storm sudden commencements, *J. Geophys. Res.*, 102, 14,087–14,102.

Hudson, M. K., S. R. Elkington, J. G. Lyon, C. C. Goodrich, and T. J. Rosenberg (1999), Simulation of radiation belt dynamics driven by solar wind variations, in *Sun-Earth Plasma Connections, Geophys. Monogr. Ser.*, vol. 109, edited by J. L. Burch, R. L. Carovillano, and S. K. Antiochos, pp. 171–182, AGU, Washington, D. C.

Hudson, M. K., B. T. Kress, J. E. Mazur, K. L. Perry, and P. L. Slocum (2004), 3D modeling of shock-induced trapping of solar energetic particles in the Earth's magnetosphere, *J. Atmos. Sol. Terr. Phys.*, 66, 1389, doi:10.1016/j.jastp.2004.03.024.

Joselyn, J. A., and B. T. Tsurutani (1990), Geomagnetic sudden impulses and storm sudden commencements: A note on terminology, *Eos Trans. AGU*, 71(47), 1808.

Judge, D. L. (1998), First solar EUV irradiance obtained from SOHO by the CELIAS/SEM, *Sol. Phys.*, 177, 161.

Kahler, S. W., N. R. Sheeley Jr., R. A. Howard, M. J. Koomen, D. J. Michels, R. E. McGuire, T. T. von Rosenvinge, and D. V. Reames (1984), Association between coronal mass ejections and solar energetic proton events, *J. Geophys. Res.*, 89, 9683–9693.

Kallenrode, M. B. (2003), Current views on impulsive and gradual energetic particle events, *J. Phys. G: Nucl. Part. Phys.*, 29, 965.

Kallenrode, M. B., E. Cliver, and G. Wibberenz (1992), Composition and azimuthal spread of solar energetic particles from impulsive and gradual flares, *Astrophys. J.*, 391, 370.

Kamide, Y., N. Yokoyama, W. Gonzalez, B. T. Tsurutani, I. A. Daglis, A. Brekke, and S. Masuda (1998), Two-step development of geomagnetic storms, *J. Geophys. Res.*, 103, 6917–6921.

Kawasaki, K., S.-I. Akasofu, F. Yasuhara, and C.-I. Meng (1971), Storm sudden commencements and polar magnetic substorms, *J. Geophys. Res.*, 76, 6781–6789.

Kelley, M. C., B. G. Fejer, and C. A. Gonzales (1979), An explanation for anomalous equatorial ionospheric electric field associated with a northward turning of the interplanetary magnetic field, *Geophys. Res. Lett.*, 6(4), 301–304.

Kelley, M. C., J. J. Makela, J. L. Chau, and M. J. Nicolls (2003), Penetration of the solar wind electric field into the magnetosphere/ionosphere system, *Geophys. Res. Lett.*, 30(4), 1158, doi:10.1029/2002GL016321.

Kennel, C. F., J. P. Edmiston, and T. Hada (1985), A quarter century of collisionless shock research, in *Collisionless Shocks in the Heliosphere: A Tutorial Review, Geophys. Monogr. Ser.*, vol. 34, edited by R. G. Stone and B. T. Tsurutani, pp. 1–36, AGU, Washington, D. C.

Kikuchi, T., and T. Araki (1979), Horizontal transmission of the polar electric field, *J. Atmos. Terr. Phys.*, *41*, 927.

Kokubun, S., R. L. McPherron, and C. T. Russell (1977), Triggering of substorms by solar wind discontinuities, *J. Geophys. Res.*, *82*, 74–86.

Kozyra, J. U., et al. (2006), Response of the upper/middle atmosphere to coronal holes and powerful high-speed solar wind streams in 2003, in *Recurrent Magnetic Storms: Corotating Solar Wind Streams, Geophys. Monogr. Ser.*, vol. 167, edited by B. T. Tsurutani et al., pp. 319–340, AGU, Washington, D. C.

Krieger, A. S., A. F. Timothy, and E. C. Roelof (1973), A coronal hole and its identification as the source of a high velocity solar wind stream, *Sol. Phys.*, *23*, 123.

Lepping, R. P., T. W. Narock, and H. Chen (2007), Unique Vector Features of the Magnetic Fields of Interplanetary Magnetic Clouds as Cylindrically Symmetric Force-Free Flux Ropes, Goddard Space Flight Center internal doc., Greenbelt, Md.

Li, X. (2006), The role of radial transport in accelerating radiation belt electrons, in *Recurrent Magnetic Storms: Corotating Solar Wind Streams, Geophys. Monogr. Ser.*, vol. 167, edited by B. T. Tsurutani et al., pp. 139–150, AGU, Washington, D. C.

Li, X., and M. A. Temerin (2001), The electron radiation belt, *Space Sci. Rev.*, *95*, 569.

Liu, Y., J. D. Richardson, J. W. Belcher, C. Wang, Q. Hu, and J. C. Kasper (2006), Constraints on the global structure of magnetic clouds: Transverse size and curvature, *J. Geophys. Res.*, *111*, A12S03, doi:10.1029/2006JA011890.

Lorentzen, K. R., J. E. Mazur, M. D. Looper, J. F. Fennell, and J. B. Blake (2002), Multisatellite observations of MeV ion injections during storms, *J. Geophys. Res.*, *107*(A9), 1231, doi:10.1029/2001JA000276.

Mannucci, A. J., B. T. Tsurutani, B. A. Iijima, A. Komjathy, A. Saito, W. D. Gonzalez, F. L. Guarnieri, J. U. Kozyra, and R. Skoug (2005), Dayside global ionospheric response to the major interplanetary events of October 29–30, 2003 "Halloween Storms", *Geophys. Res. Lett.*, *32*, L12S02, doi:10.1029/2004GL021467.

Maruyama, T., G. Ma, and M. Nakamura (2004), Signature of TEC storm on 6 November 2001 derived from dense GPS receiver network and ionosonde chain over Japan, *J. Geophys. Res.*, *109*, A10302, doi:10.1029/2004JA010451.

Mazur, J. E., J. B. Blake, P. L. Slocum, M. K. Hudson, and G. M. Mason (2006), The creation of new ion radiation belts associated with solar energetic particle events and interplanetary shocks, in *Solar Eruptions and Energetic Particles, Geophys. Monogr. Ser.*, vol. 165, edited by N. Gopalswamy, R. Mewaldt, and J. Torsti, pp. 345–352, AGU, Washington, D. C.

McComas, D. J., H. A. Elliot, J. T. Gosling, D. B. Reisenfeld, R. M. Skoug, B. E. Goldstein, M. Neugebauer, and A. Balogh (2002), Ulysses second fast-latitude scan: Complexity near solar maximum and the reformation of polar coronal holes, *Geophys. Res. Lett.*, *29*(9), 1290, doi:10.1029/2001GL014164.

Meredith, N. P., R. B. Horne, and R. R. Anderson (2001), Substorm dependence of chorus amplitudes: Implications for the acceleration of electrons to relativistic energies, *J. Geophys. Res.*, *106*, 13,165–13,178.

Meredith, N. P., R. B. Horne, R. M. Thorne, and R. R. Anderson (2003), Favored regions for chorus-driven electron acceleration to relativistic energies in the Earth's outer radiation belt, *Geophys. Res. Lett.*, *30*(16), 1871, doi:10.1029/2003GL017698.

Mitra, A. P. (1974), *Ionospheric Effects of Solar Flares*, 294 pp., Springer, New York.

Mullen, E. G., M. S. Gussenhoven, K. Ray, and M. Vilet (1991), A double peaked inner radiation belt: Cause and effect as seen on CRRES, *IEEE Trans. Nucl. Sci.*, *38*, 1713, doi:10.1109/23.124167.

Namba, S., and K.-I. Maeda (1939), *Radio Wave Propagation*, 86 pp., Corona, Tokyo.

Neupert, W. M., and V. Pizzo (1974), Solar coronal holes as sources of recurrent geomagnetic disturbances, *J. Geophys. Res.*, *79*, 3701–3709.

Nishida, A. (1968), Coherence of geomagnetic DP 2 fluctuations with interplanetary magnetic variations, *J. Geophys. Res.*, *73*, 5549–5559.

Obayashi, T. (1967), The interaction of solar plasma with geomagnetic field, disturbed condition, in *Solar Terrestrial Physics*, vol. 107, edited by J. W. King and W. S. Newman, Academic Press, London.

O'Brien, T. P., R. L. McPherron, D. Sornette, G. D. Reeves, R. Friedel, and H. J. Singer (2001), Which magnetic storms produce relativistic electrons at geosynchronous orbit?, *J. Geophys. Res.*, *106*, 15,533–15,544.

Onwumechili, A., K. Kawasaki, and S.-I. Akasofu (1973), Relationships between the equatorial electrojet and polar magnetic variations, *Planet. Space Sci.*, *21*, 1.

Paulikas, G. A., and J. B. Blake (1979), Effects of the solar wind on magnetospheric dynamics: Energetic electrons at the synchronous orbit, in *Quantitative Modeling of Magnetospheric Processes, Geophys. Monogr. Ser.*, vol. 21, edited by W. Olsen, pp. 180–202, AGU, Washington, D. C.

Pizzo, V. J. (1985), Interplanetary shocks on the large scale: A retrospective on the last decade's theoretical efforts, in *Collisionless Shocks in the Heliosphere: Reviews of Current Research, Geophys. Monogr. Ser.*, vol. 35, edited by B. T. Tsurutani and R. G. Stone, pp. 51–68, AGU, Washington, D. C.

Proelss, G. W. (1995), Ionospheric F region storms, in *Handbook of Atmospheric Electrodynamics*, vol. 2, edited by H. Volland, p. 195, CRC Press, Boca Raton, Fla.

Proelss, G. W. (1997), Magnetic storm associated perturbations of the upper atmosphere, in *Magnetic Storms, Geophys. Monogr. Ser.*, vol. 98, edited by B. T. Tsurutani et al., pp. 227–241, AGU, Washington, D. C.

Reames, V. D. (1999), Particle acceleration at the Sun and in the heliosphere, *Space Sci. Rev.*, *90*, 413.

Richardson, I. G. (2006), The formation of CIRs at stream-stream interfaces and resultant geomagnetic activity, in *Recurrent Magnetic Storms: Corotating Solar Wind Streams, Geophys. Monogr. Ser.*, vol. 167, edited by B. T. Tsurutani et al., pp. 45–58, AGU, Washington, D. C.

Richmond, A. D., and G. Lu (2000), Upper-atmospheric effects of magnetic storms: A brief tutorial, *J. Atmos. Sol. Terr. Phys.*, *62*, 1115.

Riley, P., and N. U. Crooker (2004), Kinematic treatment of CME evolution in the solar wind, *Astrophys. J., 600*, 1035.

Sahai, Y., et al. (2005), Effects of the major geomagnetic storms of October 2003 on the equatorial and low-latitude F region in two longitudinal sectors, *J. Geophys. Res., 110*, A12S91, doi:10.1029/2004JA010999.

Sanderson, T. R., R. Reinhard, P. van Nes, K.-P. Wenzel, E. J. Smith, and B. T. Tsurutani (1985), Observation of 35- to 1600-keV protons and low-frequency waves up stream of interplanetary shocks, *J. Geophys. Res., 90*, 3973–3980.

Scherliess, L., and B. G. Fejer (1997), Storm time dependence of equatorial disturbance dynamo zonal electric fields, *J. Geophys. Res., 102*, 24,037–24,046.

Schieldge, J. P., and G. L. Siscoe (1970), A correlation of the occurrence of simultaneous sudden magnetospheric compressions and geomagnetic bay onsets with selected geophysical indices, *J. Atmos. Terr. Phys., 32*, 1819.

Schwabe, H. (1843), Solar observations during 1843, *Astron. Nachr., 20*, 495.

Schwenn, R. (2006), Space weather: The solar perspective, *Living Rev. Sol. Phys., 3*, 1–76.

Smith, E. J., and J. H. Wolfe (1976), Observations of interaction regions and corotating shocks between one and five Au: Pioneers 10 and 11, *Geophys. Res. Lett., 3*, 137–140.

Smith, E. J., B. T. Tsurutani, and R. L. Rosenberg (1978), Observations of the interplanetary sector structure up to heliographic latitudes of 16°: Pioneer 11, *J. Geophys. Res., 83*, 717–724.

Soraas, F., K. Aarsnes, K. Oksavik, M. I. Sandanger, D. S. Evans, and M. S. Greer (2004), Evidence for particle injection as the cause of Dst reduction during HILDCAA events, *J. Atmos. Sol. Terr. Phys., 66*, 177.

Søraas, F., K. Aarsnes, D. V. Carlsen, K. Oksavik, and D. S. Evans (2005), Ring current behavior as revealed by energetic proton precipitation, in *The Inner Magnetosphere: Physics and Modeling, Geophys. Monogr. Ser.*, vol. 155, edited by T. I. Pulkkinen, N. A. Tsyganenko, and R. H. W. Friedel, pp. 237–247, AGU, Washington, D. C.

Southwood, D. J. (1977), The role of hot plasma in magnetospheric convection, *J. Geophys. Res., 82*, 5512–5520.

Southwood, D. J., and R. A. Wolf (1978), An assessment of the role of precipitation in magnetospheric convection, *J. Geophys. Res., 83*, 5227–5232.

St. Cyr, O. C., et al. (2000), Properties of coronal mass ejections: SOHO LASCO observations from January 1996 to June 1998, *J. Geophys. Res., 105*, 18,169–18,185.

Stone, E. C., A. M. Frandsen, R. A. Mewaldt, E. R. Christian, D. Margolies, J. F. Ormes, and F. Snow (1998), The advance composition explorer, *Space Sci. Rev., 86*, 1.

Sturrock, P. A. (Ed.) (1980), *Solar Flares: A Monograph From Skylab Solar Workshop II*, Colo. Associated Univ. Press, Boulder, Colo.

Svetska, Z. (1976), *Solar Flares*, D. Reidel, Dordrecht.

Tang, F., B. T. Tsurutani, W. D. Gonzalez, S. I. Akasofu, and E. J. Smith (1989), Solar sources of interplanetary southward B_z events responsible for major magnetic storms (1978–1979), *J. Geophys. Res., 94*, 3535–3541.

Thome, G. D., and L. S. Wagner (1971), Electron density enhancements in the E and F regions of the ionosphere during solar flares, *J. Geophys. Res., 76*, 6883–6895.

Thomson, N. R., C. J. Rodger, and R. L. Dowden (2004), Ionosphere gives the size of greatest solar flare, *Geophys. Res. Lett., 31*, L06803, doi:10.1029/2003GL019345.

Thorne, R. M. (1980), The importance of energetic particle precipitation on the chemical composition of the middle atmosphere, *Pure Appl. Geophys., 118*, 128.

Torr, M. R., et al. (1995), A far ultraviolet imaging for the International Solar Terrestrial Physics mission, *Space Sci. Rev., 71*, 329.

Trakhtengerts, V. Y., M. J. Rycroft, D. Nunn, and A. G. Demekhov (2003), Cyclotron acceleration of radiation belt electrons by whistlers, *J. Geophys. Res., 108*(A3), 1138, doi:10.1029/2002JA009559.

Tsurutani, B. T., and W. D. Gonzalez (1987), The cause of high intensity long-duration continuous AE activity (HILDCAAs): Interplanetary Alfvén wave trains, *Planet. Space Sci., 35*, 405.

Tsurutani, B. T., and W. D. Gonzalez (1995), The efficiency of "viscous interaction" between the solar wind and the magnetosphere during intense northward IMF events, *Geophys. Res. Lett., 22*, 663–666.

Tsurutani, B. T., and W. D. Gonzalez (1997), The interplanetary causes of magnetic storms: A review, in *Magnetic Storms, Geophys. Monogr. Ser.*, vol. 98, edited by B. T. Tsurutani et al., pp. 77–89, AGU, Washington, D. C.

Tsurutani, B. T., and R. P. Lin (1985), Acceleration of >47 keV ions and >2 keV electrons by interplanetary shocks at 1 AU, *J. Geophys. Res., 90*, 1–11.

Tsurutani, B. T., and E. J. Smith (1974), Postmidnight chorus: A substorm phenomenon, *J. Geophys. Res., 79*, 118–127.

Tsurutani, B. T., and E. J. Smith (1977), Two types of magnetospheric ELF chorus and their substorm dependences, *J. Geophys. Res., 82*, 5112–5128.

Tsurutani, B. T., and X. Y. Zhou (2003), Interplanetary shock triggering of substorms: WIND and POLAR, *Adv. Space Res., 31*, 1063.

Tsurutani, B. T., E. J. Smith, K. R. Pyle, and J. A. Simpson (1982), Energetic protons accelerated at corotating shocks: Pioneer 10 and 11 observations from 1 to 6 AU, *J. Geophys. Res., 87*, 7389–7404.

Tsurutani, B. T., W. D. Gonzalez, F. Tang, S. I. Akasofu, and E. J. Smith (1988), Origin of interplanetary southward magnetic fields responsible for major magnetic storms near solar maximum (1978–1979), *J. Geophys. Res., 93*, 8519–8531.

Tsurutani, B. T., W. D. Gonzalez, F. Tang, and Y. T. Lee (1992), Great magnetic storms, *Geophys. Res. Lett., 19*, 73–76.

Tsurutani, B. T., C. M. Ho, E. J. Smith, M. Neugebauer, B. E. Goldstein, J. S. Mok, J. K. Arballo, A. Balogh, D. J. Southwood, and W. C. Feldman (1994), The relationship between interplanetary discontinuities and Alfvén waves: Ulysses observations, *Geophys. Res. Lett., 21*, 2267–2270.

Tsurutani, B. T., W. D. Gonzalez, A. L. C. Gonzalez, F. Tang, J. K. Arballo, and M. Okada (1995), Interplanetary origin of geomagnetic activity in the declining phase of the solar cycle, *J. Geophys. Res., 100*, 21,717–21,733.

Tsurutani, B. T., W. D. Gonzalez, Y. Kamide, and J. K. Arballo (1997), Preface, in *Magnetic Storms, Geophys. Monogr. Ser.*, vol. 98, edited by B. T. Tsurutani et al., pp. ix–x, AGU, Washington, D. C.

Tsurutani, B. T., X. Y. Zhou, V. M. Vasyliunas, G. Haerendel, J. K. Arballo, and G. S. Lakhina (2001), Interplanetary shocks, magnetopause boundary layers and dayside auroras: The importance of a very small magnetospheric region, *Surv. Geophys.*, *22*, 101.

Tsurutani, B., S. T. Wu, T. X. Zhang, and M. Dryer (2003), Coronal mass ejection (CME)-induced shock formation, propagation and some temporally and spatially developing shock parameters relevant to particle energization, *Astron. Astrophys.*, *412*, 293, doi:10.1051/0004-6361.

Tsurutani, B. T., et al. (2004), Global dayside ionospheric uplift and enhancement associated with interplanetary electric fields, *J. Geophys. Res.*, *109*, A08302, doi:10.1029/2003JA010342.

Tsurutani, B. T., et al. (2005), The October 28, 2003 extreme EUV solar flare and resultant extreme ionospheric effects: Comparison to other Halloween events and the Bastille Day event, *Geophys. Res. Lett.*, *32*, L03S09, doi:10.1029/2004GL021475.

Tsurutani, B. T., et al. (2006a), Corotating solar wind streams and recurrent geomagnetic activity: A review, *J. Geophys. Res.*, *111*, A07S01, doi:10.1029/2005JA011273.

Tsurutani, B. T., R. L. McPherron, W. D. Gonzalez, G. Lu, N. Gopalswamy, and F. L. Guarnieri (2006b), Magnetic storms caused by corotating solar wind streams, in *Recurrent Magnetic Storms: Corotating Solar Wind Streams, Geophys. Monogr. Ser.*, vol. 167, edited by B. T. Tsurutani et al., pp. 1–18, AGU, Washington, D. C.

Tsurutani, B. T., et al. (2006c), The dayside ionospheric superfountain (DIS), plasma transport and other consequences, in *Solar Influence on the Heliosphere and Earth's Environment: Recent Progress and Prospects*, Proc. ILWS Workshop, Goa, India, February 19–24, edited by N. Gopalswamy, and A. Bhattacharyya, p. 384, Quest Pub.

Tsurutani, B. T., O. P. Verkhoglyadova, A. J. Mannucci, T. Araki, A. Sato, T. Tsuda, and, K. Yumoto (2007), Oxygen ion uplift and satellite drag effects during the October 30, 2003 daytime superfountain event, *Ann. Geophys.*, *25*, 569.

Tsurutani, B. T., E. Echer, F. L. Guarnieri, and J. U. Kozyra (2008), CAWSES Novermber 7–8, 2004, superstorm: Complex solar and interplanetary features in the post-solar maximum phase, *Geophys. Res. Lett.*, *35*, L06S05, doi:10.1029/2007GL031473.

Turner, N. E., E. J. Mitchell, D. J. Knipp, and B. A. Emery (2006), Energetics of magnetic storms driven by corotating interaction regions: A study of geoeffectiveness, in *Recurrent Magnetic Storms: Corotating Solar Wind Streams, Geophys. Monogr. Ser.*, vol. 167, edited by B. T. Tsurutani et al., pp. 113–124, AGU, Washington, D. C.

Verkhoglyadova, O. P., B. T. Tsurutani, and A. J. Mannucci (2006), Temporal development of dayside TEC variations during the October 30, 2003 superstorm: Matching modeling to observations, in *Advances in Geosciences*, vol. 8, edited by M. Duldig et al., p. 69, World Sci. Publ.

Verkhoglyadova, O. P., B. T. Tsurutani, A. J. Mannucci, A. Saito, T. Araki, D. Anderson, M. Abdu, and J. H. A. Sobral (2008), Simulation of PPEF effects in dayside low-latitude ionosphere for the October 30, 2003 superstorm, this volume.

Webb, D. F., and R. A. Howard (1994), The solar cycle variation of coronal mass ejections and the solar wind mass flux, *J. Geophys. Res.*, *99*(A3), 4201–4220.

Winterhalter, D., E. J. Smith, M. E. Burton, N. Murphy, and D. J. McComas (1994), The heliospheric plasma sheet, *J. Geophys. Res.*, *99*, 6667–6680.

Wu, C. S. (1996), Production of solar energetic ions associated with flares, *Astrophys. J.*, *472*, 818.

Zhou, X., and B. T. Tsurutani (1999), Rapid intensification and propagation of the dayside aurora: Large scale interplanetary pressure pulses (fast shocks), *Geophys. Res. Lett.*, *26*, 1097–1100.

Zhou, X., and B. T. Tsurutani (2001), Interplanetary shock triggering of nightside geomagnetic activity: Substorms, pseudo-breakups, and quiescent events, *J. Geophys. Res.*, *106*, 18,957–18,967.

Zhou, X.-Y., and B. T. Tsurutani (2004), Dawn and dusk auroras caused by gradual, intense solar wind ram pressure (GISWRP) events, *J. Atmos. Sol. Terr. Phys.*, *66*, 153.

Zhou, X.-Y., R. J. Strangeway, P. C. Anderson, D. G. Sibeck, B. T. Tsurutani, G. Haerendel, H. U. Frey, and J. K. Arballo (2003a), Shock aurora: FAST and DMSP observations, *J. Geophys. Res.*, *108*(A4), 8019, doi:10.1029/2002JA009701.

Zhou, X.-Y., H. U. Frey, J. F. Waterman, B. T. Tsurutani, S. B. Mende, and J. K. Arballo (2003b), Dawn/dusk auroras and geomagnetic fluctuations: Ionospheric effects of intense solar wind ram pressure, in *Earth's Low-Latitude Boundary Layer, Geophys. Monogr. Ser.*, vol. 133, edited by P. Newell and T. Onsager, pp. 361–370, AGU, Washington, D.C.

E. Echer, Brazilian National Institute for Space Research (INPE), Av. dos Astronautas, 1758, Sao Jose dos Campos, SP 12227-010, Brazil.

F. L. Guarnieri, University of Paraiba Valley (UNIVAP), Ave. Shishima Hifumi, 2911, Sao Jose dos Campos, SP 12244-000, Brazil.

B. Tsurutani and O. P. Verkhoglyadova, Jet Propulsion Laboratory, California Institute of Technology, 4800 Oak Grove Drive, Pasadena, CA 91109-8099, USA. (bruce.tsurutani@jpl.nasa.gov)

Ionospheric-Magnetospheric-Heliospheric Coupling: Storm-Time Thermal Plasma Redistribution

John C. Foster

MIT Haystack Observatory, Westford, Massachusetts, USA

Large-scale thermal plasma redistribution phenomena have been observed from the ground and space during the major geomagnetic storms of the recent solar cycle. Plasma redistribution is a multistep, systemwide process involving the equatorial, low, mid, auroral, and polar latitude regions. Penetration electric fields enhance the equatorial ionization anomaly peaks, whereas polarization electric field effects at the dusk terminator redistribute the low-latitude total electron content (TEC) in both longitude and latitude. Magnetic field geometry creates a preferred longitude for the enhancement of low- and mid-latitude TEC in the American sector at sunset. This TEC enhancement forms a localized source for the intense storm enhanced density (SED) erosion plumes that are observed over the Americas during major storms. A second set of storm-time processes leads to the erosion of the plasmasphere boundary layer, transporting the SED plumes to the noontime cusp in the ionosphere and to the dayside magnetopause at high altitudes. Ring current enhancements generate strong poleward-directed subauroral polarization stream (SAPS) electric fields in the evening sector as field-aligned currents close through the low-conductivity ionosphere. The SAPS electric field overlaps the outer plasmasphere, drawing out the SED/plasmasphere erosion plumes. These greatly enhanced fluxes of cold plasma traverse the cusp and enter the polar cap forming the polar tongue of ionization and providing a rich source of heavy ions for the magnetospheric injection and acceleration mechanisms that operate in these regions.

1. PREAMBLE

Thermal plasma redistribution during major geomagnetic disturbances is a multistep, systemwide process involving the equatorial, low, mid, auroral, and polar latitude regions. During the preceding solar maximum, IMAGE extreme ultraviolet (EUV) space-based imagery [*Sandel et al.*, 2001] revealed dramatic storm-time erosion of the plasmasphere and the formation of the tails or drainage plumes that had been predicted in early studies modeling the interaction of the disturbance electric field with the plasmasphere [e.g., *Chen and Grebowsky*, 1974]. Mid-latitude ground-based observations revealed the characteristics of a number of related ionospheric phenomena [e.g., *Mendillo*, 1973; *Foster*, 1993], whereas a study by *Su et al.* [2001] drew the connection between the ionospheric observations, plasmaspheric tails, and the large-scale redistribution of thermal plasma in the magnetosphere.

A series of studies addressing the characteristics and causes of thermal plasma redistribution, and their space weather effects, has been undertaken by the research team at

MIT Haystack Observatory. This focused review provides an overview of the picture of storm-time plasma redistribution developed in this prior body of work. Global imaging from the ground using distributed GPS total electron content (TEC) observations and from space by the IMAGE EUV and far ultraviolet instruments has provided two-dimensional snapshots of the plasma distribution at ionospheric and plasmaspheric heights, and of their temporal evolution during disturbed conditions. Incoherent scatter radar (ISR) and overflights with the DMSP satellites give details of plasma convection velocities and altitude/spatial distributions. A system perspective involving the role of ionosphere–magnetosphere–heliosphere coupling in shaping and controlling these important space weather phenomena is developed through a synopsis of the MIT and related publications available through the year 2007.

2. INTRODUCTION

Earth's ionosphere responds dramatically to severe geomagnetic storms. Coupled closely to the neutral thermosphere and the overlying magnetosphere, storm-time ionospheric effects appear with a wide variety of scale sizes, characteristics, time histories, and associated drivers. Recent studies highlighting the disturbance effects observed in the ionosphere and magnetosphere during geomagnetic disturbances, and their space weather effects, have heightened interest in the processes occurring in the plasmasphere boundary layer (PBL) [*Carpenter and Lemaire*, 2004], the boundary between the corotating field lines of the plasmasphere and the processes of the outer magnetosphere. During storms, the effects of magnetospheric electric field and ExB plasma convection extend deep into the mid-latitude ionosphere. At the PBL, strong electric fields contribute to the formation of the deep mid-latitude density trough that spans the nightside, whereas penetration electric fields uplift, destabilize, and perturb the low-latitude ionosphere. Equatorward of the ionospheric trough, large increases in the midlatitude ionospheric *F*-region electron density and TEC are often observed in local dusk sector during magnetic storms [e.g., *Mendillo*, 2006]. Early investigations [e.g., *Evans*, 1973; *Mendillo*, 1973] proposed that an uplifting of the *F* layer by an eastward electric field and convergence in the east–west direction might be responsible for such dusk effect mid-latitude TEC enhancements.

3. STORM ENHANCED DENSITY

Foster [1993] examined Millstone Hill ISR observations and found that recurring localized dusk-sector TEC enhancements observed at the PBL in disturbed conditions were associated with sunward convection (toward noon and poleward at ionospheric heights) of high-density plasma originating from lower latitudes. These sunward-convecting density enhancements at the equatorward edge of the dusk-sector ionospheric trough are termed storm enhanced density (SED) [*Foster*, 1993]. Plate 1 presents a cross section of one such SED feature observed with a Millstone Hill ISR elevation scan. *F*-region TEC (calculated for the ISR density–altitude profiles) approached 100 TEC unit (TECU; 1 TECU = 10^{16} el m^{-2}) within the SED plume and sharp TEC gradients (~50 TECU/deg latitude) were observed bordering the SED feature [*Vo and Foster*, 2001].

GPS TEC measures the integrated column content of cold electrons through the ionosphere and overlying plasmasphere to an altitude of ~20,000 km (~4 Re) [*Coster et al.*, 2003]. Plate 2 [from *Foster et al.*, 2005a] presents in polar projection a GPS TEC map of the spatial extent of the strong plume of SED seen during the 31 March 2001 event discussed by *Foster et al.* [2002]. The SED plume stretches from a region of enhanced TEC in southeastern United States, poleward to the limit of the GPS observations near the noontime cusp over north central Canada. At right, we have used the *Tsyganenko* [2002] magnetic field model to project the ionospheric footprint of the TEC observations into the magnetosphere equatorial plane. The SED plume maps into a narrow drainage plume reaching sunward from the greatly eroded plasmapause position near $L = 2$ to the dayside magnetopause near noon (noon is at the right of the figure). The strength of the plume, its position, and related magnetospheric and ionospheric effects depend on both the cold plasma source of the SED material and the characteristics of the electric field that transports it.

The overall storm-time enhancement of TEC at low and mid latitudes consists of two parts (as shown schematically in Plate 3 from *Foster et al.* [2005a]). First is the increase in TEC seen at and poleward of the crests of the equatorial anomalies associated with plasma uplift and redistribution from low to mid latitudes under the influence of the penetration electric field [e.g., *Tsurutani et al.*, 2004]. This occurs inside the PBL and serves as an enhanced source population for the second TEC feature, the plumes of SED that are eroded from the lower-latitude ionosphere/plasmasphere by magnetospheric disturbance electric fields. This process produces narrow plasmasphere drainage plumes [*Sandel et al.*, 2001], whose extent along magnetic field lines stretches between the plasmasphere and the ionosphere [e.g., *Chi et al.*, 2005]. Cold plasma within the plumes is streaming sunward across magnetic field lines toward the cusp (at low altitudes) and the dayside magnetopause (at high altitudes) accounting for significant storm-time thermal plasma redistribution in the coupled ionosphere–magnetosphere system.

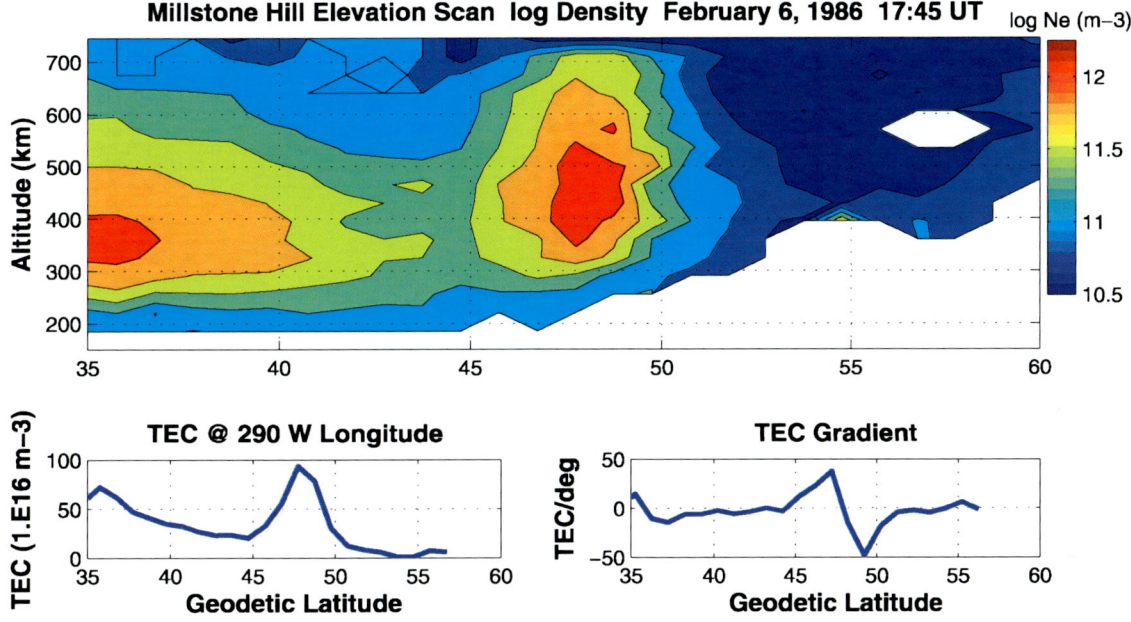

Plate 1. Isodensity contours observed by the Millstone Hill radar scanning N–S across a region of strong storm enhanced density (SED) near local noon are presented with as a function of geodetic latitude (invariant latitude = geodetic latitude + 11°). F region total electron content (TEC) and TEC gradient across the region are derived from the radar elevation scan and are shown in the bottom panels. The 8 February 1986 event represents a severe example of the ionospheric density perturbation that can occur over the continental United States associated with SED plumes. TEC near 48°N geodetic latitude is ~100 TECU with latitude gradient in TEC of ~50 TECU/deg [from *Vo and Foster*, 2001].

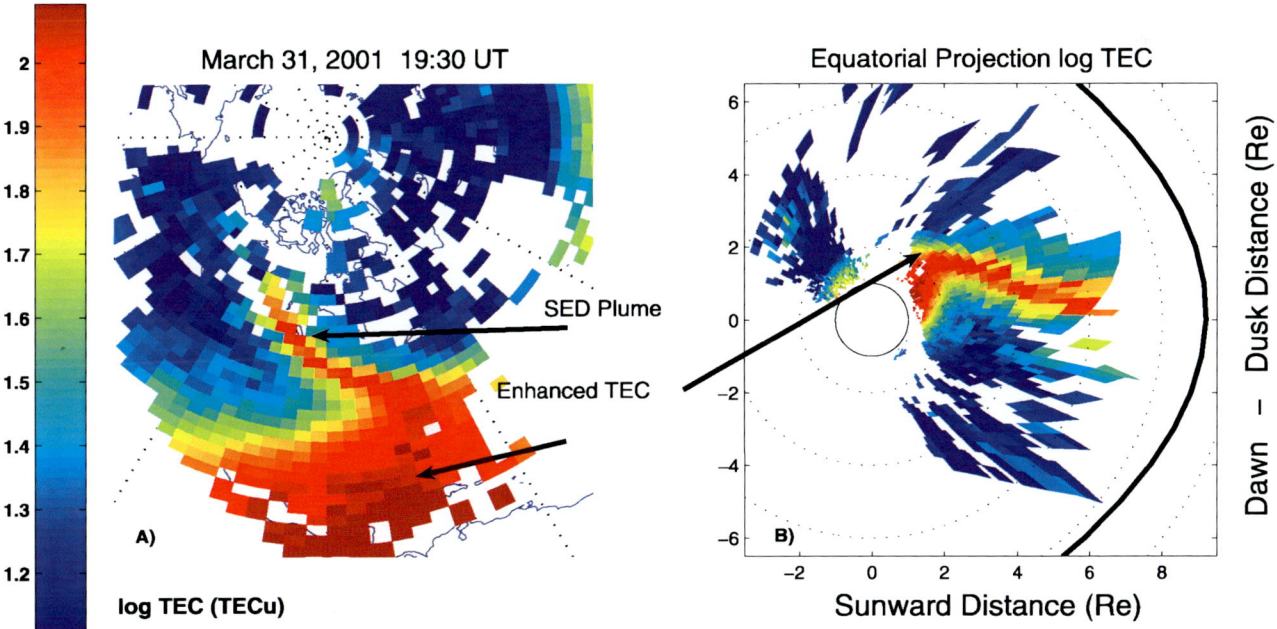

Plate 2. (A) A region of enhanced GPS TEC was observed at the base of the plume of SED seen over North America during the 31 March 2001 event. (B) Projecting the GPS TEC observations into the magnetospheric equatorial plane using Tsyganenko mapping (with the sun at the right), indicates that the enhancement at the base of the plume is field lines threading the outer plasmasphere [from *Foster et al.*, 2005a].

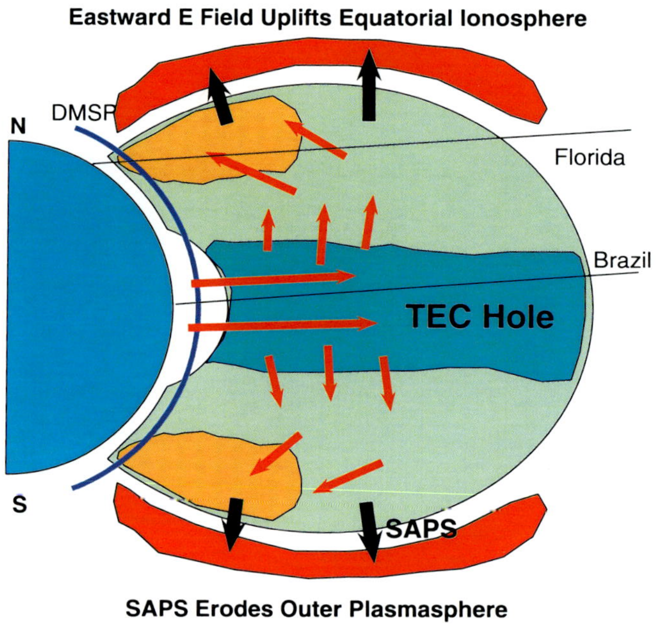

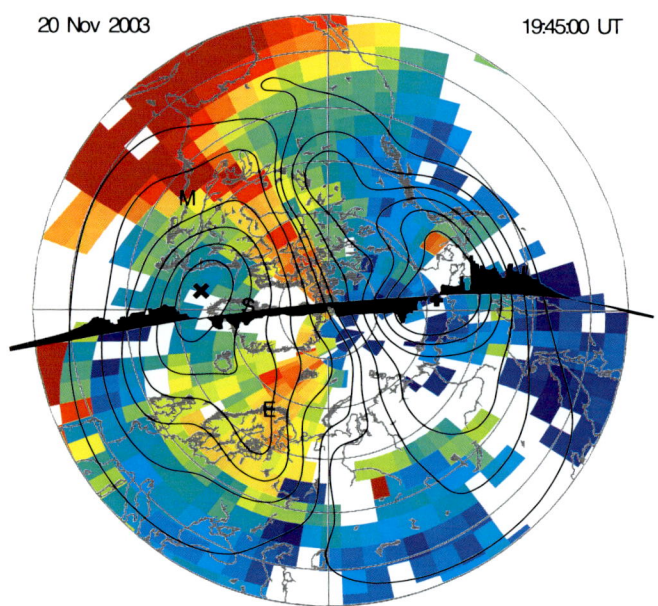

Plate 3. The dual effects of disturbance electric fields are presented schematically. Undershielded penetration electric fields uplift the equatorial ionosphere redistributing equatorial plasma poleward, whereas SAPS electric fields strip away the enhanced outer layers of the plasmasphere [from *Foster et al.*, 2005a].

Plate 4. Combined GPS TEC and convection observations are displayed in polar projection (mag lat/MLT coordinates; 10° latitude circles; with noon at the top). $\log_{10}$ vertical TEC observations binned by lat/long at 350-km altitude are displayed with the simultaneous, independent convection pattern derived from combined Super Dual Auroral Radar Network and DMSP observations. Ion drift meter cross-track velocity data from a transpolar cap DMSP pass are shown, indicating antisunward convection above 60° latitude spanning the polar region. A plume of elevated TEC follows the convection contours back across the polar cap from the dayside cusp forming the polar tongue of ionization [from *Foster et al.*, 2005b].

4. RADAR OBSERVATIONS OF SED PLUMES

The MIT Millstone Hill ISR, located at 55°Λ (invariant latitude) near the ionospheric projection of the plasmapause and the PBL, regularly observes SED in the premidnight subauroral ionosphere during the early stages of magnetic disturbances [*Foster*, 1993]. These high-TEC plumes of ionization are seen in the premidnight and afternoon sector at the equatorward edge of the main ionospheric trough and stream sunward carried by the low-latitude edge of the subauroral disturbance electric field (the subauroral polarization stream [SAPS]; *Foster and Burke* [2002]; *Foster and Vo* [2002]). Combining ground- and space-based thermal plasma imaging techniques, *Foster et al.* [2002] demonstrated that one such ionospheric SED plume mapped into the low-altitude signature of a plasmasphere drainage plume associated with the storm-time erosion of the PBL. *Foster et al.* [2004] used direct observations of the sunward ExB advection to quantify the flux of ions carried to the noontime cusp F region ionosphere by the SED plume ($\sim 10^{26}$ ions/s), amounting to a tenfold enhancement of the ionospheric source plasma available to cusp acceleration and injection mechanisms. *Foster et al.* [2005b] combined ISR observations from Millstone Hill, Sondrestrom, and European Incoherent Scatter, with Super Dual Auroral Radar Network HF radar mapping and DMSP simultaneous observations to reveal how the SED plume carrying low-latitude material to the cusp enters the polar cap forming a high-TEC tongue of ionization (TOI) spanning polar latitudes from noon to midnight (cf. Plate 4 from *Foster et al.* [2005b]).

5. ELECTRIC FIELDS

During the early phase of a geomagnetic storm, enhanced cross-tail electric fields drive plasmasheet particles inward. There is little shielding in place at the time, and the fields penetrate deep into the inner magnetosphere. Penetrating eastward electric fields are observed at mid and low latitudes in the postnoon sector [e.g., *Foster and Rich*, 1978], driving the F region plasma upward and poleward in the ExB direction. At equatorial latitudes, the ionosphere is lifted and spreads poleward along the field lines in both hemispheres under the influence of gravity and plasma pressure to form the equatorial ion fountain and the enhanced plasma of the equatorial ionospheric anomaly (EIA) peaks. During extreme uplift events (e.g., March 1989, *Greenspan et al.* [1991]; July 2000, *Basu et al.* [2001]; or 30 October 2003, *Mannucci et al.* [2005a]; *Tsurutani et al.* [2006]) the bottomside of the equatorial F region can rise above the 830-km altitude orbit of the DMSP satellites. Such drastic uplifting is seen preferentially at longitudes near the South Atlantic magnetic anomaly (SAA) and most often in the dusk sector [e.g., *Basu et al.*, 2007]).

The observed changes in low- and mid-latitude TEC in the American sector respond to the dual action of storm-time electric fields associated with inner-magnetosphere cold plasma redistribution (cf. Plate 3). An initial strong equatorial upwelling redistributes the low-latitude ionospheric plasma to higher-latitude flux tubes within the plasmasphere, resulting in enhanced levels of density on flux tubes in the PBL near the plasmapause. As the event progresses, a shielding layer is set up where the freshly injected ring current particles abut the plasmapause. The inward extent of the energetic ring current ions lies equatorward of the plasmasheet electrons, and region II currents are driven into the subauroral ionosphere. There, a strong poleward electric field is needed to drive poleward-directed Pedersen closure currents in the low-conductivity ionosphere equatorward of the precipitating auroral electrons. This SAPS electric field [*Foster and Vo*, 2002] overlaps the outer plasmasphere and draws out the SED/plasmasphere erosion plumes that stretch sunward from their dusk-sector source to the dayside cusp and magnetopause merging region [*Foster et al.*, 2002, 2004].

6. SUBAURORAL POLARIZATION STREAM

Electric fields are of prime importance in the formation and transport of storm-time ionospheric disturbances. The SAPS electric field refers to the region of enhanced poleward electric field that forms equatorward of electron precipitation in the dusk to midnight sector in disturbed conditions. Often found embedded within the broader SAPS channel are narrow, intense subauroral ion drifts (SAID) [e.g., *Anderson et al.*, 2001]. SAPS and SAID are features of the magnetospheric electric field whose characteristics are controlled by ionospheric conductance. SAPS forms as pressure gradients at the inner edge of the disturbance-enhanced ring current drive region 2 field-aligned currents into the evening-sector ionosphere. Large poleward-directed electric fields are set up to drive closure currents across the low-conductivity region equatorward of the auroral electron precipitation. *Anderson et al.* [2001] have demonstrated that the SAID electric fields are magnetically conjugate and extend along magnetic field lines into the magnetosphere. Observations of the SAPS phenomenon at ionospheric altitudes [*Yeh et al.*, 1991; *Foster and Burke*, 2002; *Foster and Vo*, 2002] describe the occurrence characteristics and persistence of the SAPS electric field whose latitude extent spans the region between the electron plasma sheet and the outer reaches of the plasmasphere (the PBL). Significant magnetospheric effects occur as the SAPS electric field maps along field lines and is observed at higher altitudes [e.g., *Rowland and Wygant*, 1998; *Burke et al.*, 1998].

SAPS is strongest in the dusk/premidnight sector where it lies equatorward of and often distinct from the region of sunward ionospheric convection at auroral latitudes. It is the SAPS electric field that overlaps the PBL at the base of the erosion plumes, resulting in the erosion of the enhanced outer layer of the storm-time plasmasphere (see discussion below and *Foster et al.* [2002, 2004]). Nearer to noon, the SAPS convection merges into the equatorward extent of the auroral convection [*Foster and Vo*, 2002], bringing the SED plume at ionospheric heights into the dayside cusp [*Foster et al.*, 2004].

In situ electric field data further characterize this subauroral feature and its extent to magnetospheric heights. One year of data from ISEE 1 was used by *Maynard et al.* [1983] to determine the average characteristics of the electric field from L of 2 to 6. Those data indicated a penetration of the convection field into $L < 4$ during disturbed conditions in the late evening and early morning sector. *Rowland and Wygant* [1998] studied in situ electric field data from CRRES for a 10-month period in 1991 and showed the presence of a localized region of enhanced electric field between $L = 3$ and $L = 6$ for $Kp > 3$.

Foster et al. [2007] used magnetic field-aligned mapping between the ionosphere and the magnetosphere to intercompare ground-based observations of SED, and plasmasphere drainage plumes imaged from space by the IMAGE EUV imager, with the enhanced inner-magnetosphere/ionosphere SAPS electric field that develops during large storms. In this detailed comparison they found that the inner edge of the SAPS electric field overlaps the erosion plume and that plume material is carried sunward in the SAPS overlap region. Figure 1, taken from *Foster et al.* [2007], describes the

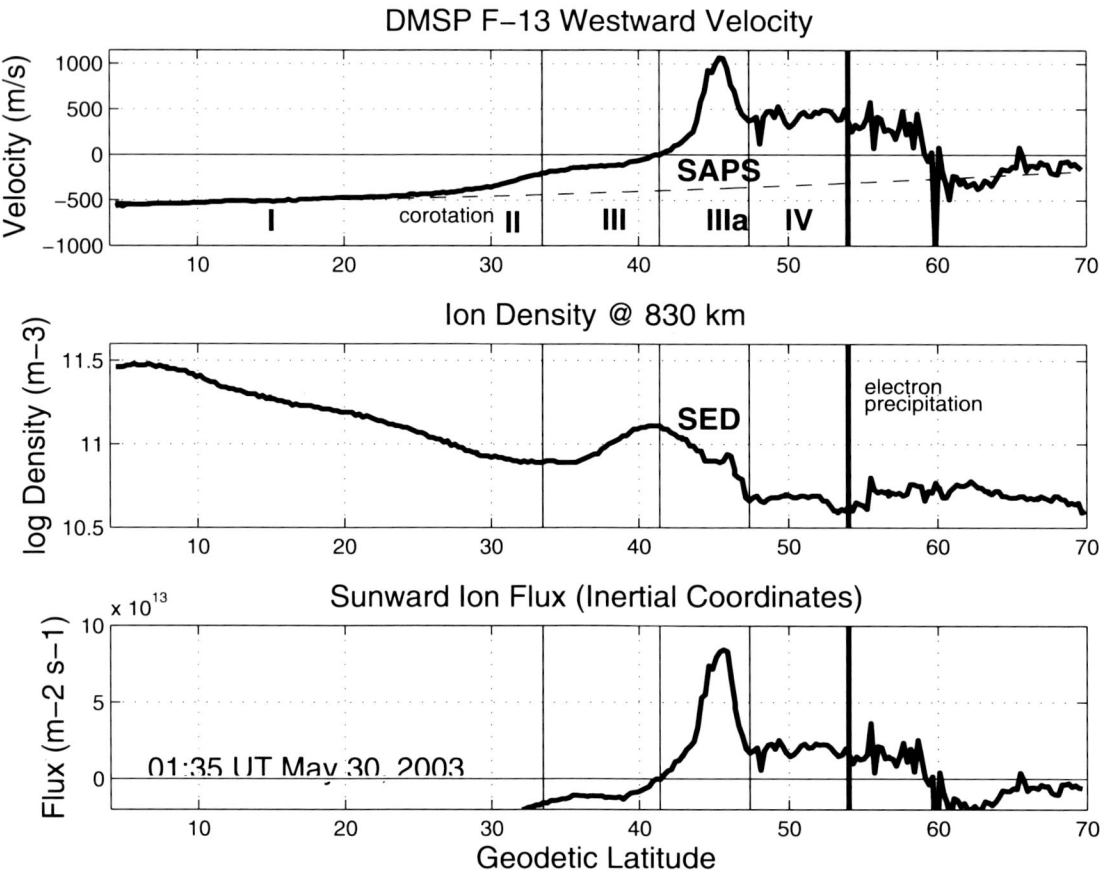

Figure 1. DMSP F-13 observations of plasma density (middle panel) and cross-track velocity (top panel; positive westward) are shown. The equatorward limit of electron precipitation is indicated by vertical lines near 54° N. An extensive region of SAPS convection extends inward to <35°. The bottom panel shows sunward ion flux calculated as the product of density velocity. The westward ion flux exceeds 8×10^{13} m^{-2} s^{-1} in the topside F region (~830 km) in the outer portion of the SED plume [from *Foster et al.*, 2007].

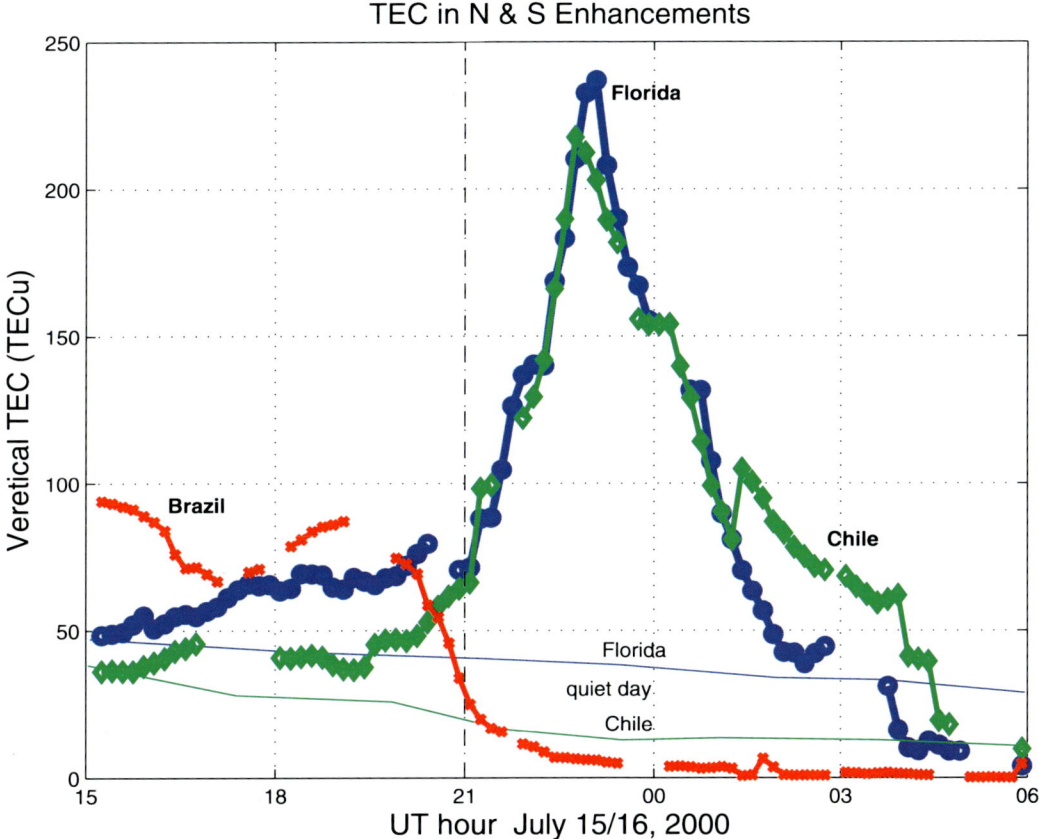

Plate 5. Temporal variation of TEC in the N–S conjugate enhancements during the July 2000 event indicates a close similarity. TEC over Florida and Chile increases sharply as the equatorial TEC over Brazil drops during storm intensification. Conjugate enhancements of ~200 TECU above the quiet-day background (13–14 July 2000) were observed [from *Foster and Coster*, 2007].

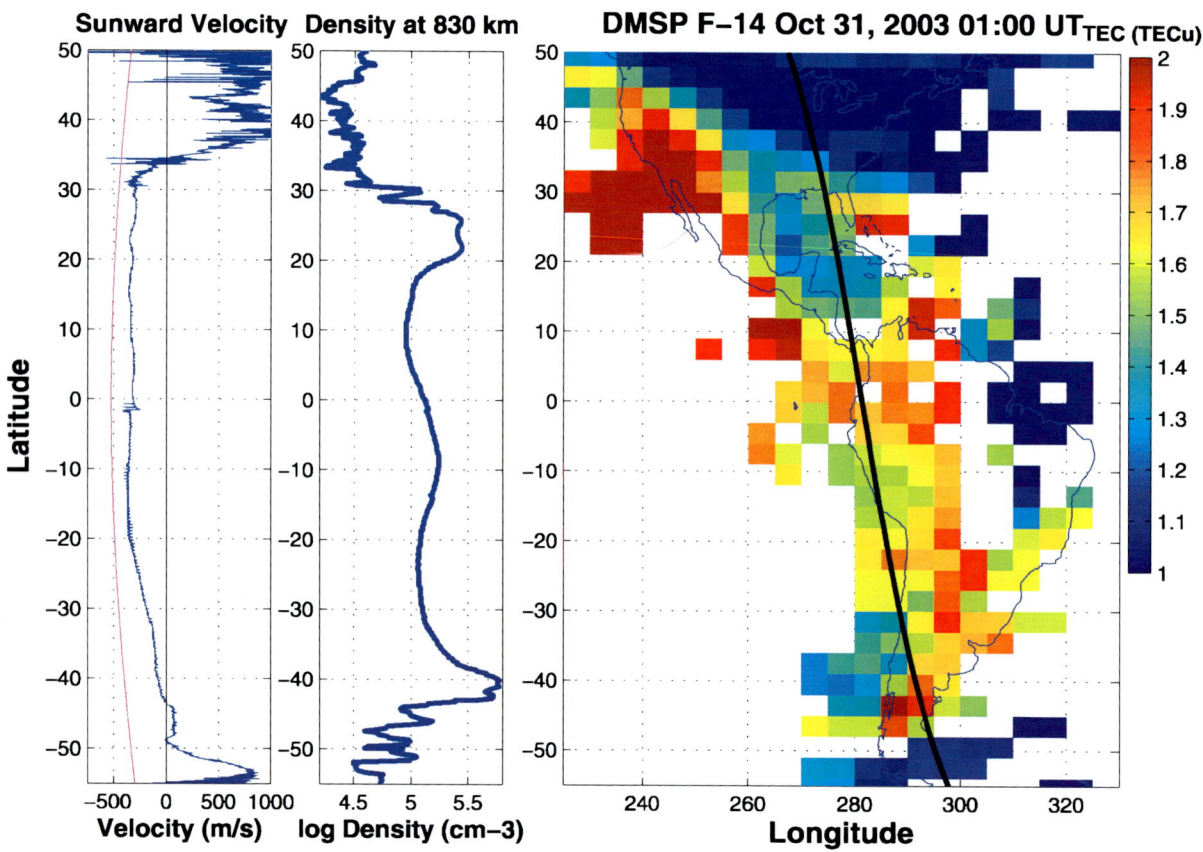

Plate 6. Sunward ion velocity in inertial coordinates and in situ density are shown at the left for a DMSP F-14 pass at 830-km altitude through remnant enhanced TEC regions. The (negative) corotation speed is shown as the smooth curve in the leftmost panel. The position of the southern TEC enhancement at ~42°S latitude is nearly fixed in local time (zero velocity), with the earth rotating eastward under it. In the north, the enhancement near 24°N latitude approximately is corotating at this time of the event (see text for discussion). The enhancements are in magnetically conjugate locations and lie immediately inside the steep density gradient at the trough boundary that also is seen in the GPS TEC map at the right [from *Foster and Coster*, 2007].

overlap regions of SAPS and SED observed in the topside ionosphere by DMSP. The two phenomena, SED in the ionosphere and the erosion plume at magnetospheric heights, define a common trajectory for sunward-propagating cold plasma fluxes in the midnight–dusk–postnoon sector. The SAPS electric field abuts and overlaps both the PBL and the plasmasphere erosion plume from premidnight through postnoon local times. The SAPS channel at ionospheric heights and its projection into the equatorial plane serve to define the sharp outer boundary of the erosion plume.

7. TEC ENHANCEMENT AT THE BASE OF THE SED PLUME

Near sunset during the expansion phase of large geomagnetic storms, a localized enhancement of TEC takes place in the American longitude sector off the coast of Florida [*Mannucci et al.*, 2005b] and in its magnetically conjugate region. These enhancements lie poleward of the crests of the equatorial ionization anomaly and on field lines mapping into the PBL. The onset of the enhanced TEC in these regions is described in Plate 5, which presents the time histories of the variation of vertical TEC in magnetically conjugate regions in the northern Caribbean SE of Florida ("Florida") and in southern Chile ("Chile") for the large storm of 15–16 July 2000 (taken from *Foster and Coster* [2007]). Also shown is the variation of TEC at the magnetic equator, in the region of the SAA over eastern Brazil. At the time of sharply decreasing Dst, TEC at the equator dropped sharply from >80 to <10 TECU and remained very low throughout the remainder of the interval. Vertical TEC over Chile and southeast of Florida track each other surprisingly well—clearly showing the simultaneity and similarity of the storm-induced enhancement in these localized regions in the two hemispheres. A dashed vertical line is shown at 21 UT, the time of the E-region sunset terminator at the Brazilian site. Both poleward and eastward components of the disturbance electric field are needed to account for the observed redistribution of cold plasma at low and mid latitudes inside the plasmapause [*Kelley et al.*, 2004; *Lin et al.*, 2005]. *Foster et al.* [2005a] present a similar figure showing an identical timing and relationship between TEC over Florida and over Brazil for the 30 May 2003 event. Those authors and the data and discussions of *Foster and Coster* [2007] make the case that such events involve the redistribution of cold plasma from low and mid latitudes in the vicinity of the SAA to produce localized enhancements of TEC on field lines mapping into the PBL in the region southeast of Florida, and in its conjugate region. Plasma uplift in the SED can retard recombination and permit additional TEC buildup if the SED region is in sunlight.

After their formation, the conjugate enhancements persist at American longitudes and corotate into the night sector. Plate 6, taken from the study of *Foster and Coster* [2007], presents observations during the pass of the DMSP F14 satellite that crossed the magnetically conjugate GPS TEC enhancements at 24°N and 42°S latitude, observing a factor of 2–5 increase in in situ plasma density. Observations of cross-track plasma drift made with the F14 ion drift meter (shown in the left panel) indicate that the enhanced TEC regions were approximately corotating in place over the positions in which they were formed earlier in the event.

The repeatability and geographic localization of these features suggest a longitudinal specificity for the process associated with their formation. Plasma redistribution from low latitudes into the base of the erosion plumes and its enhancement at longitudes near the SAA in the dusk sector suggest a UT dependence for the injection of ionospheric ions into the storm-time magnetosphere via the effects of strong plasmasphere drainage plumes. The long-lasting conjugate enhancements of TEC in the PBL at longitudes in the American sector serve as a corotating source for the erosion plumes of SED seen during such events. *Foster et al.* [2005b] demonstrate that the SED plumes pass through the cusp, filling the polar caps with dense tongues of heavy ions (TOI), and feeding processes that inject heavy ions into the magnetotail. A large flux of these cold ions will change the characteristics of the magnetosphere and may well alter the response of the magnetosphere to the developing magnetic storm—that is, they may alter the geoeffectiveness of the storm drivers. *Foster et al.* [2005a] speculate that the appearance of rich concentrations of ionospheric ions in the disturbance ring current in the latter phases of great storms indicates that plasmaspheric erosion/SED/ion injection may have preconditioned the magnetosphere before the final surges of particle injection that drive the storm-time Dst to extreme values during these events. It is possible that maximum negative Dst during the truly great storms ($Dst < -300$ nT) occur at similar UTs—that is, when the SAA has recently been in the dusk/bulge sector (20 UT–00 UT).

8. MAGNETIC CONJUGACY CHARACTERISTICS OF THE SED

The cold ionospheric and plasmaspheric material constituting the SED plumes is redistributed at ionospheric and magnetospheric heights by the actions of electric fields. *Foster and Rideout* [2007] intercompared simultaneous observations of ground-based TEC with Jason and TOPEX satellite TEC observations to investigate the magnetic conjugacy characteristics of the plumes and other storm-time density and TEC enhancements. They found that the TEC enhancements on inner-magnetospheric field lines at the base of the SED plumes exhibit localized and longitude-dependent features that are not strictly magnetically conjugate, whereas the SED plumes streaming away from these source regions closely follow magnetic conjugate paths, confirming that SED is a convection electric field dominated effect. The location and flow of the SED plumes can be used as a tracer of the extent and strength of disturbance electric fields.

Plate 7, taken from the study of *Foster and Rideout* [2007], demonstrates the close similarity in magnitude and location between the strong SED features that developed over the continental United States during the October 2003 superstorms [e.g., *Foster and Rideout*, 2005] and their magnetic-conjugate counterparts observed in the south Pacific sector by Jason TEC measurements. The figure maps ground-based GPS TEC observations in geographic coordinates. A red contour outlines the regions of elevated TEC observed in the northern hemisphere, and the magnetic conjugate projection of these contour lines is shown in the south. A heavy black line denotes the N to S orbital track of the Jason satellite. The magnetic conjugate of the Jason orbit has been calculated and is denoted by the heavy dashed black line. The conjugate orbit crosses North America, where good ground-based GPS TEC coverage was available, intersecting the regions of low-latitude TEC enhancement and the SED plume. To the left, is plotted the vertical TEC measured by Jason along its direct orbit (heavy black line) as a function of latitude along the orbital track depicted in the right panel. The conjugate to the poleward boundary of the SED plume is crossed near $-60°$ latitude. Wherever the conjugate Jason orbit (dashed curve) intersects a region where there are GPS TEC data, the underlying values of GPS TEC are plotted point by point in the left-hand panel (red diamonds) at their corresponding conjugate latitudes, on top of the direct Jason TEC observations. In those regions, there is a direct comparison of TEC observed by Jason with the GPS TEC observed at magnetic conjugate points.

Conclusions of the *Foster and Rideout* [2007] study include:

1. The SED plume occurs in magnetically conjugate regions in both hemispheres.
2. The position of the sharp poleward edge of the SED plume is closely conjugate.
3. The SAPS electric field is observed in magnetically conjugate regions (SAPS channel).
4. The strong TEC enhancement at the base of the SED plume in the North American sector is more extensive than in its magnetic conjugate region.
5. The entry of the SED plume into the polar cap near noon, forming the polar TOI, is seen in both hemispheres in magnetically conjugate regions.

9. POLARIZATION TERMINATOR

Foster et al. [2005a] observed that many of the strong enhancements of TEC in the Caribbean sector at the base of the SED plumes begin sharply near 21 UT, associated with the passage of the sunset terminator into the SAA. Foster and Erickson (J. C. Foster and P. J. Erickson, Ionospheric superstorms: Sunset terminator effects, submitted to *Geophysics Research Letters*, 2008, hereinafter referred to as Foster and Erickson, submitted manuscript, 2008) have investigated the cause of this effect. They show how the varying ionospheric conductivity at the ends of magnetic fields lines in the Atlantic sector can produce a greatly distorted dusk terminator effect. As a result, the polarization electric fields set up at the terminator in disturbed conditions sweep the storm-enhanced EIA crest plasma westward and poleward, creating magnetically conjugate TEC enhancements at the base of the SED plumes that form in the American sector.

A combination of the storm-time penetration electric fields, the effect of the reduced magnetic field strength in the SAA, and the geographic distortion of the magnetic field in the Atlantic sector contribute to the characteristics of the polarization electric fields at the sunset terminator. These combined effects lead to a strong localized enhancement of TEC at low to mid latitudes in the American sector during ionospheric superstorms. At dusk, polarization electric field effects begin on magnetic field lines when the E region at either end goes into darkness. Foster and Erickson (submitted manuscript, 2008) define polarization terminator (PT) to be the locus of points at a given altitude for which the E-region shadow height at either end of the magnetic field line equals 100 km. Electric fields associated with the charge buildup in the conductivity-gradient region due to the effects of winds or penetration electric fields are directed perpendicular to the PT and increase in magnitude as the PT is approached from the dayside.

Ground-based GPS observations of vertical TEC at 2040 UT on 15 July 2000 are shown in Plate 8 in geodetic coordinates, during the initial stages of the strong, magnetically conjugate enhancement of TEC (>200 TECU) in the American sector shown in Plate 5 (from Foster and Erickson, submitted manuscript, 2008). Superimposed on the TEC map is a family of contours, each indicating the locus of points at 800-km altitude and 20:40 UT for which the E-region shadow height in either hemisphere along the local magnetic field line first exceeds a fixed value ranging from 50 to 500 km in 50-km intervals. This family of contours indicates the two-dimensional extent and orientation of the PT conductivity-gradient region at this specific time. The distorted geometry of the magnetic field at these longitudes, and the significant differences in solar zenith angle at the

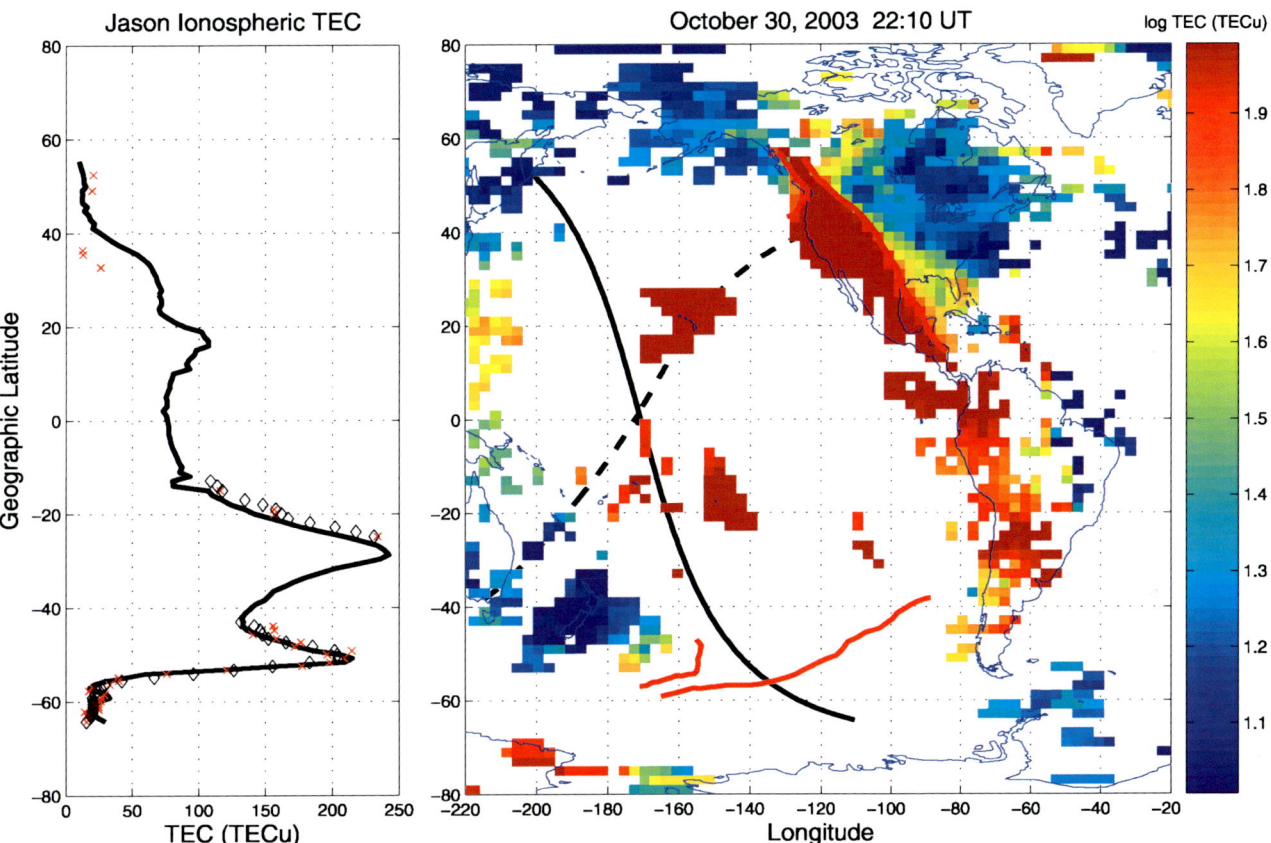

Plate 7. Characteristics of the strong SED plume (>200 TECU) seen over the U.S. mainland and at its magnetic conjugate are similar for the 30 October 2003 superstorm, indicating the conjugate actions of magnetospheric electric fields mapping along field lines between hemispheres. The closely similar magnitude of TEC in the two hemispheres suggests a similar source for the TEC enhancements at low and mid latitudes during that event [from *Foster and Rideout*, 2007].

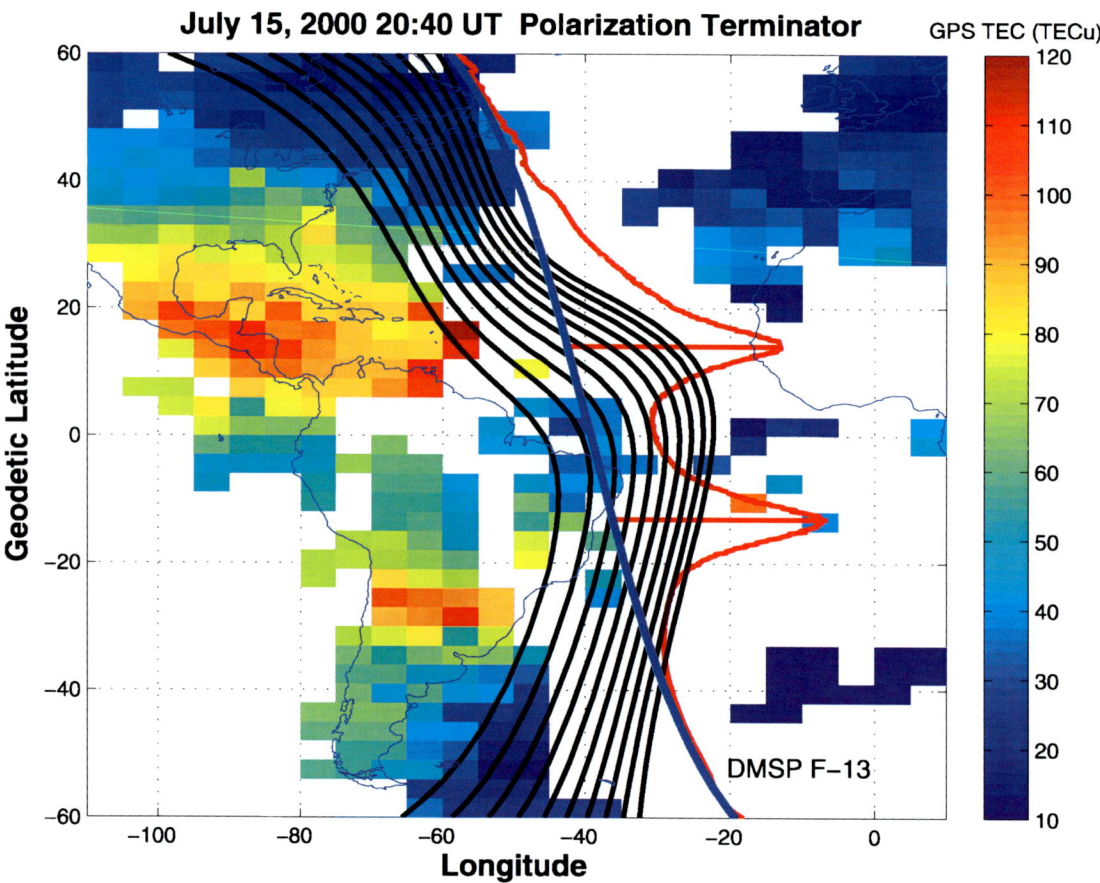

Plate 8. Ground-based GPS observations of vertical TEC at 2040 UT on 15 July 2000 are shown in geodetic coordinates, during the initial stages of a strong enhancement of TEC (>200 TECU) in the American sector. Superimposed is a family of contours that indicates the two-dimensional extent and orientation of the polarization terminator conductivity-gradient region at this time (see text). The trajectory of the DMSP F-13 satellite is shown as a heavy blue line. Superimposed on the orbital track in red is the in situ observation of plasma density at 830-km altitude (from Foster and Erickson, submitted manuscript, 2008).

S and N hemisphere ends of the field lines, make the PT vary significantly from a simple N–S orientation. ExB plasma convection is parallel to the PT contours, indicating plasma redistribution westward and poleward from the vicinity of the EIA crests to offshore the Americas. The S–N trajectory of the DMSP F-13 satellite across the region of the SAA is shown as a heavy blue line. Superimposed on the orbital track is the in situ observation of plasma density observed by DMSP at 830-km altitude. Intense, spread EIA crests are seen at 15°N and 13°S latitude as DMSP crosses the PT.

Velocity observations with DMSP F-13 reported by Foster and Erickson (submitted manuscript, 2008) indicate that the enhanced EIA plasma shown in Plate 8 is being swept westward in geographic coordinates at exactly the speed of the terminator. The PT acts like a standing wave in inertial coordinates, with the earth rotating under it, and with the plasma of the EIA crests surf-riding the terminator wave westward and poleward across the Atlantic sector. The PT effect results in the pronounced magnetically conjugate enhancement of TEC shown in Plate 5.

In this manner, the particular configuration of the magnetic field in the Atlantic sector due to the offset of the poles and declination effects near the SAA creates a preferred longitude/Universal Time sector (western Atlantic/21 UT) for the buildup of enhanced TEC on field lines inside the dusk plasmapause. The electric fields associated with the PT sweep up the plasmas of the EIA crests from the Atlantic sector and redistribute it to the base of the erosion plumes and into the mid-latitude SAPS channels, contributing to the strength of the high total content SED plumes observed during strong storms in the American sector. This effect is strongest for northern hemisphere summer conditions, as experienced during the 15–16 July 2000 superstorm.

10. SUMMARY AND CONCLUSIONS

Plasma redistribution during large storms is a multistep systemwide process involving and coupling processes and regions at equatorial, low, mid, auroral, and polar latitudes acting under the influence of storm-time drivers of the coupled system. Penetration electric fields enhance the uplift and redistribution of the equatorial ionosphere to form enhancements of the equatorial ionization anomaly peaks at low latitudes. Polarization electric field effects at the dusk terminator act to redistribute this elevated TEC further in both longitude and latitude. Plasma redistribution at the dusk PT is most pronounced in northern hemisphere summer conditions and in the western Atlantic sector. This creates a preferred longitude for the formation of TEC enhancements at low mid latitudes at the base of the erosion plumes in the PBL.

Energetic particle injection into the storm-time ring current leads to the formation of strong poleward-directed SAPS electric fields in the evening to postnoon sector as field-aligned currents close through the low-conductivity subauroral ionosphere. The SAPS electric fields overlap the outer plasmasphere and the TEC enhancements formed by low-latitude redistribution effects drawing out plumes of SED. Convection in the SAPS channel transports the eroded material to the noontime cusp in the ionosphere and to the dayside magnetopause at high altitudes. These greatly enhanced fluxes of cold plasma traverse the cusp and enter the polar cap forming the polar tongue of ionization and providing a rich source of heavy ions for the magnetospheric injection and acceleration mechanisms that operate in these regions.

Acknowledgments. Partial support for this work at the MIT Haystack Observatory was provided by NASA research Grants NAG5-12875 and NNX06AB86G, and NSF Cooperative Agreement ATM-0233230 with the Massachusetts Institute of Technology. The contributions of Anthea Coster, Phil Erickson, Bill Rideout, and Fred Rich to this research are gratefully acknowledged.

REFERENCES

Anderson, P. C., D. L. Carpenter, K. Tsuruda, T. Mukai, and F. J. Rich (2001), Multisatellite observations of rapid subauroral ion drifts (SAID), *J. Geophys. Res., 106*, 29,585–29,599.

Basu, S., Sa. Basu, K. M. Groves, H. C. Yeh, F. J. Rich, P. J. Sultan, and M. J. Keskinen (2001), Response of the equatorial ionosphere in the South Atlantic Region to the great magnetic storm of July 15, 2000, *Geophys. Res. Lett., 28*, 3577–3580.

Basu S., Su. Basu, F. J. Rich, K. M. Groves, E. MacKenzie, C. Coker, Y. Sahai, P. R. Fagundes, and F. Becker-Guedes (2007), Response of the equatorial ionosphere at dusk to penetration electric fields during intense magnetic storms, *J. Geophys. Res., 112*, A08308, doi:10.1029/2006JA012192.

Burke, W. J., N. C. Maynard, M. P. Hagan, R. A. Wolf, G. R. Wilson, L. C. Gentile, M. S. Gussenhoven, C. Y. Huang, T. W. Garner, and F. J. Rich (1998), Electrodynamics of the inner magnetosphere observed in the dusk sector by CRRES and DMSP during the magnetic storm of June 4–6, 1991, *J. Geophys. Res., 103*, 29,399–29,418.

Carpenter, D. L., and J. Lemaire (2004), The plasmasphere boundary layer, *Ann. Geophys., 22*, 4291.

Chen, A. J., and J. M. Grebowsky (1974), Plasma tail interpretations of pronounced detached plasma regions measured by OGO 5, *J. Geophys. Res., 79*, 3851–3855.

Chi, P. J., C. T. Russell, J. C. Foster, M. B. Moldwin, M. J. Engebretson, and I. R. Mann (2005), Density enhancement in the plasmasphere–ionosphere plasma during the 2003 Halloween Superstorm: Observations along the 330th meridian in

North America, *Geophys. Res. Lett., 32*, L03S07, doi:10.1029/2004GL021722.

Coster, A. J., J. Foster, and P. Erickson (2003), Monitoring the ionosphere with GPS: Space weather, *GPS World*, *14*(5), 42–49.

Evans, J. V. (1973), The causes of storm-time increases of the F-layer at mid-latitudes, *J. Atmos. Terr. Phys., 35*, 593.

Foster, J. C. (1993), Storm-time plasma transport at middle and high latitudes, *J. Geophys. Res., 98*, 1675–1689.

Foster, J. C., and W. J. Burke (2002), SAPS: A new categorization for sub-auroral electric fields, *Eos Trans. AGU, 83*, 393–394.

Foster, J. C., and A. J. Coster (2007), Conjugate localized enhancement of total electron content at low latitudes in the American sector, *J. Atmos. Sol. Terr. Phys., 69*, 1241–1252.

Foster, J. C., and F. J. Rich (1998), Prompt midlatitude electric field effects during severe geomagnetic storms, *J. Geophys. Res., 103*, 26,367–26,372.

Foster, J. C., and W. Rideout (2005), Midlatitude TEC enhancements during the October 2003 superstorm, *Geophys Res. Lett., 32*, L12S04, doi:10.1029/2004GL021719.

Foster, J. C., and W. Rideout (2007), Storm enhanced density: Magnetic conjugacy effects, *Ann. Geophys., 25*, 1791–1799.

Foster, J. C., and H. B. Vo (2002), Average characteristics and activity dependence of the subauroral polarization stream, *J. Geophys. Res., 107*(A12), 1475, doi:10.1029/2002JA009409.

Foster, J. C., A. J. Coster, P. J. Erickson, J. Goldstein, and F. J. Rich (2002), Ionospheric signatures of plasmaspheric tails, *Geophys. Res. Lett., 29*(13), 1623, doi:10.1029/2002GL015067.

Foster, J. C., A. J. Coster, P. J. Erickson, F. J. Rich, and B. R. Sandel (2004), Stormtime observations of the flux of plasmaspheric ions to the dayside cusp/magnetopause, *Geophys. Res. Lett., 31*, L08809, doi:10.1029/2004GL020082.

Foster, J. C., A. J. Coster, P. J. Erickson, W. Rideout, F. J. Rich, T. J. Immel, and B. R. Sandel (2005a), Redistribution of the storm-time ionosphere and the formation of the plasmaspheric bulge, in *Inner Magnetosphere Interactions: New Perspectives from Imaging*, edited by J. Burch, and M. Schultz, pp. 277–289, AGU Press, Washington, D. C.

Foster, J. C., A. J. Coster, P. J. Erickson, J. M. Holt, F. D. Lind, W. Rideout, M. McCready, A. van Eyken, R. J. Barnes, R. A. Greenwald, and F. J. Rich (2005b), Multiradar observations of the polar tongue of ionization, *J. Geophys. Res., 110*, A09S31, doi:10.1029/2004JA010928.

Foster, J. C., W. Rideout, B. Sandel, W. T. Forrester, and F. J. Rich (2007), On the relationship of SAPS to storm enhanced density, *J. Atmos. Sol. Terr. Phys., 69*, 303–313.

Greenspan, M. E., C. E. Rasmussen, W. J. Burke, and M. A. Abdu (1991), Equatorial density depletions observed at 840 km during the great magnetic storm of March 1989, *J. Geophys. Res., 96*, 13,931–13,942.

Kelley, M. C., M. Vlassov, J. C. Foster, and A. J. Coster (2004), A quantitative explanation for the phenomenon known as storm-enhanced density, *Geophys. Res. Lett., 31*, L19809, doi:10.1029/2004GL020875.

Lin, C. S., and H. C. Yeh (2005), Satellite observations of electric fields in the South Atlantic anomaly region during the July 2000 magnetic storm, *J. Geophys. Res., 110*, A03305, doi:10.1029/2003JA010215.

Mannucci, A. J., B. T. Tsurutani, B. A. Iijima, A. Komjathy, A. Saito, W. D. Gonzales, F. L. Guarnieri, J. U. Kozyra, and R. Skoug (2005a), Dayside global ionospheric response to the major interplanetary events of October 29–30, 2003 "Halloween Storm," *Geophys. Res. Lett., 32*, L12S02, doi:10.1029/2004GL021467.

Mannucci, A. J., S. Datta-Barua, T. Walter, A. Komjathy, L. Sparks, and B. T. Tsurutani (2005b), Anomalous nighttime plasma structure in the recovery phase of a superstorm, *Eos Trans. AGU, 86*(52), Fall Meet. Suppl., Abstract SA21A-0275.

Maynard, N. C., T. L. Aggson, and J. P. Heppner (1983), The plasmaspheric electric field as measured by ISEE 1, *J. Geophys. Res., 88*, 3991–4003.

Mendillo, M. (1973), A study of the relationship between geomagnetic storms and ionospheric disturbance at mid-latitudes, *Planet. Space Sci., 21*, 349.

Mendillo, M. (2006), Storms in the ionosphere: Patterns and processes for total electron content, *Rev. Geophys., 44*, RG4001, doi:10.1029/2005RG000193.

Rowland, D. E., and J. R. Wygant (1998), Dependence of the large scale, inner magnetospheric electric field on geomagnetic activity, *J. Geophys. Res., 103*, 14,959–14,964.

Sandel, B. R., R. A. King, W. T. Forrester, D. L. Gallagher, A. L. Broadfoot, and C. C. Curtis (2001), Initial results from the IMAGE extreme ultraviolet imager, *Geophys. Res. Lett., 28*, 1439–1442.

Su, Y.-J., M. F. Thomsen, J. E. Borovsky, and J. C. Foster (2001), A linkage between polar patches and plasmaspheric drainage plumes, *Geophys. Res. Lett., 28*, 111–113.

Tsurutani, B. T., et al. (2004), Global dayside ionospheric uplift and enhancement associated with interplanetary electric fields, *J. Geophys. Res., 109*, A08302, doi:10.1029/2003JA010342.

Tsurutani, B. T., et al. (2006), Extreme solar EUV flares and ICMEs and resultant extreme ionospheric effects: Comparison of the Halloween 2003 and the Bastille Day events, *Radio Sci., 41*, RS5S07, doi:10.1029/2005RS003331.

Tsyganenko, N. A. (2002). A model of the near magnetosphere with a dawn–dusk asymmetry: 1. Mathematical structure, *J. Geophys. Res., 107*(A8), 1179, doi:10.1029/2001JA00219.

Vo, H. B, and J. C. Foster (2001), A quantitative study of ionospheric density gradients at mid-latitudes, *J. Geophys. Res. 106*, 21,555–21,563.

Yeh, H.-C., J. C. Foster, F. J. Rich, and W. Swider (1991), Storm-time electric field penetration observed at mid-latitude, *J. Geophys. Res., 96*, 5707–5721.

Yin, P., C. N. Mitchell, P. S. J. Spencer, J. C. Foster (2004), Ionospheric electron concentration imaging using GPS over the USA during the storm of July 2000, *Geophys. Res. Lett., 31*, L12806, doi:10.1029/2004GL019899.

J. C. Foster, MIT Haystack Observatory, Off Route 40, Westford, MA 01886, USA. (jfoster@haystack.mit.edu)

The Linkage Between the Ring Current and the Ionosphere System

P. C. Brandt, Y. Zheng, and T. S. Sotirelis

The Johns Hopkins University Applied Physics Laboratory, Laurel, Maryland, USA

K. Oksavik

The University Center in Svalbard, Longyearbyen, Norway

F. J. Rich

Air Force Research Laboratory, Acton, Massachusetts, USA

The coupling between the ring current and the ionosphere is briefly reviewed and discussed. Given global energetic neutral atom (ENA) observations of the ring current, the three-dimensional current system driven by ring current plasma pressure (the region 2 system) is derived to illustrate where the ring current connects to the ionosphere. Special attention is given to how the ring current and ionospheric conductance set up the sub-auroral polarization streams (SAPS) through the closure of the region 2 current through the ionospheric trough region. Simultaneous ENA observations of the ring current and radar observations of the SAPS flow show that the onset of SAPS flow and equatorward motion is coincident with the injection and buildup of plasma pressure in the inner magnetosphere. The comprehensive ring current model is used to demonstrate how the SAPS can be generated by computing ring current pressure and allowing its pressure-driven currents to close through a model of ionospheric conductance including the trough region. The paper ends by discussing open questions and what is missing in our understanding of the generation of the large-scale electric fields of the inner magnetosphere and sub-auroral ionosphere.

1. INTRODUCTION

Phenomena in near-Earth space stem from an intrinsically coupled system of plasmas and fields. To understand phenomena in the magnetosphere, ionosphere, and even the thermosphere, we must understand how these systems work together to produce the phenomena we observe. The plasma pressure of the ring current is by far the largest source of inner magnetospheric currents (region 2 current system) that couple through the ionosphere and is therefore critical to the magnetosphere–ionosphere coupling process. The closure of the magnetospheric currents through the ionospheric conductance pattern give rise to the large-scale electric field of the magnetosphere as well as the ionosphere and can therefore be viewed as an important input to thermospheric phenomena. The picture is further complicated by the fact that the thermosphere modifies ionospheric conductance.

The purpose of this paper is to give a brief review of our knowledge on how the ring current couples to the ionosphere and the effects that arise. We begin by giving a brief background to storm-time ring current dynamics in section 2. In section 3, we discuss the global picture of the ring current–ionosphere coupling, and continue presenting observations on the effects of this coupling, as exemplified by the subauroral polarization stream (SAPS).

2. RING CURRENT DYNAMICS

Historically, the ring current has been thought of as a ring of electrical current encircling the Earth. However, recent global energetic neutral atom (ENA) observations of the ring current by the IMAGE mission [*Burch*, 2000] have noted that the storm-time ring current is highly asymmetric with its center on the night side [*Brandt et al.*, 2002]. Observations by IMAGE have enabled investigations of the global dynamics of not only the ring current through ENA imaging but also of the plasmasphere [*Goldstein et al.*, 2005], which can be used as a tracer of the electric field of the inner magnetosphere.

Plate 1 demonstrates how dynamic the ring current really is. The ENA image sequence was obtained via the high energy neutral atom (HENA) imager [*Mitchell et al.*, 2000] on 24 November 2001 during a transition from a geomagnetic storm main phase to a recovery phase. The hydrogen ENA images are integrated for 20 min and are obtained in the 60- to 119-keV range, which represents the bulk of the plasma pressure in the ring current.

The peak of the storm was reached at about 1444 UTC and the southward interplanetary magnetic field (IMF) turned northward at Earth at about 1500 UTC. Within minutes, the ring current began to respond by becoming more symmetric. This is due to the different drift patterns in the magnetosphere during northward IMF, during which the dominating drift is curvature and gradient drift. Drift trajectories then circle the Earth and energy dispersion, and charge exchange decay helps in azimuthally smoothing out spatial gradients in the ring current plasma pressure.

At 1618 UTC the ring current had already become significantly symmetric, and at 1752 UTC it was almost entirely symmetric. Note that the *SYM-H* index is very negative large throughout this transition, which justifies strong caution that *SYM-H* does not always represent the state of the ring current accurately. The clear conclusion of these observations is that the ring current is *not* a ring during the driving phase of the storm [*Brandt et al.*, 2002; *Ebihara and Ejiri*, 2000; *Ebihara et al.*, 2002].

3. COUPLING AND ITS EFFECTS

The key quantity of the ring current is the plasma pressure carried mostly by protons and O^+ as shown by *Krimigis et al.* [1985]. The plasma pressure gradients are responsible for most of the electric current in the ring current. Contrary to some literature, the *equatorial* magnetospheric ring current is *not* dominated by the ions curvature and gradient drifting westward and the electrons eastward. Although this does give rise to a net westward current, it is a negligible effect on the currents perpendicular to the magnetic field for an isotropic plasma pressure distribution. Under such condition, the perpendicular current is dominated by the gradients of the perpendicular component of the pressure (the so-called diamagnetic current). See, for example, *Parker* [1957] for a complete derivation from single particle motion to electrical currents. One can show that the basic force balance equation for the ring current is

$$\mathbf{J} \times \mathbf{B} = \nabla \cdot \mathbf{P}. \tag{1}$$

Note that this equation only gives us information about the currents perpendicular to the magnetic field. To visualize the electrical current flow lines of the inner magnetosphere, one can assume current continuity $\nabla \cdot \mathbf{J} = 0$. Combining this with the force balance equation (1), one may derive the three-dimensional system of currents connecting to the ionosphere. For a more detailed treatment, see *Roelof* [1987].

The result of this exercise is displayed in Plate 2. The storm-time ring current plasma pressure is displayed in the magnetic equator and is distributed around the midnight sector. The pressure is retrieved from HENA images by using a constrained linear inversion algorithm [*DeMajistre et al.*, 2004]. The electric current flow lines have been plotted for one iso-pressure surface and are shown as dashed lines. The complete ring current system is built up of an infinite number of such iso-pressure surfaces and is commonly known as the region 2 current system.

In the equator, the current flow lines close on themselves and there are no field-aligned currents (FAC) flowing. The magnetospheric equatorial current is large and westward on the outer edge of the ring current pressure peak, but is much smaller and eastward on the inside edge of the ring current pressure. At higher latitudes, the FAC component increase and the current flow lines start to connect to the ionosphere. Currents flow into the ionosphere on the dusk side and up through the dawn side sector.

Note that the solution does permit a constant interhemispheric FAC that we have set to zero for simplicity. Independent in situ measurements confirm that the FAC density is indeed small near the equatorial plane and that there may be an interhemispheric current present [*Vallat et al.*, 2005].

In the dusk and dawn side ionosphere, the Pedersen conductance dominates so that the region 2 FAC currents close poleward (equatorward) at dusk (dawn) through the region

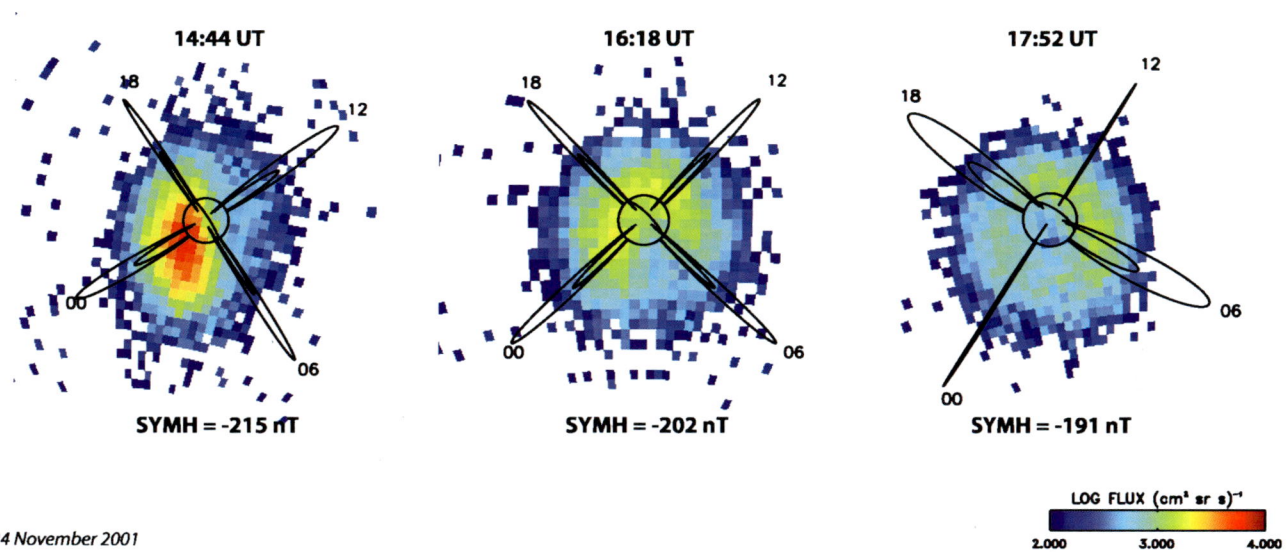

Plate 1. The ENA image sequence obtained by the high energy neutral atom (HENA) imager on board the IMAGE spacecraft. The sequence was obtained during a transition from the storm main phase to the storm recovery phase. The interplanetary magnetic field turned from southward to northward at about 1500 UTC and within minutes the ring current responded by becoming more symmetric.

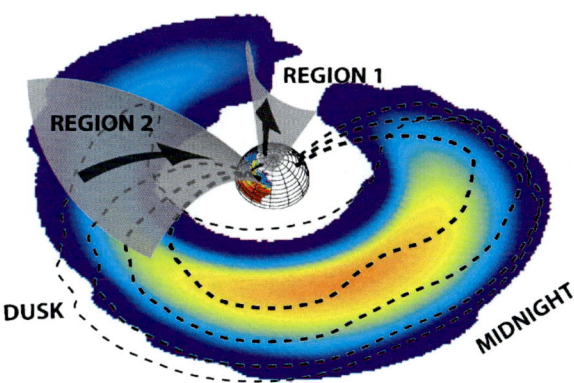

Plate 2. The ring current plasma pressure drives an electrical current system ("region 2") that closes through the dusk and dawn side ionosphere during storms. This global electrical circuit is a significant player in producing the sub-auroral polarization streams (SAPS) in the dusk side ionosphere, and is also responsible for modifying the electric field of the inner magnetosphere.

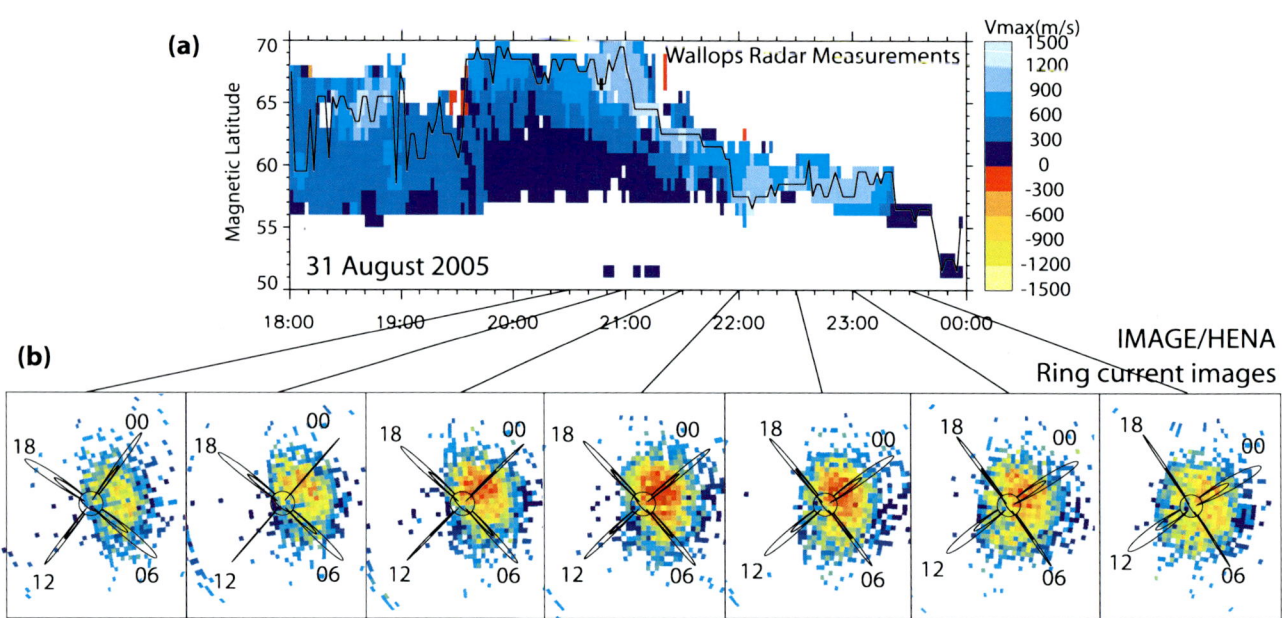

Plate 3. (a) The color coded line-of-sight (LOS) velocities as a function of magnetic latitude and UTC. The black line marks maximum LOS velocity. (b) The HENA image sequence in the 60- to 198-keV range.

1 current system. Let us here focus on what happens on the dusk side. Here, the ionospheric currents close through the low-conductance ionospheric trough region, which sets up a poleward electric field associated with enhanced ionospheric plasma flow velocities ($\mathbf{E} = -\mathbf{v} \times \mathbf{B}$) commonly referred to as SAPS [*Foster and Burke*, 2002].

The trough is present at quiet times and is formed through the stagnation of plasma in the region where the two-cell convection speed meets the corotating ionospheric plasma [*Spiro et al.*, 1978]. The result is an increased recombination rate and extremely low conductances (tenths of a mho). As convection increases due to a southward IMF, the two-cell convection pattern expands to lower latitudes and pushes the trough region with it.

There are numerous observations of the enhanced flow velocities associated with the SAPS phenomena made from both in situ spacecraft and remote radar observations [*Foster and Burke*, 2002; *Anderson et al.*, 2001]. Now with the addition of global ENA data and modeling efforts, a more complete picture of the SAPS mechanism is beginning to emerge. The ring current closure through the trough region plays a central role in producing the high-speed flows. On one hand, for a given trough conductance and location, the ring current strength determines the speed of the SAPS flow. On the other hand, *Anderson* [2004] demonstrated that the relative location of the trough and the aurora plays a significant role, which can be intuitively explained since their relative locations determine the distance the ionospheric current has to flow, which directly relates to the total resistance. Also, *Brandt et al.* [2005] demonstrated that the relative location may be the controlling effect of the SAPS velocity more than the FAC current density.

Plate 3 shows remote radar measurements of the flow velocities with simultaneous global ring current observations, suggesting that the ring current plays an active role in the formation and evolution of the SAPS. Plate 3a shows the line-of-sight (LOS) velocities of the SAPS flow during an event on 31 August 2005 (black line). The measurements were obtained by the SuperDARN radar facility on Wallops Island, VA [*Oksavik et al.*, 2006]. Plate 3a shows color-coded LOS velocities versus magnetic latitude and UTC. Plate 3b shows the corresponding HENA images for this period. All HENA images were integrated for 20 min. It is also possible to derive the actual three-dimensional current system for this period [*Roelof et al.*, 2004], but that is beyond the scope of this paper.

The ring current shows a sharp increase just before the enhancement of the SAPS flow velocities at about 2040 UTC. The SAPS velocity is maintained at a high level and the magnetic latitude of the SAPS channel decreases as ring current intensity continues to increase. *Foster and Burke* [2002] found that the magnetic latitude of the SAPS flow velocity decreases with increasing magnetic local time. We cannot rule out that the observed decrease in latitude is a spatial effect. There are other temporal effects that may cause the same phenomena. First of all, the auroral oval moves to lower latitudes during storms. Second, the ring current moves closer to Earth during storms, which would cause the region 2 FAC system to move to lower latitudes. At this point, we cannot separate these effects in the observations.

At about 2200 UTC the ring current intensity decreases, which is immediately reflected in the magnetic latitude of the SAPS, but not as much in the SAPS velocity. We cannot rule out that the decrease in magnetic latitude of the peak of SAPS flow velocity is a spatial effect as found by *Foster and Burke* [2002]. Keep in mind here that the plotted velocities are not the absolute SAPS velocity, but rather the LOS velocities derived from the radar measurements.

4. MODELING

Attempts to model the SAPS formation are not new. The Rice Convection Model (RCM) [*Wolf*, 1970; *Southwood and Wolf*, 1978] was among the first to illustrate the importance of ionospheric conductivity on magnetospheric and ionospheric flows. Other attempts include those reported by *Pintér et al.* [2006], *Lyatsky et al.* [2006], *Garner et al.* [2004], and *Harel et al.* [1981].

Today, most ring current models self-consistently compute the electric field in the ionosphere and magnetosphere by solving Poisson's equation for the pressure-driven currents closing through the ionosphere [*Fok et al.*, 2001; *Liemohn et al.*, 2006]. The equation provides the electric potential distribution in the ionosphere, which is assumed to be the same in the magnetosphere. The ionospheric and magnetospheric flow velocities are then simply computed as the $E \times B$ velocity. There are also recent important efforts to develop ring current models with a self-consistent magnetic field by computing the magnetic field deformation due to the pressure-driven currents [*Zaharia et al.*, 2006].

The comprehensive ring current model (CRCM) was developed by *Fok et al.* [2001] and is a further development of the RCM in that it computes the pitch-angle distributions by using the bounce averaged Boltzmann equation. The ionospheric conductance usually consists of three parts: (1) dayside conductance due to solar illumination, (2) nightside background conductance, and (3) auroral conductance. These conductances have usually been determined through statistical models such as the auroral conductance model by *Hardy et al.* [1985]. Figure 1a illustrates the magnetospheric electric fields resulting from a simple super position of the dawn to dusk electric field and the corotational electric field,

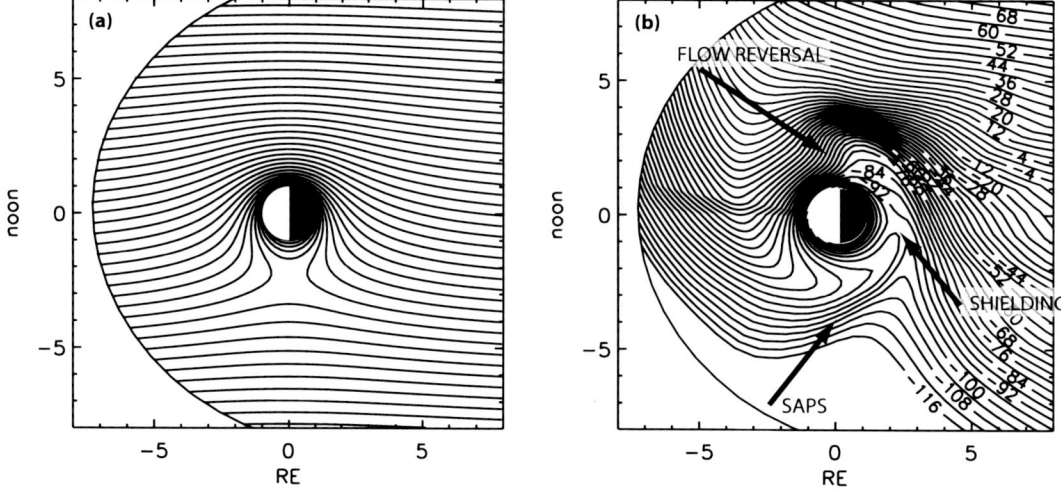

Figure 1. Closure of the pressure-driven current system through the ionosphere makes a dramatic difference on the magnetospheric and ionospheric electric field. (a) Electric potential pattern at the magnetic equator without current closure through the ionosphere. (b) Potential pattern from a typical comprehensive ring current model (CRCM) run (12 August 2000) allowing for current closure through the ionosphere.

representing the knowledge of the global distribution of storm-time electric fields of the inner magnetosphere before 1970. Figure 1b illustrates a typical storm-time CRCM run and demonstrates how a dramatically different electric field pattern arises when the model computes the electric field self-consistently. Even if the common statistical conductance models are heavily averaged and do not resolve the ionospheric trough region, the presence of an auroral oval on top of a lower background conductance reproduces a slightly enhanced SAPS flow velocity (200 m/s) at the equatorward edge of the auroral oval. Strong ionospheric and magnetospheric flows and electric fields are also known to exist on the dawn side [*Ebihara et al.*, 2005].

Zheng et al. [2008] implemented a *Kp*-driven empirical model of the ionospheric trough conductance [*Spiro et al.*, 1978] in the CRCM to determine the importance of the trough to SAPS formation for a storm period on 12 June 2005. In the present paper we have set the absolute conductance of the trough minimum to estimations from actual measurements by the Millstone Hill Incoherent Scatter Radar (P. Erickson and J. Foster, personal communication). Plate 4 summarizes the CRCM results and compares them to DMSP observations. Note that the CRCM model runs used a *Tsyganenko* [1995] model, which is strictly speaking a quiet-time magnetic field model. More accurate model runs will use a magnetic field model such as the *Tsyganenko and Sitnov* [2007] magnetic field model. We find that the following characteristics are important.

4.1. SAPS Velocity

With previous lack of realistic trough conductances, the CRCM reproduced much smaller SAPS flow velocities, and the region 2 FAC appeared slightly equatorward of the max SAPS flow. With the realistic trough conductances (dotted blue line in Plate 4), SAPS total flow velocities increase to about 2000 m/s (dotted red line), whereas the observed maximum cross-track velocity (solid red line) was 1400 m/s.

4.2. SAPS Location

The CRCM results clearly show that downward (positive) region 2 FAC (dotted black line), trough conductance (dotted blue line), and SAPS flow maximum (dotted red line) all coincide. The DSMP observations show a similar behavior where the region 2 FAC (gradient-filled boxed) coincides with the max of the SAPS flow maximum (solid red line). The background magnetic field has been subtracted to yield the plotted observed residual magnetic field (solid black line in the lower panel). We assume here that the FAC is sheet-like and extends in the east–west direction. The center of the current sheet will then be at the location where the residual magnetic field changes sign. No measured trough conductances were directly available for this date. However, from the total precipitating electron intensities (solid blue line in lower panel), we see that the observed SAPS flow maximum sits directly at the equatorward edge of the electron

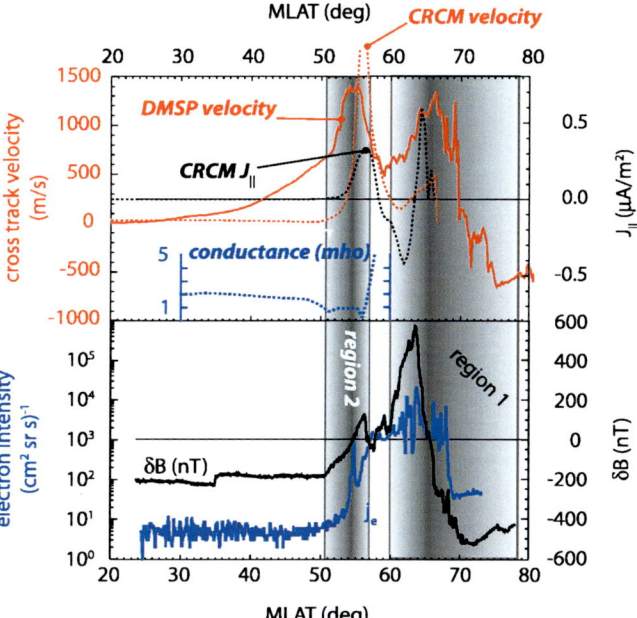

Plate 4. The location of the trough, the SAPS flow, and the region 2 field-aligned currents coincide. CRCM results are shown as dotted lines and DMSP measurements as solid lines.

precipitation, which is consistent with the location of the ionospheric trough.

5. DISCUSSION

Near-Earth space is an intrinsically coupled system of plasma and fields in the magnetosphere, ionosphere, and thermosphere. Although several sophisticated models can reproduce, at least qualitatively, phenomena such as the large-scale electric fields in the sub-auroral ionosphere and inner magnetosphere, it does not mean we understand how the different regions and regimes of space work together in reality to produce the observed phenomena. The SAPS phenomena is an excellent illustration: The modeled SAPS flow is reproduced by closing a current through an ionospheric conductance pattern with no feedback from the thermosphere. Yet, it is known that ionospheric conductance depends on the electric field [*Schunk and Walker*, 1973] through changes in thermospheric recombination rates. Therefore, the closure of the ring current across the ionospheric trough modifies the conductance, which in turn modifies the magnetospheric electric field. There are ongoing efforts to couple the inner magnetosphere, ionosphere, and thermosphere to understand this intrinsic coupling.

Furthermore, existing models of the SAPS flows are computed from the closure of currents through the ionosphere. However, we must not forget that the solar–wind magnetosphere interaction results in circulation of magnetized plasma in the magnetosphere, which imposes the two-cell convection pattern in the ionosphere. The ionospheric flow speeds depend on the ionospheric conductance. For a high ionospheric conductance, the convecting magnetic flux tubes will move slowly through the ionosphere (and speed up the plasma flow). At the lack of any significant conductance, flux tubes are free to move at high speeds (determined by the flow speeds in the magnetosphere). Therefore, the mere presence of an imposed flow and a low trough conductance will naturally result in a region with enhanced flows such as the SAPS flow. The SAPS electric field is then simply a consequence of the flow [*Vasyliūnas*, 2001], and the current system is set up to maintain the proper charge balance in accordance with the flow-induced electric field. This illustrates that the fundamental generation mechanism of the SAPS flow is a result of the complicated interaction between the magnetosphere and the ionosphere. Our interpretation, as manifested through the CRCM, represents a simplification of the entire system.

We have discussed ionospheric closure of the ring current as a way to explain the generation of SAPS flows. However, there are other important factors that impact the generation of the SAPS flow that we have not discussed in detail here. In summary, there are at least three factors that work together to create the large-scale electric field (or flow) of the inner magnetosphere/sub-auroral ionosphere:

1. *Flows*. The flows imposed on the ionosphere by magnetospheric convection depend on the ionospheric conductance. There are also thermospheric (neutral) wind-induced plasma flows that contribute to the ionospheric flows. It is still an open question how SAPS is affected by these flows.

2. *Currents*. The strength and morphology of the region 2 current system and where it closes through the ionosphere is critical in determining the ionospheric electric fields and flows. At present, this is the way most models compute the ionospheric electric fields due to ring current coupling. In a steady state, the currents, conductance, fields, and flows have to be consistent regardless of what drives what.

3. *Conductance*. It is clear that ionospheric conductance is of critical importance to the electric field, especially if it the conductance is low, as in the ionospheric trough region. In turn, changes in the thermosphere alter the ionospheric conductance in response to the flows and fields.

Acknowledgment. This work was supported by NASA grant NNX06AC29G.

REFERENCES

Anderson, P. C. (2004), Subauroral electric fields and magnetospheric convection during the April 2002 geomagnetic storms, *Geophys. Res. Lett.*, *31*, L11801, doi:10.1029/2004GL019588.

Anderson, P. C., D. L. Carpenter, K. Tsuruda, T. Mukai, and F. J. Rich (2001), Multisatellite observations of rapid subauroral ion drifts (SAID), *J. Geophys. Res.*, *106*(A12), 29,585–29,599.

Brandt, P. C., S. Ohtani, D. G. Mitchell, M.-C. Fok, E. C. Roelof, and R. DeMajistre (2002), Global ENA observations of the storm mainphase ring current: Implications for skewed electric fields in the inner magnetosphere, *Geophys. Res. Lett.*, *29*(20), 1954, doi:10.1029/2002GL015160.

Brandt, P. C., J. Goldstein, P. C. Anderson, B. J. Anderson, R. DeMajistre, E. C. Roelof, and D. G. Mitchell (2005), On the relation between sub-auroral electric fields, the ring current and the plasmasphere, *AGU Geophys. Monogr. Ser.*, *155*, 163–172.

Burch, J. L. (Ed.) (2000), *The IMAGE Mission*, Kluwer Academic, reprinted from *Space Sci. Rev.*, *91*(1–2).

DeMajistre, R., L. J. Paxton, D. Morrison, J.-H. Yee, L. P. Goncharenko, and A. B. Christensen (2004), Retrievals of nighttime electron density from thermosphere ionosphere mesosphere energetics and dynamics (timed) mission global ultraviolet imager (GUVI) measurements, *J. Geophys. Res.*, *109*, A05305, doi:10.1029/2003JA010296.

Ebihara, Y., and M. Ejiri (2000), Simulation study on the fundamental property of the storm-time ring current, *J. Geophys. Res.*, *105*, 15,843–15,859.

Ebihara, Y., M. Ejiri, H. Nilsson, I. Sandahl, A. Milillo, M. Grande, J. F. Fennell, and J. L. Roeder (2002), Statistical distribution of the storm-time proton ring current: POLAR measurements, *Geophys. Res. Lett.*, *29*(20), 1969, doi:10.1029/2002GL01540.

Ebihara, Y., M.-C. Fok, S. Sazykin, M. F. Thomsen, M. R. Hairston, D. S. Evans, F. J. Rich, and M. Ejiri (2005), Ring current and the magnetosphere–ionosphere coupling during the superstorm of 20 November 2003, *J. Geophys. Res.*, *110*, A09S22, doi:10.1029/2004JA010924.

Fok, M. C., R. A. Wolf, R. W. Spiro, and T. E. Moore (2001), Comprehensive computational model of Earth's ring current, *J. Geophys. Res.*, *106*(A5), 8417–8424.

Foster, J. C., and W. J. Burke (2002), SAPS: A new categorization for sub-auroral electric fields, *Eos Trans. AGU*, *83*(36), 393.

Garner, T. W., R. A. Wolf, R. W. Spiro, W. J. Burke, B. G. Fejer, S. Sazykin, J. L. Roeder, and M. R. Hairston (2004), Magnetospheric electric fields and plasma sheet injection to low L-shells during the 4–5 June 1991 magnetic storm: Comparison between the Rice Convection Model and observations, *J. Geophys. Res.*, *109*, A02214, doi:10.1029/2003JA010208.

Goldstein, J., J. L. Burch, B. R. Sandel, S. B. Mende, P. C. Brandt, and M. R. Hairston (2005), Coupled response of the inner magnetosphere and ionosphere on 17 April 2002, *J. Geophys. Res.*, *110*, A03205, doi:10.1029/2004JA010712.

Hardy, D. A., M. S. Gussenhoven, and E. Holeman (1985), A statistical model of auroral electron precipitation, *J. Geophys. Res.*, *90*(A5), 4229–4248.

Harel, M., R. A. Wolf, P. H. Reiff, R. W. Spiro, W. J. Burke, F. J. Rich, and M. Smiddy (1981), Quantitative simulation of a magnetospheric substorm: I. Model logic and overview, *J. Geophys. Res.*, *86*, 2217–2241.

Krimigis, S. M., G. Gloeckler, R. W. McEntire, T. A. Potemra, F. L. Scarf, and E. G. Shelley (1985), Magnetic storm of September 4, 1984: A synthesis of ring current spectra and energy densities measured with AMPTE/CCE, *Geophys. Res. Lett.*, *12*(5), 329–332.

Liemohn, M. W., A. J. Ridley, J. U. Kozyra, D. Gallagher, M. Thomsen, M. Henderson, M. Denton, P. C. Brandt, and J. Goldstein (2006), Analyzing electric field morphology through data-model comparisons of the geospace environment modeling inner magnetosphere/storm assessment challenge events, *J. Geophys. Res.*, *111*, A11S11, doi:10.1029/2006JA011700.

Lyatsky, W., A. Tan, and G. V. Khazanov (2006), A simple analytical model for subauroral polarization stream (SAPS), *Geophys. Res. Lett.*, *33*, L19101, doi:10.1029/2006GL025949.

Mitchell, D. G., et al. (2000), High energy neutral atom (HENA) imager for the IMAGE mission, *Space Sci. Rev.*, *91*, 67–112.

Oksavik, K., R. A. Greenwald, J. M. Ruohoniemi, M. R. Hairston, L. J. Paxton, J. B. H. Baker, J. W. Gjerloev, and R. J. Barnes (2006), First observations of the temporal/spatial variation of the sub-auroral polarization stream from the SuperDARN Wallops HF radar, *Geophys. Res. Lett.*, *33*, L12104, doi:10.1029/2006GL026256.

Parker, E. N. (1957), Newtonian development of the dynamical properties of ionized gases of low density, *Phys. Rev.*, *107*(4), 924–933.

Pintér, B., S. D. Thom, R. Balthazor, H. Vo, and G. J. Bailey (2006), Modeling subauroral polarization streams equatorward of the plasmapause footprints, *J. Geophys. Res.*, *111*, A10303, doi:10.1029/2005JA011457.

Roelof, E. C. (1987), Energetic neutral atom image of a storm-time ring current, *Geophys. Res. Lett.*, *14*, 652–655.

Roelof, E. C., P. C. Brandt, and D. G. Mitchell (2004), Derivation of currents and diamagnetic effects from global pressure distributions obtained by IMAGE/HENA, *Adv. Space Res.*, *33*(5), 747–751, doi:10.1016/S0273-1177(03)00633-1.

Schunk, R. W., and J. C. G. Walker (1973), Theoretical ion densities in the lower ionosphere, *Planet. Space Sci.*, *21*, 1875–1896, doi:10.1016/0032-0633(73)90118-9.

Southwood, D. J., and R. A. Wolf (1978), An assessment of the role of precipitation in magnetospheric convection, *J. Geophys. Res.*, *83*, 5227–5232.

Spiro, R. W., R. A. Heelis, and W. B. Hanson (1978), Ion convection and the formation of the mid-latitude F region ionization trough, *J. Geophys. Res.*, *83*, 4255–4264.

Tsyganenko, N. A. (1995), Modeling the Earth's magnetospheric magnetic field confined within a realistic magnetopause, *J. Geophys. Res.*, *100*, 5599–5612.

Tsyganenko, N. A., and M. I. Sitnov (2007), Magnetospheric configurations from a high-resolution data-based magnetic field model, *J. Geophys. Res.*, *112*, A06225, doi:10.1029/2007JA012260.

Vallat, C., et al. (2005), First current density measurements in the ring current region using simultaneous multi-spacecraft CLUSTER-FGM data, *Ann. Geophys.*, *23*, 1849–1865.

Vasyliūnas, V. M. (2001), Electric field and plasma flow: What drives what?, *Geophys. Res. Lett.*, *28*, 2177–2180, doi:10.1029/2001GL013014.

Wolf, R. A. (1970), Effects of ionospheric conductivity on convective flow of plasma in the magnetosphere, *J. Geophys. Res.*, *75*(25), 4677–4698.

Zaharia, S., V. K. Jordanova, M. F. Thomsen, and G. D. Reeves (2006), Self-consistent modeling of magnetic fields and plasmas in the inner magnetosphere: Application to a geomagnetic storm, *J. Geophys. Res.*, *111*, A11S14, doi:10.1029/2006JA011619.

Zheng, Y., P. Brandt, A. Lui, and M.-C. Fok (2008), On ionospheric trough conductance and subauroral ion drift: Simulation results, *J. Geophys. Res.*, *113*, A04209, doi:10.1029/2007JA012532.

P. C. Brandt, T. S. Sotirelis, and Y. Zheng, The Johns Hopkins University Applied Physics Laboratory, 11100 Johns Hopkins Rd, Laurel, MD 20723, USA. (pontus.brandt@jhuapl.edu)

K. Oksavik, The University Centre in Svalbard, NO-9171 Longyearbyen, Norway.

F. J. Rich, Air Force Research Laboratory, 15 Juniper Ridge Road, Acton, MA 01720, USA.

Storm Phase Dependence of Penetration of Magnetospheric Electric Fields to Mid and Low Latitudes

Takashi Kikuchi,[1,2] Kumiko K. Hashimoto,[3] and Kenro Nozaki[2]

Penetration of the magnetospheric electric fields to the equatorial ionosphere was examined using magnetometer data from high-equatorial latitudes for three geomagnetic storms characterized by the equatorial DP2 current during the main phase and the counter electrojet (CEJ) during the early recovery phase. The equatorial DP2 started simultaneously with the onset of the ring current, and continued for 2–3 h during the main phase, indicating instantaneous transmission of the convection electric field to the equator for the period of ring current development. However, the equatorial DP2 decreased its magnitude concurrently with increase in the auroral electrojet (AEJ) during the late main phase, and changed into the CEJ when the AEJ moved rapidly poleward at the beginning of the recovery phase. It is suggested that the electric field associated with the DP2 current may play a role in driving the ring current, and that the overshielding responsible for the CEJ contributed to reduce electric fields responsible for ring current development.

1. INTRODUCTION

It is well known that the convection electric field causes ionospheric currents responsible for the quasiperiodic DP2 magnetic fluctuations at high latitude and at the dayside geomagnetic equator [*Nishida et al.*, 1966; *Nishida*, 1968]. *Kikuchi et al.* [1996] demonstrated that the DP2 fluctuation occurred simultaneously at these latitude regions within the temporal resolution of 25 s, and suggested that the convection electric field was instantaneously transmitted to the equatorial ionosphere via the mid latitude. During a geomagnetic storm, strong DP2 current flowed into the mid to equatorial latitude ionosphere [*Wilson et al.*, 2001; *Tsurutani et al.*, 2004; *Huang et al.*, 2005]. *Wilson et al.* [2001] demonstrated that intensified DP2 currents were observed at mid latitudes during a major geomagnetic storm, when a significant electric field was detected by Combined Release and Radiation Effects Satellite (CRRES) inside the ring current. *Wilson et al.* [2001] suggested that the ionospheric electric field responsible for the DP2 current contributed to the development of the storm ring current.

On the other hand, the enhanced convection drives a partial ring current and the field-aligned current (FAC) builds up an electric field with an opposite direction to that of the convection electric field at low latitude [*Vasyliunas*, 1972; *Jaggi and Wolf*, 1973; *Southwood*, 1977; *Senior and Blanc*, 1984]. The time constant of this shielding electric field has been estimated as 17–20 min from magnetometer observations [*Somajajulu et al.*, 1987; *Kikuchi et al.*, 2000] and 20–30 min from theoretical calculations [*Senior and Blanc*,

---

[1] Solar–Terrestrial Environment Laboratory, Nagoya University, Nagoya, Japan.
[2] National Institute of Information and Communications Technology, Koganei, Tokyo, Japan.
[3] Kibi International University, Takahashi, Japan.

Midlatitude Ionospheric Dynamics and Disturbances
Geophysical Monograph Series 181
Copyright 2008 by the American Geophysical Union.
10.1029/181GM14

1984; *Peymirat et al.*, 2000]. During the storm, however, the shielding is not effective for many hours as suggested by *Huang et al.* [2005]. After the shielding electric field grows, the electric field at mid and low latitudes is often reversed when the convection electric field is decreased abruptly because of the northward turning of the interplanetary magnetic field (IMF) [*Rastogi and Patel*, 1975; *Kelley et al.*, 1979; *Fejer et al.*, 1979; *Gonzales et al.*, 1979; *Kobea et al.*, 2000; *Kikuchi et al.*, 2000, 2003]. The reversal of the penetrated electric field was identified as the overshielding electric field [*Kelley et al.*, 1979; *Gonzales et al.*, 1979; *Fejer et al.*, 1979] and the reversed current at the equator appears as the counter electrojet (CEJ) [*Rastogi*, 1977, 1997; *Kobea et al.*, 1998, 2000; *Kikuchi et al.*, 2000, 2003]. The reversed electric field was observed in the inner magnetosphere by CRRES during the recovery phase of the storm [*Wygant et al.*, 1998]. The reversed electric field associated with the storm was explained by means of the disturbance dynamo [*Huang et al.*, 2001].

Three questions can be raised on the relationship between the storm time electric field and ring current development; (1) Does the DP2 current play a role in ring current evolution? (2) Does the shielding work or not during the main phase of the storm? (3) What is the role of the overshielding in storm evolution? To answer these questions, we analyzed three geomagnetic storms characterized by concurrent development of the ring current and equatorial DP2, which were initiated by a solar wind shock accompanied by the southward IMF, and therefore, their onsets were determined within the temporal resolution of a few minutes. The recovery of these storms was clearly related to reduction in the southward IMF [e.g., *Daglis et al.*, 2003]. We used magnetometer data from the geomagnetic equator (Yap, −0.3° GML) and the low latitude (Okinawa, 14.47° GML), to derive the equatorial DP2.

2. OBSERVATIONS

2.1. Selected Geomagnetic Storms

We analyzed geomagnetic storms on April 18, 2001, November 6, 2001, and September 4, 2002. These three storms were characterized by sudden commencement (SC) immediately followed by ring current development as seen in the SYM-H, which were caused by the solar wind shock accompanied by the southward IMF. On the other hand, the recovery of the three storm events was initiated by the northward turning of the IMF, reduction in the southward IMF, and an impulsive northward deflection embedded in the prolonged southward IMF. To detect the electric field penetrated to low latitudes, we used the equatorial DP2 defined as a difference between H-component magnetic fields at the geomagnetic equator, Yap (YAP, 0.3°S GML), and at low latitude, Okinawa (OKI, 14.47°N GML). In deriving the equatorial DP2, we assumed that these two stations are under the same effects of the magnetospheric currents because of their short latitudinal distance, and that the DP2 at low latitude is considerably less than the equatorial DP2. We used the B_y at Cambridge Bay [CBB; 77.21° Corrected Geomagnetic Latitude (CGML), magnetic local time (MLT) = UT − 8] or B_x at Thule (THL; 85.22° CGML, MLT = UT − 3) to infer variations in the polar cap potential (PCP), and used contour maps of the intensity of the westward auroral electrojet (AEJ) derived from the International Monitor for Auroral Geomagnetic Effects (IMAGE) magnetometer array data to infer the location and motion of the auroral oval during the main and recovery phases of the storm. Positive deflections of the B_x (THL) and negative deflections of the B_y (CBB) responded well to the southward IMF, which, therefore, represent variations in PCP.

2.2. April 18, 2001 Storm

The first storm was caused by the southward IMF of magnitude 17 nT accompanied by the solar wind shock as observed by WIND located at (5.6, −227, −132 Re at 01 UT) (Figure 1). The storm ring current developed simultaneously with increases in PCP (upper panel) and AEJ (middle panel), immediately after the SC at 0046 UT (Figure 2), and the development of the ring current continued for 210 min as expressed with the SYM-H (Figure 1). The simultaneous development of the PCP and ring current implies near-instantaneous transmission of the convection electric field to the inner magnetosphere.

PCP and AEJ developed during the early main phase (0046–0310 UT), and the AEJ moved equatorward from 68° to 60° CGML concurrently with the increase in PCP during the late main phase (0330–0415 UT). On the other hand, the equatorial DP2 increased simultaneously with PCP and AEJ, and remained positive during the main phase. This indicates continuous penetration of the convection electric field for more than 3 h during the main phase of the storm, in agreement with the results presented by *Huang et al.* [2005]. It should be noted, however, that the equatorial DP2 decreased conversely to the increase in PCP and AEJ during the late main phase. This converse behavior of the DP2 must be caused by a shielding electric field developed equatorward of the AEJ. The AEJ then moved poleward at 0415 UT, and reached the latitude of 72° CGML at 0500 UT at the beginning of the storm recovery phase. At this time, the DP2 changed into the CEJ at the equator, and the storm changed into the recovery phase at 0430 UT.

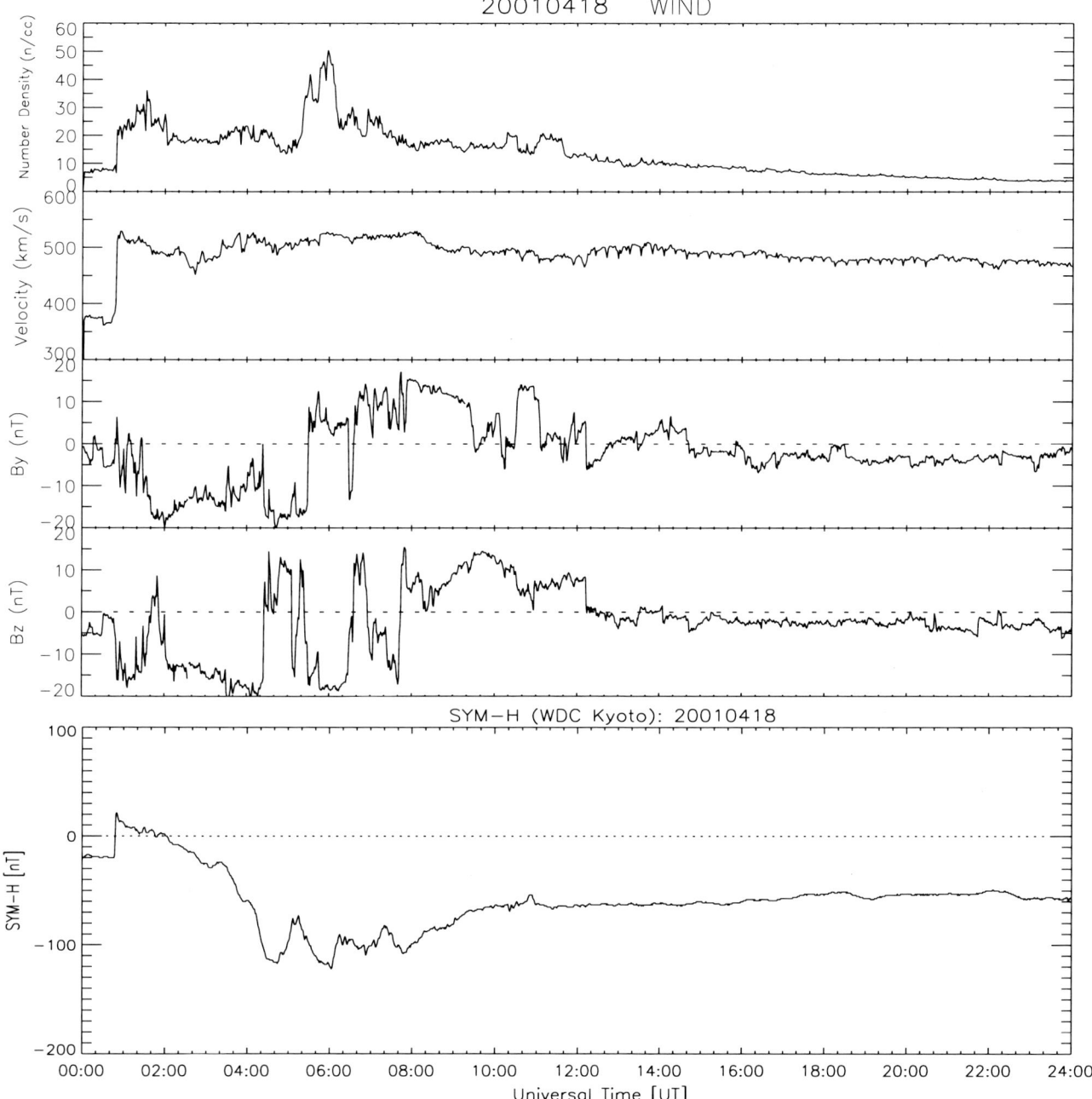

Figure 1. From top to bottom, the solar wind number density, velocity, the IMF B_y and B_z observed by WIND at (5.6, -227, -132 Re), and the SYM-H for the April 18, 2001 storm event, are shown.

2.3. November 6, 2001 Storm

Figure 3 shows the IMF observed by WIND at (44.1–75.0, 22.0 Re at 02 UT) and SYM-H for the second storm event. The SC started at 0152 UT with the amplitude of 89 nT, and the ring current developed immediately after the SC, being caused by the southward IMF of -55 nT. The ring current continued to develop for 80 min to reach the minimum of -330 nT at 0310 UT, and decayed after 0400 UT, when the southward IMF was decreasing. The PCP increased over

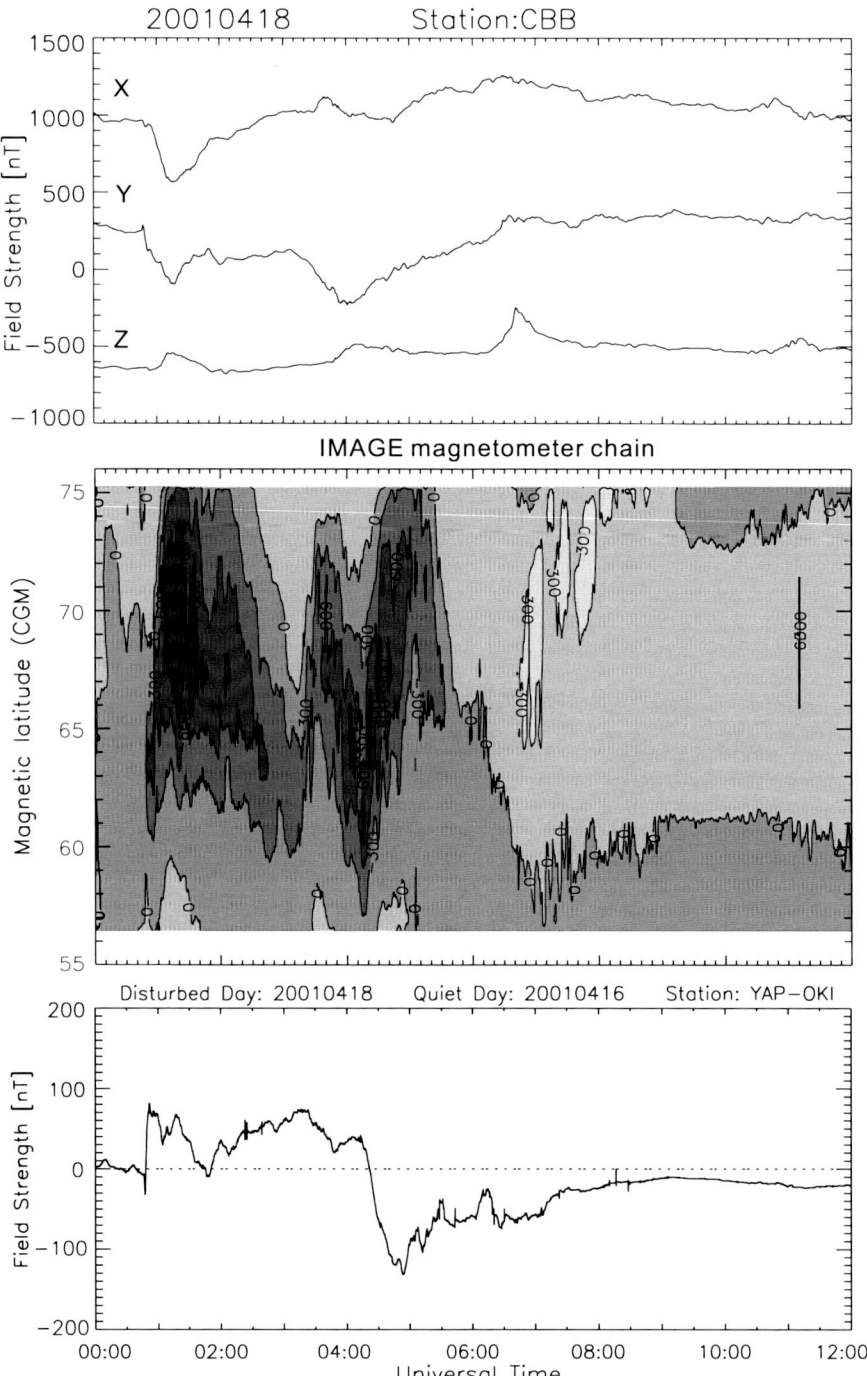

Figure 2. *X*-, *Y*-, and *Z*-component magnetic fields at Cambridge bay (CBB) (top panel), contour map of the westward auroral electrojet (AEJ) intensity derived from the IMAGE magnetometer array (middle panel), and the equatorial DP2 derived from the *H*-component magnetic fields at the geomagnetic equator (Yap) and low latitude (Okinawa) (bottom panel) for the April 18, 2001 storm event.

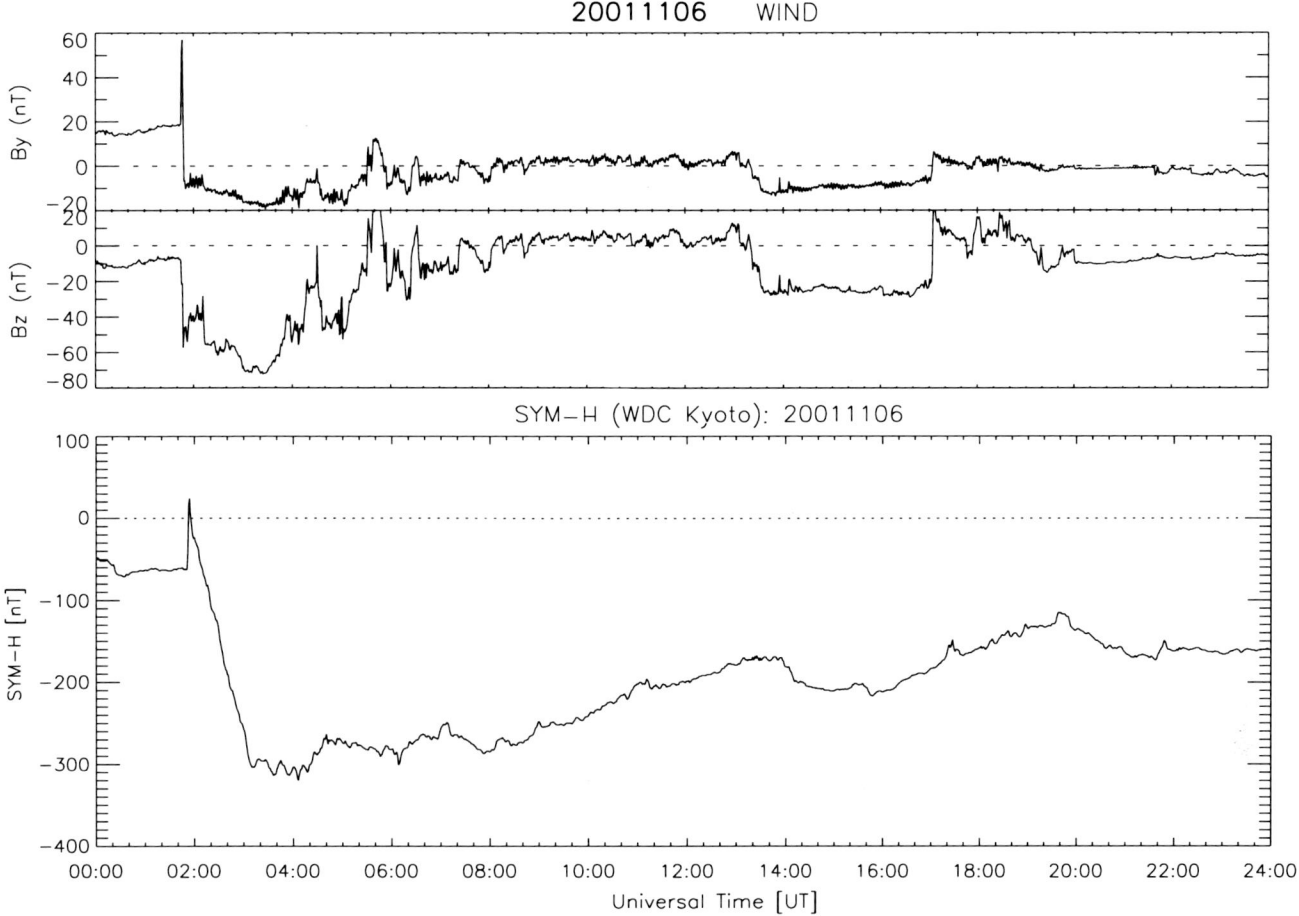

Figure 3. From top to bottom, the IMF B_y and B_z observed by WIND at (44.1–75.0, 22.0 Re), and the SYM-H for the November 6, 2001 storm event, are shown.

two time intervals, 0150–0340 UT and 0410–0550 UT, as seen in the X-component of the magnetic field at THL (upper panel, Figure 4). In correspondence to the first PCP increase, the AEJ developed immediately after the SC at mid latitudes (55–60° CGML) centered at 57° CGML (middle panel, Figure 4), and remained strong with a magnitude of 2000 nT for the first time interval. The AEJ then moved rapidly poleward to the auroral latitude centered at 67°, and remained high with a magnitude of 2000 nT for the second time interval. The equatorial DP2 developed simultaneously with the increase in PCP and AEJ, and remained positive with a peak amplitude of 280 nT until 0340 UT, and then the DP2 turned into the CEJ (lower panel, Figure 4). It should be noted that the equatorial DP2 started to decrease at 0240 UT, whereas the AEJ was strengthened with the peak at 0300 UT and remained high until 0340 UT. The decrease in equatorial DP2 indicates growth of the shielding electric field during the late growth phase, and the CEJ occurred due to overshielding during the early recovery phase. The rapid poleward motion of the AEJ implies a contraction of the auroral oval and would decrease the convection electric field at low latitudes. As a result, the overshielding occurred at lower latitudes and caused the CEJ at the equator. It should be stressed that the storm went into the recovery phase, when the overshielding occurred.

2.4. September 4, 2002 Storm

Figure 5 shows the solar wind parameters observed by WIND located at (63.8, 40.5, –6.5 Re at 02 UT) and SYM-H. The SC started at 0150 UT, and the ring current developed at 0210 UT, which was caused by the southward IMF of –19 nT. The ring current developed for 140 min, and started to decay at the time of an impulsive positive deflection of the

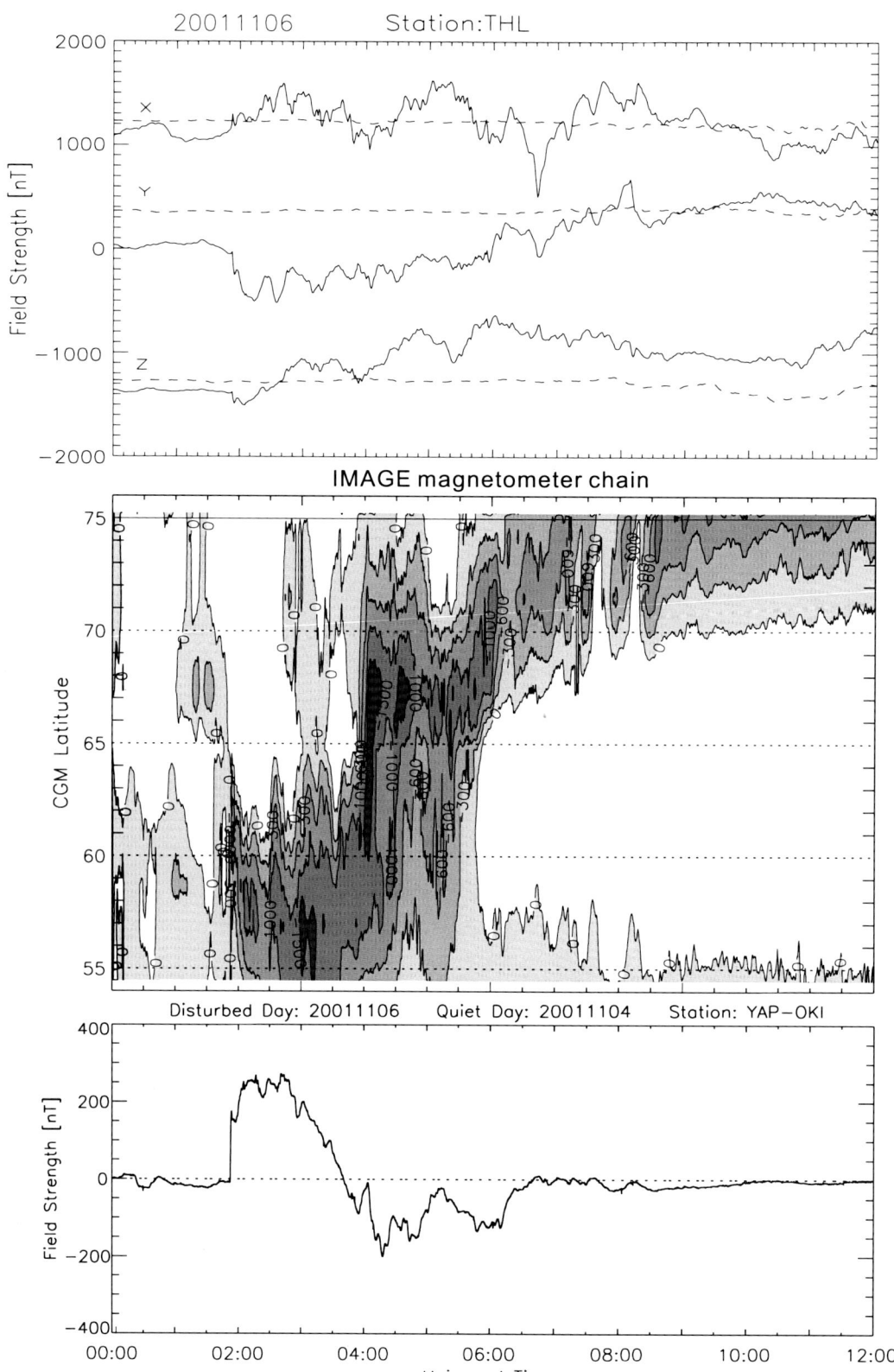

Figure 4. *X*-, *Y*-, and *Z*-component magnetic fields at Thule (THL) (top panel), contour map of the westward AEJ intensity derived from the IMAGE magnetometer array (middle panel), and the equatorial DP2 derived from the *H*-component magnetic fields at the geomagnetic equator (Yap) and low latitude (Okinawa) (bottom panel) for the November 6, 2001 storm event.

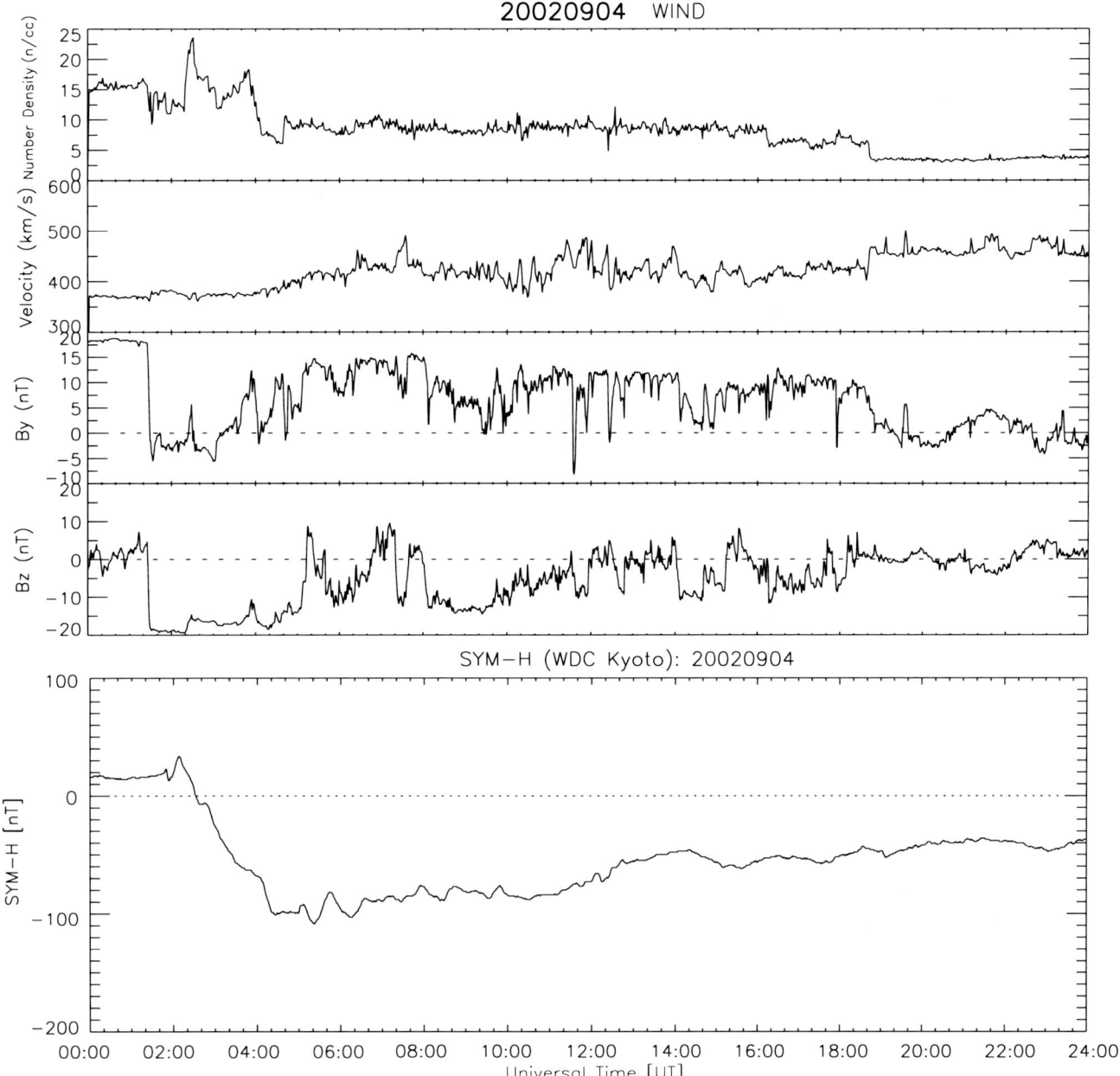

Figure 5. From top to bottom, the solar wind number density, velocity, the IMF B_y and B_z observed by WIND at (63.8, 40.5, −6.5 Re), and the SYM-H for the September 4, 2002 storm event, are shown.

IMF at 0400 UT. The southward IMF increased again for 30 min, but the ring current did not develop again as seen in the SYM-H (Figure 5).

The AEJ developed at auroral latitudes (65–70° CGML) during the early main phase (0200–0250 UT) (middle panel, Figure 6). On the other hand, the AEJ intensified at lower latitudes (58–64° CGML) during the late main phase (0250–0420 UT), and then intensified at auroral latitude (62–67° CGML) during the early recovery phase. These temporal and latitudinal behaviors of the AEJ are similar to those of the first and second events.

The equatorial DP2 increased at 0150 UT from the negative level that might be caused by previous disturbances, and remained positive until 0410 UT (lower panel, Figure 6). The

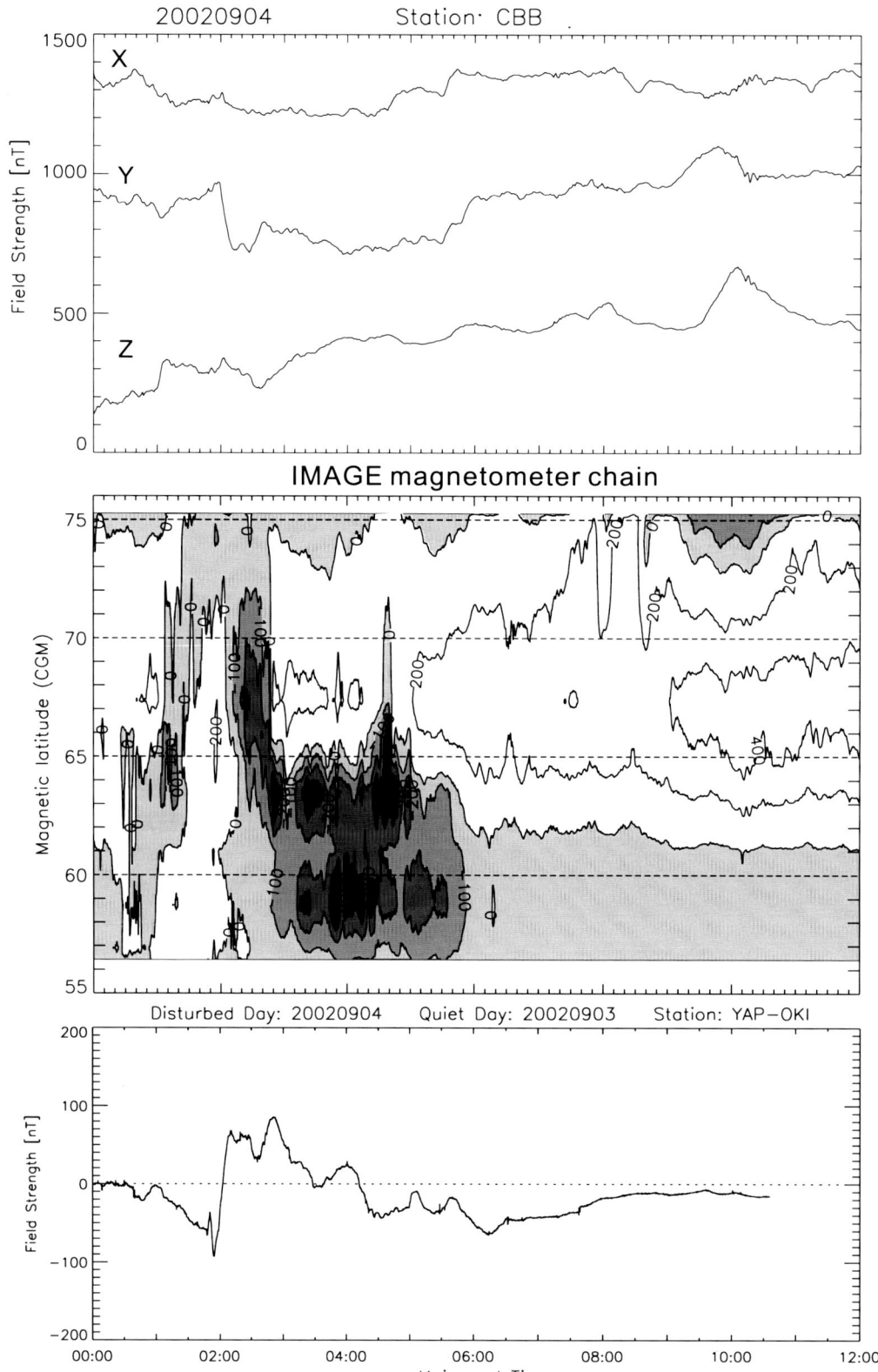

Figure 6. *X*-, *Y*-, and *Z*-component magnetic fields at Cambridge bay (CBB) (top panel), contour map of the westward AEJ intensity derived from the IMAGE magnetometer array (middle panel), and the equatorial DP2 derived from the *H*-component magnetic fields at the geomagnetic equator (Yap) and low latitude (Okinawa) (bottom panel) for the September 4, 2002 storm event.

equatorial DP2 decreased gradually with some fluctuations over the period of the equatorward shift of the AEJ (0250–0420 UT) during the late main phase. The CEJ then occurred at 0410 UT, a little earlier than the onset of the poleward shift of the AEJ, and continued for about 3 h during the early recovery phase. These latitudinal and temporal variations are very similar to those of the first and second events, suggesting growth of the shielding electric field during the late main phase. It is remarkable that the southward IMF became strong again after the impulsive northward deflection, but it did not increase the ring current again, probably because of the significant growth of the overshielding electric field. It is suggested that the overshielding contributed to the decay of the ring current.

3. DISCUSSION AND CONCLUSION

We have shown three storm events initiated by the shock-associated southward IMF and recovered by a decrease in the southward IMF. The equatorial DP2 started to increase simultaneously with the ring current development as well as the PCP and AEJ. These facts imply that the convection electric field penetrated simultaneously into the equatorial ionosphere and inner magnetosphere. The fast mode wave propagates across the magnetic field line, carrying an inductive electric field to the equatorial ionosphere. However, the electric field associated with the fast mode wave never causes currents responsible for the ground magnetic perturbations, but tends to shield the incoming magnetic perturbations [*Kikuchi and Araki*, 1979a]. Furthermore, the fast mode in the ionospheric F region suffers great attenuation because of the dominant electron-neutral particle collision frequency [*Strangeway et al.*, 2001]. As a result, the ionosphere behaves as an incompressible medium for the ultralow frequency range perturbations [*Kivelson and Southwood*, 1988]. The instantaneous transmission of the convection electric field to the equator has been explained by means of the zeroth-order transverse magnetic (TM_0) mode wave in the Earth–ionosphere waveguide [*Kikuchi et al.*, 1978; *Kikuchi and Araki*, 1979b], which propagates horizontally at the speed of light. The TM_0 mode wave accompanies electric currents in the ionosphere and on the ground, which are connected by the displacement current at the wave front of the TM_0 mode wave. The ionospheric current generates an electric field in the ionospheric finite conductivity, which is mapped upward along the field lines by the Alfven waves, with no attenuation under a condition of large ionospheric conductance to Alfven conductance ratio [*Kikuchi*, 2005]. Indeed, strong electric field penetrated to the inner magnetosphere as observed by CRRES and Akebono satellites during geomagnetic storms [*Burke et al.*, 1998; *Wilson et al.*, 2001; *Shinbori et al.*, 2005]. The electric field would drive the plasma convection in the inner magnetosphere, causing the development of the storm ring current. *Wilson et al.* [2001] suggested close relationship between the DP2 currents and the electric field in the inner magnetosphere. We are able to suggest that the development of the polar–equatorial ionospheric currents has an impact on the ring current evolution.

The convection electric field penetrates instantaneously to low latitudes, but suffers from shielding in about 20 min for substorm events [*Somajajulu et al.*, 1987; *Kikuchi et al.*, 2000]. *Huang et al.* [2005], however, suggested that the penetration continued for many hours during the main phase of the storm. Indeed, the equatorial DP2 continued for 2–3 h during the main phase of the storms analyzed in this paper (Figures 2, 4, and 6). It should be noted, however, that shielding electric field became effective during the late main phase, for example, 1 h after the onset of the main phase of the November 6, 2001 storm event. The growth of the shielding electric field resulted in the overshielding when the convection electric field decreased because of the reduction in the southward IMF at the beginning of the recovery phase. It should be noted that the overshielding occurred when the IMF remained southward during the storm events, whereas it occurs when the IMF turns northward during the substorm [*Rastogi and Patel*, 1975; *Kelley et al.*, 1979; *Kikuchi et al.*, 2003]. The distinct feature of the stormtime overshielding may be due to the fact that the ring current is much stronger and the location of the R1 and R2 FACs is far equatorward from the auroral latitude.

The growth of the shielding electric field accompanied the equatorward shift of the AEJ, and the AEJ moved rapidly poleward at the beginning of the recovery phase. These latitudinal motions of the AEJ may be a signature of the substorm. The substorm may play a crucial role in initiating the storm recovery phase as suggested by *Iyemori and Rao* [1996] and *Ohtani et al.* [2001]. Continuous reduction in the southward IMF is needed for the recovery of the storm [e.g., *Ohtani et al.*, 2001], which is valid for the first two storm events analyzed above. On the other hand, the southward IMF increased again after the impulsive northward deflection in the third event, but it never intensified the ring current again. The overshielding may have contributed to end the development of the ring current, and initiate the storm recovery phase.

In conclusion, there are two types of equatorial electrojet during the storm. One is the DP2 current driven by the penetrated convection electric field during the main phase, which occurs concurrently with the enhancement in PCP and AEJ. The other is the CEJ caused by the overshielding electric field during the recovery phase, which was accompanied by the rapid poleward shift of the AEJ. The substorm may have played a role in the transition of the storm into the recovery phase, as has been suggested by *Iyemori and Rao* [1996] and *Ohtani et al.* [2001]. In addition, we suggest that overshielding electric fields may have played a role in the storm recovery, reducing the electric fields that influence ring current development in the inner magnetosphere. From the electric current viewpoint, the R1 and R2 FACs flowed into the dayside equatorial ionosphere via the polar ionosphere, driving the equatorial DP2 and CEJ, during the main and recovery phases, respectively.

Acknowledgments. We would like to thank the Finnish Meteorological Institute for the IMAGE magnetometer data, and the U.S. NOAA/Weather Service Office at Yap and Ryukyu University for their help in operating the NICT space weather monitoring magnetometers at Yap and Okinawa, respectively. The SYM-H was provided by the WDC for Geomagnetism, Kyoto. Magnetometer data at Thule and Cambridge Bay were provided by the INTERMAGNET data site (http://www.intermagnet.org/Data_e.html). The WIND data were obtained through the Coordinated Data Analysis Web (CDAWeb).

REFERENCES

Daglis, I. A., J. U. Kozyra, Y. Kamide, D. Vassiliadis, A. S. Sharma, M. W. Liemohn, W. D. Gonzalez, B. T. Tsurutani, and G. Lu (2003), Intense space storms: Critical issues and open disputes, *J. Geophys. Res.*, *108*(A5), 1208, doi:10.1029/2002JA009722.

Fejer, B. G., C. A. Gonzales, D. T. Farley, M. C. Kelley, and R. F. Woodman (1979), Equatorial electric fields during magnetically disturbed conditions, 1. The effect of the interplanetary magnetic field, *J. Geophys. Res.*, *84*, 5797–5802.

Gonzales, C. A., M. C. Kelley, B. G. Fejer, J. F. Vickrey, and R. F. Woodman (1979), Equatorial electric fields during magnetically disturbed conditions, 2. Implications of simultaneous auroral and equatorial measurements, *J. Geophys. Res.*, *84*, 5803–5812.

Huang, C.-S., J. C. Foster, and M. C. Kelley (2005), Long-duration penetration of the interplanetary electric field to the low-latitude ionosphere during the main phase of magnetic storms, *J. Geophys. Res.*, *110*, A11309, doi:10.1029/2005JA011202.

Huang, C. Y., W. J. Burke, J. S. Machuzak, L. C. Gentile, and P. J. Sultan (2001), DMSP observations of equatorial plasma bubbles in the topside ionosphere near solar maximum, *J. Geophys. Res.*, *106*(A5), 8131–8142.

Iyemori, T., and D. R. K. Rao (1996), Decay of the D_{st} field of geomagnetic disturbances after substorm onset and its implication to storm–substorm relation, *Ann. Geophys.*, *14*, 608–618.

Kelley, M. C., B. G. Fejer, and C. A. Gonzales (1979), An explanation for anomalous equatorial ionospheric electric fields associated with a northward turning of the interplanetary magnetic field, *Geophys. Res. Lett.*, *6*, 301–304.

Kikuchi, T. (2005), Transmission line model for driving plasma convection in the inner magnetosphere, in *The Inner Magnetosphere: Physics and Modeling*, Geophys. Monogr. Ser., vol. 155, edited by T. I. Pulkkinen, N. A. Tsyganenko, and R. H. W. Friedel, pp. 173–179, AGU, Washington, D. C.

Kikuchi, T., and T. Araki (1979a), Transient response of uniform ionosphere and preliminary reverse impulse of geomagnetic storm sudden commencement, *J. Atmos. Terr. Phys.*, *41*, 917–925.

Kikuchi, T., and T. Araki (1979b), Horizontal transmission of the polar electric field to the equator, *J. Atmos. Terr. Phys.*, *41*, 927–936.

Kikuchi, T., T. Araki, H. Maeda, and K. Maekawa (1978), Transmission of polar electric fields to the equator, *Nature*, *273*, 650–651.

Kikuchi, T., H. Lühr, T. Kitamura, O. Saka, and K. Schlegel (1996), Direct penetration of the polar electric field to the equator during a *DP* 2 event as detected by the auroral and equatorial magnetometer chains and the EISCAT radar, *J. Geophys. Res.*, *101*, 17,161–17,173.

Kikuchi, T., H. Liihr, K. Schlegel, H. Tachihara, M. Shinohara, and T.-I. Kitamura (2000), Penetration of auroral electric fields to the equator during a substorm, *J. Geophys. Res.*, *105*, 23,251–23,261.

Kikuchi, T., K. K. Hashimoto, T.-I. Kitamura, H. Tachihara, and B. Fejer (2003), Equatorial counterelectrojets during substorms, *J. Geophys. Res.*, *108*(A11), 1406, doi:10.1029/2003JA009915.

Kivelson, M. G., and D. J. Southwood (1988), Hydromagnetic waves and the ionosphere, *Geophys. Res. Lett.*, *15*, 1271–1274.

Kobea, A. T., C. Amory-Mazaudier, J. M. Do, H. Lühr, E. Houginou, J. Vassal, E. Blanc, and J. J. Curto (1998), Equatorial electrojet as part of the global circuit: a case-study from the IEEY, *Ann. Geophys.*, *16*, 698–710.

Kobea, A. T., A. D. Richmond, B. A. Emery, C. Peymirat, H. Lühr, T. Moretto, M. Hairston, and C. Amory-Mazaudier (2000), Electrodynamic coupling of high and low latitudes: Observations on May 27, 1993, *J. Geophys. Res.*, *105*(A10), 22,979–22,989.

Nishida, A. (1968), Coherence of geomagnetic *DP* 2 magnetic fluctuations with interplanetary magnetic variations, *J. Geophys. Res.*, *73*, 5549–5559.

Nishida, A., N. Iwasaki, and T. Nagata (1966), The origin of fluctuations in the equatorial electrojet; A new type of geomagnetic variation, *Ann. Geophys.*, *22*, 478–484.

Ohtani, S., M. Nosé, G. Rostoker, H. Singer, A. T. Y. Lui, and M. Nakamura (2001), Storm-substorm relationship: Contribution of the tail current to *Dst*, *J. Geophys. Res.*, *106*(A10), 21,199–21,210.

Peymirat, C., A. D. Richmond, and A. T. Kobea (2000), Electrodynamic coupling of high and low latitudes: Simulations of shielding/overshielding effects, *J. Geophys. Res.*, *105*(A10), 22,991–23,003.

Rastogi, R. G. (1977), Geomagnetic storms and electric fields in the equatorial ionosphere, *Nature*, *268*, 422–424.

Rastogi, R. G. (1997), Midday reversal of equatorial ionospheric electric field, *Ann. Geophys.*, *15*, 1309–1315.

Rastogi, R. G.., and V. L. Patel (1975), Effect of interplanetary magnetic field on ionosphere over the magnetic equator, *Proc. Indian Acad. Sci.*, *82*, 121–141.

Senior, C., and M. Blanc (1984), On the control of magnetospheric convection by the spatial distribution of ionospheric conductivities, *J. Geophys. Res.*, *89*, 261–284.

Somayajulu, V. V., C. A. Reddy, and K. S. Viswanathan (1987), Penetration of magnetospheric convective electric field to the equatorial ionosphere during the substorm of March 22, 1979, *Geophys. Res. Lett.*, *14*, 876–879.

Southwood, D. J. (1977), The role of hot plasma in magnetospheric convection, *J. Geophys. Res.*, *82*, 5512–5520.

Strangeway, R. J., and J. Raeder (2001), On the transition from collisionless to collisional magnetohydrodynamics, *J. Geophys. Res.*, *106*, 1955–1960.

Tsurutani, B., et al. (2004), Global dayside ionospheric uplift and enhancement associated with interplanetary electric fields, *J. Geophys. Res.*, *109*, A08302, doi:10.1029/2003JA010342.

Vasyliunas, V. M. (1972), The interrelationship of magnetospheric processes, in *Earth's Magnetospheric Processes*, edited by B. M. McCormac, pp. 29–38 D. Reidel, Norwell, Mass.

Wilson, G. R., W. J. Burke, N. C. Maynard, C. Y. Huang, and H. J. Singer (2001), Global electrodynamics observed during the initial and main phases of the July 1991 magnetic storm, *J. Geophys. Res.*, *106*(A11), 24,517–24,539.

Wygant, J., D. Rowland, H. J. Singer, M. Temerin, F. Mozer, and M. K. Hudson (1998), Experimental evidence on the role of the large spatial scale electric field in creating the ring current, *J. Geophys. Res.*, *103*(A12), 29,527–29,544.

T. Kikuchi, Solar–Terrestrial Environment Laboratory, Nagoya University, Nagoya, Furo-cho, Chikusa-ku, Aichi 464-8601, Japan. (kikuchi@stelab.nagoya-u.ac.jp)

T. Kikuchi and K. Nozaki, National Institute of Information and Communications Technology (NICT), 4-2-1 Nukui-Kitamachi, Koganei, Tokyo 184-8795, Japan.

K. K. Hashimoto, Kibi International University, 8 Igamachi, Takahashi, Okayama 716-8508, Japan.

Relating the Interplanetary-Induced Electric Fields With the Low-Latitude Zonal Electric Fields Under Geomagnetically Disturbed Conditions

Adela Anghel,[1,2] David Anderson,[1,2] Jorge Chau,[3] Kiyohumi Yumoto,[4] and Archana Bhattacharyya[5]

The overall ionospheric variability with periods ranging from long-term, secular changes to days, hours, and even minutes and seconds, is influenced by the solar activity, geomagnetic activity, and processes originating in the lower atmospheric layers. Using a wavelet transform approach, in this paper, we study the short-term (minutes to hours) and day-to-day variability of the ionospheric low-latitude zonal electric fields (LLZEF) at three longitude sectors, Peruvian, Philippine, and Indian, during time intervals of increased geomagnetic activity and relate the LLZEF variability to changes in the dawn-to-dusk component of the interplanetary electric field (IEF). Continuous Morlet wavelet and cross-wavelet amplitude spectra with reduced and increased frequency resolutions were obtained to analyze and compare the oscillation activity in the LLZEF and IEF spectra, in the 10-min to 10-h and 1.25- to 12-d period ranges. For the 1.25- to 12-d period range, periodicities in the LLZEF spectrum were compared with similar periodicities in the IEF spectrum over 9 February to 9 June 2001, with our wavelet results indicating the geomagnetic activity as an important driver of LLZEF variability in this period range. For the 10-min to 10-h period range, four case studies were examined when concurrent observations of Jicamarca incoherent scatter radar zonal electric field and IEF, as calculated from the ACE satellite solar wind velocity and interplanetary magnetic field data, were available. We show that the wavelet transform represents a powerful tool to study the frequency dependence of the two specific mechanisms of ionospheric electric field variability, which are dominant during geomagnetic storms, namely penetration and disturbance dynamo.

---

[1] Cooperative Institute for Research in Environmental Sciences, University of Colorado, Boulder, Colorado, USA.

[2] Space Weather Prediction Center, NOAA Boulder, Colorado, USA.

[3] Radio Observatorio de Jicamarca, Instituto Geofisico del Peru, Jicamarca, Peru.

[4] Space Environment Research Center, Kyushu University, Fukuoka, Japan.

[5] Indian Institute of Geomagnetism, New Panvel, Navi Mumbai, India.

Midlatitude Ionospheric Dynamics and Disturbances
Geophysical Monograph Series 181
Copyright 2008 by the American Geophysical Union.
10.1029/181GM15

1. INTRODUCTION

Stemming from the need to understand and predict the ionosphere behavior and its deviations from the normal climatological mean, under both quiet and disturbed conditions, recent modeling and observational studies have shown an increased interest in the short-term (minutes to hours) and day-to-day variability of the upper atmosphere and ionosphere [e.g., *Mendillo and Schatten*, 1983; *Parish et al.*, 1994; *Forbes et al.*, 2000; *Pancheva et al.*, 2002]. The ionosphere–thermosphere system is a complex system, its complexity being determined by (1) inherent internal interactions occurring inside the system, (2) interactions with the magnetosphere above, where space plasma processes in the magnetosphere caused by its coupling to the solar wind provide an interface with highly variable inputs of electrodynamic energy and energetic particles, (3) interactions with the middle atmosphere below, itself modulated by tropospheric weather and surface topology, and (4) variability of the external sources driving the system. As a result, the ionosphere displays variations from its normal patterns that affect the ionospheric predictions on timescales ranging from secular to days, hours, and even minutes and seconds. These variations have been observed in different ionospheric parameters [e.g., *Parish et al.*, 1994; *Rishbeth and Mendillo*, 2001; *Pancheva et al.*, 2002; *Fagundes et al.*, 2005; *Abdu et al.*, 2006] possibly induced by wave activity originating in the lower regions of the atmosphere, quasiperiodic oscillations in the geomagnetic activity, or other triggers such as the periodic variability of the solar radiation flux. The solar radiation influences mostly the long-term (months to years) variability, while the geomagnetic activity and the lower atmosphere processes can induce oscillations with periods ranging from about few seconds or minutes to several days or weeks [e.g., *Laštovička*, 2006].

In this paper, the main focus is on the variability of the ionospheric low-latitude zonal electric field (LLZEF) in response to changes in the dawn-to-dusk component of the interplanetary electric field (IEF) during time intervals of increased geomagnetic activity. During storm times, large ionospheric electric field and current perturbations travel from high to equatorial latitudes, changing the ionization distribution over large areas and controlling the storm-time dynamics and electrodynamics of the low- and mid-latitude ionosphere. The most important sources of low-latitude electric field perturbations during storm times are (1) the prompt penetration electric fields of solar wind/magnetospheric origin and (2) the atmospheric disturbance dynamo electric fields due to auroral Joule heating and ion-drag acceleration. The penetration electric fields are associated with changes in the field-aligned current system responsible for shielding the inner magnetosphere and mid- and low-latitude ionosphere from the high-latitude magnetospheric convection electric fields and propagate instantaneously to equatorial latitudes in response to changes in the magnetospheric convection [e.g., *Fejer and Scherliess*, 1997; *Huang et al.*, 2005]. On the other hand, the disturbance dynamo electric fields are associated with enhanced deposition of energy and momentum in the auroral zone that causes winds to develop, producing predominantly westward electric fields on the dayside and eastward on the nightside at equatorial latitudes [e.g., *Blanc and Richmond*, 1980; *Scherliess and Fejer*, 1997].

Equatorial electric field measurements are rather sporadic, but recently, *Anderson et al.* [2004] developed a neural network-based plasma drift model, for both quiet and disturbed conditions, that uses ground-based magnetometer observations from pairs of equatorial stations to infer realistic, daytime equatorial vertical E × B plasma drift velocities wherever appropriately placed magnetometers exist. Using their neural network drift model, *Anderson et al.* [2006] and *Anghel et al.* [2007] calculated seasonally averaged patterns of quiet-time vertical E × B drifts at the Peruvian, Philippine, and Indian longitude sectors, showing that there is a very good agreement with drift patterns obtained with the global vertical drift model developed by Scherliess and Fejer [1999]. In addition, *Anghel et al.* [2007] also conducted an investigation on the variability of the daytime LLZEF with respect to changes in the IEF conditions. Expanding on their studies, in our paper, we use wavelet and cross-wavelet spectral analyses to compare the oscillation activity in the LLZEF and IEF spectra in the 10-min to 10-h and 1.25- to 12-d period ranges. Therefore, the purpose of this paper is twofold: (1) to study the variability of the daytime LLZEF in the 1.25- to 12-d period range at three longitudes, Peruvian, Philippine, and Indian, over a time interval of relatively increased geomagnetic activity, 9 February to 9 June 2001, and relate this variability to similar changes in the IEF and (2) to analyze and relate the oscillation activity in the LLZEF and IEF in the 10-min to 10-h period range for three case studies characterized by enhanced geomagnetic activity, the 17–19 April 2001, 15–18 April 2002, and 9–12 November 2004 storm events, using concurrent observations of Jicamarca incoherent scatter radar (ISR) zonal electric field and IEF data in a wavelet analysis approach. A fourth quiet-time case, 29 March to 2 April 2004, is also considered for comparison purposes.

The paper is organized as follows: in the next section, we briefly describe the data sets and the wavelet analysis approach, then we present the wavelet results for the 1.25- to 12-d period range, followed by an examination of the four case studies for periodicities in the 10-min to 10-h range, and in the last section, we present succinct our conclusions.

2. DATA SETS AND ANALYSIS METHODS

2.1. Data Sets

A direct measure of the strength of the equatorial electrojet current and of the magnitude of the F region vertical E × B drifts is provided by the difference in the horizontal H components, ΔH, between a magnetometer placed on the magnetic equator and one displaced 6–9° away, after subtracting the nighttime baseline at each station [e.g., *Anderson et al.*, 2004]. For our study, we used magnetometer data with a 5-min time resolution from three pairs of equatorial stations located in Peru, Philippines, and India. Magnetometer observations at the Peruvian sector were obtained from Jicamarca (geographic coordinates, 11.9°S, 283.1°E, geometric latitude 0.8°N) and Piura (geographic coordinates, 5.2°S, 279.4°E, geometric latitude 6.8°N), at the Philippine sector, from Davao (geographic coordinates, 7°N, 125.4°E, geometric latitude 1.32°S) and Muntinlupa (geographic coordinates, 14.37°N, 121.02°E, geometric latitude 6.39°N), and at the Indian sector, from Tirunelveli (geographic coordinates, 8.7°N, 76.9°E, geometric latitude 0.5°S) and Alibag (geographic coordinates, 18.6°N, 72.9°E, geometric latitude 10°N). The neural network drift model described by *Anderson et al.* [2004] and the calculated ΔH values at each longitude sector were then used to estimate the daytime vertical E × B drifts and, in turn, the equatorial zonal electric fields knowing that 1 mV/m corresponds to a vertical drift of ~40 m/s at the Peruvian sector, ~28 m/s at the Philippine sector, and ~25 m/s at the Indian sector.

In our analysis, the interplanetary conditions are described using 64-s averages of merged ACE Magnetic Field Experiment (MAG)–Solar Wind Electron, Proton, and Alpha Monitor (SWEPAM) Level 2 interplanetary magnetic field and solar wind velocity data, where the two parameters are important in establishing the dawn-to-dusk component of the IEF, IEF Ey. First, the calculated IEF-Ey at the spacecraft position L1 (~1.4 million km) is time-shifted to the magnetopause position using the radial component of the solar wind velocity, in addition to a 10-min time delay. Then, the time-shifted IEF Ey is smoothed by using a sliding-window average procedure, which is equivalent to a low-pass filtering of the data. By using the time-shifting and smoothing procedures, we obtain a very good correlation between the IEF and the equatorial zonal electric fields at all three longitude sectors [e.g., *Kelley et al.*, 2003; *Anghel et al.*, 2007].

2.2. Continuous Morlet Wavelet

Over the last few years, the continuous wavelet transform [e.g., *Kumar and Foufoula-Georgiou*, 1997; *Torrence and Compo*, 1998] has become a favored tool to analyze periodicities that occur in the atmosphere [e.g., *Pancheva*, 2000; *Abdu et al.*, 2006]. The wavelet method targets nonstationary signals with variable frequency content like the ones we deal with here. For our analysis, we favored the continuous Morlet wavelet transform to relate the variability in the LLZEF and IEF. The Morlet wavelet is a complex-valued function consisting of a plane wave modulated by a Gaussian envelope:

$$\psi(t) = \frac{1}{(\pi \cdot \sigma^2)^{1/4}} \cdot \exp\left(-\frac{t^2}{2\sigma^2}\right) \cdot \exp(j\omega_o t), \quad (1)$$

where the parameters σ and ω_o control the tradeoff between the time and frequency resolution. Here, we chose $\sigma = 1$, and for ω_o, we selected different values throughout the paper, where large ω_o values correspond to increased frequency resolution in the wavelet domain. An important advantage of the wavelet method described here consists in its ability to extract amplitude information about the periodicities present in the signal spectrum. The method can also be used to calculate "instantaneous" frequency response functions in the wavelet domain, defined in a similar way like in the Fourier domain at each time instant. The "classical" frequency response function in the Fourier domain can then be obtained by averaging over time the "instantaneous" frequency response functions, with the specific consequences of averaging. Also, to obtain information about the simultaneous presence of similar periodicities in different signals, we used a type of cross-wavelet analysis defined as a geometric mean of the wavelet amplitude spectra of the signals [e.g., *Manson et al.*, 2005]. In all our wavelet plots, the statistical significance levels were calculated based on a first-order autoregressive parametric spectral estimate of the power spectrum [*Roberts and Mullis*, 1987] considered as a background spectrum, multiplied by the desired percentile value for the χ^2 distribution with two degrees of freedom.

In the following, three examples of simulated wavelet spectra are presented to familiarize the reader with the method, for a better understanding of our results. The first example refers to Plate 1a and shows the wavelet amplitude spectra of a signal obtained by superimposing harmonics of amplitude 1 at different time intervals. As shown in Plate 1a (left), for $\omega_o = 6$ there is a good time resolution but reduced frequency resolution in the wavelet domain, with strong beatings between relatively closed spectral components, appearing as amplitude modulations. In Plate 1a (right), ω_o is 12, and the spectral components appear as distinct spectral lines well localized in time. The contour lines in the wavelet spectra represent the 95% significance levels, and the two slant lines define the cone-of-influence where the edge effects become significant.

Our second example refers to an amplitude modulation case and is presented in Plate 1b. In this case, a 30-d-long signal of 1-d period is modulated by a 5-d period:

$$\begin{aligned} x(t) &= \cos(2\pi t/T_{p1})[1 + \cos(2\pi t/T_{p2})] \\ &= \cos(2\pi t/T_{p1}) + 0.5\cos(2\pi t/T_s) + 0.5\cos(2\pi t/T_d) \end{aligned} \quad (2)$$

In equation (2), the primary periods are $T_{p1} = 1$ d and $T_{p2} = 5$ d, and the sum and difference secondary periods are $T_s = T_{p1}T_{p2}/$

$(T_{p1}+T_{p2}) = 0.833$ d and $T_d = T_{p1} \cdot T_{p2}/(T_{p2} - T_{p1}) = 1.25$ d. For $\omega_o = 6$ (Plate 1b, left), the two secondary periods beat with the primary 1-d period producing a 1-d period modulated in amplitude by a 5-d period. For $\omega_o = 30$ (Plate 1b, middle), distinct spectral lines corresponding to the two secondary periods and to the primary 1-d period can be distinguished. Plate 1 (right) shows the three spectral components at day 15. In general, in a wavelet domain with reduced frequency resolution, two simultaneous and relatively closed periods, T_1 and T_2, may beat with each other with the beating period $T = T_1 T_2 / |T_1 - T_2|$ and produce an amplitude modulation effect that may appear as distinct and unrelated bursts of oscillation activity.

In our third example, illustrated in Plate 1c, the signal is generated as:

$$x(t) = 2\cos(2\pi t/T_{p1})\cos(2\pi t/T_{p2}) \qquad (3)$$
$$= \cos(2\pi t/T_s) + \cos(2\pi t/T_d),$$

where $T_{p1} = 1$ d, $T_{p2} = 10$ d, $T_s = 0.9091$ d, and $T_d = 1.11$ d. For $\omega_o = 6$ (Plate 1c, left), the two secondary periods beat with each other with a 5-d beating period, producing, like in our second example, a 1-d period modulated in amplitude by a 5-d period, although different spectral components are present in this case. For $\omega_o = 30$, two distinct spectral lines of amplitude 1 are clearly noticed in the wavelet spectrum (Plate 1c, middle) and in the corresponding plot for day 15 (Plate 1c, right).

We conclude that, in most cases, a time domain analysis of nonstationary signals by itself, which is very similar with our wavelet analysis for $\omega_o = 6$, might not be sufficient. At the other extreme, a Fourier analysis might not help much either since the time information in this case is completely lost. However, as shown here, the wavelet analysis can provide an accurate way of analyzing nonstationary signals in the time–frequency domain over large frequency bands.

3. LLZEF PERTURBATIONS OF GEOMAGNETIC ORIGIN

3.1. LLZEF Perturbations in the 1.25- to 12-d Period Range

Previous studies have shown that the geomagnetic activity is an important driver of planetary wavelike oscillations (periods in the 2- to 30-d range) in the ionosphere [*Altadill and Apostolov*, 2001; *Forbes et al.*, 2000; *Rishbeth and Mendillo*, 2001; *Pancheva*, 2002]. Therefore, in this section, we relate periodicities in the 1.25- to 12-d range that are present in the LLZEF spectra corresponding to the Peruvian, Philippine, and Indian longitude sectors, to similar periodicities in the IEF spectrum using the continuous wavelet approach described above. The zonal electric fields at the three longitude sectors were obtained from magnetometer ΔH observations via a neural network-based vertical drift model [*Anderson et al.*, 2004]. Plate 2a shows the daytime ΔH observations from the Peruvian, Philippine, and Indian sectors as a function of local time and day of the year, for the entire year 2001, indicating a large day-to-day variability in ΔH and seasonal changes, with peaks at equinox. The wavelet amplitude spectra of the ΔH observations at the three sectors, in the 0.2- to 1.8- and 1.5- to 33-d period ranges, are shown in Plates 2b and 2c. Similar wavelet amplitude spectra of the ΔH-inferred LLZEF at the three sectors are displayed in Plate 3. It can be seen in Plates 2 and 3 that the main spectral features of the ΔH data in the 0.2- to 33-d period range are in general preserved by the neural network processing procedure of inferring the LLZEF data. Referring to Plate 3b, some general characteristics are worth mentioning: (1) oscillations with periods in the 1.5- to 33-d range are present in the LLZEF spectra at all three longitudes over the entire year 2001, (2) continuous and strong bursts of oscillation activity with periods less than about 5 d are observed throughout the year at all three locations, (3) there is an enhanced oscillation activity over a large range of periods, mostly at equinoxes, and especially at the Indian sector, (4) overall, in this period range, the oscillation activity is more intense in the Indian sector whereas the least activity is recorded at the Philippine sector, and (5) referring to Plate 3a, there are stronger 8-, 12-, and 24-h periodicities at the Philippine sector than at the other two sectors.

Plate 4 shows the wavelet amplitude spectrum of the IEF/15 for $\omega_o = 6$, over the entire year 2001. A scaling factor of 15 was used to obtain comparable amplitudes for the IEF and LLZEFs in the wavelet plots. The main features distinguished in Plate 4 are (1) the presence of continuous and strong bursts of oscillation activity with periods less than about 5 d present throughout the year, (2) enhanced oscillation activity over a large range of periods during the November and December months and at equinoxes, (3) during January, February, and summer solstice, the periodicities are less than about 5 d and have smaller amplitudes than for the rest of the year, and (4) the significant peaks in the spectrum have periods less than about 10 d.

Subsequently, we focus our discussion on periods in the 1.25- to 12-d range in the LLZEF and IEF spectra, over the 9 February to 9 June 2001 (40–160) interval characterized by relatively increased geomagnetic activity. The wavelet amplitude spectra of the LLZEF at the three longitudes and of the IEF/15 are shown in Plate 5 (left, $\omega_o = 6$) and Plate 5 (middle, $\omega_o = 30$) and their cross-spectra in Plate 5 (right,

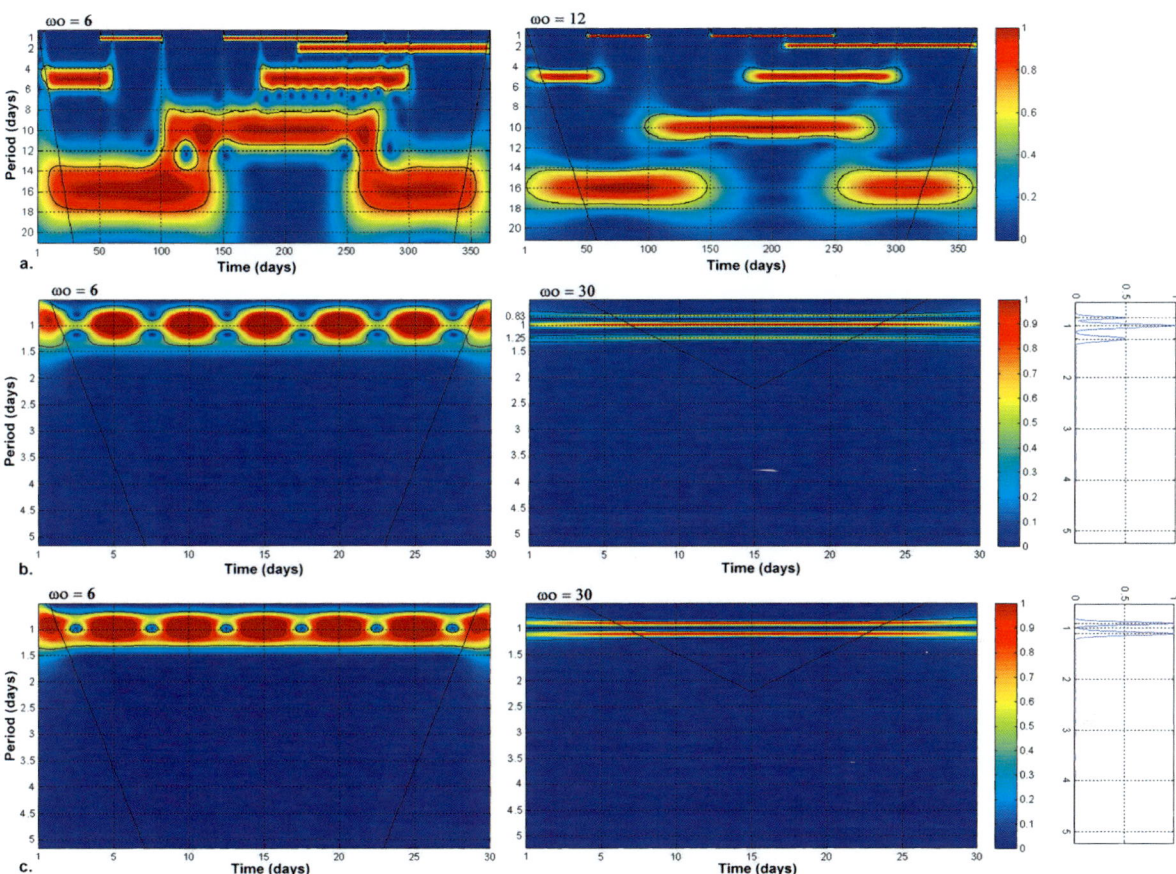

Plate 1. Wavelet amplitude spectra for the (a) first, (b) second, and (c) third examples.

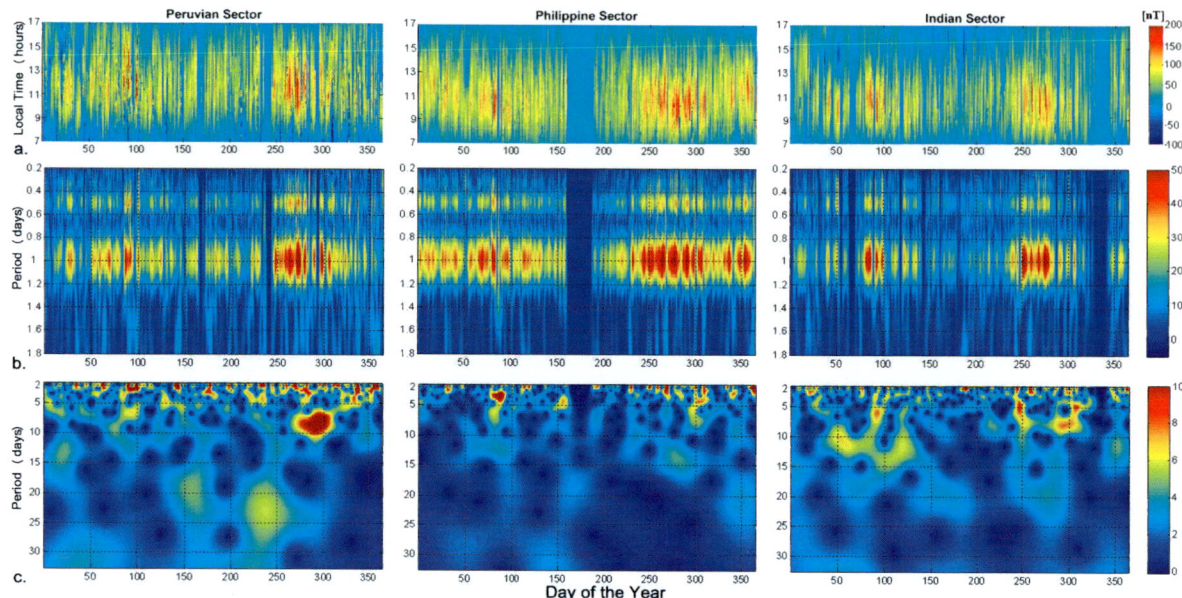

Plate 2. (a) ΔH observations at the three longitude sectors over the entire year of 2001 as a function of local time, (b) wavelet amplitude spectra for the 0.2- to 1.8-d period range, and (c) wavelet amplitude spectra for the 1.5- to 33-d period range ($\omega_o = 6$).

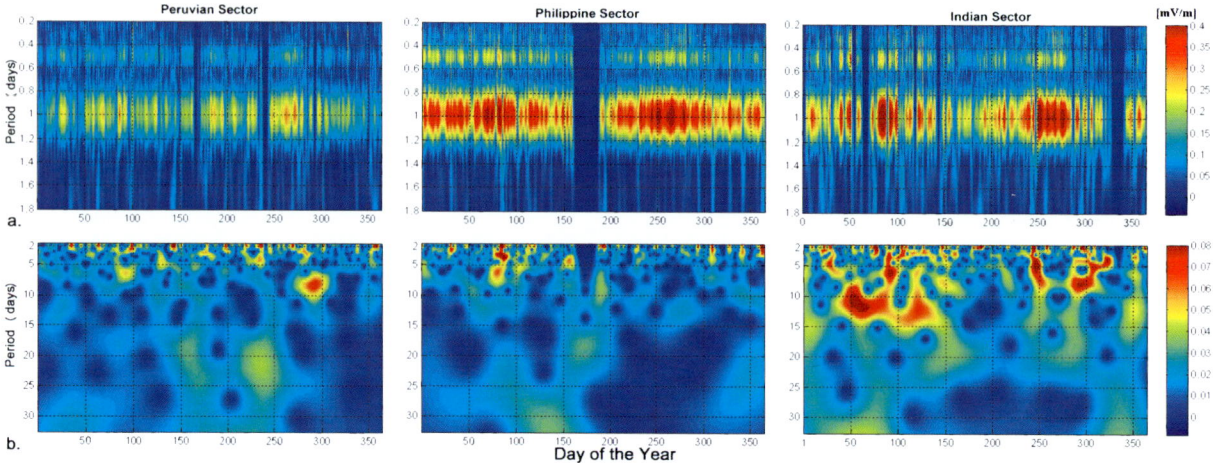

Plate 3. Wavelet amplitude spectra ($\omega_o = 6$) of the ΔH-inferred LLZEF for the 0.2- to 1.8- and 1.5- to 33-d period ranges, at the three longitude sectors.

$\omega_o = 30$). In Plate 5a (left), in the 70- to 120-d time interval, the IEF spectrum displays ongoing bursts of oscillation activity, which appear as an amplitude modulation of the 1.25- to 5-d periods by a 10-d period. By representing the IEF spectrum for $\omega_o = 30$ (Plate 5, middle), distinct spectral lines can be distinguished, revealing a relative constant spectral content of the IEF over this entire time interval.

Plates 5b, 5c, and 5d show the LLZEF wavelet amplitude spectra corresponding to the Peruvian, Philippine, and Indian longitude sectors, respectively. The LLZEF spectra ($\omega_o = 30$) at the three sectors present some common features, with more pronounced similarities between the Philippine and Indian sectors, which are separated by 0300 UT hours. To relate the oscillation activity in the LLZEF and IEF, we used a cross-wavelet analysis as described in the previous section. The cross-wavelet spectra are shown in Plate 5 (right) and indicate similar periodicities that are present simultaneously in the IEF and LLZEF spectra. In Plate 5 (top), the global cross-wavelet spectrum of the IEF/15 and of the three LLZEFs is displayed. The periodicities present in the cross-wavelet spectra are periodicities in the equatorial zonal electric fields most probably of geomagnetic origin and most probably associated with the electric field penetration mechanism.

3.2 LLZEF Perturbations in the 10-min to 10-h Period Range

In this section, we analyze the oscillation activity in the LLZEF and IEF wavelet spectra in the 10-min to 10-h period range for three storm case studies and a quiet-time case. The three storm cases examined are the 17–19 April 2001, 15–18 April 2002, and 9–12 November 2004 storm events, when concurrent observations of Jicamarca ISR zonal electric field and IEF data were available for at least 3 days. The quiet-time example 29 March to 2 April 2004 is only included for comparison purposes. Time and frequency domain analyses of the three storm events have been reported previously in the literature [e.g., *Kelley et al.*, 2007; *Nicolls et al.*, 2007; *Maruyama et al.*, 2007; *Anghel et al.* 2007], but this is the first time, to our knowledge, when wavelet analysis in the form presented here has been ever used to study the relationship between the equatorial zonal electric field and IEF data.

It is known that the two most important sources of ionospheric electric fields are (1) electric fields induced by solar tidal and thermospheric winds through the E and F region dynamo mechanisms and (2) electric fields of solar wind/magnetospheric origin generated at high altitudes by circulation of the magnetospheric plasma as a result of solar wind interaction with the Earth's magnetic field [*Gonzales et al.*, 1979]. While the atmospheric electric field sources are dominant at low and mid latitudes during quiet times, the magnetospheric sources become dominant at these latitudes during disturbed magnetic conditions. Thus, during storm times, the most important sources of ionospheric electric field disturbances are the prompt penetration electric fields and the disturbance dynamo electric fields, although significant contributions from other sources, such as substorms, neutral atmospheric waves, or spread F enhanced electric fields during nighttime, might also be involved. Previous studies have used different Fourier-based spectral analysis techniques to identify the sources of ionospheric electric field variability during disturbed magnetic conditions. *Earle and Kelley* [1987] performed Fourier analyses in the 1- to 10-h period range to identify the sources of ionospheric electric field at equatorial latitudes and study the frequency dependence of the penetration of the high-latitude magnetospheric convection electric fields to low latitudes. They reported that for Kp larger than 3 and periods less than about 5 h, the magnetospheric electric field sources dominate the atmospheric sources, the entire system acting as a high-pass filter with a peak in the 3- to 5-h period range and a roll-off near the 10-h period [*Vasyliunas*, 1972]. More recently, *Nicolls et al.* [2007], using large data sets of concurrent IEF and ΔH-inferred LLZEF data, studied the frequency dependence of the penetration mechanism using a frequency response function approach, showing that their filter, with IEF as input and LLZEF as output, peaks near the 2-h period, passes periodicities in the 30-min to 5-h range, attenuates the longer periodicities, and drops off for periodicities shorter than about 30 min. In this section, we show that, to some extent, the wavelet analysis allows us to better identify different sources of periodicities in the equatorial zonal electric fields, and that the filtering process is more complex than predicted by a time or a Fourier domain analysis, the wavelet analysis revealing in each of our case studies some peculiarities regarding the relationship between the IEF and LLZEF.

Plate 6a shows the wavelet results for the 18 April 2001 storm event. Plate 6a (top) displays the LLZEF (red line) and IEF (blue line), scaled by a factor of 5, as a function of local time for 17–19 April 2001 (107–109), when concurrent Jicamarca ISR zonal electric field and IEF data were available. The storm event was characterized by a daily Ap of 50 and commenced and developed mostly during the nighttime at Jicamarca on 17–18 April, which explains the strong anticorrelation between the two time series. The days before and after the event were quiet with a daily Ap of 7. This storm event was analyzed in more details by *Anghel et al.* [2007], and here we compare the spectra of the IEF and LLZEF in a wavelet domain with increased frequency resolution. In Plate 6a (middle and bottom), the wavelet amplitude spectra ($\omega_o = 30$) of the IEF and LLZEF, respectively, for periods ranging from 10 min to 10 h, are plotted as functions of local time and period, with the contour lines representing the 95% significance levels. The ratio between the maximum amplitudes in the two spectra is about 7, but for plotting purposes, we used a scaling factor of 5. A strong and highly oscillating IEF was recorded by the ACE satellite during the nighttime hours at Jicamarca. The wavelet amplitude spectrum of the IEF shows (1) strong periodicities less than about 1 h confined to a time interval of increased IEF, (2) significant periodicities in the 1.5- to 3-h range developing few hours prior to the onset of increased IEF activity, persisting over the entire time interval of increased IEF, and then fading away about 4 h after the increased activity in the IEF ceased, and (3) a less significant 4-h period of long time extent. As seen in the IEF wavelet spectrum, periodicities longer than about 1.5 h form a background spectrum on which high-frequency components, with periods less than about 1 h, superimpose during the interval of increased IEF activity. The LLZEF spectrum shows some significant frequency components with periods less than about 3 h that occur simultaneously in the IEF spectrum. Significant in the LLZEF spectrum are two spectral lines at about 1-h period and one spectral line near the 2-h period, each having a corresponding periodicity in the IEF spectrum. Of less significance are five spectral lines around the 3-, 4-, 5-, and 8-h periods. The 2.75-h period seems to be associated with a strong similar periodicity in the IEF spectrum, while the 3.2- and 5-h periods cannot be directly linked with similar periodicities in the IEF spectrum. They extend over long time intervals and might be associated with a disturbance dynamo effect. Although very weak and narrow-banded, the 4-h period has a corresponding periodicity in the IEF spectrum, while the 8-h period in the LLZEF spectrum has a relatively constant amplitude over the entire time interval and might be attributed to the terdiurnal tide.

The results for the 17 April 2002 storm event are presented in Plate 6b. Plate 6b (top) shows the Jicamarca zonal electric field (red line) and the IEF (blue line), scaled by a factor of 10, as a function of local time for 15–18 April 2002 (105–108). During this time interval, the daily Ap varied from 7, on 15 April, to 41 and 54, on 17 and 18 April. The storm commenced and developed on 17 April during the daytime hours at Jicamarca, which explains the strong correlation between the two signals. In this case study, we have a scenario when both the IEF and LLZEF evolved from quieter to more fluctuating values. The wavelet spectra for the IEF and LLZEF are shown in Plate 6b (middle and bottom, respectively). The ratio between the maxima of the two spectra is about 7. The IEF spectrum shows periods less than about 1 h during daytime on 17 April, when the IEF was highly fluctuating. Some significant spectral components with periods longer than 1 h can also be distinguished in the IEF spectrum: 1.5-, 2-, 3.5- to 4-, 6-, and 9-h periods. The 1.5-h period was present in the spectrum for about 24 h, between the midnights on 17 and 18 April and the 2-h period persisted between noontime on 16 April and midnight on 18 April, both periodicities being very strong in these time intervals. The other periodicities in the IEF spectrum have smaller amplitudes and extend from early morning on 16 April to noontime on 18 April. The LLZEF spectrum is dominated by a strong 8-h period over the entire time interval and, on 17 April, periodicities less than 1 h correlate very well with similar periodicities in the IEF spectrum. In addition, 1.5-, 2-, and 4-h periods also appear as significant in the LLZEF spectrum having very narrow bands compared with similar periodicities in the IEF spectrum but extending over the same time intervals.

The third case study, presented in Plate 6c, refers to the 9 November 2004 storm event, which is part of a more complex event that commenced on 7 November 2004. Plate 6c (top) displays the LLZEF (red line) and IEF (blue line), scaled by a factor of 10, as a function of local time for 9–12 November 2001 (314–317). Daily Ap values of 120 and 181 were registered on 9 and 10 November, respectively. The ratio between the maxima of the two spectra is about 7. The wavelet amplitude spectrum of the IEF is displayed in Plate 6c (middle) and shows significant periodicities less than about 2 h mostly confined to an interval of strong and highly fluctuating IEF. A strong 3-h period is also observed developing earlier on 9 November and persisting till about noon on 10 November. Other spectral lines with significant amplitudes are also observed in the IEF spectrum extending over long time intervals of more than 4–5 days, which possibly developed even earlier, maybe on 6 or 7 November. The LLZEF spectrum for this storm event is not as spectacular as the IEF spectrum, and there are no significant periodicities in the 3- to 8-h period band. Some significant periodicities less then about 2 h and a strong 3-h period similar with those observed in the IEF spectrum are distinguished, and two less significant and narrow-banded 4- and 6-h periods can also be noticed. The strong 9.5-h period might be associated with a disturbance dynamo effect since it does not have a direct correspondent in the IEF spectrum. During the entire interval, a strong 8-h period can be observed in the equatorial zonal electric field spectrum. The 8-h periodicity, most probably attributed to the terdiurnal tide, is present in all three storm events and also in the quiet-time example in Plate 6d but with different amplitudes in each case, suggesting that its magnitude might be dependent on the strength of the storm.

For two of the storm events presented here, 18 April 2001 and 17 April 2002, with daily Ap values of about 50, the IEF wavelet spectra are characterized by distinct period bands

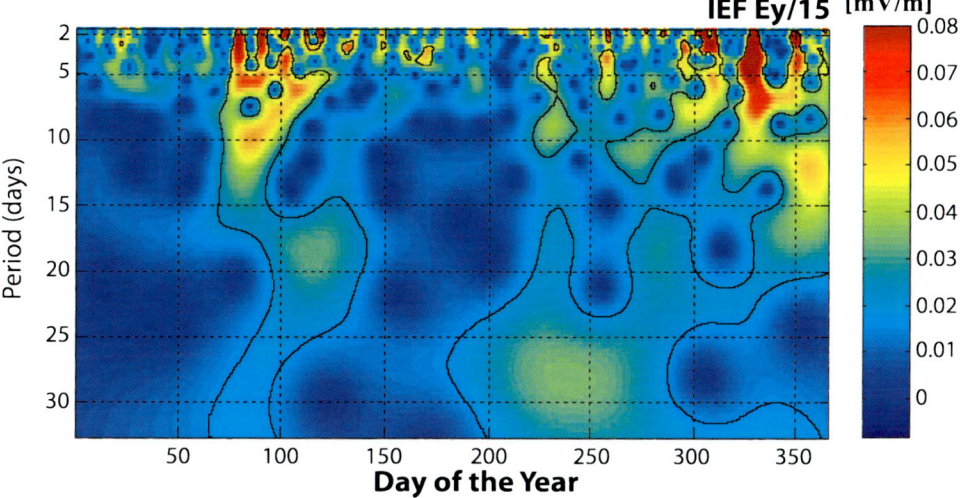

Plate 4. Wavelet amplitude spectrum of the IEF for the 1.5- to 33-d period range.

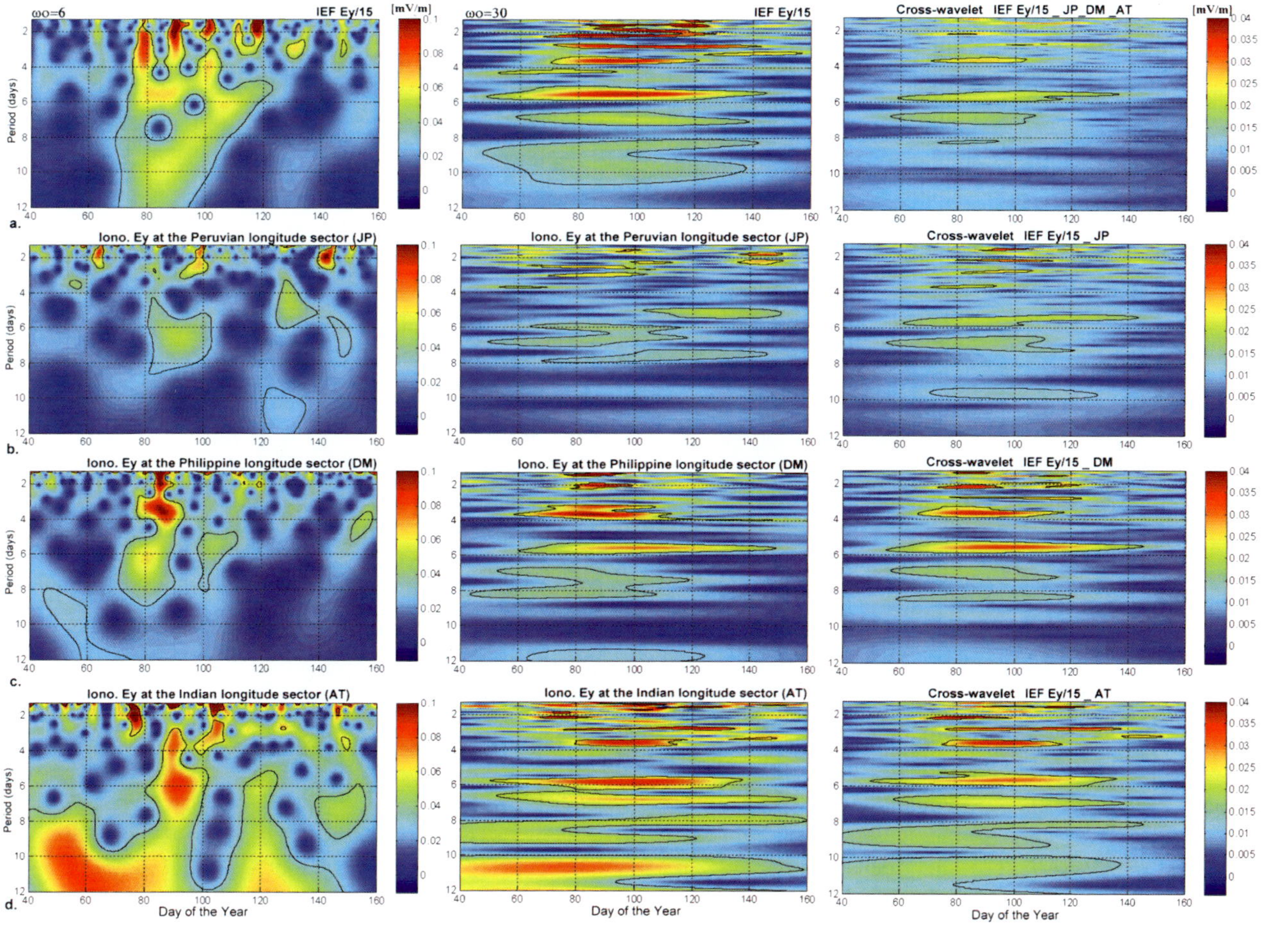

Plate 5. Wavelet amplitude spectra of the LLZEF at the three longitude sectors and of the IEF/15 in the 1.25- to 12-d period range and over 40- to 160-d time interval for (left) $\omega_o = 6$ and (middle) $\omega_o = 30$ and (right) the cross-wavelet spectra for $\omega_o = 30$.

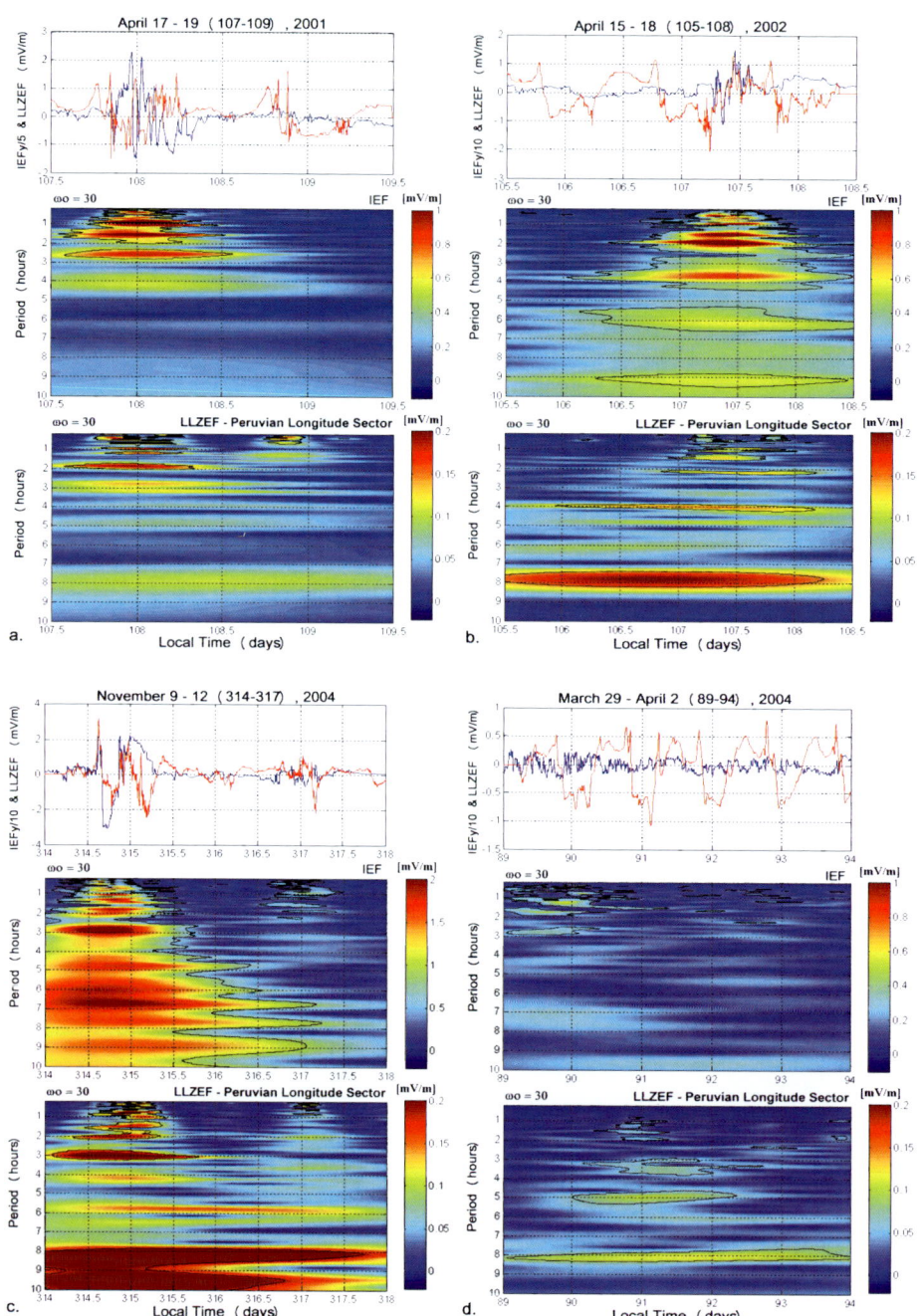

Plate 6. Time series of the (red line) LLZEF and (blue line) IEF and their wavelet amplitude spectra for the 10-min to 10-h period range, for (a) 17–19 April 2001, (b) 15–18 April 2002, (c) 9–12 November 2004, and (d) 29 March to 2 April 2004.

less than about 4 h that emerged from a noiselike background spectrum. Also, during the 9 November 2004 event, characterized by daily Ap values greater than 120 for three consecutive days, distinct period bands of large amplitudes are observed in the IEF spectrum, over the entire 10-min to 10-h period range, superimposed on a noiselike background spectrum. In each of the three storm cases, some of the significant periodicities in the IEF wavelet spectrum, especially those less than about 4 h, are also observed in the LLZEF spectrum as very narrow bands relative to their counterparts in the IEF spectrum, extending over the same time interval, but with a certain degree of attenuation [*Earle and Kelley*, 1987; *Nicolls et al.*, 2007]. In the 9 November 2004 case, the LLZEF wavelet spectrum does not show any significant periods in the 4- to 8-h range, only two weak and narrow spectral bands, although there is a quite strong oscillation activity in the IEF spectrum in this period range. We also showed that periodicities longer than about 1.5 h are present in both IEF and LLZEF spectra hours in advance of a visible onset of the storm, as in the case of the 17 April 2002 event, and many hours after the main phase of the storm, as in the 18 April 2001 and 9 November 2004 events.

We conclude that in the 10-min to 10-h period range, the system, with IEF as input and LLZEF as output, behaves like a highly nonlinear time-varying filter affecting more or less each periodicity in the spectrum, with possibly more attenuation on periodicities longer than about 4 h, as seen in the 9 November 2004 event. Similar and simultaneous periodicities present in both LLZEF and IEF wavelet spectra are most probably associated with penetration effects, while periodicities longer than about 3 h, which are present in the LLZEF spectrum but do not have a correspondent in the IEF spectrum, might be attributed to disturbance dynamo effects, although some other processes may also be accounted for.

4. CONCLUSION

In this paper, we used the continuous Morlet wavelet transform to relate the oscillation activity in the LLZEF and IEF spectra, in the 10-min to 10-h and 1.25- to 12-d period ranges, during time intervals of increased geomagnetic activity. The wavelet method described here can be easily tuned to different frequency resolutions in the wavelet domain and can provide accurate amplitude values of the periodicities present in the spectrum. Three examples of simulated wavelet spectra were presented to familiarize the reader with the method.

For periods in the 10-min to 10-h range, we showed that the wavelet method represents a powerful tool to study the frequency dependence of the two specific mechanisms of equatorial electric field variability, which are dominant during disturbed conditions, namely penetration and atmospheric disturbance dynamo, without the need to subtract the quiet-time components from the LLZEF and IEF data. In general, in separating out the two contributions, we associate a periodicity longer than about 3 h in the LLZEF wavelet spectrum with disturbance dynamo effects when it is not simultaneously present in the IEF wavelet spectrum, although some other processes may also be accounted for, and consider that a periodicity in the LLZEF wavelet spectrum is due to penetration effects most probably when it is simultaneously present in both LLZEF and IEF wavelet spectra. Previous studies associated a periodicity in the LLZEF Fourier spectrum with the penetration effects if it were present in both LLZEF and IEF Fourier spectra [e.g., *Earle and Kelley*, 1987; *Nicolls et al.*, 2007], although there is no time information about the occurrence of the periodicities in the two spectra. Also, since the wavelet method provides accurate amplitude values of the periodicities in the spectrum, it makes the wavelet transform suitable for studying the penetration efficiency and its time dependency by developing "instantaneous" frequency response functions in the same way like in the Fourier domain [*Kelley et al.*, 2003; *Huang et al.*, 2007; *Nicolls et al.*, 2007].

For periods in the 1.25- to 12-d range, the wavelet and cross-wavelet analyses of the IEF and LLZEF data over the 9 February to 9 June 2001 (40–160) time interval, indicate that there are significant periodicities in the LLZEF wavelet spectrum of possible geomagnetic origin, which are well correlated with similar and simultaneous periodicities in the IEF wavelet spectrum. We consider that these periodicities might be associated with penetration effects since we do not impose a limit on the maximum period that can be passed by the system, although it is not excluded that other processes may also be involved.

It is possible that the wavelet analysis in conjunction with physics-based models and ground-based and satellite data sets will bring more insight about the sources of periodicities in the LLZEF and about the system that links the IEF and the equatorial zonal electric fields. Here, we provide a new method of analysis and present a few case studies, but separating out different sources of ionospheric electric field variability still remains a task that requires further investigations.

Acknowledgment. Funding to carry out this study came from an NSF Space Weather grant (ATM#0207992).

REFERENCES

Abdu, M. A., T. K. Ramkumar, I. S. Batista, C. G. M. Brum, H. Takahashi, B. W. Reinisch, and J. Sobral (2006), Planetary wave signatures in the equatorial atmosphere–ionosphere system and mesosphere-E- and F-region coupling, *J. Atmos. Sol. Terr. Phys.*, *68*, 509–522.

Altadill, D., and E. Apostolov (2003), Time and scale size of planetary wave signatures in the ionospheric F region: Role of the geomagnetic activity and mesosphere/lower thermosphere winds, *J. Geophys. Res.*, *108*(A11), 1403, doi:10.1029/2003JA010015.

Anderson, D., A. Anghel, J. L. Chau, and O. Veliz (2004), Daytime vertical E × B drift velocities inferred from ground-based magnetometer observations at low latitudes, *Space Weather*, *2*, S11001, doi:10.1029/2004SW000095.

Anderson D., A. Anghel, J. L. Chau, and K. Yumoto (2006), Global, low-latitude, vertical E × B drift velocities inferred from daytime magnetometer observations, *Space Weather*, *4*, S08003, doi:10.1029/2005SW000193.

Anghel, A., D. Anderson, N. Maruyama, J. Chau, K. Yumoto, A. Bhattacharyya, and S. Alex (2007), Interplanetary electric fields and their relationship to low-latitude electric fields under disturbed conditions, *J. Atmos. Sol. Terr. Phys.*, *69*, 1147–1159.

Blanc, M., and A. D. Richmond (1980), The ionospheric disturbance dynamo, *J. Geophys. Res.*, *85*(A4), 1669–1686.

Earle, G. D, and M. C. Kelley (1987), Spectral studies of the sources of ionospheric electric fields, *J. Geophys. Res.*, *92*(A1), 213–224.

Fagundes, P. R., V. G. Pillat, M. Bolzan, Y. Sahai, F. Becker-Guedes, J. R. Abalde, S. L. Aranha, and J. A. Bittencourt (2005), Observations of *F* layer electron density profiles modulated by planetary wave type oscillations in the equatorial ionospheric anomaly region, *J. Geophys. Res.*, *110*, A12302, doi:10.1029/2005JA011115.

Fejer, B. G., and L. Scherliess (1997), Empirical models of storm-time equatorial zonal electric fields, *J. Geophys. Res.*, *102*, 24,047–24,056.

Forbes, J. F., S. E. Palo, and X. Zhang (2000), Variability of the ionosphere, *J. Atmos. Sol. Terr. Phys.*, *62*, 685–693.

Gonzales, C. A., M. C. Kelley, B. G. Fejer, J. F. Vickrey, and R. F. Woodman (1979), Equatorial electric fields during magnetically disturbed conditions, 2. Implications of simultaneous auroral and equatorial measurements, *J. Geophys. Res.*, *84*(A10), 5803–5812.

Huang, C.-S., J. C. Foster, and M. C. Kelley (2005), Long-duration penetration of the interplanetary electric field to the low-latitude ionosphere during the main phase of magnetic storms, *J. Geophys. Res.*, *110*, A11309, doi:10.1029/2005JA011202.

Huang, C.-S., S. Sazykin, J. Chau, N. Maruyama, and M. C. Kelley (2007), Penetration electric fields: Efficiency and characteristic time scale, *J. Atmos. Sol. Terr. Phys.*, *69*, 1135–1146.

Kelley, M. C., J. J. Makela, J. L. Chan, and M. J. Nicolls (2003), Penetration of the solar wind electric field into the magnetosphere/ionosphere system, *Geophys. Res. Lett.*, *30*(4), 1158, doi:10.1029/2002GL016321.

Kelley, M. C., M. J. Nicolls, D. Anderson, A. Anghel, J. Chau, R. Sekar, K. Subbarao, and A. Bhattacharyya (2007), Multi-longitude case studies comparing the interplanetary and equatorial ionospheric electric fields using an empirical model, *J. Atmos. Sol. Terr. Phys.*, *69*, 1174–1181.

Kumar, P., and E. Foufoula-Georgiou (1997), Wavelet analysis for geophysical applications, *Rev. Geophys.*, *35*(4), 385–412.

Laštovička, J. (2006), Forcing of the ionosphere by waves from below, *J. Atmos. Sol. Terr. Phys.*, *68*, 479–497.

Manson, A. H., C. E. Meek, T. Chshyolkova, S. K. Avery, D. Thorsen, J. W. MacDougall, W. Hocking, Y. Murayama, and K. Igarashi (2005), Wave activity (planetary, tidal) throughout the middle atmosphere (20–100km) over the CUJO network: Satellite (TOMS) and medium frequency (MF) radar observations, *Ann. Geophys.*, *23*, 305–323.

Maruyama, N., S. Sazykin, R. Spiro, D. Anderson, A. Anghel, R. Wolf, F. Toffoletto, T. J. Fuller-Rowell, M. Codrescu, and A. D. Richmond (2007), Modeling storm-time electrodynamics of the low-latitude ionosphere–thermosphere system: Can long lasting disturbance electric fields be accounted for?, *J. Atmos. Sol. Terr. Phys.*, *69*, 1182–1199.

Mendillo M., and K. Schatten (1983), Influence of solar sector boundaries on ionospheric variability, *J. Geophys. Res.*, *88*, 9145–9153.

Nicolls, M. J., M. C. Kelley, J. Chau, O. Veliz, D. Anderson, and A. Anghel (2007), The spectral properties of low latitude daytime electric fields inferred from magnetometer observations, *J. Atmos. Sol. Terr. Phys.*, *69*, 1160–1173.

Pancheva, D. (2000), Evidence for nonlinear coupling of planetary waves and tides in the lower thermosphere over Bulgaria, *J. Atmos. Sol. Terr. Phys.*, *62*(2), 115–132.

Pancheva, D., N. Mitchell, R. Clark, J. Drobjeva, and J. Lastovicka (2002), Variability in the maximum height of the ionospheric F2-layer over Millstone Hill (September 1998–March 2000): Influence from below and above, *Ann. Geophys.*, *20*(11), 1807–1819.

Parish, H. F., J. M. Forbes, and F. Kamalabadi (1994), Planetary wave and solar emission signatures in the equatorial electrojet, *J. Geophys. Res.*, *99*, 355–368.

Rishbeth, H., and M. Mendillo (2001), Patterns of F2-layer variability, *J. Atmos. Sol. Terr. Phys.*, *63*, 1661–1680.

Roberts, R. A., and C. T. Mullis (1987), *Digital Signal Processing*, Addison-Wesley, Boston, Mass.

Scherliess, L., and B. G. Fejer (1997), Storm time dependence of equatorial disturbance dynamo zonal electric fields, *J. Geophys. Res.*, *102*(A11), 24,047–24,056.

Scherliess, L., and B. G. Fejer (1999), Radar and satellite global equatorial F-region vertical drift model, *J. Geophys. Res.*, *104*(A4), 6829–6842.

Torrence, C., and G. P. Compo (1998), A practical guide to wavelet analysis, *Bull. Am. Meteorol. Soc.*, *79*, 61–78.

Vasyliunas, V. M. (1972), The interrelationship of magnetospheric processes, in *Earth's Magnetospheric Processes*, edited by B. M. McCormac, pp. 29–38, D. Reidel, Dordrecht, Netherlands.

D. Anderson and A. Anghel, Cooperative Institute for Research in Environmental Sciences, University of Colorado, Campus Box 216, Boulder, CO 80309, USA. (adela.anghel@noaa.gov)

A. Bhattacharyya, Indian Institute of Geomagnetism, Kalamboli Highway, New Panvel, Navi Mumbai 410218, India.

J. Chau, Radio Observatorio de Jicamarca, Instituto Geofisico del Peru, Jicamarca, Peru.

K. Yumoto, Space Environment Research Center, Kyushu University, 53 6-10-1 Hakozaki, Higashi-ku, Fukuoka 812-8581, Japan.

Simulation of PPEF Effects in Dayside Low-Latitude Ionosphere for the October 30, 2003, Superstorm

Olga P. Verkhoglyadova,[1,2] Bruce T. Tsurutani,[2] Anthony J. Mannucci,[2] Akinori Saito,[3] Tohru Araki,[3] David Anderson,[4] M. Abdu,[5] and J. H. A. Sobral[5]

One of the important signatures during strong magnetic storms is prompt penetrating electric fields (PPEFs) into the ionosphere, which causes the dayside ionospheric superfountain (DIS). Interplanetary-ionosphere coupling for the October 30, 2003, superstorm is analyzed by using ACE and ground-based measurements. The relationships between the interplanetary magnetic field B_z component, ionospheric vertical velocities above Jicamarca, and horizontal magnetic field components measured at Huancayo are presented. DIS is associated with uplift, displacement, and enhancement of the equatorial ionospheric anomalies. We apply an extended SAMI-2 ionospheric model to simulate DIS effects above Jicamarca for this superstorm. An agreement between our results and observed f_oF_2 during the main phase of the storm is reported. It is shown that the PPEF approach and corresponding modeling results capture the main physics of the dayside low-latitude ionospheric response during the first couple hours of the magnetic superstorm.

1. INTRODUCTION

The extreme disturbances of the ionosphere that occurred during the superstorms of October 29 and 30, 2003, are well documented in a special issue of *Geophysical Research Letters* (2005). It is believed that prompt penetrating electric fields (PPEFs) in the dayside ionosphere caused extreme $\mathbf{E} \times \mathbf{B}$ uplift of the equatorial ionospheric anomalies (EIAs) and dayside ionospheric superfountains (DISs) [*Tsurutani et al.*, 2004, 2006, 2008]. This uplift, latitudinal displacement, and photoionization dramatically increased total electron content (TEC) above 400 km for the near-equatorial to midlatitude regions [*Mannucci et al.*, 2005; *Verkhoglyadova et al.*, 2006]. In this paper, we will focus on two features of the DIS that have not been previously explored. First, we will discuss high time resolution interplanetary-ionospheric coupling during the superstorm of October 30, 2003. Second, we will apply an extended SAMI-2 ionospheric model to simulate motion of the ionospheric F layer over Jicamarca (−11.9° geographic latitude, 283.1° geographic longitude, 0.8° dip latitude). Ionospheric vertical velocities derived from ground magnetometer observations during 0700 to 1700 LT [*Anderson et al.*, 2006] will be used as input into the model. Results from the model will be directly compared with the observed f_oF_2. Advantages and limitations of our current modeling approach will be addressed in Section 4.

---

[1]CSPAR, University of Alabama in Huntsville, Huntsville, Alabama, USA.
[2]Jet Propulsion Laboratory, California Institute of Technology, Pasadena, California, USA.
[3]Department of Geophysics, Kyoto University, Kyoto, Japan.
[4]NOAA, Boulder, Colorado, USA.
[5]INPE, Sao Paulo, Brazil.

Midlatitude Ionospheric Dynamics and Disturbances
Geophysical Monograph Series 181
Copyright 2008 by the American Geophysical Union.
10.1029/181GM16

2. INTERPLANETARY AND GROUND-BASED MAGNETIC FIELD MEASUREMENTS RELATED TO THE OCTOBER 30, 2003, SUPERSTORM

The magnetic storm dynamics is indicated by geomagnetic indices shown in Figure 1a. The PPEF event studied is superposed on the recovery phase of a previous strong storm of October 29. The superstorm of October 30 starts with a supersubstorm at ~1700 UT (seen in AU/AL) that presumably causes a moderate level of asymmetric ring current disturbances (represented by ASY-H). The magnetic storm main phase commences at ~1840 UT, reaching local maximum intensity (D_{st} ~ −320 nT) at ~2100 UT, as seen from the ASY-H profile. Peak D_{st} excursion of ~ −390 nT (not shown) was recorded at 2315 UT [*Mannucci et al.*, 2005]. The B_z component of the interplanetary magnetic field (IMF) is plotted in Figure 1b. IMF measurements during the magnetic storm are taken from ACE, which was located in the

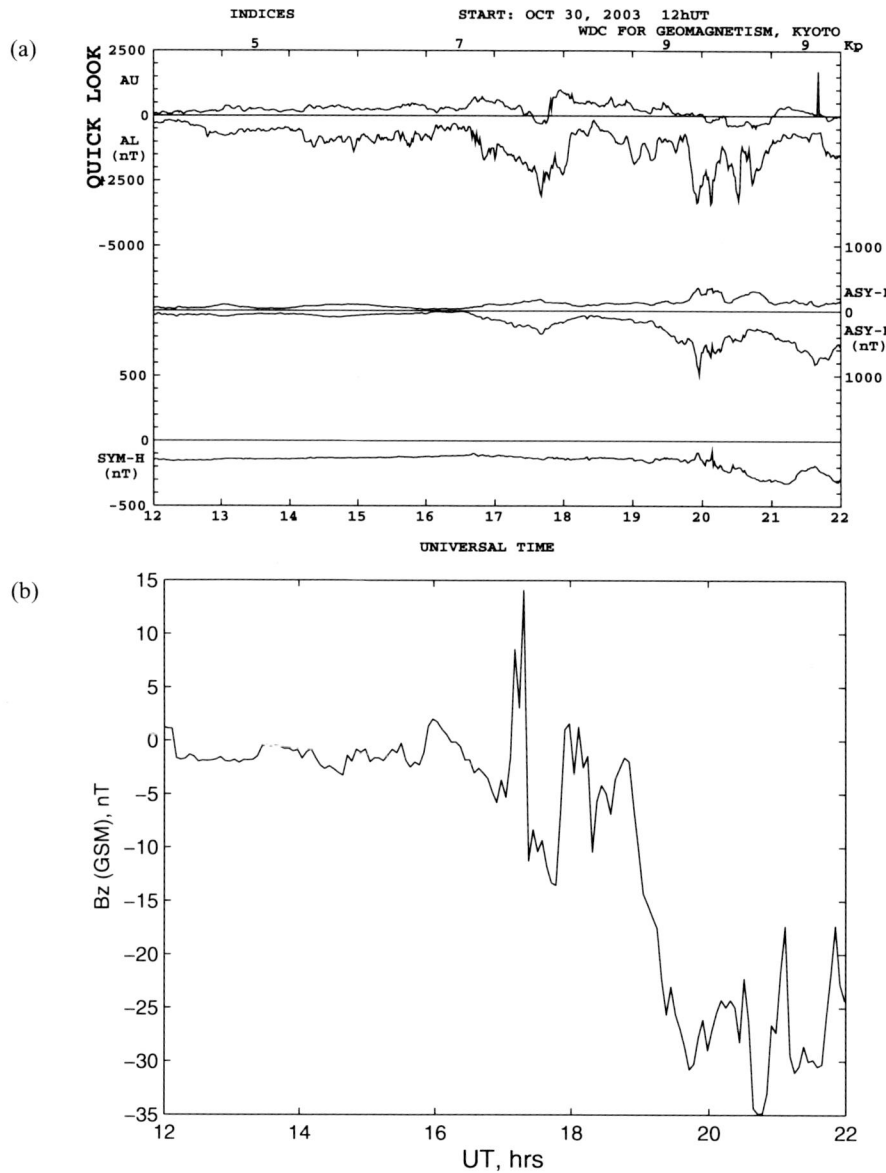

Figure 1. (a) Geomagnetic indices provided by WDC Kyoto for October 30, 2003, superstorm and (b) IMF B_z taken from ACE. The ACE data have been shifted by ~35 min in the plot due to a solar wind convectional delay. Difference between UT and local time at Jicamarca location is 5 h.

solar wind at (230, 41.3, −20.3)Re in GSE coordinates. We assume a measured speed of 700 km/s to derive a solar wind convection delay of ~35 min. The solar wind data have been shifted by this amount in the plot. There is a clear correspondence between southward IMF B_z and ring current disturbances (presented by SYM/ASY-H) as the magnetic storm progresses, as has been noted for all magnetic storms [*Gonzalez and Tsurutani*, 1987; *Gonzalez et al.*, 1994]. The asymmetric ring current is enhanced during the start of the storm main phase (seen in ASY-H), whereas SYM-H is a good indicator of an enhanced symmetric current during the main and recovery phases [*Liemohn et al.*, 2001; *Wanliss and Showalter*, 2006].

What are the global ionospheric consequences of a strong magnetic storm at low latitudes in the daytime ionosphere? In this paper, we will focus on several phenomena associated with penetration of magnetospheric electric fields into the low-latitude ionosphere under storm-time conditions (see *Abdu et al.* [1991], *Abdu* [1997], *Sobral et al.* [2001], *Tsurutani et al.* [2004, 2006], *Batista et al.* [2006] for a complete review of PPEF-associated ionospheric disturbances). The interplanetary motional electric field is defined by vector product of the solar wind flow velocity (V_{SW}) and IMF B as $V_{SW} \times B$. If the IMF has a southward component (B_z <0), the corresponding interplanetary electric fields are in the dawn-to-dusk direction or eastward (westward) on the dayside (nightside). Observations show that these electric fields partially penetrate into the ionosphere [*Nishida*, 1968; *Sastri*, 1988; *Kelley et al.*, 1979, 2003] during intense magnetic storms as PPEFs with an efficiency of ~5 to 10%. These ionospheric electric fields are associated with vertical drift velocities of the plasma. In the case of southward IMFs, the PPEFs lead to plasma uplift of the dayside ionosphere causing DIS and a positive ionospheric storm. The main properties of positive ionospheric storms at low and middle latitudes have been reviewed by *Prölss* [1993, 1995]. It should be noted that this $\mathbf{E} \times \mathbf{B}$ storm-time drift is, in addition to the vertical quiet-time drift, associated with diurnal variations due to the ionospheric dynamo zonal electric field. However, $\mathbf{E} \times \mathbf{B}$ drift velocities during superstorms may exceed the velocities associated with quiet-time diurnal variations by almost 1 order of magnitude [*Tsurutani et al.*, 2004], and the quiet-time variations may be neglected, to first order. Another point is that strictly speaking, vertical velocities are the result of $\mathbf{E} \times \mathbf{B}$ drift orthogonal to the magnetic field lines and of meridional winds along magnetic field lines. The contribution to the total velocity vector from these two processes depends on the local magnetic inclination [see, e.g., *Abdu*, 1997]. Because the magnetic fields are horizontal at the magnetic equator, meridional winds do not cause vertical drifts. Away from the magnetic equator, meridional winds could contribute to the vertical drift but, depending on the magnetic inclination, this is considered to be only of secondary importance to the process of our interest and will not be discussed further. Hereafter in our discussion of PPEF effects, we assume that storm-time $\mathbf{E} \times \mathbf{B}$ drift is the predominant contribution to the vertical drift velocity (V_{drift}) in the low-latitude ionosphere.

The PPEF mechanism of ionospheric uplift implies dependence between the IMF B_z and V_{drift}, in terms of $\mathbf{E} \times \mathbf{B}$ drift, of the dayside low-latitude ionosphere. To study this relationship, we use experimental data of the ionospheric vertical drift velocity derived from the measurements above Jicamarca, Peru (0.8° dip latitude), during 0700 to 1700 LT on October 30, 2003. For details of the technique used, we refer to Figure 2a in *Anderson et al.* [2006] and related discussion. The drift velocities inferred from ground magnetometer observations with 5-min cadence are plotted in Figure 2. Values of $-B_z$ from ACE measurements in the solar wind near the Earth are shown by a solid line. The ACE measurements have been shifted by ~35 min and converted to Jicamarca local times. At the beginning of the time interval up to ~1100 LT, the V_{drift} of ~ −17 m/s is in the negative direction (plasma downdraft). This interval is associated with a weak southward IMF (B_z ~ −3 nT) in the recovery phase of the strong magnetic storm of October 29. At 1300 LT the

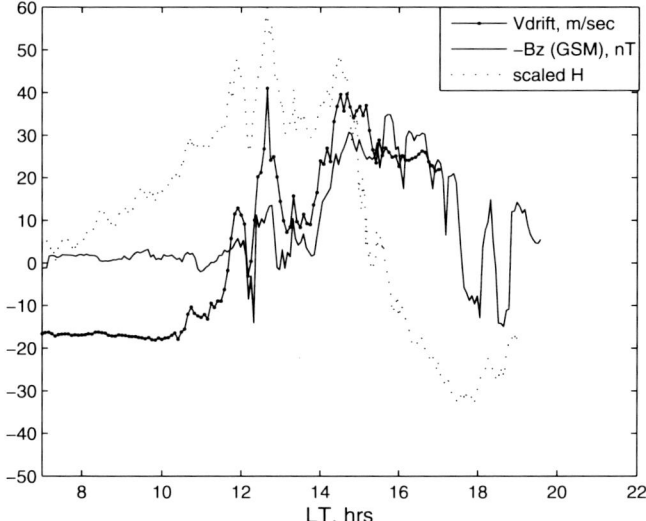

Figure 2. The vertical drift velocity derived from ground magnetometer observations above Jicamarca [*Anderson et al.*, 2006] (solid dot line), $-B_z$ measured by ACE (solid line) and scaled H component of the geomagnetic field from Huancayo (dotted line) for the October 30, 2003, superstorm. The ACE data have been shifted by ~35 min in the plot due to a solar wind convectional delay. Time is shown in LT of the Jicamarca location.

ionospheric drift velocity increases dramatically to ~40 m/s. At the same time, the IMF B_z decreases from ~0 to -12 nT. Another large increase in V_{drift} to ~40 m/s occurs at ~1500 LT and corresponds to IMF B_z ~ -30 nT and the main phase of the magnetic storm. Note that there is an ~5-h difference between UT and LT for the Jicamarca location. There is a correlation between IMF $-B_z$ (inverted B_z) and V_{drift} in Figure 2, especially at ~1200, 1300, 1330, 1400, 1500, and 1700 LT. These times correspond to local maxima of V_{drift}. This correlation indicates that strong vertical ionospheric uplift is presumably driven by the southward component of IMF.

We obtained 1-min measurements of the horizontal H component of the magnetic field at Huancayo at approximately the same geographical location as Jicamarca (-11.9° geographic latitude, 283.1° geographic longitude, 0.8° dip latitude) and plot them in Figure 2 (dotted line). The data were shifted along the Y axis and scaled by the factor of 10^5 to be plotted within the same data limits as V_{drift} and $-B_z$. The horizontal magnetic field disturbances are primarily produced by enhanced equatorial electrojets (EEJs), since the station is located almost at the magnetic equator. We suggest that the EEJ variations are mostly caused by storm-time PPEFs, which in turn are caused by southward and northward turnings of IMF B_z. The ring current enhancement has a lesser contribution and is also controlled by the IMF B_z. Thus, the observed correlation between the ground magnetic field H component and V_{drift} suggests a strong link between the electrodynamics of the strong magnetic storm and the ionospheric uplift. Figure 2 shows no significant delay (<1 h) between IMF B_z, ground-based measurements of the disturbed geomagnetic field near the magnetic equator and the ionospheric response. This result is consistent with prompt transmission of the horizontal electric field from the polar ionosphere to the equatorial region [*Araki*, 1977; *Kikuchi and Araki*, 1979; *Tsunomura and Araki*, 1984; *Araki*, 1994]. Traveling ionospheric disturbances, on the other hand, can be initiated by the magnetic storm and propagate from high to low latitudes for 3 to 4 h, thus contributing to the ionospheric storm at later times [*Prölss*, 1995]. This multihour time delay is too long to account for the observations presented here, however.

3. MODELING OF IONOSPHERIC ELECTRON DENSITY DYNAMICS AND THE CRITICAL FREQUENCY OF F_2 LAYER OVER JICAMARCA DURING THE OCTOBER 30, 2003, EVENT

In this section, we will present simulation results of ionospheric dynamics in the Peruvian sector during the first hours of the southward turning of the IMF. The numerical approach is based on the NRL SAMI-2 ionospheric model [*Huba et al.*, 2000]. This model has been extended to include enhanced vertical drifts due to PPEFs. We use the drift velocities inferred from ground magnetometer observations (V_{drift}) [*Anderson et al.*, 2006] (Figure 2) as input to the code to model the ionospheric storm-time effects. In the previous section, we have discussed that positive ionospheric vertical drifts above Jicamarca are related to southward IMFs. We assume that the upward vertical drifts are driven by storm-time PPEFs. Hereafter in the paper, the numerical model with V_{drift} input will be referred to as the PPEF model. For comparison, we use two quiet-time ionospheric vertical drift models: the diurnal "sine" model and the Fejer–Scherliess model [*Scherliess and Fejer*, 1999; *Huba et al.*, 2000]. The ionospheric vertical drift in the "sine" model is defined by $V_0 \times \sin[(t - 7)/24]$, where the magnitude of V_0 is 15 m/s and t is the local time in hours. In Figure 3, input vertical drifts of the "sine" model (dashed) and V_{drift} (solid) are plotted against local time at the Jicamarca location. This modeling effort with SAMI-2 is an extension of the approach presented in *Anderson et al.* [2006].

Figure 4 presents "temporal snapshots" of the electron density distribution before the storm main phase. The results are obtained with the PPEF model (right) and with the "sine" or the quiet-time model (left). The electron densities at ~1049 LT are relatively low (see Figure 4b) as compared to the quiet-time values (Figure 4a). This can be explained by negative values of the vertical drift velocity from 0700 until 1130 LT (Figure 3, solid line) and the resulting downdraft of the ionosphere. A strong southward turning of the

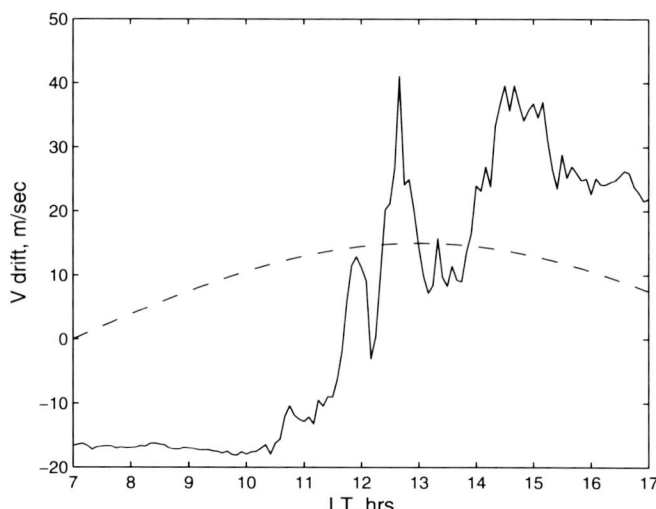

Figure 3. The vertical drift velocity above Jicamarca on October 30, 2003 [*Anderson et al.*, 2006] (solid line), and vertical drift from the "sine" model (dashed line) plotted versus LT at Jicamarca.

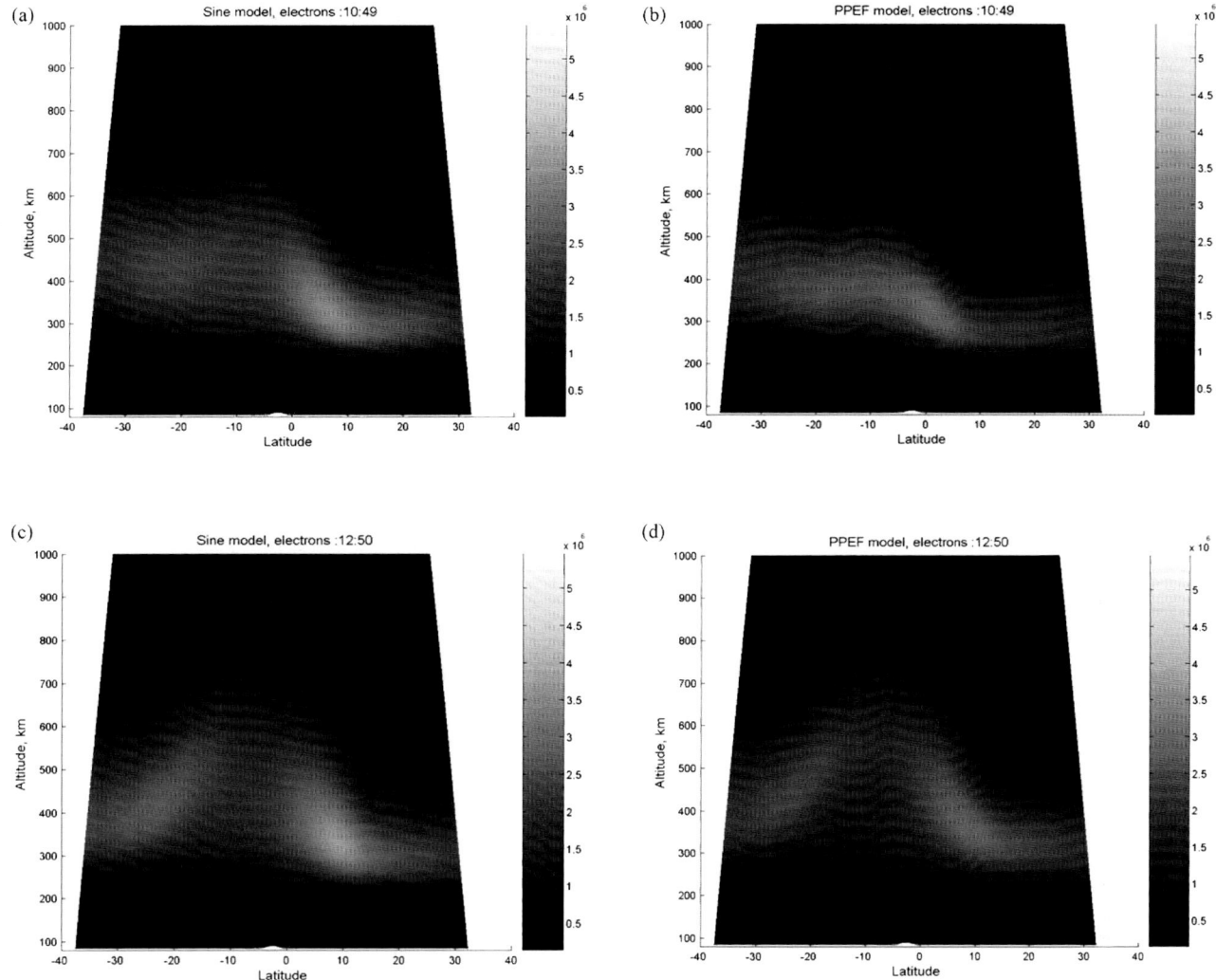

Figure 4. Modeled electron density distribution at the beginning of the ionosphere uplift at using "sine" model (left) and V_{drift} over Jicamarca (right). Time is shown in LT: (top) ~1049 LT and (bottom) ~1250 LT.

IMF B_z occurred around ~1130 LT as seen in Figure 1b (southward field corresponds to $-B_z > 0$). After ~1130 LT, the ionosphere starts moving upward, and V_{drift} exceeds the quiet-time vertical velocity after ~1220 LT (see Figure 3). There is a small difference between the two model results at ~1250 LT (see Figures 4c and 4d) and the EIA crests are even more enhanced for the "sine" model, probably because the storm-time uplift was suppressed at earlier times. At ~1606 LT, EIAs are uplifted, enhanced, and displaced to higher latitudes (see Figure 5b). Quiet-time EIAs created by the regular ionospheric fountain are shown in Figure 5a for comparison. At the end of the superfountain event at ~1850 LT, the uplifted plasma is noted to descend along the local magnetic field lines, which causes enhancement of the electron density in northern hemisphere at ~350–600 km in altitude and around −25° and 20° in geographic latitude. This effect is shown in Figure 5d. The quiet-time fountain also reverses its direction at dusk causing plasma downflow shown in Figure 5c. However, this effect is much smaller in terms of electron density enhancements and the EIAs are located around −20° and 15° in geographic latitude. Modeling results for the local times beyond 1700 LT are based on the quiet-time "sine" model, since no drift velocity data are available after that time. Our results reflect a general scenario of plasma downflow at the end of a superfountain event [*Verkhoglyadova et al.*, 2006; *Tsurutani et al.*, 2007].

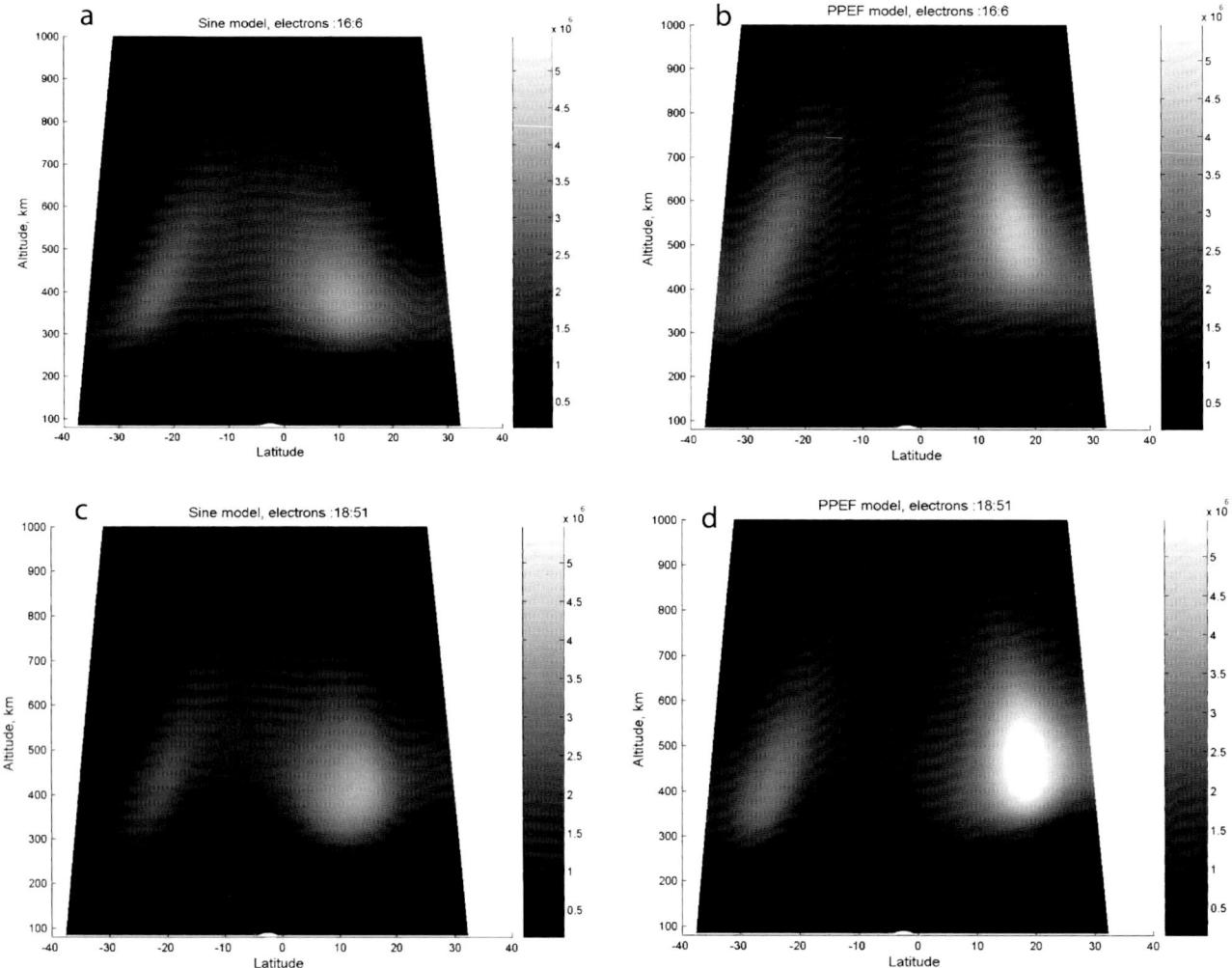

Figure 5. Modeled electron density distribution during ionosphere uplift and downdraft using "sine" model (left) and V_{drift} over Jicamarca (right; up to 1700 LT). Time is shown in LT: (top) ~1606 LT and (bottom) ~1851 LT.

Figure 6a shows results of modeling of the critical frequency of the F layer (f_oF_2) over Jicamarca (shown by triangles) using the PPEF model based on the V_{drift} inferred from ground magnetometer observations. Observed values of f_oF_2 are shown by solid line. There is a reasonably good agreement between the modeled and the observed values from 0700 until 1400 LT. Unfortunately, data are not available from 1010 to 1245 LT. The PPEF model predicts the peak value ~14.8 MHz after ~1100 LT, which is close in time to the strong southward turning of IMF B_z (Figure 2) discussed above. The "sine" model (shown by circles) gives a definite underestimation of the uplift. The quiet time model cannot explain the ionospheric observations during the strong positive ionospheric storm. Our PPEF model also shows poor agreement with the observations at the end of the uplift interval.

Figure 6b presents a comparison among the observed f_oF_2 (solid line), PPEF model (triangles), and the Fejer–Scherliess model (diamonds). There is poor agreement between observations and the Fejer–Scherliess results. We attribute the discrepancy to the fact that the Fejer–Scherliess model [*Scherliess and Fejer*, 1999] is a statistical quiet-time model, thus one cannot use it to reproduce ionospheric effects during strong magnetic storms and disturbed conditions.

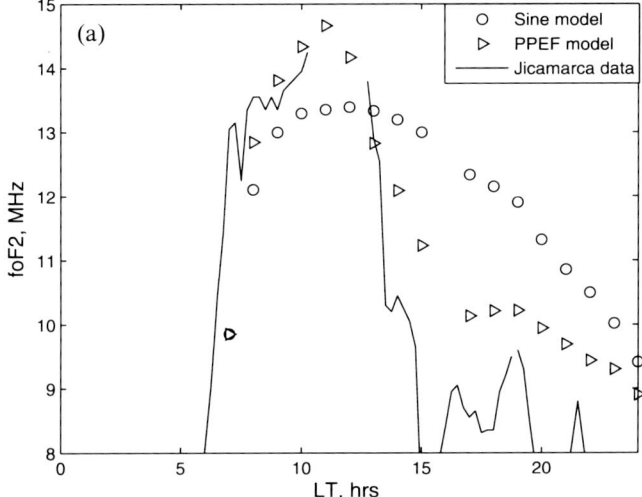

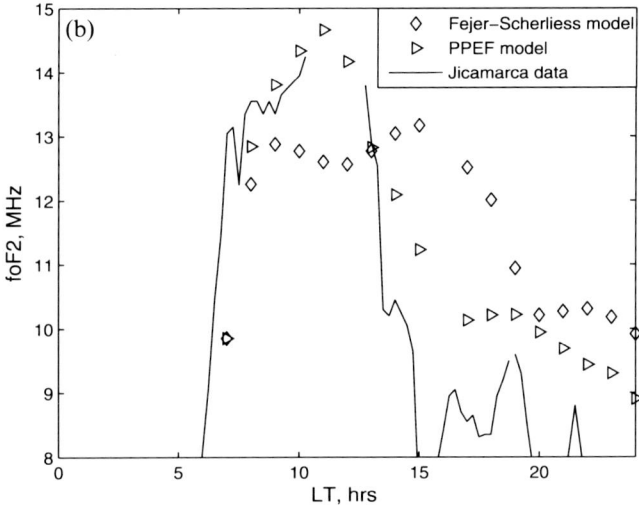

Figure 6. Results of f_oF_2 modeling using the storm-time PPEF model (triangles). Observations over Jicamarca are shown by solid line. Two quiet-time ionospheric models are plotted for comparison: (a) "sine" model (circles) and (b) Fejer–Scherliess model (diamonds).

4. DISCUSSION AND CONCLUSIONS

The paper presents evidence of a close relationship between the southward IMF component, H component of the horizontal magnetic field, and local vertical drift velocities of the equatorial ionosphere during the main phase of the October 30, 2003, magnetic storm. The results indicate that the dayside ionospheric uplift is storm-time electric field-driven. There is very little delay (minutes?), if any, between the interplanetary driver and the ionospheric response, once the solar wind convection delay from the satellite to the magnetosphere has been removed.

We used the vertical velocities over Jicamarca inferred from ground magnetometer observations as input into the SAMI-2 ionospheric model to simulate electron density dynamics during this superstorm. The PPEFs directly drive ionospheric uplift. Electron density decreased in comparison to the quiet-time model at the beginning of the event presumably due to initial overshielding from the previous storm. Later, the density became enhanced and the EIAs moved upward and toward high latitudes as the storm progressed. The uplifted plasma started to descend along the magnetic field lines during the recovery phase of the storm. This evolution of the electron density is consistent with the DIS scenario. Peak magnitudes of EIAs enhancements, maximum altitude, and latitude of their displacements are dependent on timing of the storm main phase. Thus, the morning or post-dawn effects of the PPEFs will be strongly diminished, partly because of low initial ion densities supplied by photoionization at daytime and weaker eastward electric fields at these times (see below).

We modeled the critical frequency of F_2 layer over Jicamarca with the PPEF model. Results show reasonably good agreement with direct radar measurements during the first several hours of the f_oF_2 enhancement, from 0700 LT until ~1400 LT. The reference quiet-time models underestimated the true values. Our PPEF model shows poor agreement with the observations at the end of the uplift. This discrepancy might be due to the effects of ionospheric heating and resultant winds in the equatorial ionosphere–thermosphere system caused by the superfountain [*Abdu*, 1997]. We expect that these effects become important after several hours and should be incorporated into future models.

As the next step, more sophisticated model of PPEFs at local dawn should be developed. Eastward electric fields applied to the polar ionosphere produce afternoon and morning current vortices of the DP2 type current system [*Nishida*, 1968; *Araki et al.*, 1985]. The afternoon current vortex is always larger than the morning vortex and produces strong eastward EEJs along the dayside dip equator. Morning vortices dominate from midnight to dawn local time and impose electric fields in the opposite direction [*Tsunomura and Araki*, 1984]. As a result, ionospheric electric fields at the dip equator reverse their direction from nightside to dayside. Storm-time ionospheric uplift after dawn is possible if the afternoon vortex rotates to the nightside during strong ionospheric disturbances. The combined effect of two current vortices, however, produces a much weaker eastward electric field during morning hours than that around noon. Thus, PPEF magnitude for the morning ionospheric superfountain should be reduced compared to that of at noon local time. Extended

PPEF model should also take into account altitudinal and local time dependencies of the ionospheric conductivity.

The PPEF approach and the modeling results capture the main physics of the low-latitude ionospheric response during first couple hours of the October 30, 2003, magnetic storm. However, other effects will become important as the storm develops. Further research efforts should include:

- ionosphere heating and disturbance dynamo into a general model;
- altitude and local time dependence of the PPEF.

The study is currently in progress.

Acknowledgments. Portion of this work was performed at the Jet Propulsion Laboratory, California Institute of Technology, under contract with NASA. B. T. Tsurutani and O. P. Verkhoglyadova wish to thank RISH, Kyoto University, for hosting us during our visit and while this paper was being written. This work uses the SAMI-2 ionosphere model written and developed by the Naval Research Laboratory. The ACE magnetic field data were provided by N. Ness through the CDAWeb. We thank WDC Kyoto for support and providing the data.

REFERENCES

Abdu, M. A. (1997), Major phenomena of the equatorial ionosphere–thermosphere system under disturbed conditions, *J. Atmos. Sol. Terr. Phys.*, *59*(13), 1505.

Abdu, M. A., I. S. Batista, G. O. Walker, J. H. A. Sobral, N. B. Trivedi, and E. R. de Paula (1995), Equatorial ionospheric electric field during magnetospheric disturbances: Local time/longitude dependences from recent EITS campaigns, *J. Atmos. Terr. Phys.*, *57*, 1065.

Anderson, D., A. Anghel, E. Araujo, V. Eccles, C. Valladares, and C. Lin (2006), Theoretically modeling the low-latitude, ionospheric response to large geomagnetic storms, *Radio Sci.*, *41*, RS5S04, doi:10.1029/2005RS003376.

Araki, T. (1977), Global structure of geomagnetic sudden commencements, *Planet. Space Sci.*, *25*, 373.

Araki, T. (1994), A physical model of the geomagnetic sudden commencement, in *Solar Wind Sources of Magnetospheric Ultra-Low-Frequency Waves, Geophys. Monogr. Ser.*, vol. 81, edited by M. J. Engebretson, K. Takahashi, and M. Scholer, pp. 183–200, AGU, Washington, D. C.

Araki, T., J. H. Allen, and Y. Araki (1985), Extension of a polar ionospheric current to the nightside equator, *Planet. Space Sci.*, *33*(1), 11–16.

Batista, I. S., M. A. Abdu, J. R. Souza, F. Bertoni, M. T. Matsuoka, P. O. Camargo, and G. J. Bailey (2006), Unusual early morning development of the equatorial anomaly in the Brazilian sector during the Halloween magnetic storm, *J. Geophys. Res.*, *111*, A05307, doi:10.1029/2005JA011428.

Gonzalez, W. D., and B. T. Tsurutani (1987), Criteria of interplanetary parameters causing intense magnetic storms ($D_{st}<-100$ nT), *Planet. Space Sci.*, *35*, 1101.

Gonzalez, W. D., J. A. Joselyn, Y. Kamide, H. W. Kroehl, G. Rostoker, B. T. Tsurutani, and V. M. Vasyliunas (1994), What is a geomagnetic storm?, *J. Geophys. Res.*, *99*, 5771–5792.

Huba, J. D., G. Joyce, and J. A. Fedder (2000), Sami2 is another model of the ionosphere (SAMI2): A new low-latitude ionosphere model, *J. Geophys. Res.*, *105*(A10), 23,035–23,053.

Kelley, M. C., B. G. Fejer, and C. A. Gonzales (1979), An explanation for anomalous equatorial ionospheric electric fields associated with a northward turning of the interplanetary magnetic field, *Geophys. Res. Lett.*, *6*(4), 301–304.

Kelley, M. C., J. J. Makela, J. L. Chau, and M. J. Nicolls (2003), Penetration of the solar wind electric field into the magnetosphere/ionosphere system, *Geophys. Res. Lett.*, *30*(4), 1158, doi:10.1029/2002GL016321.

Kikuchi, T., and T. Araki (1979), Transient response of uniform ionosphere and preliminary reverse impulse of geomagnetic storm sudden commencement, *J. Atmos. Terr. Phys.*, *41*, 917.

Liemohn, M. W., J. U. Kozyra, M. F. Thomsen, J. L. Roeder, G. Lu, J. E. Borovsky, and T. E. Cayton (2001), Dominant role of the asymmetric ring current in producing the stormtime Dst^*, *J. Geophys. Res.*, *106*(A6), 10,883–10,904.

Mannucci, A. J., B. T. Tsurutani, B. A. Iijima, A. Komjathy, A. Saito, W. D. Gonzalez, F. L. Guarnieri, J. U. Kozyra, and R. Skoug (2005), Dayside global ionospheric response to the major interplanetary events of October 29–30, 2003 "Halloween storms," *Geophys. Res. Lett.*, *32*, L12S02, doi:10.1029/2004GL021467.

Nishida, A. (1968), Coherence of geomagnetic *DP* 2 fluctuations with interplanetary magnetic variations, *J. Geophys. Res.*, *73*, 5549–5559.

Prölss, G. W. (1993), Common origin of positive ionospheric storms at middle latitudes and the geomagnetic activity effect at low latitudes, *J. Geophys. Res.*, *98*, 5981–5991.

Prölss, G. W. (1995), Ionospheric *F*-region storms, in *Handbook of Atmospheric Electrodynamics*, vol. 2, edited by H. Volland, pp. 195–248, CRC Press, Boca Raton.

Sastri, J. H. (1988), Equatorial electric fields of the disturbance dynamo origin, *Ann. Geophys.*, *6*, 635.

Scherliess, L., and B. G. Fejer (1999), Radar and satellite global equatorial *F* region vertical drift model, *J. Geophys. Res.*, *104*, 6829–6842.

Sobral, J. H. A., M. A. Abdu, W. D. Gonzalez, C. S. Yamashita, A. L. Clua de Gonzalez, I. Batista, and C. J. Zamlutti (2001), Responses of the low latitude ionosphere to very intense geomagnetic storms, *J. Atmos. Sol. Terr. Phys.*, *63*, 965.

Tsunomura, S., and T. Araki (1984), Numerical analysis of geomagnetic sudden commencement, *Planet. Space Sci.*, *32*, 599.

Tsurutani, B., et al. (2004), Global dayside ionospheric uplift and enhancement associated with interplanetary electric fields, *J. Geophys. Res.*, *109*, A08302, doi:10.1029/2003JA010342.

Tsurutani, B. T., et al. (2006), The Dayside Ionospheric "Superfountain" (DIS), Plasma transport and other consequences, in *Solar Influence on the Heliosphere and Earth's Environment:*

Recent Progress and Prospects, Proc. ILWS Workshop, Goa, India, February 19–24, edited by N. Gopalswamy and A. Bhattacharyya, pp. 384–387, Quest Publ., Mumbai, India.

Tsurutani, B. T., O. P. Verkhoglyadova, A. J. Mannucci, T. Araki, A. Saito, T. Tsuda, and K. Yumoto (2007), Oxygen ion uplift and satellite drag effects during the 30 October 2003 daytime superfountain event, *Ann. Geophys.*, *25*, 569.

Tsurutani, B. T., et al. (2008), Prompt penetration electric fields (PPEFs) and their ionospheric effects during the great magnetic storm of 30–31 October 2003, *J. Geophys. Res.*, *113*, A05311, doi:10.1029/2007JA012879.

Verkhoglyadova, O. P., B. T. Tsurutani, and A. J. Mannucci (2006), Temporal development of dayside TEC variations during the October 30, 2003, superstorm, edited by M. Duldig et al., World Scientific Co., Pte. Ltd., Singapore, *Adv. Geosci.*, *8*, 69.

Wanliss, J. A., and K. M. Showalter (2006), High-resolution global storm index: *Dst* versus SYM-H, *J. Geophys. Res.*, *111*, A02202, doi:10.1029/2005JA011034.

M. Abdu and J. H. A. Sobral, INPE, São José dos Campos, Sao Paulo, SP 12201-97D, Brazil.

D. Anderson, NOAA, Boulder, CO 80305-3328, USA.

T. Araki and A. Saito, Department of Geophysics, Kyoto University, Kyoto 606-8502, Japan.

A. J. Mannucci, B. T. Tsurutani, and O. P. Verkhoglyadova, Jet Propulsion Laboratory, California Institute of Technology, 4800 Oak Grove Drive, Pasadena, CA 91109, USA. (Olga.Verkhoglyadova@jpl.nasa.gov)

Impact of the Neutral Wind Dynamo on the Development of the Region 2 Dynamo

T. W. Garner

Space and Geophysics Laboratory, Applied Research Laboratories, University of Texas at Austin, Austin, Texas, USA

Geoff Crowley

Atmospheric and Space Technology Research Associates, San Antonio, Texas, USA

R. A. Wolf

Department of Physics and Astronomy, Rice University, Houston, Texas, USA

Three separate dynamos create the electric fields that impact the Earth's ionosphere. To first order, the currents and electric fields generated by these dynamos can be computed separately, and the superposition of these electric fields used to drive the ionospheric dynamics. However, this methodology ignores important second-order effects, in which the electric field from one dynamo system drives the plasma within another dynamo. This paper investigates the second-order impact of the neutral-wind dynamo on the Region 2 Birkeland current dynamo. The Region 2 dynamo is calculated by the Rice Convection Model with and without an input, non-self-consistent neutral-wind dynamo electric field from the Thermosphere–Ionosphere–Mesosphere Electrodynamics-General Circulation Model. Comparisons between these two model runs show that the wind-generated electric field alters the Region 2 Birkeland currents and the resultant electric field. The additional electric field alters the near-Earth plasma pressure gradients, which in turn drive the Region 2 dynamo to produce an "overshielding" electric field. The potential drop at 10° magnetic latitude is more than 4.5 times greater in the RCM run using the wind-generated electric field. Hence, the second-order influence of the combined dynamos is important to the Region 2 dynamo system.

1. INTRODUCTION

The dynamics of the midlatitude ionosphere is a complex superposition of various forces and their interactions with the Earth's magnetic field. The two dominant forcing terms are neutral wind drag and the electric field. While numerous studies have examined the role of neutral wind dynamics on the ionosphere [e.g., *Prölss*, 1995; *Rishbeth*, 1998; *Lin et al.*, 2005], much of the focus of recent studies on midlatitude dynamics focuses on the influence of the electric field [e.g., *Foster et al.*, 2005; *Huba et al.*, 2005; *Lin et al.*, 2005; *Mannucci et al.*, 2005].

The midlatitude electric field is generated through three separate dynamos. The first dynamo is associated with the

Region 1 or driving Birkeland currents. It is frequently the strongest dynamo, and the electric field generated is called the convection electric field. On the dayside, the Region 1 currents are related to Alfvén waves in the cusp and the physics of magnetic reconnection. On the nightside, these currents are generated by pressure gradients in the far tail. The second dynamo is associated with the Region 2 or shielding Birkeland currents. The electric field generated by this dynamo opposes the convection electric field equatorward of the Region 2 currents and strengthens the convection field on the poleward side. The pressure gradient along the inner edge of the plasma sheet provides the energy to drive this dynamo [*Vasyliunas*, 1970]. The final dynamo is generated by thermospheric winds preferentially pushing ions across magnetic field lines. To first order, these dynamo systems can be computed separately, and the resultant electric fields superimposed. Because magnetic field lines generally act as approximate electrical equipotentials, the potential electric fields generated in the ionosphere (where the load exists) map along magnetic field lines into the magnetosphere and change the particle dynamics in both regions. Thus, it can be expected that the neutral wind dynamo will alter the particle dynamics in the inner magnetosphere and the pressure gradients that drive the Region 2 dynamos.

The relative role of these dynamos in driving ionospheric and magnetospheric plasmas has been investigated by several authors. *Peymirat et al.* [1998] developed the Magnetosphere–Thermosphere–Ionosphere Electrodynamics General Circulation Model (MTIEGCM) to calculate both the neutral wind and Region 2 dynamos self-consistently. MTIEGCM has been used to study a variety of features including over- and undershielding events [*Peymirat et al.*, 2000] and the neutral wind dynamo impact on MI coupling [*Peymirat et al.*, 2002]. In addition, *Maruyama et al.* [2005] have coupled the Region 2 dynamo electric field with an ionospheric and neutral wind dynamo model to determine how the Region 2 dynamo alters the neutral wind dynamo. This study demonstrated that the addition of the Region 2 potentials alter the ionosphere in such a way that the resultant electric field is different than the superposition of the two separate fields. This paper investigates this same phenomenon, but from the other direction. It looks at how the neutral wind dynamo changes the Region 2 dynamo.

This study examines the second-order influence of the neutral wind dynamo electric field on the Region 2 Birkeland current dynamo. Two separate runs of the Rice Convection Model (RCM) are compared to isolate the influence of the wind-generated electric field on the Region 2 dynamo. The RCM is a theoretical model that calculates the Region 2 dynamo and the inner magnetospheric particle distribution. The first RCM run was conducted for an idealized magnetic storm, corresponding to constant $Kp = 6$ conditions ignoring the impact of the neutral wind-dynamo. The second run is the same as the first, but includes the wind-generated electric field, which is calculated from a thermosphere–ionosphere–mesosphere electrodynamics-general circulation model (TIME-GCM) for the same storm conditions. TIME-GCM is a theoretical model of the upper atmosphere that calculates the neutral wind dynamo electric field and is run independently from the RCM with no self-consistent information exchange. A comparison of the two RCM runs demonstrates a meaningful change in the Region 2 dynamo caused by the interaction with the neutral wind dynamo.

2. MODEL DESCRIPTION

The RCM and the TIME-GCM are used to simulate the plasma structures and the electric fields in the inner magnetosphere and the ionosphere, respectively. Both models have a long history of successfully studying their respective regions of the near-Earth geospace.

2.1. Rice Convection Model

The RCM is a first-principles model that calculates the inner magnetospheric particle distribution, the Region 2 Birkeland currents, and the electric field generated from the Region 2 current system [*Jaggi and Wolf*, 1973; *Wolf*, 1983; *Toffoletto et al.*, 2003]. The effects of the Region 1 dynamo are included as a boundary condition. Using a $\kappa = 6$ particle distribution [*Christon et al.*, 1991] with a typical plasma sheet pressure (~0.4 nPa at midnight, $R_{SM} = -15R_E$; see *Kauffman et al.* [2001]) boundary and initial particle distribution, the RCM evolves the inner magnetospheric particle distribution by calculating the bounced-averaged guiding center drift and tracing particle forward in time. Initially, the plasma sheet population is placed at the outer boundary and run to equilibrium in order to fill the lower L shells. As the magnetospheric particles move, they alter the behavior of the Region 2 dynamo. The relationship between the Region 2 Birkeland currents, the inner magnetospheric particle distribution, and the electric field generated by this dynamo was first mathematically enunciated by *Vasyliunas* [1970] as the equation of magnetosphere–ionosphere coupling. It is written as

$$1/\sin(I)\nabla \cdot (-\Sigma \cdot \nabla \Phi) = j_{\|,n} - j_{\|,s}$$
$$= B_{ion}/B_{eq} \, \underline{b} \cdot \nabla V \times \nabla p, \quad (1)$$

where I is the magnetic dip angle, Σ is the field-aligned integrated conductance tensor, Φ is the electric potential, $j_{\|,n}$ and $j_{\|,s}$ are the field-aligned current (FAC) flowing into the northern hemisphere and southern hemisphere, respectively,

B is the magnetic field strength in the ionosphere (*ion*) and the magnetic equatorial plane (*eq*), $\underline{b}$ is the magnetic field unit vector, V is the volume of a magnetic flux tube (with 1 Wb of flux), and p is the magnetospheric particle pressure. In this equation, the left-hand side is calculated in the ionosphere (including all gradients), and the right-hand side is computed on the magnetic equatorial plane. This equation is a simple current conservation equation with the left-hand side representing the divergence of horizontal ionospheric current, the center representing the divergence of the FACs and the right-hand side representing the divergence of the gradient/curvature drift currents in the inner magnetosphere. Note that the free energy source is the inner magnetospheric pressure gradient, which provides the work that drives this dynamo. The RCM solves a different, but equivalent, equation with the particle distribution expressed in terms of the bounced-averaged adiabatic energy and density invariants assuming strong, elastic pitch-angle scattering.

The RCM needs only a few inputs to solve the bounced-averaged drift and the Region 2 dynamo. The first and most significant is the magnetic field model. These runs use the *Tsyganenko* [1989] magnetic field model for $Kp = 6$ conditions. It should be noted that the magnetic field perturbations generated by the Region 2 currents are not self-consistently calculated. The second RCM input is the solar-produced ionospheric conductance, which is calculated from International Reference Ionosphere (IRI-2000) [*Bilitza et al.*, 2001] and the Mass Spectrometer Incoherent Scatter (MSIS) [*Hedin*, 1991] model. For these runs, the input models used solar maximum $F10.7$ cm solar radio flux = 250 SFU), magnetic storm ($Kp = 6$, $ap = 80$) conditions. The redistribution of the ionospheric plasma and their related changes in the ionospheric conductance caused by the Region 2 dynamo field are not included in the RCM's calculations. The auroral conductances are taken from the *Robinson et al.* [1987] formula using a partially self-consistent auroral particle precipitation [*Sazykin*, 2000]. For the present runs, the location of the precipitating particles is consistent with the location of the magnetospheric source populations; however, to increase consistency with TIME-GCM, the magnitude of the fluxes is adjusted for consistency with appropriate latitude integrals of the *Roble and Ridley* [1987] auroral precipitation model, for a hemispheric power 65.4 GW, interplanetary magnetic field (IMF) $By = 0$, and the appropriate time-dependent polar cap potential (PCP). The PCP is determined for $Kp = 6$ conditions using the *Boyle et al.* [1997] formula, which gives a PCP of 109.5 kV. The final input is the electric potential distribution on the poleward boundary of the modeling region. The electric potential model of *Heelis et al.* [1982] is used because it is the PCP model used by TIME-GCM. The *Heelis* model is an empirical model based on ion drift measurements from the Dynamics Explorer 2 data and corresponds to both the convection electric field and the polar cap neutral wind dynamo field. It is driven by the fixed PCP. The RCM treats the Region 1 dynamo as a voltage source with a magnitude given by the PCP. When the RCM includes the TIME-GCM electric field, the *Boyle et al.* PCP is used. But the neutral wind dynamo imposes a roughly 21.5-kV potential drop across the RCM's boundary so that the effective PCP is ~131 kV. This is the PCP value used as the input to the RCM when it is run without the neutral wind dynamo electric field. A failure to do so would confuse the matter since the base run and wind run would experience different convection fields within the RCM's architecture.

The RCM is run on an irregular ionospheric grid in solar-fixed magnetic coordinates. Hence, the coordinates are invariant latitude and magnetic local time (MLT). In the auroral regions, the latitude grid is ~0.25° apart. Near the equator, it is roughly ~1°. The grid points are separated by 15 min in MLT. The grid is mapped along magnetic field lines to the magnetic equatorial plane. For these runs, 5-s time steps are used in the RCM. These runs are run for a complete day starting from a precomputed steady state $Kp = 6$ RCM run. In addition, these runs neglect field-aligned potential drops, but use ion charge exchange and flux tube refilling modules.

To isolate the influence of the wind-generated electric field, two RCM runs were conducted. The first run, known hereafter as the base run, is a standard RCM run without the inclusion of the wind-generated electric field. The second run, called the wind run, includes the neutral wind-generated electric field as a model input. This electric potential is added to the Region 2 dynamo electric field calculated at each time step. The combined electric field is then used to calculate the magnetospheric particle drift. These runs are compared below.

2.2. Thermosphere–Ionosphere–Mesosphere Electrodynamics General Circulation Model

The TIME-GCM is a first-principles model that solves the basic chemical and physical for the ionosphere, thermosphere, and mesosphere [*Roble and Ridley*, 1994; *Crowley et al.*, 1999]. The primary inputs to the TIME-GCM are the auroral precipitation, the high-latitude electric field, the solar EUV flux, and the upflowing energy and momentum from atmospheric tides. The auroral precipitation model is the *Roble and Ridley*'s [1982] model, which uses the hemispheric power and PCP values given above as inputs. The PCP field is the Heelis model described above. The solar EUV fluxes are specified by the $F10.7$ cm flux = 250 SFU. The final input is an atmospheric tidal model, which was developed for the TIME-GCM.

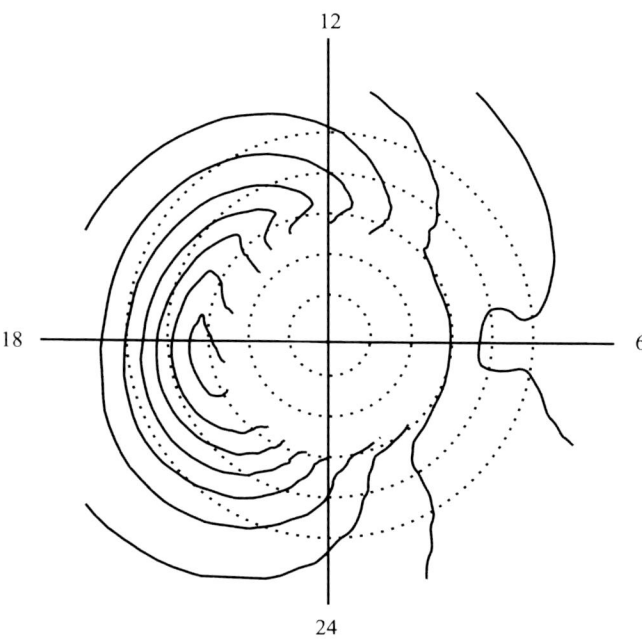

Figure 1. A plot of the electric potential pattern generated by the neutral wind dynamo as calculated TIME-GCM. The potentials are shown in the northern polar cap of the ionosphere with the magnetic latitude lines drawn every 10° (dashed lines). The equipotentials are 5 kV apart. The equipotentials poleward of the RCM's modeling region are not plotted.

The TIME-GCM grid uses geographic coordinates with a regular grid spacing of 5° latitude by 5° longitude. Unlike the Region 2 dynamo system primarily controlled by the magnetic field, the neutral wind dynamo is significantly controlled by solar heating and is best described on a geographic coordinate system. The RCM's magnetic coordinate system is offset from the TIME-GCM's geographic. This creates a wobble to the electric field input to the RCM, which corresponds to the rotation of the geographic pole around the magnetic pole.

The electric field from the neutral wind dynamo is calculated by the TIME-GCM and used as an input to the RCM. Figure 1 shows the equipotentials of this electric field in the RCM's modeling regime. This additional electric field is mapped along the magnetic field line into the magnetic equatorial plane and alters the drift of the magnetospheric plasma.

3. COMPARISON OF MODEL RESULTS

As previously mentioned, each RCM run lasted for a full day. However, the RCM inputs are held constant over the day, except for the wobble in the wind-generated electric field. The wobble has a rather minor impact on the total potential distribution (not shown) and can generally be ignored. Because the model essentially reaches steady state, only one model time step is shown here.

3.1. Changes in the Inner Magnetospheric Plasma

As an additional forcing term, the wind-generated electric field alters the particle distribution in the inner magnetosphere. Figure 2 shows a comparison of the inner magnetospheric plasma pressure between the two RCM runs. In this figure, the logarithm of the total plasma pressure is shown in the magnetic equatorial plane. Figure 2 (left) and Figure 2 (center) show the plasma distribution in the base and wind runs, respectively. Figure 2 (right) shows the difference in the plasma pressure between the two runs (base–wind). The wind-generated electric field pulls plasma from the dawnside into the dusk region. This effect occurs predominately at low L, where the wind-generated electric field is strongest. In addition, the pressure difference is most noticeable on the dawnside where the ion gradient and curvature drifts oppose the $\mathbf{E} \times \mathbf{B}$ drift. The pressure at dawn is 0.37 nPa at $L = 2$ in the wind run compared to $1.9 \cdot 10^{-4}$ nPa in the base run. Similarly, the pressure at $L = 3$ is 162 nPa in the wind run, but only 66 nPa in the base run. However, the pressure is the nearly the same in both runs at $L = 4$ and 5. Hence, the wind-generated electric field is most effective near the local null of the purely magnetospheric drifts.

3.2. Changes in the Region 2 Birkeland Currents

As Figure 2 shows, the pressure gradients in the inner magnetosphere change because of addition of the wind-generated electric field. The altered pressure gradients in turn modify the Region 2 Birkeland currents. Plate 1 shows a comparison between the Region 2 currents from the base run and the wind run. As with Figure 2, the panels correspond (from left to right) to the Region 2 currents in the base run, those in the wind run, and difference between the runs. While these Birkeland currents are rotated with respect to the classic Birkeland current pattern from *Iijima and Potemra* [1978], similar patterns have recently been observed by the magnetometers onboard the Iridium satellites (see Figure 3 of *Korth et al.* [2004]). The downward currents (represented in blue) are pushed equatorward by the wind-generated field with a downward peak FAC density of 3.15 μA/m^2 at 55.21° magnetic latitude and 16.75 h MLT compared with 2.67 μA/m^2 at 63.2°, 12.75 h. The region of upward current in the difference plot indicates the area where the currents in the wind run have moved equatorward of the location of the base run's current. The neutral wind dynamo has a different impact on the dawnside currents, where it strengthens the

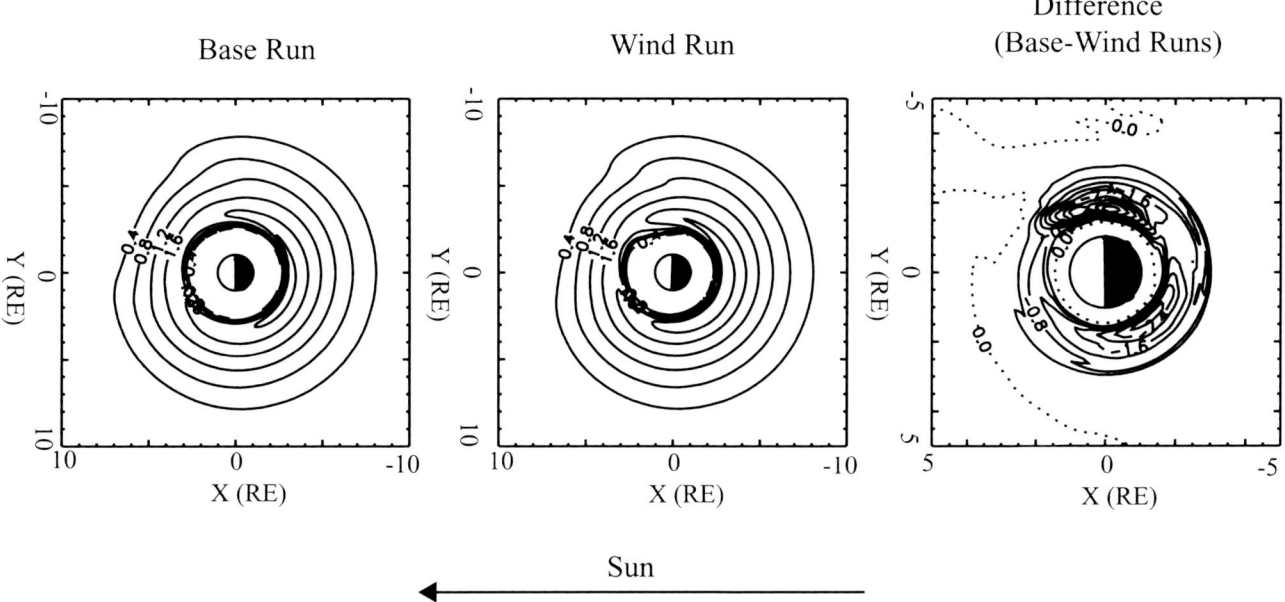

Figure 2. A comparison of the distribution of plasma pressures in the inner magnetosphere. (left) Base-10 logarithm of the plasma pressure in nPa for the base run. (center) Pressure distribution for the wind run. The contour interval is 0.4 $\log_{10}(\text{nPa})$, which corresponds to a factor of ~2.5. Difference between the two runs (base–wind pressures) in a zoomed in box. The pressures are plotted on the magnetic equator in SM coordinates with the Sun to the left.

currents more than pushing them to different L shells. The peak upward current is 3.99 $\mu\text{A/m}^2$ at 61.4°, 6.5 h in the wind run as opposed to 2.83 $\mu\text{A/m}^2$ at 62.2°, 6.75 h. This is a reflection of the relative impacts that the wind-generated electric fields have on the dynamics of the two regions. On the duskside, the wind-generated electric fields added to the gradient-curvature and $\mathbf{E} \times \mathbf{B}$ drifts and pushed the plasma sheet particles earthward causing the Birkeland currents to

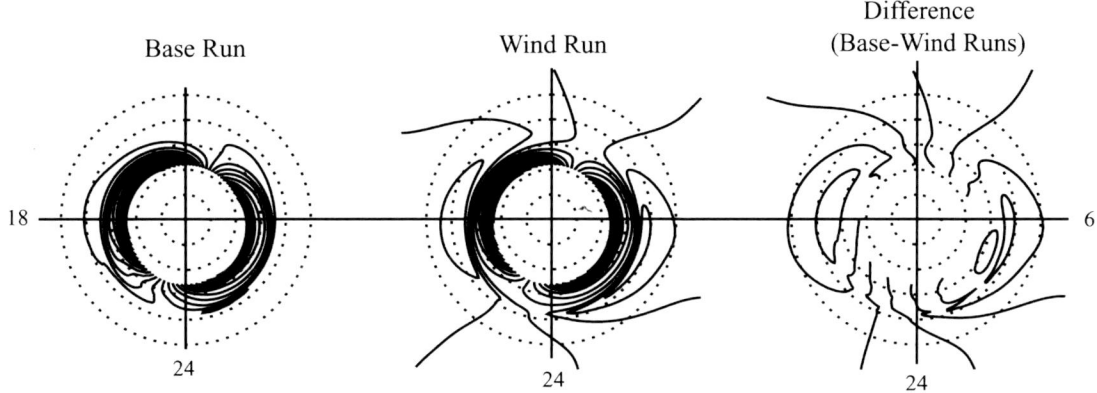

Figure 3. A comparison of the electric potential pattern generated by the input convection field and the Region 2 dynamo with and without the wind-generated electric field. The potentials are shown in the northern polar cap of the ionosphere with the magnetic latitude lines drawn every 10° (dashed lines). (left) The base run is well shielded. (center) The wind run shows large electric fields at low latitudes. (right) Difference between these runs. The equipotentials are 5 kV apart. *The equipotentials are 5 kV apart.* The equipotentials poleward of the RCM's modeling region are not plotted.

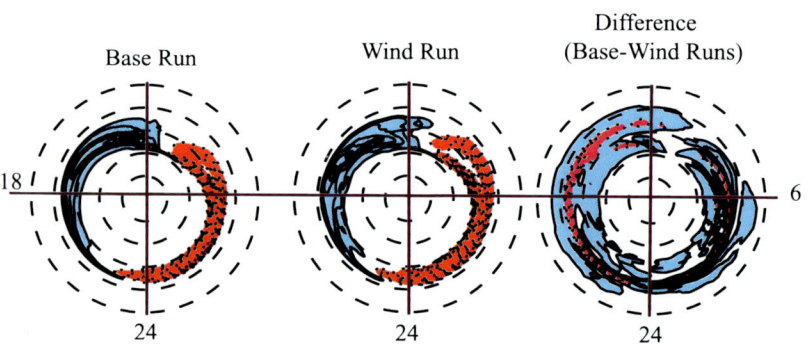

Plate 1. A comparison of the distribution of Region 2 field-aligned currents in the ionosphere. (left) Region 2 currents from the base run. (center) Pressure distribution for the wind run. (right) Difference between these two runs. The contours are 0.5 µA/m² apart. The blue shaded region represents downward currents, whereas upward currents are shaded in red. The dashed lines represent lines of constant magnetic latitude and are 10° apart.

move equatorward. On the dawnside, the plasma piles up in the null region of the electric field and generates stronger FAC densities.

3.3. Changes in the Mid- and Low-Latitude Electric Field

The electric field generated by the Region 2 dynamo differs between these two runs as shown in Figure 3. These potential patterns are the sum of the inputted dawn–dusk convection pattern and the purely Region 2 potentials. As before, Figure 3 (left) shows the electric potential pattern in the ionosphere for the base run. Figure 3 (center) shows the potential pattern from the wind run after subtracting off the wind-driven potentials. Finally, Figure 3 (right) shows the difference between the runs. These plots show nearly total shielding in the base run. The potential drop in the base run is 2.86 kV at 10° magnetic latitude, 2.87 kV at 20°, and 2.95 kV at 30°. This is in contrast to the wind run, which shows a strong electric field reaching the equator, with a potential drop of 13.03 kV at 10°, 13.22 kV at 20°, and 14.35 kV at 30°. The electric field pattern in the wind run is generated by the new low L-shell pressure gradients (Figure 2).

4. SUMMARY

This study has examined the effects of the neutral wind dynamo electric field on the Region 2 dynamo system. The wind-generated electric fields were included as a non-self-consistent input to the RCM for this study. These model runs demonstrate that the wind-generated electric fields make significant changes in the inner magnetospheric particle pressure, the Region 2 Birkeland currents, and the mid- and low-latitude electric field. The wind-generated electric fields are most effective in changing the plasma pressure near drift null in the thermal ion velocities. At lower L shells, plasma is removed from the magnetosphere by the wind-generated field. These changes in the plasma pressure alter the Region 2 Birkeland currents, particularly in the dawn sector. In general, the Birkeland currents are stronger in the wind run. The stronger Birkeland currents lead to a different electric field structure with a stronger electric field at lower latitudes, where the potential drop at 10° magnetic latitude increased by a factor of ~4.5 (13.03 kV in the wind run, compared to 2.86 kV in the base run). The new electric field pattern resembles a typical overshielding equipotential pattern.

Acknowledgments. This work was supported by NASA's Living with a Star program under grant NNG-GN35-1. The authors would also like to thank S. Sazykin for his helpful advice and useful comments.

REFERENCES

Bilitza, D. (2001), International Reference Ionosphere 2000, *Radio Sci.*, *36*, 261.

Boyle, C. B., P. H. Reiff, and M. R. Hairston (1997), Empirical polar cap potentials, *J. Geophys. Res.*, *102*, 111–125.

Christon, S. P., D. J. Williams, D. G. Mitchell, C. Y. Huang, and L. A. Frank (1991), Spectral characteristics of plasma sheet ion and electron populations during disturbed geomagnetic conditions, *J. Geophys. Res.*, *96*, 1–22.

Crowley, G., C. Freitas, A. Ridley, D. Winningham, R. G. Roble, and A. D. Richmond (1999), Next generation space weather specification and forecasting model, in *Proceedings of the 1999 Ionospheric Effects Symposium*, edited by J. M. Goodman, pp. 34–41. JMG Associates Ltd., Alexandria, Va.

Foster, J. C., A. J. Coster, P. J. Erickson, W. Rideout, F. J. Rich, T. J. Immel, and B. R. Sandel (2005), Redistribution of the storm-time ionosphere and the formation of a plasmaspheric bulge, in *Inner Magnetosphere Interactions: New Perspectives from Imaging, Geophys. Monogr. Ser.*, vol. 159, edited by J. Burch, M. Schulz, and H. Spence, pp. 277–290. AGU, Washington, D. C.

Hedin, A. E. (1991), Extension of the MSIS thermospheric model into the middle and lower atmosphere, *J. Geophys., Res.*, *96*, 1159–1172.

Heelis, R. A., J. K. Lowell, and R. W. Spiro (1982), A model of the high-latitude ionospheric convection pattern, *J. Geophys. Res.*, *87*, 6339–6345.

Huba, J. D., G. Joyce, S. Sazykin, R. Wolf, and R. Spiro (2005), Simulation study of penetration electric field effects on the low- to mid-latitude ionosphere, *Geophys. Res. Lett.*, *32*, L23101, doi:10.1029/2005GL024162.

Iijima, T., and T. A. Potemra (1976), The amplitude distribution of field-aligned currents at northern high latitudes observed by TRIAD, *J. Geophys. Res.*, *81*, 2165–2174.

Jaggi, R. K., and R. A. Wolf (1973), Self-consistent calculation of the motion of a sheet of ions in the magnetosphere, *J. Geophys. Res.*, *78*, 2852–2866 (Correction, *J. Geophys. Res.*, *80*, 4109).

Lin, C. H., A. D. Richmond, R. A. Heelis, G. J. Bailey, G. Lu, J. Y. Liu, H. C. Yeh, and S.-Y. Su (2005), Theoretical study of the low- and midlatitude ionospheric electron density enhancement during the October 2003 superstorm: Relative importance of the neutral wind and the electric field, *J. Geophys. Res.*, *110*, A12312, doi:10.1029/2005JA011304.

Kaufmann, R. L., B. M. Ball, W. R. Paterson, and L. A. Frank (2001), Plasma sheet thickness and electric currents, *J. Geophys. Res.*, *106*, 6179–6193.

Korth, H., B. J. Anderson, M. J. Wiltberger, J. G. Lyon, and P. C. Anderson (2004), Intercomparison of ionospheric electrodynamics from the Iridium constellation with global MHD simulations, *J. Geophys. Res.*, *109*, A07307, doi:10.1029/2004JA010428.

Mannucci, A. J., B. T. Tsurutani, B. A. Iijima, A. Komjathy, A. Saito, W. D. Gonzalez, F. L. Guarnieri, J. U. Kozyra, and R. Skoug (2005), Dayside global ionospheric response to the major interplanetary events of October 29–30, 2003 "Halloween Storms," *Geophys. Res. Lett.*, *32*, L12S02, doi:10.1029/2004GL021467.

Maruyama, N., A. D. Richmond, T. J. Fuller-Rowell, M. V. Codrescu, S. Sazykin, F. R. Toffoletto, R. W. Spiro, and G. H. Milward (2005), Interaction between direct penetration and disturbance dynamo electric fields in the storm-time equatorial ionosphere, *Geophys. Res. Lett., 32*, L17105, doi:10.1029/2005GL023763.

Peymirat, C., A. D. Richmond, B. A. Emery, and R. G. Roble (1998), A Magnetosphere–Thermosphere–Ionosphere Electrodynamics General Circulation Model, *J. Geophys. Res., 103*, 17,467–17,477.

Peymirat, C., A. D. Richmond, and A. T. Kobea (2000), Electro-dynamic coupling of high and low latitudes: Simulations of shielding/overshielding effects, *J. Geophys. Res., 105*, 22,991–23,003.

Peymirat, C., A. D. Richmond, and R. G. Roble (2002), Neutral wind influence on the electrodynamic coupling between the ionosphere and the magnetosphere, *J. Geophys. Res., 107*, doi:10.1029/2001JA900106.

Prölss, G. W. (1995), Ionospheric F-region storms, in: *Handbook of Atmospheric Electrodynamics*, edited by H. Volland, pp. 195–248, CRC Press, Boca Raton, Fla.

Rishbeth, H. (1998), How the thermospheric circulation affects the ionospheric F2-layer, *J. Atmos. Sol. Terr. Phys., 60*, 1385–1402.

Roble, R. G., and E. C. Ridley (1987), An auroral model for the NCAR Thermospheric General Circulation Model (TGCM), *Ann. Geophys., 5*, 369–382.

Roble, R. G., and E. C. Ridley (1994), Thermosphere–Ionosphere–Mesosphere-Electrodynamics General Circulation Model (TIME-GCM): Equinox solar cycle minimum simulations (300–500 km), *Geophys. Res. Lett., 22*, 417–420.

Robinson, R. W., R. R. Vondrak, K. Miller, T. Dabbs, and D. Hardy (1987), On calculating ionospheric conductances from the flux and energy of precipitation electrons, *J. Geophys. Res., 92*, 2565–2569.

Sazykin, S. (2000), Theoretical studies of penetration electric fields to the ionosphere, Ph.D. thesis, Utah State Univ., Logan.

Toffoletto, F. R., S. Sazykin, R. W. Sprio, and R. A. Wolf (2003), Inner magnetospheric modeling with the Rice Convection Model, *Space Sci. Rev., 107*, 175–196.

Tsyganenko, N. A. (1989), A magnetospheric magnetic field model with a warped tail current sheet, *Planet. Space Sci., 37*, 5–20.

Vasyliunas, V. M. (1970), Mathematical models of magnetospheric convection and its coupling to the ionosphere, in: *Particles and Fields in the Magnetosphere*, edited by B. M. McCormac, pp. 60–71, D. Reidel, Norwell, Mass.

Wolf, R. A. (1983), The quasi-static (slow-flow) region of the magnetosphere, in *Solar–Terrestrial Physics: Principles and Theoretical Foundations*, edited by R. L. Carovillano and J. M. Forbes, pp. 303–368, D. Reidel, Norwell, Mass.

G. Crowley, Atmospheric and Space Technology Research Associates, 11118 Quail Pass, San Antonio, TX 78249, USA

T. W. Garner, Applied Research Laboratories, University of Texas at Austin, P.O. Box 8029, Austin, TX 78713, USA. (garner@arlut.utexas.edu)

R. A. Wolf, Department of Physics and Astronomy, Rice University, 6100 S Main, Houston, TX 77005, USA

Global Modeling of Storm-Time Thermospheric Dynamics and Electrodynamics

T. J. Fuller-Rowell

CIRES University of Colorado and NOAA Space Weather Prediction Center, Boulder, Colorado, USA

A. D. Richmond

High Altitude Observatory, NCAR, Boulder, Colorado, USA

N. Maruyama

CIRES University of Colorado and NOAA Space Weather Prediction Center, Boulder, Colorado, USA

Understanding the neutral dynamic and electrodynamic response of the upper atmosphere to geomagnetic storms, and quantifying the balance between prompt penetration and disturbance dynamo effects, are two of the significant challenges facing us today. This paper reviews our understanding of the dynamical and electrodynamic response of the upper atmosphere to storms from a modeling perspective. After injection of momentum and energy at high latitude during a geomagnetic storm, the neutral winds begin to respond almost immediately. The high-latitude wind system evolves quickly by the action of ion drag and the injection of kinetic energy; however, Joule dissipation provides the bulk of the energy source to change the dynamics and electrodynamics globally. Impulsive energy injection at high latitudes drives large-scale gravity waves that propagate globally. The waves transmit pressure gradients initiating a change in the global circulation. Numerical simulations of the coupled thermosphere, ionosphere, plasmasphere, and electrodynamic response to storms indicate that although the wind and waves are dynamic, with significant apparent "sloshing" between the hemispheres, the net effect is for an increased equatorward wind. The dynamic changes during a storm provide the conduit for many of the physical processes that ensue in the upper atmosphere. For instance, the increased meridional winds at mid latitudes push plasma parallel to the magnetic field to regions of different composition. The global circulation carries molecular rich air from the lower thermosphere upward and equatorward, changing the ratio of atomic and molecular neutral species, and changing loss rates for the ionosphere. The storm wind system also drives the disturbance dynamo, which through plasma transport modifies the strength and location of the equatorial ionization anomaly peaks. On a global scale, the increased equatorward meridional winds, and the generation of zonal winds at mid latitudes via the Coriolis effects, produce a current system opposing the normal quiet-time Sq current system. At the equator, the storm-time zonal electric fields reduce or reverse the normal upward and downward plasma drift on the dayside and nightside, respectively. In the numerical simulations, on the dayside, the disturbance dynamo appears fairly uniform, whereas at night a stronger local time dependence is apparent with increased upward drift between midnight and dawn. The simulations also indicate the possibility for a rapid dynamo response at the equator, within 2 h of storm onset, before the arrival of the large-scale gravity waves. All these wind-driven processes can result in dramatic ionospheric changes during storms. The disturbance dynamo can combine and interact with the prompt penetration of magnetospheric electric fields to the equator.

1. INTRODUCTION

The neutral thermosphere comprises more than 99% of the mass of the upper atmosphere at altitudes above 100 km. It is therefore not surprising that the plasma state, the ionosphere, is dependent on the dynamics and composition of the thermosphere. Similarly, although only a relatively minor species in the weakly ionized plasma of the upper atmosphere, the ionosphere has a significant, and sometimes dominant impact on the neutral gas, at all latitudes. The coupling and interaction between the neutral and plasma state is therefore crucial to a thorough understanding of each component.

At high latitude, ions respond directly to the strong magnetospherically imposed electric fields, which cause ion drifts of many hundred, if not thousands, of m/s. Although collisions between ions and the neutral gas are infrequent above ~160 km, they are sufficient to accelerate the thermosphere at high latitudes to many hundreds of m/s over periods of tens of minutes or more [*Killeen et al.*, 1984, 1988]. Plate 1 shows the ion drift and neutral wind, during moderate geomagnetic activity ($Kp \sim 3$) poleward of 40° latitude geographic latitude, in response to a fairly typical two-cell pattern of magnetospheric convection [*Weimer*, 1995]. In the upper figure, the vectors represent plasma drift velocity in the upper thermosphere; where the ion drift motion is close to the $\mathbf{E} \times \mathbf{B}/B^2$, since the collisions with the thermosphere are relatively infrequent. The color contours represent plasma density near the F region peak, close to 300 km altitude. The effect of transport on the plasma density is clearly visible with the formation of the sub-auroral trough, and advection of dayside solar-produced plasma being drawn toward and across the polar region.

In the bottom panel of Plate 1, the neutral wind and temperature response over the same region is shown. Winds driven by solar heating alone would typically be antisunward, and reach ~150 m/s at 300 km in the polar region. With the imposition of magnetospheric convection, even with the infrequent collisions at this altitude there is a sufficient momentum source to accelerate the medium close to 400 m/s. The first reaction of the neutrals is to follow the ion convection, but other inertial and viscous forces act to introduce asymmetries in the circulation, or cell, structure. The clockwise cell excited in the dusk sector is particularly strong due to an inertial resonance between the ion and neutral convection [*Fuller-Rowell et al.*, 1984; *Fuller-Rowell*, 1995]. The natural motion of the neutral gas is to form a clockwise vortex due to the action of the Coriolis force, so there is a natural tendency for this cell to develop and resonate. In contrast, the dawn cell is less well formed with weak, or virtually nonexistent, sunward winds in the dawn sector auroral oval in response to the plasma convection. This anticlockwise cell does not have a resonance with the ion convection since momentum is continually being transported out of the cell by the tendency for the vortex to diverge. The neutral temperature structure at this altitude, shown in the bottom panel, has a fairly modest impact from Joule heating, but is more controlled by transport of the neutral gas by the elevated circulation. The warmer dayside temperatures tend to be advected poleward by the neutral circulation, in much the same way as the plasma density is transported poleward by the ion drift. The numerical simulation results depicted in Plate 1 are taken from the coupled thermosphere–ionosphere–plasmasphere (CTIPe) model with self-consistent mid- and low-latitude electrodynamics [*Fuller-Rowell et al.*, 1996b; *Millward et al.*, 1996]. A similar presentation in the lower thermosphere, where collisions between the neutrals and ions are more dominant, would show a quite different pattern.

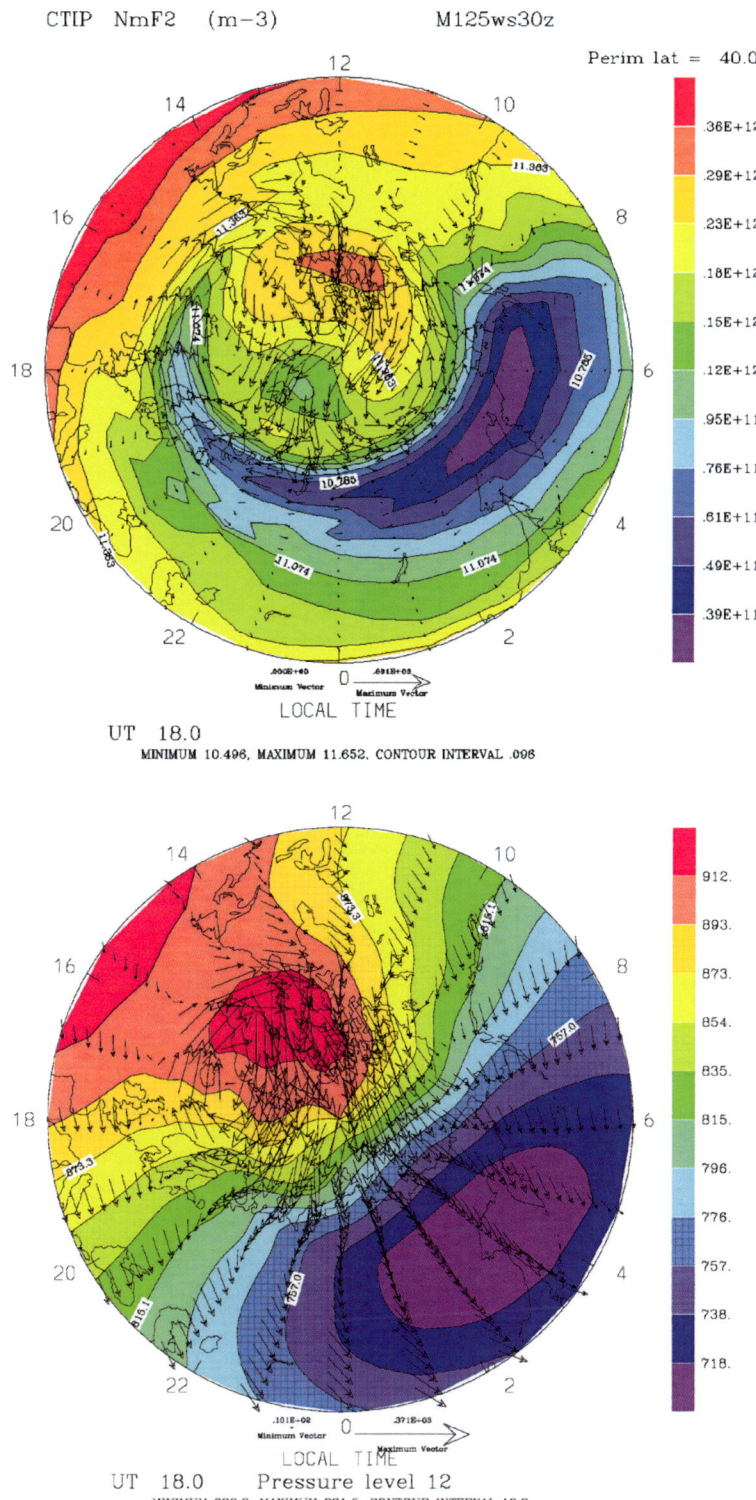

Plate 1. (top) Simulation of plasma drift and electron density at the altitude of the F-region peak in the upper thermosphere for moderate geomagnetic activity ($Kp \sim 3$). The region poleward of 40° latitude is shown at 18 UT. (bottom) Neutral wind and temperature over the same region at 300 km altitude in response to the plasma drift. The maximum plasma drift velocity is ~ 700 m/s, and the peak wind speed is ~380 m/s.

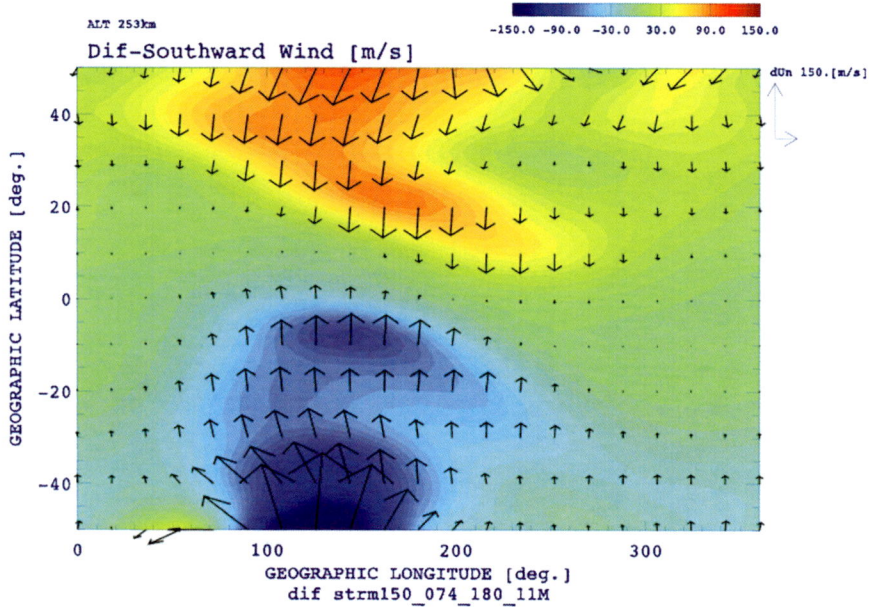

Plate 2. Simulation of the response of the neutral winds at mid and low latitudes at 250 km altitude, shortly after a sudden increase in high-latitude Joule heating. The region within 50° of the geographic equator is shown at 15 UT, 3 hours after the increase in high-latitude magnetospheric forcing, equivalent to a $Kp \sim 7$. Wind surges of ~150 m/s are produced, mainly on the nightside.

During more active storm conditions the dynamics at high latitudes are more disturbed with winds approaching 1 km/s. In these cases, Joule heating becomes more prominent and can elevate the temperature at high latitudes by hundreds of Kelvin in the upper thermosphere. The high-latitude Joule heating has global consequences because it is responsible for the launch of large-scale gravity wave or wind surges toward the equator, and is the source of subsequent changes in the global circulation. The large-scale waves have typical wavelength of 1000 km or more and phase propagation speeds ranging from 400 to 1000 m/s [*Hunsucker*, 1982; *Shiokawa et al.*, 2002].

Changes in neutral dynamics during storms provide the conduit for a series of changes in neutral composition, electrodynamics, and plasma density. The basic physical processes operating during a storm are the same at all phases of the solar cycle; however, the balance between the solar driven and storm circulation, and the ensuing composition changes, will vary. Large storms can occur at any phase of the solar cycle but are more likely when the Sun is more magnetically active. The impact of changes in thermospheric circulation on neutral composition is treated extensively by *Crowley et al.* [2008, this issue], and so will not be the focus of this chapter. It is also important to appreciate that the high-latitude magnetospheric electric fields can also penetrate globally [*Fejer et al.*, 1997; *Kelley et al.*, 2003], the so-called prompt penetration electric field, and can have a direct and often dramatic influence on the plasma restructuring at mid and low latitudes [*Foster and Rideout*, 2005; *Foster and Coster*, 2007; *Mannucci et al.*, 2005]. One of the current challenges facing us today is to be able to separate and quantify the influence of the various physical processes in geospace during storms. Numerical models can be useful tools in this endeavor to unravel the physical processes [e.g., *Richmond and Roble*, 1987; *Fuller-Rowell et al.*, 2002; *Fesen et al.*, 2000; *Peymirat et al.*, 2002; *Ridley et al.*, 2006]. The focus of this chapter is on modeling thermospheric dynamics during storms and quantifying its impact on electrodynamics.

2. MODELING GLOBAL THERMOSPHERE DYNAMICS

During geomagnetic storms the global dynamics of the upper atmosphere changes dramatically [*Buonsanto*, 1999; *Fuller-Rowell et al.*, 1994, 1997; *Fejer et al.*, 2002; *Emmert et al.*, 2001, 2002]. The response is complex even during the simplest of events. The thermosphere, although thought of as a sluggish medium, can respond quite quickly (in tens of minutes) and can support high-speed, large-scale, gravity waves that propagate globally initiated by impulsive forcing at high latitudes [*Richmond and Matsushita*, 1975]. Gravity waves propagate at close to sound speeds, typically ~700 m/s, so waves launched by auroral heating can reach mid latitudes in an hour, and can reach the equator and penetrate into the opposite hemisphere within 3 h. Waves launched from both hemispheres can interact to form a quite complex wave train [*Shiokawa et al.*, 2002; *Lu et al.*, 2008], even for the simplest forcing time histories. Real events with complex time histories are more difficult to unravel.

Large-scale gravity waves provide the mechanism for transmitting changes in pressure gradients around the globe. A new global circulation can therefore be imposed on the same time-scale as gravity wave propagation; it does not rely on, or require, the bulk physical transport of mass by the wind field, which is typically much slower at mid latitude, 100–200 m/s.

Plate 2 shows the changes in neutral wind at mid and low latitudes at 250 km, 3 h into a numerical simulation of a step function increase in high-latitude forcing in the auroral oval (65–75° geomagnetic latitude). The wind response is shown within 50° latitude of the geographic equator, to allow for a scale that clearly shows the mid- and low-latitude dynamic response. Whereas at auroral latitude the peak neutral winds would be close 700 m/s, at mid and low latitudes the winds are much more modest, with 100–150 m/s wind surges above the background circulation. At this time, 3 h into the simulation, the disturbance winds have reached the equator and are beginning to penetrate the other hemisphere and interact with the opposing wave front from the other pole. The arrival of the wave front at the geographic equator within 3 h indicates a propagation speed of about 600 m/s, in this case. The dependence on longitude, or local time, is quite prominent with the strongest intensity of propagation in the 100–200° longitude sector, which for the Universal Time of the image (15 UT) is on the nightside. The peak response appears to be more dependent on day or night difference, rather than the longitude sector of the magnetic pole. Stronger nightside wave propagation can be attributable to reduced ion drag [*Fuller-Rowell et al.*, 1994]. In the simulation, the forcing is limited to the increase in magnetospheric electric field and auroral precipitation at high latitudes. The affect of prompt penetration of magnetospheric electric fields to low latitudes are not included in the simulation, although self-consistent changes in the mid- and low-latitude electric fields from the disturbance dynamo are included.

A vertical cut through the thermosphere would reveal a tilted wave front with wave propagating more slowly at the lower altitudes [*Richmond and Matsushita*, 1975]. Two hours later in the numerical simulation, the wave surges penetrate the opposite hemisphere and drive poleward winds at mid latitude, at a time when the high-latitude forcing is still at its strongest. The complex wave train of equatorward and poleward winds during geomagnetic disturbances is a typical characteristic of neutral wind observations [*Shiokawa et al.*, 2002].

The CTIPe numerical simulations described above and presented by *Fuller-Rowell et al.* [2002] exhibit many of the features of observations from the Wind Imaging Interferometer (WINDII) instrument on the Upper Atmosphere Research Satellite (UARS). Plate 3 shows a comparison of meridional and zonal winds from CTIPe with the empirical wind model of *Emmert et al.* [2004], which is based on the WINDII observations. The CTIM results are a snapshot at a single storm time in geographic coordinates, whereas the WINDII results are averaged over longitude, season, and storm time, and are presented in magnetic coordinates. Although sampling of the physical model is not exactly the same as the data, many of the modeled features are consistent with the observations—for instance, the increase in the meridional winds particularly on the nightside, the increase in the equatorial zonal winds after midnight, and the reduction before midnight.

Although model simulations and observations reveal a "sloshing" of winds between hemispheres, the net integrated wind effect is for an increase in the global circulation from pole to equator in both hemispheres [*Fuller-Rowell et al.*, 1994]. This global circulation induces upwelling at high latitudes and transport of molecular rich air (O_2, N_2) from the mid and lower thermosphere, upward and equatorward. The circulation during prolonged storms can transport neutral composition to low latitudes, which has been observed by space based composition measurements (see review by *Crowley et al.* [2008, this issue]). The same can happen during solstice during even quite modest storms due to transport by the prevailing summer to winter circulation [*Fuller-Rowell et al.*, 1996a].

It is well known that changes in neutral composition can impact the ionosphere by changing the ion loss rate. A decrease in the O/N_2 ratio can cause substantial decreases in plasma density [*Strickland et al.*, 2001], often referred to as a "negative phase" ionospheric storm [*Prölss*, 1997; *Rodger et al.*, 1989].

3. ELECTRODYNAMIC MODELING

At low latitude the structure of the ionosphere is strongly controlled by electrodynamics. During quiet times, the electric fields are driven by a combination of the E- and F-region dynamo processes [*Fesen et al.*, 2000; *Millward et al.*, 2001; *Heelis*, 2004]. The net result at the magnetic equator is eastward electric fields, or upward plasma drift, during the day, downward drift at night, and a prereversal enhancement (PRE) just after sunset. The upward plasma transport induced by the electrodynamics on the dayside, generates the equatorial ionization anomaly (EIA). A strong PRE after sunset can also raise the ionosphere giving rise to the Rayleigh–Taylor instability, producing conditions conducive to the generation of ionospheric irregularities. The latter are notoriously difficult to predict on a day-to-day basis. The red curve in Plate 4 shows the diurnal variation in the vertical plasma drift at the magnetic equator during quiet geomagnetic conditions from a simulation of CTIPe.

For a given wind system and in the absence of magnetospheric penetration electric fields, the ionospheric electric fields **E** and current density **J** are determined by the dynamo equations [*Blanc and Richmond*, 1980]:

$$\mathbf{J} = \sigma(\mathbf{E} + \mathbf{U} \times \mathbf{B}) \quad (1)$$

$$\mathbf{E} = -\nabla \phi \quad (2)$$

where σ is the conductivity tensor, **U** is the neutral wind, **B** is the Earth's magnetic field, and ϕ is the electrostatic potential. To understand the storm-time response, it is useful to divide the current density **J** into components driven by the wind field alone (subscript u) and by the electric field alone (subscript E), as was done by Blanc and Richmond. The horizontal components of **J** in the magnetic equatorward (θ) and eastward (ϕ) direction are then given by

$$\begin{aligned}
J_{\theta u} &= -\frac{\sigma_1}{\sin I} u_\phi B + \sigma_2 u_\theta B \\
J_{\phi u} &= \sigma_1 \sin I u_\theta B + \sigma_2 u_\phi B \\
J_{\theta E} &= \frac{\sigma_1}{\sin I} E_\varepsilon + \frac{\sigma_2}{\sin I} E_\phi \\
J_{\phi E} &= -\sigma_2 E_\varepsilon + \sigma_1 E_\phi
\end{aligned} \quad (3)$$

where I is the magnetic inclination below the horizontal and σ_1 and σ_2 are the Pedersen and Hall conductivities, respectively.

During geomagnetic storms, the dynamo electric fields are altered because the normal quiet-day thermospheric neutral winds are disrupted. *Blanc and Richmond* [1980] were the first to describe the characteristics of the storm-time disturbance dynamo, and their results are strongly supported by observations [*Scherliess and Fejer*, 1997; *Fejer and Emmert*, 2003]. The Blanc and Richmond theory relies on the buildup of zonal winds at mid latitude under the action of the Coriolis force, in response to the increased equatorward winds. The meridional winds are forced by high-latitude heating. The dynamo action of the zonal winds drives an equatorward Pedersen current. Positive charge builds up

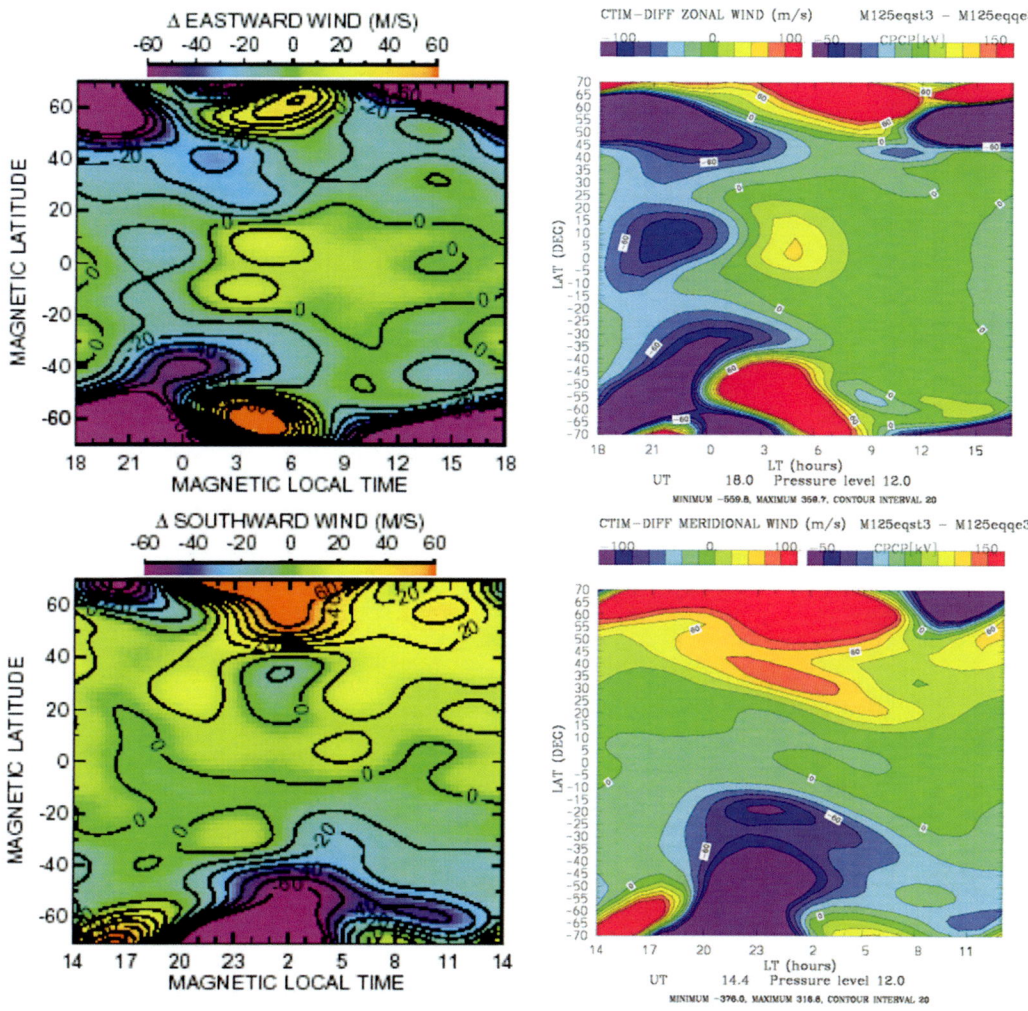

Plate 3. Comparison of a numerical simulation using the CTIPe physical model with average WINDII results in the upper thermosphere from *Emmert et al.* [2004]. The CTIPe figures are reprinted from *Fuller-Rowell et al.* [2002], with permission from Elsevier, and depict snapshots of the global disturbance wind field (top right) 6 hours (zonal component) and (bottom right) 2.5 hours (meridional component) after the onset of a $Kp \sim 7$ storm.

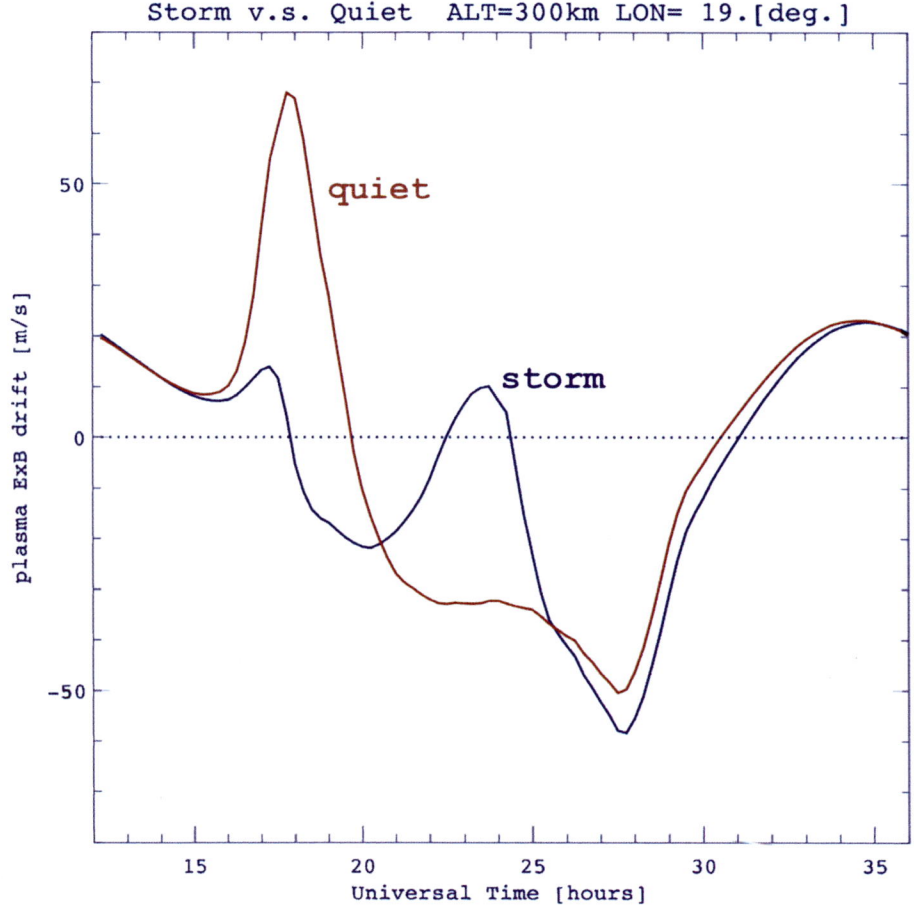

Plate 4. Simulation of vertical plasma drift at the magnetic equator during quiet time (red curve) and during a generic storm (blue curve), at 19°E longitude. The generic storm is an increase in high-latitude magnetospheric sources equivalent to $Kp \sim 7$, for a period of 12 hours, commencing at 12 UT. Pate 4 covers 24 hours of local time from 13 LT at the start of the interval; local midnight is at 23 UT.

at the equator producing a poleward-directed electric field, which balances the wind-driven equatorward current. The poleward electric field subsequently drives an eastward Hall current, which cause positive charge buildup at the dusk terminator and negative charge buildup at dawn. The zonal electric field driven by the disturbance dynamo opposes the normal dayside eastward, and nightside westward, quiet-time dynamo electric field and magnetic perturbations. The disturbance dynamo therefore act as a reverse *Sq* current vortex, reducing or even reversing the eastward electric field on the dayside, and reducing or reversing the normal westward electric field on the nightside.

The blue curve in Plate 4 shows the consequence on the nightside of the disturbance dynamo in CTIPe in response to the simplified 12-h step-function storm, as would be experienced in the 19°E longitude sector on the magnetic equator. The basic theory presented by Blanc and Richmond for the nightside is consistent with the CTIPe simulation results, but the three-dimensional model contains significantly more local time structure. In particular, rather than being a uniform reduction in the downward plasma drift on the nightside, the response is much more localized in local time, and even reverses the direction of the drift to upward in the postmidnight or predawn sector. The other significant feature in the simulations is the apparent reduction in the magnitude of the PRE. Simulations of real events [*Maruyama et al.*, 2005, 2007] shows that this nightside response is fairly typical, and is in reasonable agreement with some of the observations of storm-time response seen by the Jicamarca incoherent scatter radar facility on the magnetic equator in Peru [*Fejer and Scherliess*, 1997].

Plate 5 shows the simulation of the disturbance dynamo for a real event, during the storm in November 2004. Four days of simulation are presented beginning on 6 November 2004. The storm had two main driven phases, when the interplanetary magnetic field (IMF) Bz component was directed strongly southward. The first occurred on 7 November, shown by the orange line, and the second on 9 November, shown in blue. The dotted curve shows the normal quiet-day variation from the numerical simulation. On 7 November, the normal downward drift is reversed to strongly upward in the postmidnight sector. On 8 November, shown by the green line, the upward drift begins to abate in this sector, as the forcing from the first pulsed begins to decline. At the same time, the strength of the PRE is reduced. On 9 November (blue line), the second pulse in the solar wind further reduces the PRE and a second surge in the upward drift on the nightside is produced. The response on the dayside is more uniform with a tendency for reduced upward drift in response to the storm, which is consistent with the *Blanc and Richmond* [1980] theory.

The interesting feature about this 4-day event is that the response characteristics, although varying in magnitude, have a surprising similar local time dependence, which is consistent from day to day. So the response in the generic storm of Plate 4 turns out to be a typical response. This was also true for the simulations presented by *Maruyama et al.* [2005] for the storm in March 2001. It is interesting to note that one of the documented storm responses is that irregularities that are normally associated with postsunset enhancement in vertical plasma drift during quiet times, often appear postmidnight or predawn during a storm. This is consistent with the results of the numerical simulation. The height of the ionosphere will be raised in the predawn sector by the upward drift leading to conditions that are ripe for the initiation of plasma bubbles, or irregularities, from the Rayleigh–Taylor instability mechanism. These irregularities cause scintillation in ground-to-satellite radio signals at a range of wavelength including UHF and GPS L-band frequencies [*Basu et al.*, 1996; *Fejer and Kelley*, 1980; *Fejer et al.*, 1999; *Groves et al.*, 1997].

The Blanc and Richmond theory predicts that the disturbance dynamo is slow to develop, due to the gradual buildup of the zonal winds, and also slow to abate. An additional mechanism was noted by Blanc and Richmond, and was explored by *Fuller-Rowell et al.* [2002] in numerical simulations. The new mechanism appears to provide a means of generating a disturbance dynamo response about an hour or two after the onset of a geomagnetic storm. This second mechanism is driven by the meridional wind surges that respond within an hour or two of the high-latitude heating. The mechanism for the rapid disturbance dynamo onset is a combination of two effects. The first follows the Blanc and Richmond theory. The meridional wind surges in the geographic frame have components in both the meridional and zonal magnetic frame. The zonal component produces the same response as the Blanc and Richmond theory, except that it does not require the slow buildup of the zonal wind via the Coriolis force. The second effect arises from a direct effect of the meridional wind at mid latitudes. Equation (3) shows that an equatorward wind in the magnetic frame drives an eastward-directed zonal Pedersen current at mid latitudes. The strength of the eastward current will depend on the magnitude of the height-integrated, conductivity-weighted equatorward wind. The divergence in the eastward current produced by the local time variation of the winds (e.g., Plate 2) and the diurnal variation in conductivity produce a local time variation in the zonal electric field. In both cases, the electrodynamic response is to the wind surge, which drives the dynamo at mid latitudes within 1–2 h of storm onset at high latitudes, and is experienced at the equator on the same timescale.

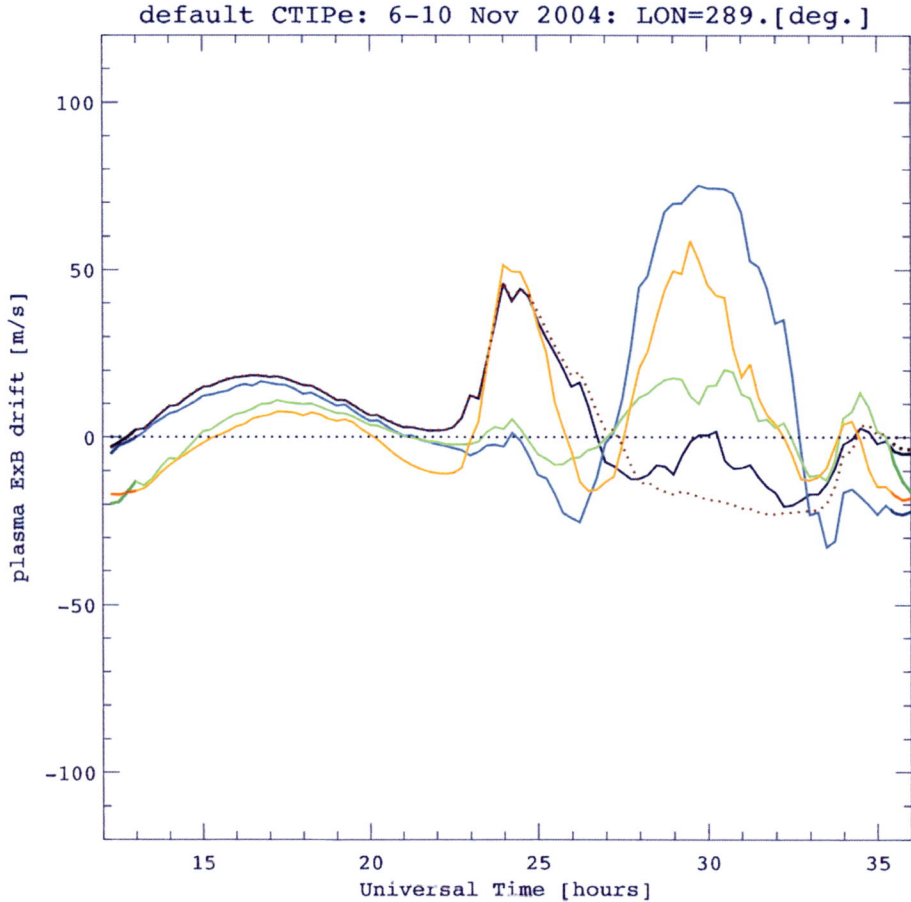

Plate 5. Simulation of a vertical plasma drift at the magnetic equator during a real storm covering the period from 6 November to 9 November 2004. Four days of simulation are presented beginning on 6 November (purple line) at 12 UT. Each 24 hour period of universal time covers local times from 7 LT at the start, with noon at 17 UT and local midnight labeled as 29 UT (i.e., 5 UT). The storm had two main driven phases, when the interplanetary magnetic field Bz component was directed strongly southward. The first occurred on 7 November (orange line), and the second occurred on 9 November (blue line). The dotted curve shows the normal quiet-day variation from the numerical simulation. On 7 November the normal downward drift is reversed to strongly upward in the postmidnight sector. On 8 November (green line) the upward drift begins to abate in this sector, as the forcing from the first pulsed begins to decline. At the same time, the strength of the prereversal enhancement is reduced. On 9 November (blue line), the second pulse in the solar wind further reduces the prereversal enhancement, and a second surge in the upward drift on the nightside is produced.

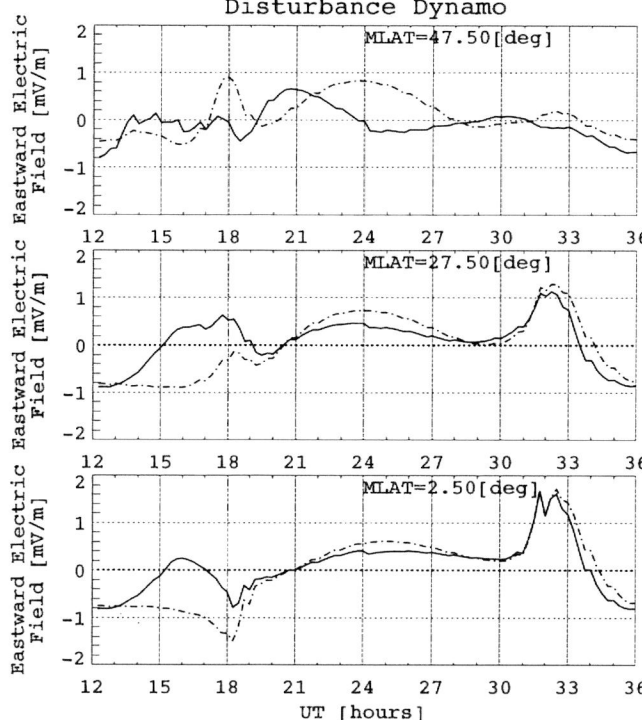

Figure 1. Illustration of the rapid initiation of the disturbance dynamo in response to the generic storm commencing at 12 UT, shown previously in Plates 2 and 3. The eastward electric field at three magnetic latitudes: 2.5°, 27.5°, and 47.5° is shown for 162°E longitude, a sector that passes through the postmidnight period in the first few hours of the simulation. Local midnight is at 13 UT, and local noon is labeled 25 UT (i.e., 1 UT). The dot-dashed lines are the quiet-time simulation, and the solid line is the storm response. Notice that the storm electrodynamics departs from the quiet curve within 2 hours at all latitudes, even at the equator, before the wind surge would have reached the low latitudes.

Figure 1 shows the electrodynamic response to the generic storm previously shown in Plates 2 and 3, for ground stations along the 162°E longitude meridian; locations that pass through the postmidnight sector in the first few hours of the simulation. The figure shows the eastward electric field at three magnetic latitudes: 2.5°, 27.5°, and 47.5°. The dot-dashed lines are the quiet-time simulation and the solid line is the storm response. Notice that the storm electrodynamics departs from the quiet curve within 2 h at all latitudes, even at the equator, before the wind surge would have reached the low latitudes.

4. DISCUSSION

In addition to the dynamo fields, prompt penetration electric fields are also a major source of disruption of the low-latitude ionosphere during geomagnetic storms. When the high-latitude magnetospheric convection increases, usually associated with a southward turning of the IMF, the high-latitude electric fields are unshielded as the magnetospheric plasma begins to respond. As a result, the electric fields can penetrate directly to the equator [*Kelley et al.*, 1979; *Spiro et al.*, 1988; *Fejer et al.*, 1990].

The observed electrodynamic response can therefore be a complex combination of prompt penetration and disturbance dynamo effects. The empirical model of *Fejer and Scherliess* [1997] attempts to separate the prompt penetration and disturbance dynamo electric fields using the time history of the *AE* geomagnetic index. The early results from the Rice magnetospheric convection model (RCM) were in good agreement with the empirical penetration electric field model [*Fejer and Scherliess*, 1997]. The magnitude of the penetration fields can be significantly larger than dynamo fields, but their duration tends to be shorter, typically lasting less than an hour. There has been speculation recently that penetration electric fields can remain unshielded for several hours [*Huang et al.*, 2005] and that the magnetospheric response to "superstorms" is somehow different from that during more modest disturbances. Whether this is a fundamental change in the magnetospheric "system response" time is difficult to determine. It is interesting to note that recent simulation by the RCM [*Maruyama et al.*, 2005] appear to support the idea of long-duration penetration events. The numerical simulations use a more complex and time-varying magnetic field model in the inner magnetosphere [*Sazykin et al.*, 2005].

Two other possibilities exist. The first is that the system response is the same but that an increasing solar wind electric field driver can continue to force the magnetosphere beyond the more typical shielding response time. The second is that penetration fields can be confused with rapid onset dynamo fields that can appear within 2 h of an event and before wave surges have actually reached the equator [*Fuller-Rowell et al.*, 2002]. Whether of short or long duration, penetration fields can be intense and can cause significant redistribution of plasma at low latitudes.

Penetration electric fields are also generated in response to a decrease in high-latitude convection, often associated with a northward turning of the IMF, but with opposite sign. Even though penetration and dynamo fields can become confused in this overshielded case, the *Fejer and Scherliess* [1997] model is able to separate the two effects. If the solar wind is highly variable, a series of prompt penetration electric fields will be initiated at low latitudes. The penetration field will interfere constructively or destructively with the dynamo component depending on the particular time history of the solar wind.

Satellite observations of the dramatic changes that can occur at low latitude in response to geomagnetic storms were presented by *Basu et al.* [2001]. They showed data from the DMSP polar orbiting satellite at 850 km altitude during the Bastille Day storm in July 2000 indicating that a wide swath of plasma had disappeared over tens of degrees in latitude. A similar event occurred during the March 1989 storm [*Greenspan et al.*, 1991] when vertical drift measurements exceeded 100 m/s. Modeling this event using the observed drifts raised the *F*-region ionosphere to more than 800 km [*Batista et al.*, 1991].

This huge restructuring of the ionosphere at low latitudes, as depicted in the DMSP observations, has never been successfully modeled. Numerical simulations are able to produce similar effects at 400 km altitude from the action of the various dynamo processes but this in no way compares to the extreme changes that are observed. The most likely scenario is that the dynamo and prompt penetration fields are acting together, due to the particular time history of the solar wind drivers. Coupled physics-based models of the inner magnetosphere and thermosphere/ionosphere are currently being developed to simulate the effects of both dynamo and penetration electric fields during storms, and their interaction [*Maruyama et al.*, 2005, 2007; *Huba et al.*, 2005].

5. CONCLUSION

It is clear that the change in thermospheric winds during a storm can be a conduit through which many of the global ionospheric and thermospheric storm-time changes flow. For instance, increased mid-latitude equatorward *F*-region winds push plasma up field lines to regions of different neutral composition and so change ion loss rates [*Prölss*, 1997]. The increase in the global circulation from pole to equator redistributes neutral composition, changing the ratio between atomic oxygen and molecular nitrogen and oxygen, and again affects ion loss rates [*Burns et al.*, 1991, 1995]. The disturbance dynamo [*Blanc and Richmond*, 1980; *Fejer and Scherliess*, 1997] drives a redistribution of plasma at low latitude, changes the strength of the EIA, and the likelihood of ionospheric scintillations. Zonal wind changes during storms can also affect the likelihood of the formation of ionospheric irregularities [*Basu et al.*, 1996].

The dynamical and physical processes involved in the response of the thermosphere to a geomagnetic storm are reasonably well understood, and simulation models have illustrated how the complexity of the response is influenced by interactions among dynamics, composition, and electrodynamics. For instance, numerical simulations indicate that winds slosh between the hemispheres having implications for the strength of the global circulation, meridional transport, and the disturbance dynamo response. However, the reality of the modeled dynamical changes has yet to be comprehensively validated. The detailed comparison of numerical simulations with observations is a task at least equal in weight to the development of the models themselves. The requirement for comprehensive datasets to complement the model predictions is evident. The recent availability of fairly comprehensive neutral composition observations and the increasing availability of global-scale electron density observations demonstrate the value of the detailed comparison with model simulations. Similar availability of data for neutral dynamics remains a pressing need, and is essential in order to come to closure on the reality of the model dynamics, and reveal a deeper understanding and appreciation of the balance between the physical processes.

Some fundamental issues remain with regard to the connection between dynamics and electrodynamics. The numerical simulations appear to show a distinct difference in the disturbance dynamo on the dayside and nightside in response to the wind surges and changes in global circulation. The dayside disturbance electric fields appear quite uniform in local time in response to the dynamo, possibly due to the influence of the stronger *E* region conductivity where the storm-time dynamical response is much weaker than in the *F* region. At night, the *F* region winds and dynamo are more dominant. Local time variation in conductivity and in the strength of the winds appears to introduce a strong local time variation in the dynamo electric fields and the vertical plasma drift at the magnetic equator. The reversal of electric field and vertical plasma drift in the postmidnight sector from downward to upward predicted by the simulations is a feature that apparently agrees with some equatorial observations, but the results of the numerical simulations have yet to be generally accepted.

The numerical simulations also reveal that the disturbance dynamo can respond quickly to the changing wind field. In fact, the wind surges can apparently begin to drive changes in current flow at mid latitudes within 2 h of a storm onset. The resulting redistribution of charge at mid latitudes maps to low latitudes virtually instantaneously, so that equatorial electric fields can change at the equator even before the actual wind surges arrive. The prompt penetration component of the storm-time changes in mid- and low-latitude electric fields is also undoubtedly an important source of plasma drift and a cause for plasma redistribution at low latitudes. The situation is now more complex (1) due to the appreciation of a rapid dynamo response and (2) due to the recent suggestion that penetration electric fields can last for many hours. The separation, balance, and interaction between the two components of storm-time electric fields at low latitudes, and their relative importance during the different phases of a storm on the day and night sides, are fundamental outstanding questions.

REFERENCES

Basu, S., E. Kudeki, S. Basu, C. E. Valladares, E. J. Weber, H. P. Zengingonul, S. Bhattacharyya, R. Sheehan, J. W. Meriwether, M. A. Biondi, H. Kuenzler, and J. Espinoza (1996), Scintillations, plasma drifts, and neutral winds in the equatorial ionosphere after sunset, *J. Geophys. Res.*, *101*, 26,795–26,810, doi:10.1029/96JA00760.

Basu, S., S. Basu, K. M. Groves, H.-C. Yeh, S.-Y. Su, F. J. Rich, P. J. Sultan, and M. J. Keskinen (2001), Response of the equatorial ionosphere in the South Atlantic region to the great magnetic storm of July 15, 2000, *Geophys. Res. Lett.*, *28*, 3577–3580.

Batista, I. S., E. R. de Paula, M. A. Abdu, N. B. Trivedi, and M. E. Greenspan (1991), Ionospheric effects of the March 13, 1989, magnetic storm at low and equatorial latitudes, *J. Geophys. Res.*, *96*, 13,943–13,952, doi:10.1029/91JA01263.

Blanc, M., and A. D. Richmond (1980), The ionospheric disturbance dynamo, *J. Geophys. Res.*, *85*, 1669–1686.

Buonsanto, M. J. (1999), Ionospheric storms—A review, *Space Sci. Rev.*, *88*, 563–601.

Burns, A. G., T. L. Killeen, and R. G. Roble (1991), A theoretical study of thermospheric composition perturbations during a impulsive geomagnetic storm, *J. Geophys. Res.*, *96*, 14,153–14,167, doi:10.1029/91JA00678.

Burns, A. G., T. L. Killeen, W. Deng, G. R. Carignan, and R. G. Roble (1995), Geomagnetic storm effects in the low- to middle-latitude upper thermosphere, *J. Geophys. Res.*, *100*(A8), 14,673–14,692, doi:10.1029/94JA03232.

Burns, A. G., T. L. Killeen, W. Wang, and R. G. Roble (2004), The solar-cycle-dependent response of the thermosphere to geomagnetic storms, *J. Atmos. Sol. Terr. Phys.*, *66*, 1–14.

Emmert, J. T., B. G. Fejer, C. G. Fesen, G. G. Shepherd, and B. H. Solheim (2001), Climatology of middle- and low-latitude daytime F region disturbance neutral winds measured by Wind Imaging Interferometer (WINDII), *J. Geophys. Res.*, *106*, 24,701–24,712, doi:10.1029/2000JA000372.

Emmert, J. T., B. G. Fejer, G. G. Shepherd, and B. H. Solheim (2002), Altitude dependence of middle and low-latitude daytime thermospheric disturbance winds measured by WINDII, *J. Geophys. Res.*, *107*(A12), 1483, doi:10.1029/2002JA009646.

Emmert, J. T., B. G. Fejer, G. G. Shepherd, and B. H. Solheim (2004), Average nighttime F region disturbance neutral winds measured by UARS WINDII: Initial results, *Geophys. Res. Lett.*, *31*, L22807, doi:10.1029/2004GL021611.

Fejer, B. G., and J. T. Emmert (2003), Low-latitude ionospheric disturbance electric field effects during the recovery phase of the 19–21 October 1998 magnetic storm, *J. Geophys. Res.*, *108*(A12), 1454, doi:10.1029/2003JA010190.

Fejer, B. G., and M. C. Kelley (1980), Ionospheric irregularities, *Rev. Geophys. Space Phys.*, *18*, 401–454.

Fejer, B. G., and L. Scherliess (1997), Empirical models of storm time equatorial zonal electric fields, *J. Geophys. Res.*, *102*(A11), 24,047–24,056.

Fejer, B. G., et al. (1990), Low- and mid-latitude ionospheric electric fields during the January 1984 GISMOS campaign, *J. Geophys. Res.*, *95*(A3), 2367–2378, doi:10.1029/89JA01555.

Fejer, B. G., L. Scherliess, and E. R. de Paula (1999), Effects of the vertical plasma drift velocity on the generation and evolution of equatorial spread F, *J. Geophys. Res.*, *104*, 19,859–19,870, doi:10.1029/1999JA900271.

Fejer B. G., J. T. Emmert, and D. P. Sipler (2002), Climatology and storm time dependence of nighttime thermospheric neutral winds over Millstone Hill, *J. Geophys. Res.*, *107*(A5), 1052, doi:10.1029/2001JA000300.

Fesen, C. G., G. Crowley, R. G. Roble, A. D. Richmond, and B. G. Fejer (2000), Simulations of the pre-reversal enhancement in the low latitude vertical ion drifts, *Geophys. Res. Lett.*, *27*, 1851–1854.

Foster, J. C., and A. J. Coster (2007), Conjugate localized enhancements of total electron content at low latitudes in the American sector, *J. Atmos. Sol. Terr. Phys.*, *69*, 1241–1252.

Foster, J. C., and W. Rideout (2005), Midlatitude TEC enhancements during the October 2003 superstorm, *Geophys. Res. Lett.*, *32*, L12S04, doi:10.1029/2004GL021719.

Fuller-Rowell, T. J. (1995), The dynamics of the lower thermosphere, in *The Upper Mesosphere and Lower Thermosphere: A Review of Experiment and Theory*, Geophys. Monogr. Ser., vol. 87, edited by R. M. Johnson and T. L. Killeen, pp. 23–36, AGU, Washington, D. C.

Fuller-Rowell, T. J., S. Quegan, D. Rees, R. J. Moffett, and G. J. Bailey (1984), The effect of realistic conductivities on the high-latitude neutral thermospheric circulation, *Planet. Space Sci.*, *32*, 469.

Fuller-Rowell, T. J., M. V. Codrescu, R. J. Moffett, and S. Quegan (1994), Response of the thermosphere and ionosphere to geomagnetic storms, *J. Geophys. Res.*, *99*, 3893–3914, doi:10.1029/93JA02015.

Fuller-Rowell, T. J., M. V. Codrescu, H. Risbeth, R. J. Moffett, and S. Quegan (1996a), On the seasonal response of the thermosphere and ionosphere to geomagnetic storms, *J. Geophys. Res.*, *101*, 2343–2354, doi:10.1029/95JA01614.

Fuller-Rowell, T. J., D. Rees, S. Quegan, R. J. Moffett, M. V. Codrescu, and G. H. Millward (1996b), A coupled thermosphere ionosphere model (CTIM), in *Handbook of Ionospheric Models*, STEP Report, edited by R. W. Schunk, pp. 217–238.

Fuller-Rowell, T. J., M. V. Codrescu, R. G. Roble, and A. D. Richmond (1997), How does the thermosphere and ionosphere react to a geomagnetic storm?, in *Magnetic Storms*, Geophys. Monogr. Ser., vol. 98, edited by B. T. Tsurutani et al., pp. 203–225, AGU, Washington, D. C.

Fuller-Rowell, T. J., G. H. Millward, A. D. Richmond, and M. V. Codrescu (2002), Storm-time changes in the upper atmosphere at low latitudes, *J. Atmos. Sol. Terr. Phys.*, *64*, 1383–1391.

Greenspan, M. E., C. E. Rasmussen, W. J. Burke, and M. A. Abdu (1991), Equatorial density depletions observed at 840 km during the great magnetic storm of March 1989, *J. Geophys. Res.*, *96*, 13,931–13,942, doi:10.1029/91JA01264.

Groves, K. M., S. Basu, E. J. Weber, M. Smitham, H. Kuenzler, C. E. Valladares, R. Sheehan, E. MacKenzie, J. A. Secan, P. Ning, W. J. McNeill, D. W. Moonan, and M. J. Kendra (1997), Equatorial scintillation and systems support, *Radio Sci.*, *32*(5), 2047–2064, doi:10.1029/97RS00836.

Heelis, R. A. (2004), Electrodynamics in the low and middle latitude ionosphere: A tutorial, *J. Atmos. Sol. Terr. Phys.*, *66*, 825–838.

Huang, C.-S., J. C. Foster, and M. C. Kelley (2005), Long-duration penetration of the interplanetary electric field to the low-latitude ionosphere during the main phase of magnetic storms, *J. Geophys. Res.*, *110*, A11309, doi:10.1029/2005JA011202.

Huba, J. D., G. Joyce, S. Sazykin, R. Wolf, and R. Spiro (2005), Simulation study of penetration electric field effects on the low- to mid-latitude ionosphere, *Geophys. Res. Lett.*, *32*, L23101, doi:10.1029/2005GL024162.

Hunsucker, R. D. (1982), Atmospheric gravity waves generated in the high-latitude ionosphere: A review, *Rev. Geophys.*, *20*, 293–315.

Kelley, M. C., B. G. Fejer, and C. A. Gonzales (1979), An explanation of anomalous ionospheric electric fields associated with a northward turning of the interplanetary magnetic field, *Geophys. Res. Lett.*, *6*, 301–304.

Kelley, M. C., J. J. Makela, J. L. Chau, and M. J. Nicolls (2003), Penetration of the solar wind electric field into the magnetosphere/ionosphere system, *Geophys. Res. Lett.*, *30*(4), 1158, doi:10.1029/2002GL016321.

Killeen, T. L., P. B. Hays, G. R. Carignan, R. A. Heelis, W. B. Hanson, N. W. Spencer, and L. H. Brace (1984), Ion-neutral coupling in the high latitude *F*-region: Evaluation of ion-neutral heating terms from the Dynamics Explorer 2, *J. Geophys. Res.*, *89*, 7495–7509.

Killeen, T. L., J. D. Craven, L. A. Frank, J.-J. Ponthieu, N. W. Spencer, R. A. Heelis, L. H. Brace, R. G. Roble, P. B. Hays, and G. R. Carignan (1988), On the relationship between dynamics of the polar thermosphere and morphology of the aurora: Global-scale observations from Dynamics Explorers 1 and 2, *J. Geophys. Res.*, *93*, 2675–2692, doi:10.1029/88JA01097.

Lu, G., L. P. Goncharenko, A. D. Richmond, R. G. Roble, and N. Aponte (2008), A dayside ionospheric positive storm phase driven by neutral winds, *J. Geophys. Res.*, *113*, A08304, doi:10.1029/2007JA012895.

Mannucci, A. J., B. T. Tsurutani, B. A. Iijima, A. Komjathy, A. Saito, W. D. Gonzalez, F. L. Guarnieri, J. U. Kozyra, and R. Skoug (2005), Dayside global ionospheric response to the major interplanetary events of October 29–30, 2003 "Halloween Storms," *Geophys. Res. Lett.*, *32*, L12S02, doi:10.1029/2004GL021467.

Maruyama, N., A. D. Richmond, T. J. Fuller-Rowell, M. V. Codrescu, S. Sazykin, F. R. Toffoletto, R. W. Spiro, and G. H. Millward (2005), Interaction between direct penetration and disturbance dynamo electric fields in the storm-time ionosphere, *Geophys. Res. Lett.*, *32*, L17105, doi:10.1029/2005GL023763.

Maruyama, N., S. Sazykin, R. W. Spiro, B. G. Fejer, R. Wolf, D. Anderson, A. Anghel, F. R. Toffoletto, T. J. Fuller-Rowell, M. Codrescu, A. D. Richmond, and G. H. Millward (2007), Modeling storm-time electrodynamics of the low-latitude ionosphere–thermosphere system: Can long lasting disturbance electric fields be accounted for?, *J. Atmos. Sol. Terr. Phys.*, *69*, 1182–1199.

Millward, G. H., R. J. Moffett, S. Quegan, and T. J. Fuller-Rowell (1986), A coupled thermosphere ionosphere plasmasphere model (CTIP), in *Handbook of Ionospheric Models*, STEP Report, edited by R. W. Schunk, pp. 239–279.

Millward, G. H., I. C. F. Müller-Wodarg, A. D. Aylward, T. J. Fuller-Rowell, A. D. Richmond, and R. J. Moffett (2001), An investigation into the influence of tidal forcing on *F* region equatorial vertical ion drift using a global ionosphere-thermosphere model with coupled electrodynamics, *J. Geophys. Res.*, *106*, 24,733–24,744.

Peymirat, C., A. D. Richmond, and R. G. Roble (2002), Neutral wind influence on the electrodynamic coupling between the ionosphere and the magnetosphere, *J. Geophys. Res.*, *107*(A1), 1006, doi:10.1029/2001JA900106.

Prölss, G. W. (1997), Magnetic storm associated perturbations of the upper atmosphere, in *Magnetic Storms, Geophys. Monogr. Ser.*, vol. 98, edited by B. T. Tsurutani et al., pp. 227–241, AGU, Washington, D. C.

Richmond, A. D. (1995), Ionospheric electrodynamics, in *Handbook of Atmospheric Electrodynamics*, vol. II, edited by H. Volland, pp. 249–290, CRC Press, Boca Raton, Fla.

Richmond, A. D., and S. Matsushita (1975), Thermospheric response to a magnetic substorm, *J. Geophys. Res.*, *80*, 2839–2850.

Richmond, A. D., and R. G. Roble (1987), Electrodynamic effects of thermospheric winds from the NCAR thermospheric general circulation model, *J. Geophys. Res.*, *92*, 12,365–12,376.

Ridley, A. J., Deng, Y., and Toth, G. (2006), The global ionosphere–thermosphere model, *J. Atmos. Sol. Terr. Phys.*, *68*(8), 839–864.

Rodger, A. S., G. L. Wrenn, and H. Rishbeth (1989), Geomagnetic storms in the Antarctic *F* region, II, physical interpretation, *J. Atmos. Terr. Phys.*, *51*, 851–866.

Sazykin, S., R. W. Spiro, R. A. Wolf, F. R. Toffoletto, N. Tsyganenko, J. Goldstein, and M. Hairston (2005), Modeling inner magnetopsheric electric fields: Latest self-consistent results, in *The Inner Magnetosphere: Physics and Modeling, Geophys. Monogr. Ser.*, vol. 115, pp. 263–269, AGU, Washington, D. C.

Scherliess, L., and B. G. Fejer (1997), Storm time dependence of equatorial disturbance dynamo zonal electric fields, *J. Geophys. Res.*, *102*, 24,037–24,046.

Shiokawa, K., Y. Otsuka, T. Ogawa, N. Balan, K. Igarashi, A. J. Ridley, D. J. Knipp, A. Saito, and K. Yumoto (2002), A large-scale traveling ionospheric disturbance during the magnetic storm of 15 September 1999, *J. Geophys. Res.*, *107*(A6), 1088, doi:10.1029/2001JA000245.

Spiro, R. W., R. A. Wolf, and B. G. Fejer (1988), Penetration of high-latitude-electric-field effects to low latitudes during SUNDIAL 1984, *Ann. Geophys.*, *6*, 39–50.

Strickland, D. J., R. E. Daniell, and J. D. Craven (2001), Negative ionospheric storm coincident with DE 1-observed thermospheric disturbance on October 14, 1981, *J. Geophys. Res.*, *106*, 21,049–21,062, doi:10.1029/2000JA000209.

Weimer, D. R. (1995), Models of high-latitude electric potentials derived with a least error fit of spherical harmonic coefficients, *J. Geophys. Res.*, *100*, 19,595–19,608, doi:10.1029/95JA01755.

T. J. Fuller-Rowell and N. Maruyama, CIRES University of Colorado and NOAA Space Weather Prediction Center, 325 Broadway, Boulder, CO 80305, USA. (tim.fuller-rowell@noaa.gov)

A. D. Richmond, High Altitude Observatory, NCAR, 3450 Mitchell Lane, Boulder, CO 80305, USA.

Thermospheric Dynamics at Low and Mid-Latitudes During Magnetic Storm Activity

J. W. Meriwether

Department of Physics and Astronomy, Clemson University, Clemson, South Carolina, USA

The coupled ionosphere-thermosphere-plasmasphere system is very complex. The study of the interrelationships of these components during geomagnetically disturbed conditions for low- and mid-latitude regions is especially challenging. This is partly because observations of the neutral atmosphere response are sparse and partly because the lack of coordination of satellite observations with ground-based measurements makes the proper identification of temporal and spatial variations in measurements difficult to achieve. In disturbed geomagnetic conditions, thermospheric neutral winds exhibit large deviations from the climatological behavior resulting in large corresponding changes in the ionospheric plasma density, composition, temperature, and electrodynamics. The general outline of the neutral thermosphere response to magnetic storm activity is clear. The mid-latitude thermospheric wind response is characterized by enhanced and expanded wind circulation cells in the polar regions (caused by enhanced ion convection) and divergent flows originating at high latitudes (caused by Joule heating/particle heating sources). Increased thermospheric temperatures and significant deviations of zonal and meridional winds from quiet-time climatology behavior are observed for the mid-latitude thermosphere. The low-latitude response of the thermosphere may be significantly less than at mid-latitudes. What remains unclear is the step-by-step sequence of the thermosphere response once the magnetic activity disturbance has been initiated. Also unclear is how this sequence might be modified as the season and the phase of the solar cycle change.

1. INTRODUCTION

The aeronomy literature regarding geomagnetic storm effects is quite extensive, with numerous papers describing aspects of the ionospheric and thermospheric phenomenology with regard to the equatorial, mid-latitudes, and polar attributes of the energy transfer processes that take place globally [*Gold*, 1962; *Mendillo et al.*, 1992; *Rees*, 1996; *Fuller-Rowell et al.*, 1997; *Prolss*, 1997; *Buonsanto*, 1999; *Foster and Jakowski*, 2000; *Fuller-Rowell et al.*, 2002; *Burke et al.*, 2007; *Maruyama et al.*, 2007). However, and perhaps for good reason, these papers have generally emphasized the ionospheric aspects of the global response for periods of strong magnetic disturbances. This is partly because the global network of ionospheric instrumentation using ionosondes, GPS sensors, and total electron content monitors, which are sensitive to ionospheric plasma density changes, has grown substantively over the past two decades. Data from these networks have provided significant information regarding the ionospheric plasma behavior in all regions of the globe. In contrast, the instrument that is most often applied to measure neutral winds and temperatures directly, the Fabry–Perot interferometer (FPI), uses optics to observe the nightglow emission spectral profile at high resolution in the visible spectrum. Consequently, useful data are collected only in periods of good weather and nighttime hours. Thus, less information has been collected regarding the response of the thermosphere to intense magnetic storms, particularly at low latitudes. A series of papers by Hernandez and Roble published during the late 1970s [*Hernandez and Roble*, 1976, 1978, 1979] provided significant characterization of the thermospheric response to geomagnetic activity at midlatitudes for the North American sector. The recent analysis of satellite optical observations (discussed in the next section) has helped to overcome this lack of information by providing an excellent picture of the statistical behavior of thermospheric winds as averaged over selected levels of magnetic disturbances and averaging bins of 2 h local time and 20° latitude bands.

Unfortunately, because the mechanism of the low-latitude and mid-latitude ion-neutral thermospheric coupling during periods of intense geomagnetic storms is not very well understood, it is not possible to rely on physics-based modeling predictions to forecast the thermosphere response even with knowledge of the global ionospheric plasma distribution. Aside from quantifying the transport of thermospheric air from the polar latitudes into middle- and low-latitude regions generated by the high-latitude disturbance, there is also a need to consider the impact of the penetration of the magnetospheric electric field into the low-latitude and mid-latitude regions. These fields may generate neutral wind disturbance effects through the ion-neutral coupling.

An excellent summary of the processes that are believed to enter into the development of the "neutral thermosphere storm" was presented by *Buonsanto* [1999] and is referred to in this review as the Buonsanto "picture." Particle and Joule heating sources at high latitudes cause expansion of the polar neutral atmosphere. Consequent upwelling and deviations from diffusive equilibrium occur that are accompanied by increases in the mean molecular mass. The resulting heating also modifies the day-to-night pressure gradient causing the global thermospheric circulation to change. In addition, enhanced equatorward winds induced by antisunward plasma convection within the polar cap [e.g., *Straus and Schulz*, 1976; *Babcock and Evans*, 1979; *Fuller-Rowell et al.*, 1997] transport the composition changes to lower latitudes, creating the formation of a composition disturbance zone of increased mean molecular mass reaching from high to middle latitudes [e.g., *Prolss*, 1987; *Fuller-Rowell et al.*, 1997].

This composition disturbance reaches the lowest latitudes at night, but then rotates with the Earth into the morning sector. Equatorward of this zone and during the afternoon hours, poleward neutral winds induced by the Coriolis force may occur. The resulting convergence of these winds results in downwelling that decreases the mean molecular weight at constant pressure levels [*Risbeth*, 1989; *Burns et al.*, 1991, 1995; *Fuller-Rowell et al.*, 1997].

Equatorward surges or traveling atmospheric disturbances may be generated when the polar heating events are impulsive [e.g., *Richmond and Matsushita*, 1975; *Roble et al.*, 1978; *Hernandez and Roble*, 1978; *Burns and Killeen*, 1992; *Hocke and Schlegel*, 1996]. These large-scale waves can penetrate to low latitude or even to the opposite hemisphere. The corresponding effect in the ionospheric plasma is seen as a large-scale traveling ionospheric disturbance, which has been detected as sequential rises and falls in h_mF_2 along north–south chains of ionosondes [e.g., *Hajkowicz*, 1990, 1991].

2. SATELLITE MEASUREMENTS

The picture of the development of neutral disturbances as described by *Buonsanto* [1999] is supported by the optical satellite measurements of thermospheric winds for low- and mid-latitude regions that have emerged from the Wind Imaging Interferometer (WINDII) observations on the UARS satellite [*Fejer et al.*, 2000; *Emmert et al.*, 2003, 2004]. Plate 1 shows the results found for the statistical analysis of the WINDII observations published by *Emmert et al.* [2004] for periods of geomagnetic activity. The quiet-time winds (i.e., $Kp \leq 3$) have been removed from the observations so that the residual values plotted represent the response of the thermosphere at low- and mid-latitude regions to three levels of geomagnetic activity.

Plate 2 summarizes the behavior of the nighttime and daytime WINDII disturbance winds for $Kp \sim 4$. Inspection of this shows a strong similarity of the disturbance wind responses for the northern and southern hemispheres. *Emmert et al.* [2004] suggested that these results support the conclusion that heating-induced pressure gradients developed at high latitudes generated the largest equatorward disturbance winds at high latitudes near 03 MLT. As this air moved equatorward, the flow turned westward and then poleward at mid-latitudes, possibly as a result of Coriolis forcing but also in part because of ion drag forcing caused by the penetration of the high-latitude electric field into the mid-latitude region.

The nighttime zonal disturbance winds are generally westward with the strongest effects extending from evening at 70° to midnight at 45°. The speeds of the meridional dis-

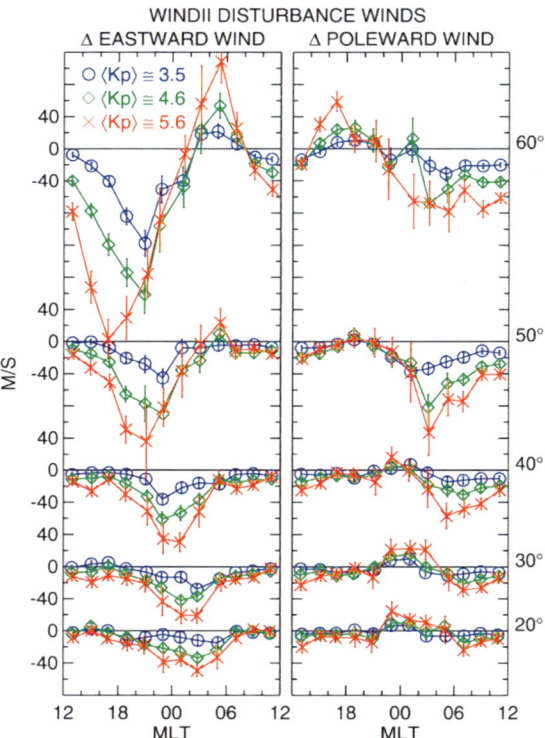

Plate 1. WINDII measurements of disturbed thermospheric winds averaged over the altitude range of 225–275 km for low- and mid-latitude regions. Results from three Kp bins are superimposed: $3 \leq Kp \leq 4.3$ (blue circles), $4 \leq Kp \leq 6$ (green diamonds), and $Kp \geq 5$ (red crosses). Local time bins are 2 h wide. Results from different 10 magnetic latitude bins, centered on the indicated values and with northern/southern hemispheres combined, are shown. (left) Zonal component, positive eastward. (right) Meridional component, positive poleward [from *Emmert et al.*, 2004].

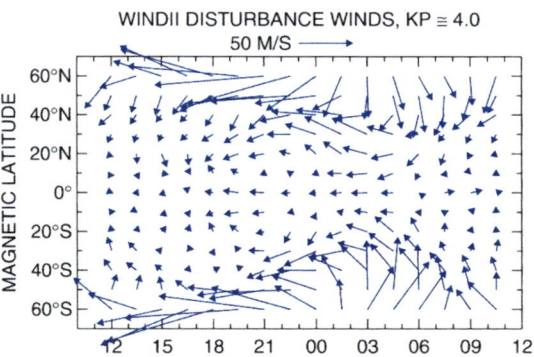

Plate 2. Vector map of disturbance winds generated by the WINDII empirical model for low- and mid-latitude regions averaged over the altitude range of 225–275 km [from *Emmert et al.*, 2004].

turbance winds shown in Plate 2 are generally small in the premidnight sector. The direction of the disturbance wind is equatorward in the postmidnight sector, and the effects appear at earlier local times with increasing latitude. The largest equatorward values vary from ~40 m s^{-1} for Kp ~3.5 to ~120 m s^{-1} for Kp ~5.6. Poleward disturbances are observed before 1930 MLT at 60° and also around midnight below 50°.

Westward disturbance winds are also observed throughout most of the night at low latitudes. Eastward perturbations occur in the postmidnight sector above 50°. The meridional disturbance winds are primarily equatorward above 40° and after 03 MLT. In the midnight sector during low and moderate solar flux conditions, poleward winds are observed below 40°. During solar maximum, the perturbations are largely equatorward throughout the night.

Figure 1 presents the height-dependent climatology of the seasonally averaged zonal and meridional thermospheric daytime disturbance winds measured by WINDII. The results are presented in three local time sectors for an average of Kp that is increased by 3.7 above the average quiet-time Kp value of 1.7. The results presented show that, in general, the zonal and meridional perturbation winds have largest amplitudes at the higher latitudes. Both components decrease sharply below 120 km and become insignificant below 100 km.

At geomagnetic latitudes between 45° and 25°, the zonal disturbance winds are westward at all local times. The largest magnitudes are seen in the early morning sector (peaking near 120 km). These winds do not change much with height above 140 km. The meridional disturbed winds are generally equatorward at all daytime hours and their magnitudes increase with increasing latitude. They are largely height-independent above 150 km with a sharp decrease seen below 120–140 km.

These results compare favorably with what was described in Section 1 as the Buonsanto "picture."

3. EXAMPLES OF THE NEUTRAL ATMOSPHERIC RESPONSE IN INDIVIDUAL STORM CASE HISTORIES

We consider now the neutral disturbance effects that have been seen with ground-based instruments at mid-latitudes during periods of large geomagnetic storms. Many reports have been published, but we concentrate on four mid-latitude studies. We also provide two examples of the observed low-latitude response. The overall impression from these cases is that our understanding of the thermospheric response in geomagnetic storm disturbances is constrained by the inability to obtain a broader picture of the global response beyond what is represented by the set of results for one geographic location. The difficulty in connecting the small-scale response to a global view remains a major challenge to the successful application of physics-based models to space weather forecasting applications.

The geomagnetic activity for individual storm events is portrayed by the variations in the Kp index and the Dst index. The former is more characteristic of the polar region level of disturbance, and the latter is taken to be representative of the equatorial region. Following the convention introduced by *Fagundes et al.* [1996], storms are classified in regards to the peak Dst value seen during the storm activity: intense, ≤ -100 nT; moderate, -50 nT $\geq Dst \geq -100$ nT; weak, -30 nT $\geq Dst \geq -50$ nT.

The ground-based instrument most commonly used to measure thermospheric wind speed and direction as well as thermospheric temperature is the FPI. This instrument observes at high spectral resolution the 630-nm spectral lineshape to determine the Doppler shifts and Doppler broadenings for a sequence of viewing directions. These are typically zonal and meridional cardinal points at a zenith angle of 60°. Because the thermosphere is highly viscous, the variation in thermospheric wind speed within the altitude range of the 630-nm volume emission profile from 200 to 300 km is small. Thus, the determined line-of-sight speed is an average of the thermosphere winds within this range with a centroid height of ~240–250 km. The Doppler reference required for the determination of the Doppler shift is the peak position of the 630-nm spectral profile for the zenith direction. This analysis approach assumes that vertical winds are small, that is, less than 1 m s^{-1}, and the evidence is that such an assumption is generally valid for low-latitude and mid-latitude thermospheric regions.

3.1. Observations of Converging Thermospheric Winds During Magnetic Storm Activity

Interestingly, two examples of mid-latitude convergence of thermospheric winds as represented by the Buonsanto "picture" are available from the FPI observations at Fritz Peak, CO (39.9°N, 105.5°W), that were reported by *Hernandez et al.* [1982]. Figure 2 shows the two cases of simultaneous poleward and equatorward winds with speeds of 200–600 m s^{-1} that was observed by the Fritz Peak FPI for two intense magnetic storms near solar maximum. The maximum Kp index for the 21 February 1979 storm was 7– and for the 25 April 1979 storm 8+. The observed zonal winds were also convergent with speeds between 100 and 200 m s^{-1}. The period of convergent winds lasted for 4–5 h. The temperature data showed an increase in thermospheric temperature of 400–500 K to a maximum of ~1700–1800 K

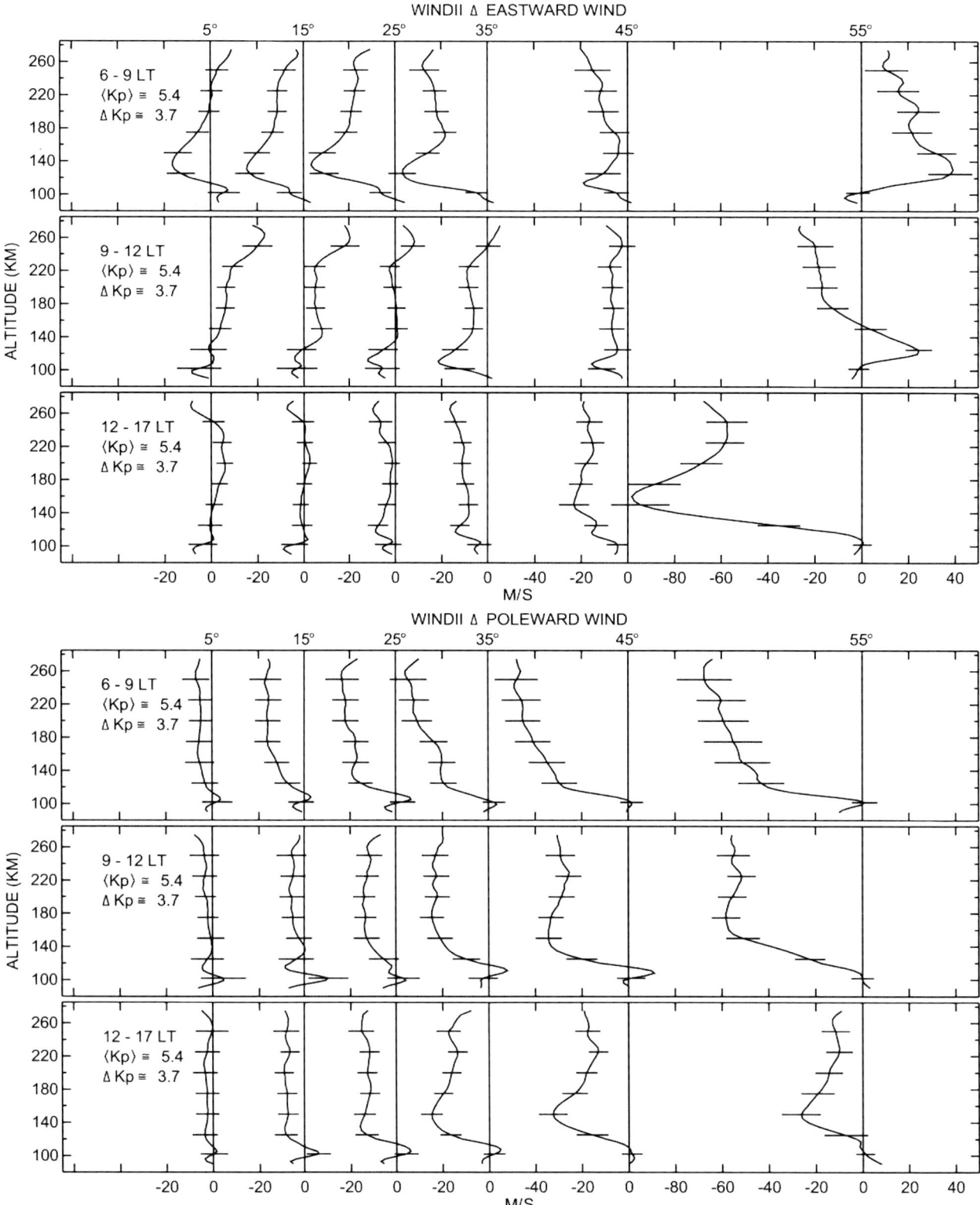

Figure 1. Seasonally averaged disturbance winds as a function of height for a disturbed condition of Kp ~4.7. The top and bottom sets of three panels show the zonal and meridional components for three different local time sectors, respectively. Each panel contains results from several 10 geomagnetic latitude bins (centered on the indicated latitudes). Error bars denote the estimated uncertainty of the means [from *Emmert et al.*, 2002].

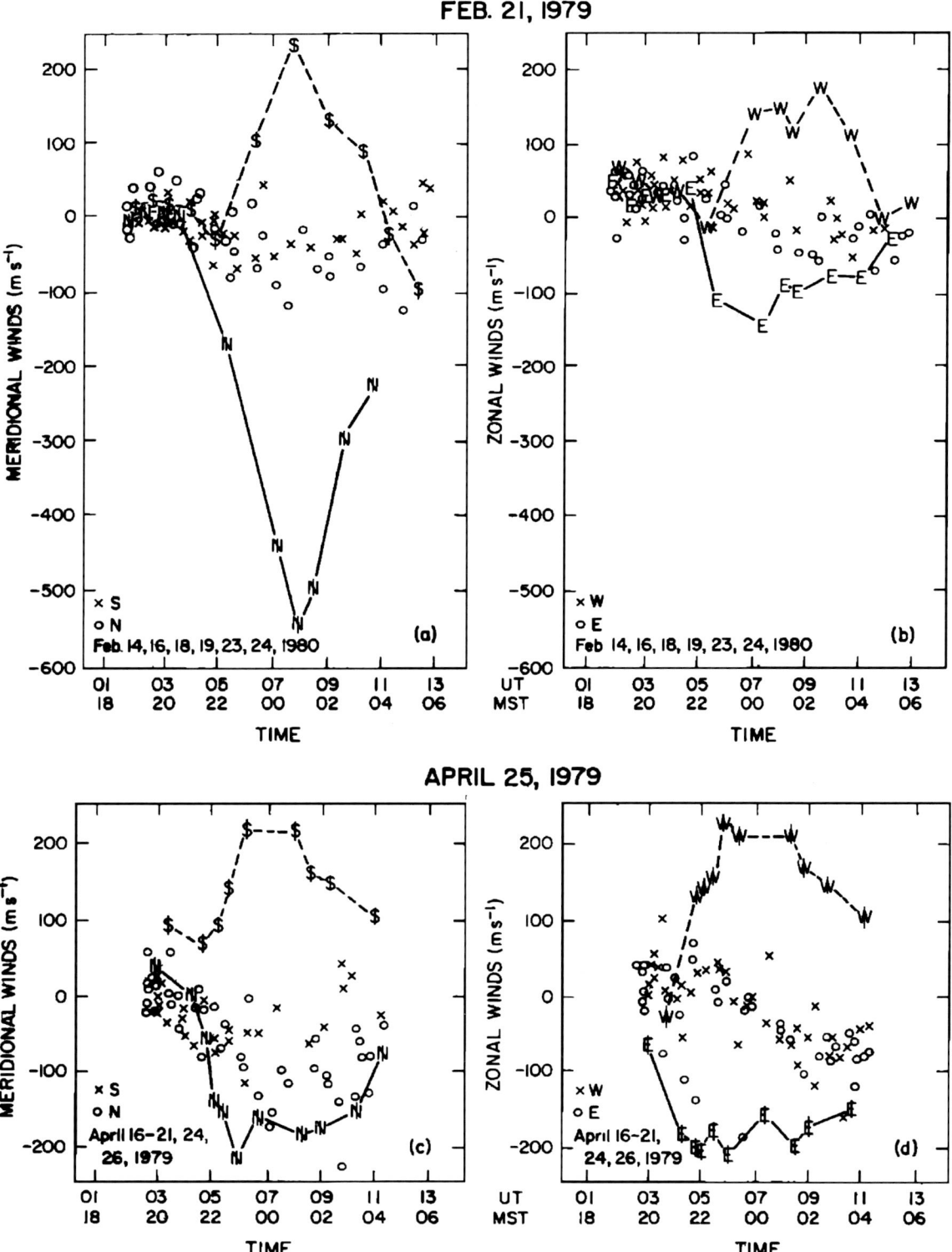

Figure 2. The meridional and zonal winds measured during the storm periods of (a, b) 21 February 1979 and (c, d) 25 April 1979. Positive values are northward and eastward for the meridional and zonal winds, respectively. The circles and crosses indicate the geomagnetic-quiet measured winds on days in February and April before and after the storm period. The nights during which the measurements were made are indicated in the lower left corner of each figure [from *Hernandez et al.*, 1982].

during the peak magnetic activity period for both storms. In the storm of 21 February 1979, the temperature was found to be warmer by ~300 K in the south rather than in the north as would normally be the case.

The source of the poleward wind seen to the south of the Fritz Peak observatory was not identified, but might indeed be the poleward flow identified in the Buonsanto description. A scenario to explain these results might be the following. In the region to the south of Fritz Peak, the flow to the north came from the storm activity of the previous night producing wind disturbances that behaved as described in the Buonsanto picture. In the region to the north, the flow to the south is a result of the nighttime dynamical activity taking place driven by the magnetic storm activity. Figure 3 selected from the *Hernandez et al.* [1982] paper illustrates how the two flow streams would need to interact to represent the results seen at Fritz Peak. One interpretation is that the southern and northern flow streams have zonal components that are driven by the Coriolis force acting on the meridional flow. An extensive shear in the north–south direction would exist at the boundary between the two flow streams. Clearly, further studies of such cases would benefit from a broader perspective of the dynamics of the mid-latitude thermosphere as might be achieved with a network of several FPI observatories [*Meriwether*, 2006].

3.2. Brazilian Observations During July 1991

It is generally accepted that the thermospheric temperature increases during geomagnetic activity [*Biondi and Meriwether*, 1986; *Tinsley and Meriwether*, 1988; *Burnside et al.*, 1991; *Fagundes et al.*, 1995, 1996]. The results presented in Figure 4 reported by *Fagundes et al.* [1995, 1996] show wind and temperature measurements obtained at Cachoeira Paulista (23°S, 45°W) for the magnetic storm of 7–11 July. The Kp index increased from 1 at 8 UT on 7 July to a maximum of 8 at 16 UT on 8 July and recovered to Kp of 2 at 16 UT on 9 July, which coincides with the

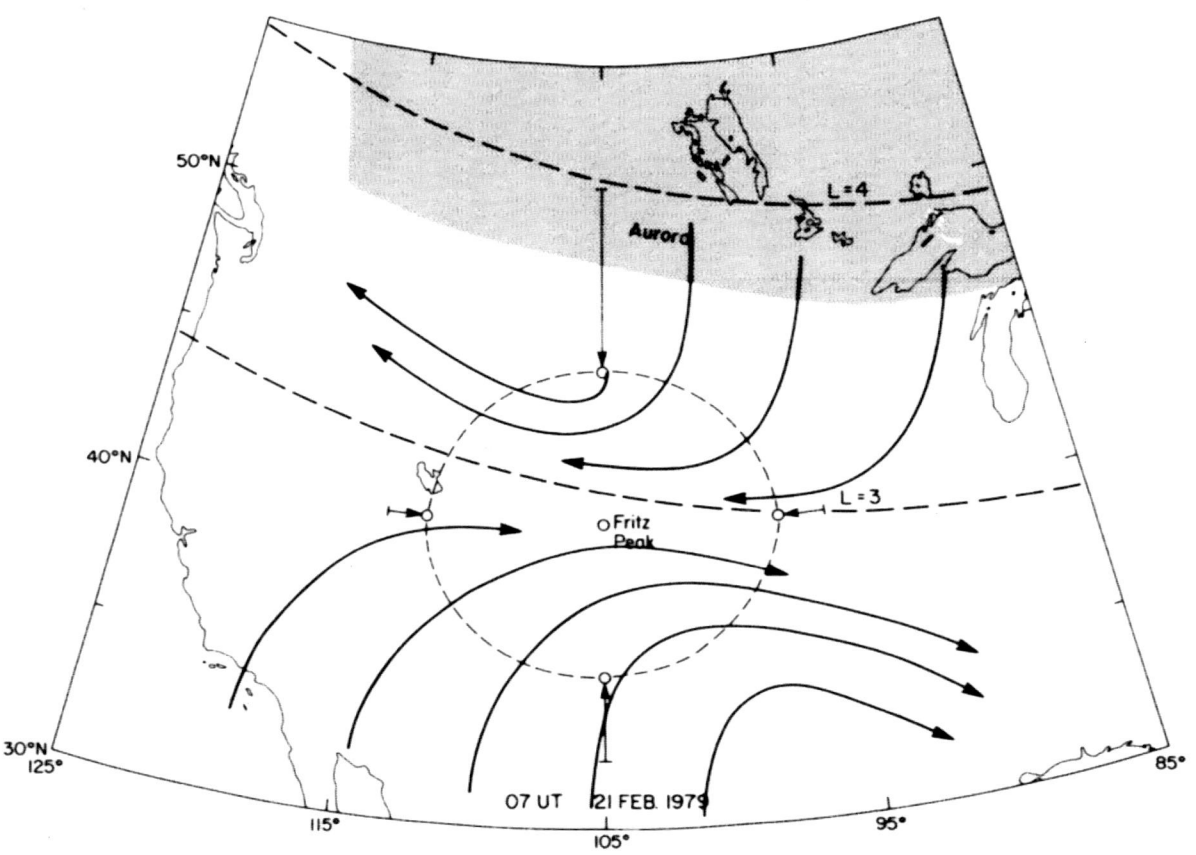

Figure 3. Schematic of two flow streams interacting near Fritz Peak to produce converging winds as observed in Figure 2 [from *Hernandez et al.*, 1982].

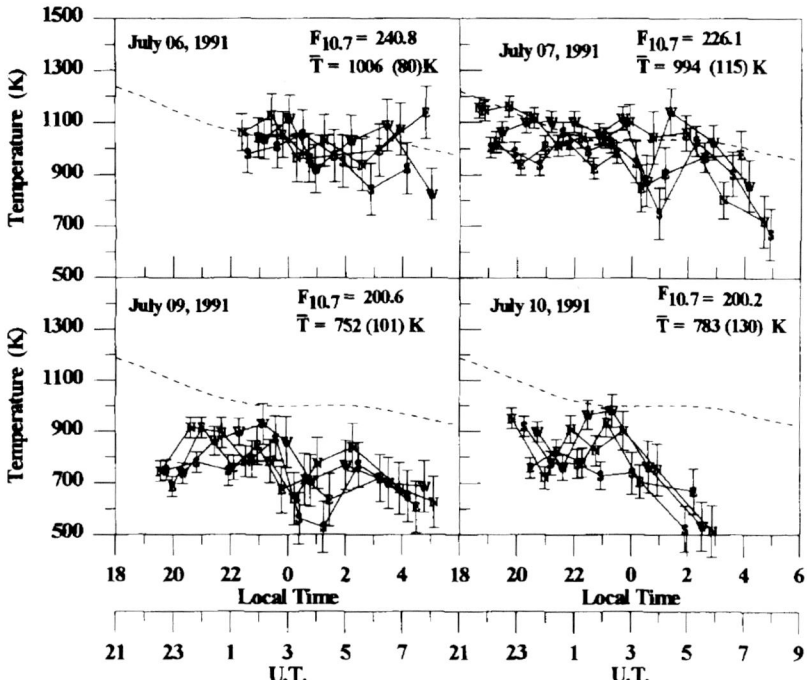

Figure 4. FPI temperature measurements for Cachoeira Paulista (23°S, 45°W) shown for the magnetic storm activity of 6–10 July 1991. The dashed line in each panel represents the MSIS-86 model temperatures calculated at 240 km for the averaged $F_{10.7}$ solar flux level indicated in the upper right corner. Also included is the averaged nighttime temperature and its standard deviation. The letters indicate the direction of observation (N, S, E, W), and the error bars for each measurement are also indicated [from *Fagundes et al.*, 1996].

minimum value of –200 gammas observed for the *Dst* index at the same time. The *Dst* index recovered to –50 by the beginning of 11 July. It was unfortunate that there were no FPI measurements available at the peak of the storm activity at 16 UT, 8 July, but there were measurements for the two nights before the peak and for the two nights after the peak. Presumably, because the Fritz Peak FPI measurements have consistently shown heating to occur during times of peak magnetic activity, the Brazilian FPI instrument would have recorded enhanced temperatures during the peak of the storm had measurements been available.

It is interesting that these results for the two nights after the magnetic storm peak activity indicate that the thermosphere during the recovery period was significantly cooler than the quiet-time temperatures seen before the storm. The analysis also showed that these temperatures were also lower than the levels predicted by the MSIS model predictions. *Fagundes et al.* [1995] found the meridional and zonal winds to be significantly modified during the peak phase of this storm relative to the initial period. They found a reversal in sign of the meridional wind from equatorward to poleward combined with a significant reduction of the eastward wind in the early evening, followed by a westward wind of 50 m s^{-1} in the predawn hours.

An explanation for the depression in the Brazilian FPI thermospheric temperatures below quiet-time levels has been advanced by *Dobbin et al.* [2006], who suggested that the observed temperature reduction is a result of a compositional change of the equatorial thermosphere. This work reported on the application of the University College London (UCL) 3-D Coupled Thermospheric and Middle Atmosphere General Circulation Model to calculate the auroral production of NO for the immense geomagnetic storm of late November 2003. Their analysis found that significant spatial transport of NO to lower latitudes would ensue with substantive equatorial NO thermospheric density increases taking place 1 or 2 days after the peak of a geomagnetic storm development. These authors noted that increased thermospheric cooling through 5.3-μm NO emission would moderate temperature increases arising from the increased geomagnetic activity and reduce the thermal relaxation time scales during the recovery phase of the storm. Thus, the Brazilian results suggest that the auroral activity in the Southern Hemisphere produced enhanced auroral thermospheric NO densities during the peak of the storm activity on 8 July. The FPI observations for 10 and 11 July showing thermospheric temperatures depressed relative to predicted MSIS-86 temperatures at 240 km and to the quiet time levels observed before the July storm started are consistent with a cooling introduced by the NO gas transported toward the Brazilian latitudes. The same behavior of depressed Fabry–Perot temperatures relative to the MSIS-86 model predictions for the altitude of 240 km was reported by *Fagundes et al.* [1996] for two other storms taking place in August 1991 and in June 1992.

3.3. Arecibo Optical and Radar Observations for Two Magnetic Storm Periods

Burnside et al. [1991] also found westward winds to occur during disturbed conditions. Figure 5 presents the results observed for the January 1988 storm. In contrast to the "Buonsanto" picture, they concluded that the penetration of the disturbed electric field to low latitudes was such that the momentum transfer from the westward-moving plasma observed by the Arecibo radar to the neutral thermosphere generated the westward wind observed. It was also determined that the electron densities were significantly increased. An extensive

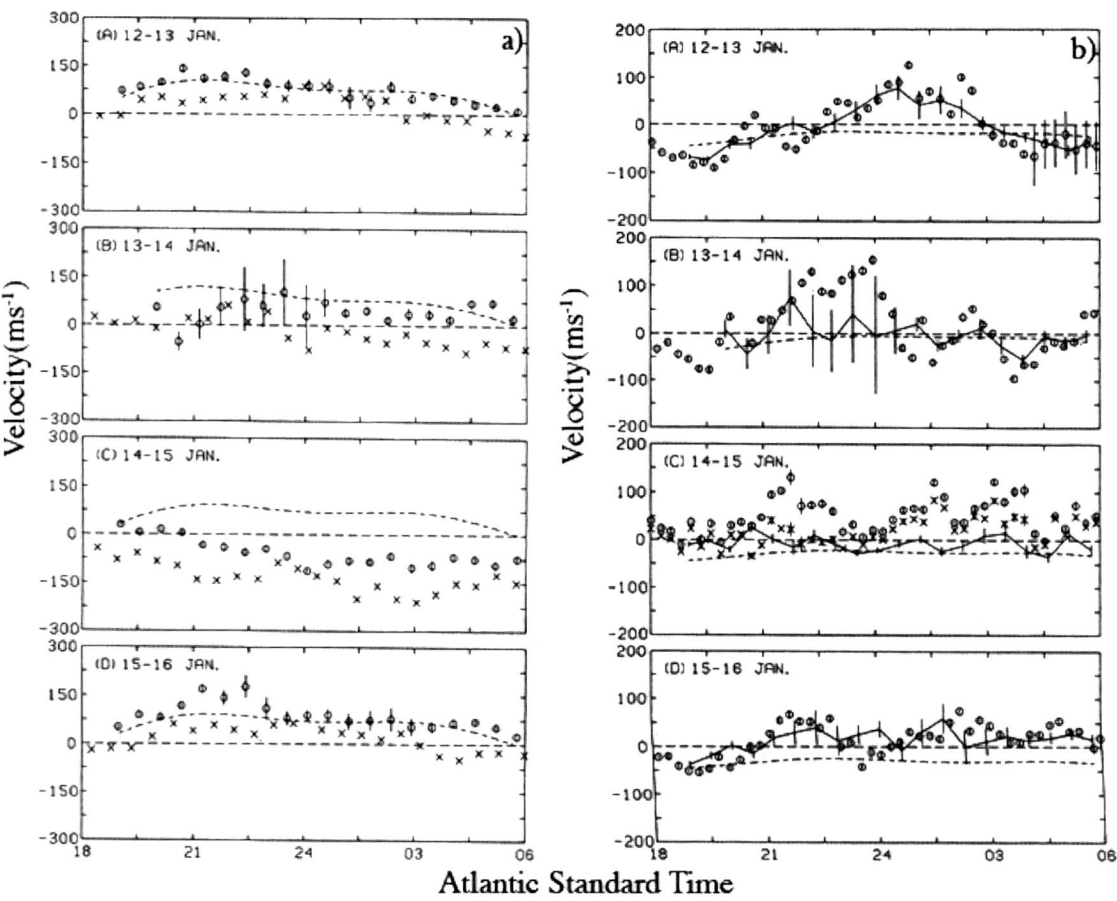

Figure 5. Arecibo radar and FPI wind and temperature measurements for Arecibo Observatory (18.5°S, 73.2°W) shown for the magnetic storm activity of 12–16 January 1988. (a) FPI (circles) and radar (crosses) measurements (300 km) of the zonal winds and ion drifts, respectively, for each of the four nights. The dashed line in each panel represents the HWM-87 empirical model wind calculated at 240 km. (b) Arecibo FPI measurements (solid line) and radar-calculated results of the meridional wind toward geomagnetic south at 240 km. Circles in each panel show the wind velocities derived from the incoherent scatter radar data when the MSIS-86 model is used. The crosses in the third panel are the velocities derived when the MSIS O densities are arbitrarily scaled upward by a factor of 3. The dashed line in each panel is the HWM-87 empirical wind model calculated at 240 km [from *Burnside et al.*, 1991].

analysis of the radar and optical observations concluded that the [O] thermospheric density had increased by nearly a factor of 2. These results suggest that westward ion convection contributed through ion-drag forcing substantive momentum transfer to help force the neutral flow toward the west. Thus, the westward neutral flow observed in the neutral disturbance reported by *Fagundes et al.* [1995] and by the two Arequipa examples presented below may not necessarily be a result of the Coriolis forcing of the meridional flow of air from the polar region as suggested by *Emmert et al.* [2004].

3.4. Beveridge FPI Measurements in March 1995

Figures 6 and 7 show Fabry–Perot observations of thermospheric winds and temperatures combined with electron densities and F_2 layer heights obtained at the mid-latitude site of Beveridge (38°S, 149°E) for the Southern Hemisphere over the six consecutive nights of 1–6 March 1995 [*Richards et al.*, 1998]. The magnetic activity was significantly active on 1 and 5 March with peak values of 6. These two periods of magnetic activity were separated by a quiet night on 3 March and these observations ended with a quiet night on 6 March. Also included in Figure 6 are the day-to-day sequence of h_mF_2 and the F_2 peak electron densities derived from the analysis of the digisonde data. Figure 7 shows that nighttime temperatures varied from the quiet time values of ~800–900 K observed on 3 March and 6 March, and higher temperatures of 1200–1250 K seen for 5 March after the peak *Kp* value of 6 reached at the end of 4 March. The peak thermospheric meridional wind of ~300 m s^{-1} equatorward was seen on 4 March.

The work described by *Richards et al.* [1998] performed a comparison between modeled and measured electron densities, winds, neutral temperatures (T_n), and emission rates at Beveridge, Australia, using wind and temperature measurements obtained by the Fabry–Perot instrument combined with the digisonde observations of *F*-region electron densities. The modeling work was based on the field-aligned interhemispheric plasma (FLIP) model tuned to calculate thermospheric winds and temperatures using as input the topside electron density profiles measured with the digisonde soundings. This study found that the measured T_n was generally higher than the MSIS T_n at night, and the temporal variation did not agree for the nighttime period after local midnight. By modifying the neutral atmosphere composition during the disturbed periods with increases in the molecular densities by 20–30%, the agreement between the modified MSIS temperatures and the observed temperatures was much improved for most nights. The FLIP model electron density was still about 25% higher than the measurements during the daytime.

This paper illustrates the potential of combining a comprehensive set of digisonde and FPI measurements with detailed theoretical modeling. A point noted in the paper is that further advances in understanding the variations of the thermospheric temperature and *F*-region electron density would benefit from multistation sets of observations including incoherent scatter radars that can measure the full profiles of electron densities and plasma temperatures.

3.5. Arequipa Equatorial Observations of Thermospheric Winds and Temperatures

The Arequipa FPI observatory (16.5°S, 71°W) has provided an extended series of data on the winds and temperature of the thermosphere over the past two decades since 1983 [*Meriweather et al.*, 1986, 1997; *Faivre et al.*, 2006]. Two examples from these observations are presented in Plates 3 and 4 to illustrate for a site at low geomagnetic latitude (8° dip angle) the thermospheric disturbance effects seen at 250 km caused by two intense geomagnetic storms occurring in 1997 and 2000.

To help characterize the thermospheric response for each storm, the observed thermospheric winds and temperatures are compared with the climatological average as established by observations during quiet periods. The observations are also compared with the results of the nominal predictions made by the National Center for Atmospheric Research Thermosphere–Ionosphere Electrodynamics General Circulation Model (NCAR TIEGCM) evaluated for the sequence of $F_{10.7}$ fluxes and *Kp* values for each case. For temperature measurements, the MSIS00 model is also shown as a reference in addition to the NCAR model calculated temperatures.

The first one observed in mid-May 1997 represents an example in which initially the magnetic activity was very quiet. This quiet period was followed by a rapid increase of *Kp* from 0 to 6+ as the *Dst* index decreased to −125 nT. The second example shows a sustained period of activity followed by a sudden recovery to quiet-time levels. Thus, these two storms illustrate the thermospheric response near the magnetic equator for two rather different temporal profiles of magnetic activity.

The measured FPI zonal and meridional winds and temperatures observed during these intense storm periods are compared with three reference curves: (1) mean zonal and meridional FPI averaged quiet-time wind variations for the month of the storm (blue dotted line), (2) the NCAR TIEGCM model results for both winds and temperatures (red lines), and (3) scintillation ion drift speed results from the Ancon system (Dr. Cesar Valladares, private communication, black line; details are provided by *Valladares et al.*

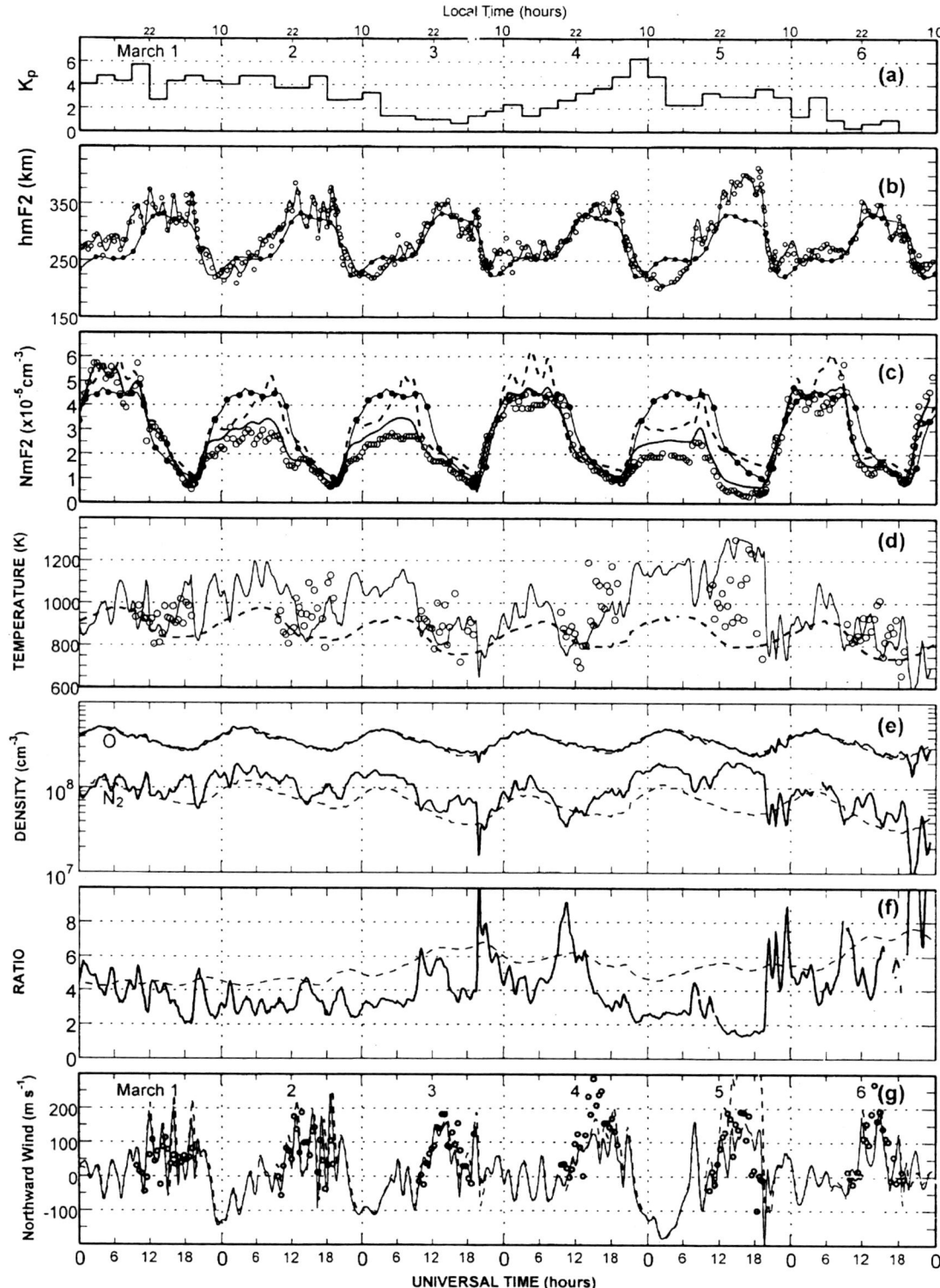

Figure 6. Time variation of (a) Kp, (b) h_mF_2, (c) N_mF_2, (d), temperatures, (e) densities, (f) O to N_2 density ratio, and (g) winds at Beveridge (38°S, 149°E) for 1–6 March 1995. The open circles in the bottom panel indicate FPI magnetic meridional wind speeds, thick solid lines with dots are for the IRI model N_mF_2 and the measured median N_mF_2, the dashed lines are the FLIP model output using the standard MSIS model, and the solid lines are the output values determined using the modified MSIS model. The neutral density and temperature are for 300-km altitude whereas equivalent wind altitudes are at h_mF_2 [from *Richards et al.*, 1998].

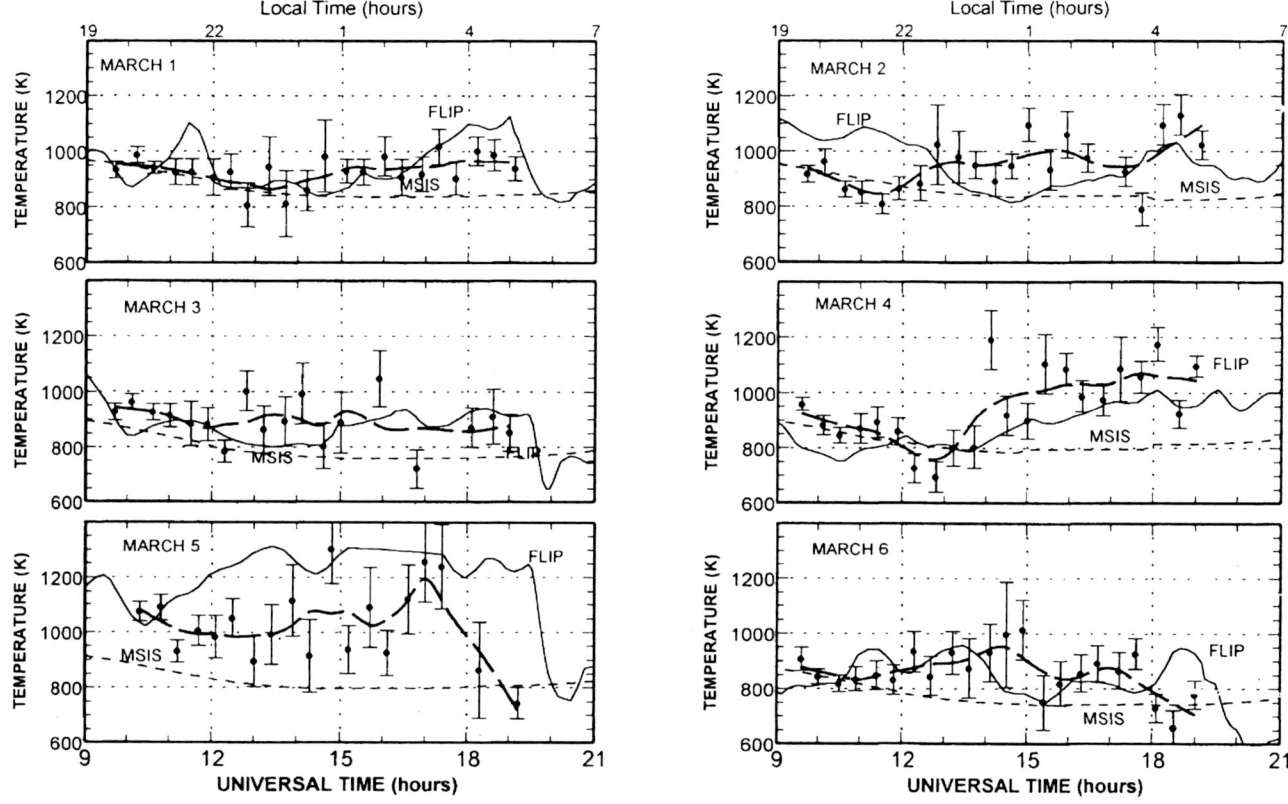

Figure 7. Temporal variations of the FPI thermospheric temperature at Beveridge (38°S, 149°E) for each night from 1 March to 6 March, 1995. The dots with error bars represent the FPI data, the dashed lines are from the standard MSIS model, and the solid lines are the modified MSIS output. The thick dash line is a minimum least-square fit through the data [from *Richards et al.*, 1998].

[2002]). The geomagnetic activity is represented by the *Dst* (bottom graphics, red dots) and by the *Kp* (black histogram) indices. The mean temperature for each night for all directions averaged together is indicated at midnight. The NCAR TIEGCM calculations were not fine-tuned for the conditions of each storm, but represent a generic type of calculation that was applied for the solar and magnetic conditions prevailing at the time. These results are included more to serve as a reference for the zonal and meridional winds and thermospheric temperatures.

3.5.1. Case 1: 25–29 August 1998. The *Kp* and *Dst* indices increased to a maximum level of 8 and to ~–150 gammas, respectively, for the early part of 27 August with the period of peak activity lasting nearly a day for both indices.

3.5.1.1. 630-nm intensities. The 630-nm nightglow showed normal behavior for the first and last two nights with a peak seen near midnight that is caused by the midnight temperature maximum (MTM) *Faivre et al.* [2006]. On the third night (Aug 27) the 630-nm intensity is ~5 times weaker than for the other nights.

3.5.1.2. Meridional winds. The behavior for the first two nights between evening twilight and midnight is similar to what is presented by the NCAR model and the climatological average (dash–dot line), that is, weakly poleward by 20–30 m s^{-1}, which may be attributed to the interhemispheric flow from the summer to the winter hemisphere. On the third and fourth nights (27 and 28 August), the speed substantially increased equatorward (75–100 m s^{-1} during the early morning hours (*Dst* ~–150) relative to the quiet-time reference and NCAR curves. A significant gradient between the north and south directions (difference of 50 m s^{-1}) can be seen for the data of 28 and 29 August. During the recovery phase (*Dst* ~–50), the speed continues to be slightly more equatorward (29 August) than the model NCAR curve but is in agreement with the quiet-time curve.

3.5.1.3. Zonal winds. For the first two nights (25–26 August), the speed was the same as the quiet-time winds, but on 27 August, the zonal wind speeds show a reduction of 100 m s^{-1} relative to the quiet-time curve and in good agreement with the Ancon scintillation drifts indicating strong thermospheric ion-neutral coupling [*Valladares et al.*, 2002]. The zonal winds for the fourth night (28 August) show a period of significant increase of 50 m s^{-1} eastward relative to the quiet-time curve in the early morning hours. During the last night (29 August), the mean zonal wind speeds were in close agreement with the climatological reference by 50–100 m s^{-1}.

3.5.1.4. Temperatures. The averaged temperatures for each of the five nights were 940, 921, 995, 1021, and 941 K. The temperature increased ~75 K during the storm and recovered slowly over the last two nights with the data for the last night indicating an increase of ~75 K relative to the MSIS00 model predicted values. The NCAR model temperatures are, in general, found to be approximately below the data.

In summary, the neutral atmosphere response to the magnetic storm activity features increased heating of 100–150 K at the peak of the storm activity, a reduction of the zonal wind eastward relative to the quiet-time curve, suppressed nightglow 630-nm intensities, and enhanced poleward winds after midnight on the fourth night. The zonal wind is increased eastward near midnight on the fourth night during the onset of storm recovery.

3.5.2. Case 2: 3–7 October 2000. This period featured strong magnetic activity with *Dst* varying from −50 to −200 nT over the first 3 days followed by a recovery to quiet-time levels on 7 October.

3.5.2.1. 630-nm intensities. The temporal variations of the 630-nm intensities behaved in much the same way from one night to the next with values between a few Rayleighs in the early evening to a maximum of ~100 Rayleighs near midnight. There is no particular departure from this nominal behavior associated with the magnetic storm activity except that the 630-nm intensity is very weak in the early evening hours of the last night after the recovery of the magnetic activity to quiet levels.

3.5.2.2. Meridional winds. For this magnetic storm there is seen no strong indication of meridional wind response to the magnetic activity. For the first three nights during the early evening, speeds are increased poleward relative to the predicted NCAR model curve during the early disturbed period of the storm interval by ~50–75 m s^{-1}. On the fourth and fifth nights, 6 October, during the period of recovery the meridional winds are increased by 100 m s^{-1} toward the south (poleward) relative to the predicted NCAR behavior. This behavior was also seen for the last night. For each of these five nights, there is a period of 2–3 h near midnight during which the winds are equatorward by ~50–75 m s^{-1}. This is likely caused by the semidiurnal tidal wave that is known to be equatorward during the midnight hours.

3.5.2.3. Zonal winds. On the first three nights there is what appears to be a significant increase of ~50 m s^{-1} relative to the quiet-time speeds of 100–150 m s^{-1} that are generally seen for early evening and what the NCAR model calculated for these nights. The zonal wind behavior for the other nights shows no unusual features although there is a period of westward drifts after midnight seen for the fourth night in the scintillation drift measurements.

3.5.2.4. Temperatures. The variation observed from night to night is in phase with the main phase of the storm activity with a strong cooling trend coincident with the reduction of magnetic activity. The measured temperatures are generally higher than the NCAR model temperatures (+100 K), except for the fourth night. In the early morning hours of the fifth night, the temperatures in the postmidnight period are significantly lower than the MSIS00 predictions. The MTM peak is seen near midnight for the 5–7 October nights with approximate amplitudes of 50, 25, and 75 K, respectively.

In summary, no significant response of the thermosphere is observed for this particular storm in spite of the fact that *Kp* reached a level of 8 and the *Dst* parameter was −150 at the peak of the *Kp* activity. There was no strong variation in temperature until the last night, which shows a reduction of ~200 K from the peak of 1156 K seen on 5 October 2000. This cooling may be another indication of the neutral atmosphere response to the transport of NO to low latitudes described by *Dobbin et al.* [2007]. The zonal or meridional winds remained much the same throughout the period of the storm activity.

4. DISCUSSION

The physics of the recovery process to the thermal and dynamical disturbances induced by the development of the geomagnetic storm is complex, so a simple conceptual model cannot be expected to be sufficient to completely describe the details of any one particular storm. The results summarized here indicate that the Buonsanto "picture" outlined in Section 1 provides a comprehensive overview of the important processes. There is a continual interplay between the polar heating and momentum sources (Joule, particle, and ion drag) and the effects of the penetration of the magnetospheric electric field into the low latitude and mid-latitude regions [*Buonsanto and Foster*, 1993]. These effects are manifested within the mid-latitude thermosphere and are

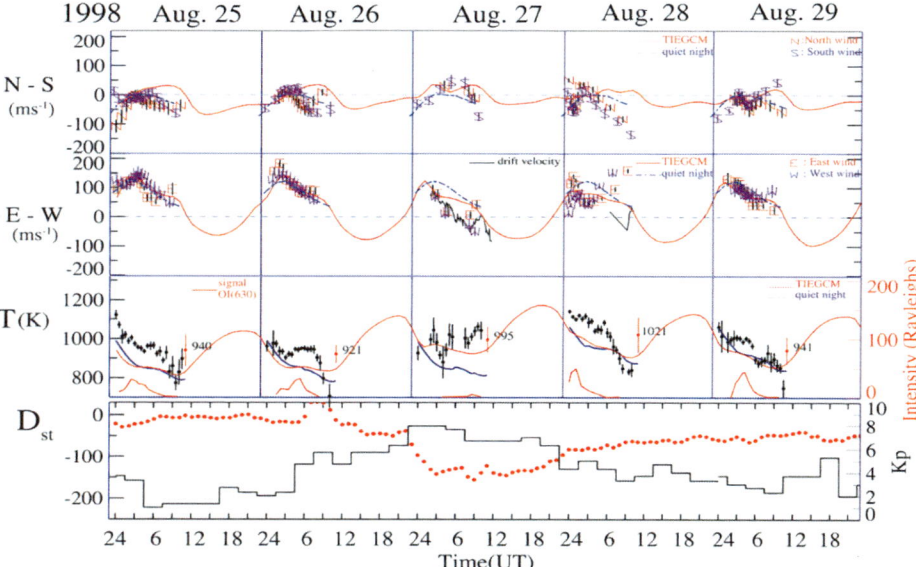

Plate 3. Arequipa FPI results shown for the intense magnetic storm activity observed for five nights (25–29 August 1998). The last row of panels shows the temporal variations of the *Dst* index (left scale, red dots) and *Kp* (right scale, histogram) for these 5 days. The blue dash–dot line in the first and second rows of panels represents the thermospheric wind climatological behavior (see text) for quiet periods of magnetic activity. In the third row of panels, the thin red, the thick red, and the blue lines plotted show the nighttime variations of the 630-nm intensity observed, the NCAR model 630-nm temperature, and the MSIS00 calculated temperature, respectively. The temperature values plotted in the third row are averaged over all directions for each 30-min period. The red line for the first and second rows of panels represents the calculated NCAR TIEGCM model wind results. The averaged nocturnal temperature and its standard deviation are indicated at midnight for each night in the third row. The black line in the zonal wind plots (second row) represents the scintillation drift results provided (courtesy of C. Valladares).

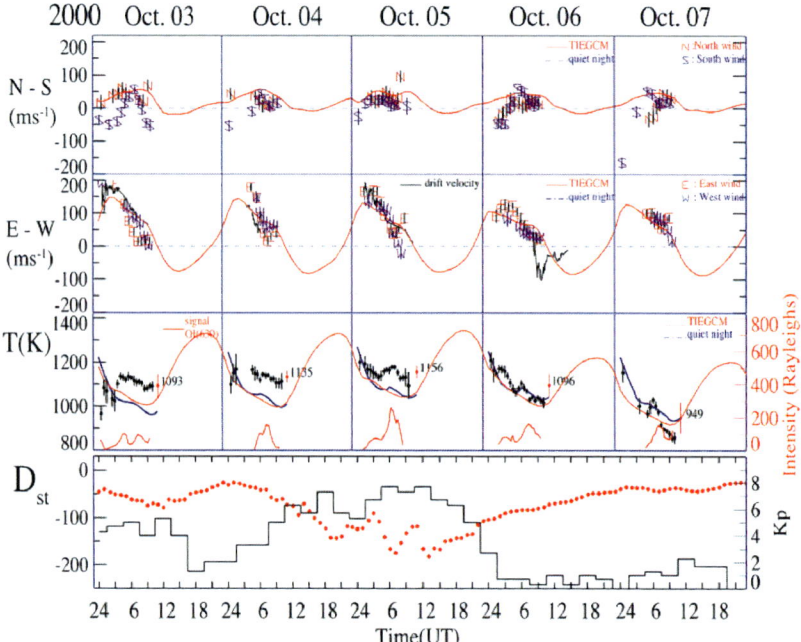

Plate 4. Arequipa FPI results shown for magnetic storm activity observed for the period of 3–7 October 2000. See Plate 3 caption for details. There is no climatological curve plotted for the winds and temperatures in the top three rows because little quiet-time FPI data were available for this period.

difficult to study in quantitative detail without knowledge of the complete global thermospheric circulation. As seen in Figure 2, there is one example showing that the mid-latitude region might be even hotter than the auroral region. The heating observed in this case may be attributed in part to the convergence of the meridional and zonal flows as described in the Buonsanto picture. Results such as those shown in Figures 2 and 3 illustrate how the interpretation of such results is hampered by the lack of additional thermospheric wind observations that expands the spatial scale of the dynamical structure being observed.

Rees et al. [1996] discussed two primary mechanisms that can operate to support the recovery of the thermosphere to thermal disturbances: (1) thermal conduction to the mesosphere where CO_2 radiates the heat and (2) radiation of nitric oxide generated within the auroral region. The modeling work described by *Dobbin et al.* [2006] points to the importance of the transport of auroral-produced NO as a significant thermospheric cooling mechanism. Again, the shortcoming of interpreting observations such as those of *Fagundes et al.* [1996] is the lack of additional measurements to help verify the different aspects of this scenario projected to apply to this case.

The satellite results that emerge after application of statistical averaging support the details of the Buonsanto "picture" in the sense of the meridional transport of hot air that is redistributed westward by the Coriolis force as the air moves toward the equator. However, the six geomagnetic storm examples presented illustrate that the Buonsanto "picture" does not represent a complete description of what happens in individual storm cases. The Arequipa results show that even for a reasonably intense magnetic storm, the low-latitude thermosphere may not be strongly affected because the storm-induced winds do not reach sufficiently far south toward the equator. The Arecibo results for two intense storms do suggest that ion-neutral coupling is important in forcing westward winds through ion drag, thus indicating that the storm-time penetration of the low-latitude plasma by the storm-induced electric field may indeed be important. However, the exact division of the neutral forcing between the ion-drag momentum source and the diversion of meridional flow by the Coriolis force is difficult to assess without a quantitative knowledge of the ion drift forcing. This lack of detail regarding this division and its temporal and spatial history represents a source of significant uncertainty in interpreting observations of mid-latitude thermospheric winds.

Analysis of the measurements from the FPIs located in Brazil, Australia, and Arecibo pointed to the need to allow for compositional changes that take place in the course of a magnetic storm. The analysis of the Beveridge FPI and digisonde results as compared with the FLIP model predictions was improved once the molecular composition of the MSIS model was substantively increased. Similarly, the Arecibo measurements demonstrated that the [O] density had to be increased a factor of 2 during the peak period of magnetic activity to achieve the agreement of the predicted MSIS temperatures with the FPI data. Finally, the smaller FPI temperatures seen for the Brazil and Arequipa observations in the recovery phase of the magnetic storm as compared with climatological-averaged data indicate that thermospheric cooling introduced by the transport of NO into low latitudes may contribute significantly to the rapid recovery of the thermosphere temperature during the recovery phase.

These results illustrate the complexity of the ion-neutral coupling for the equatorial and mid-latitude thermosphere regions. There is no simple means of describing the response to the impulsive forcing represented by development of the magnetic storm. The seasonal variation in the mid-latitude thermospheric response to storm periods represents a further complication. Because the extent of momentum transfer is closely connected to the plasma density, and this varies with the season, the timing of the storm occurrence relative to the season (whether summer, winter, or equinox) would certainly play an important role in the development of the thermospheric response at low- and mid-latitude regions.

In conclusion, in future work on the study of the equatorial and mid-latitude thermosphere response to magnetic storm activity, it is essential that a strategy be followed that emphasizes the coordination of simultaneous measurements by a network of FPI observatories combined with satellite coverage of thermospheric winds (e. g., as advocated by *Meriwether* [2006]). The combination of the global view that the satellite data provide with the local time coverage of a network of ground-based stations would greatly improve the detailed understanding of the temporal and spatial distribution of the thermospheric neutral response to geomagnetic disturbances. Modeling analysis with a general circulation global model is also crucial as part of this strategy. The assimilation of the measurements with the model predictions provides a powerful means by which the sparseness of the observations of thermospheric winds from ground and space can be surmounted.

Acknowledgments. Dr. M. Faivre's work in constructing Plates 3 and 4 is much appreciated. Dr. C. Fesen provided the NCAR modeling results for these two figures. Funding support was provided by the National Science Foundation Aeronomy program.

REFERENCES

Babcock, R. R., Jr., and J. V. Evans (1979), Effects of geomagnetic disturbances on neutral winds and temperatures in the thermosphere observed over Millstone Hill, *J. Geophys. Res.*, *84*, 5349–5354.

Buonsanto, M. J. (1999), Ionospheric storms – a review, *Space Sci. Rev.*, *88*, 563–601, doi:10.1023/A:1005107532631.

Buonsanto, M. J., and J. Foster (1993), Effects of magnetospheric electric fields and neutral winds on the low-middle latitude ionosphere during the March 20–21, 1990, storm, *J. Geophys. Res.*, *98*, 19,133–19,140.

Burke, W. J., C. Y. Huang, F. A. Marcos, J. O. Wise (2007), Interplanetary control of thermospheric densities during large magnetic storms, *J. Atmos. Sol. Terr. Phys.*, *69*, 279–287.

Burns, A. G., and T. L. Killeen (1992), The equatorial neutral thermospheric response to geomagnetic forcing, *Geophys. Res. Lett.*, *19*, 977–980.

Burns, A. G., T. L. Killeen, and R. G. Roble (1991), A theoretical study of thermospheric composition perturbations during an impulsive geomagnetic storm, *J. Geophys. Res.*, *96*, 14,153–14,167.

Burns, A. G., T. L. Killeen, G. R. Carignan, and R. G. Roble (1995), Large enhancements in the O/N2 ratio in the evening sector of the winter hemisphere during geomagnetic storms, *J. Geophys. Res.*, *100*, 14,661–14,671.

Burnside, R. G., C. A. Tepley, M. P. Sulzer, T. J. Fuller-Rowell, D. G. Torr, and R. G. Roble (1991), The neutral thermosphere at Arecibo during geomagnetic storms, *J. Geophys. Res.*, *96*, 1289–1301.

Dobbin, A. L., E. M. Griffin, A. D. Aylward, and G. H. Millward (2006), 3-D GCM modelling of thermospheric nitric oxide during the 2003 Halloween storm, *Ann. Geophys.*, *24*, 2403–2412.

Emmert, J. T., B. G. Fejer, C. G. Fesen, G. G. Shepherd, and B. H. Solheim (2001), Climatology of middle- and low latitude F region disturbance neutral winds measured by wind imaging interferometer (WINDII), *J. Geophys. Res.*, *106*, 24,701–24,712.

Emmert, J. T., B. G. Fejer, G. G. Shepherd, and B. H. Solheim (2002), Altitude dependence of mid and low latitude daytime thermospheric disturbance winds measured by WINDII, *J. Geophys. Res.*, *107*(A12), 1483, doi:10.1029/2002JA009646.

Emmert, J. T., B. G. Fejer, G. G. Shepherd, and B. H. Solheim (2004), Average nighttime F region disturbance neutral winds measured by UARS WINDII: Initial results, *Geophys. Res. Lett.*, *31*, L22807, doi:10.1029/2004GL021611.

Fagundes, P. R., Y. Sahai, H. Takahashi, D. Gobbi, and J. A. Bittencourt (1996), Thermospheric and mesospheric temperatures during geomagnetic storms at 23 S, *J. Atmos. Terr. Phys.*, *58*, 1963–1972.

Faivre, M., J. Meriwether, C. Fesen, and M. Biondi (2006), Climatology of the midnight temperature maximum at Arequipa, Peru, *J. Geophys. Res.*, *111*, A06302, doi:10.1029/2005JA011321.

Fejer, B. G., J. T. Emmert, G. G. Shepherd, and B. H. Solheim (2000), Average daytime F region disturbance neutral winds measured by UARS: Initial results, *Geophys. Res. Lett.*, *27*, 1859–1862.

Forster, M., and N. Jakowski, (2000), Geomagnetic storm effects on the topside ionosphere and plasmasphere: A compact tutorial and new results, *Surv. Geophys.*, *21*, 47–87.

Fuller-Rowell, T. J., M. V.Codrescu, R. J. Moffett, and S. Quegan (1994), Response of the thermosphere and ionosphere to geomagnetic storms, *J. Geophys. Res.*, *99*, 3893–3914.

Fuller-Rowell, T. J., M. V. Codrescu, R. G. Roble, and A. D. Richmond (1997), How does the thermosphere and ionosphere react to a geomagnetic storm?, in *Magnetic Storms*, Geophys. Monogr. Ser., vol. 98, edited by B. T. Tsurutani et al., pp. 203–225, AGU, Washington, D. C.

Fuller-Rowell, T. J., G. H. Millward, A. D. Richmond, and M. V. Codrescu (2002), Storm-time changes in the upper atmosphere at low latitudes, *J. Atmos. Sol. Terr. Phys.*, *64*, 1383.

Gold, T. (1962), Magnetic storms, *Space Sci. Rev.*, *1*, 100–114.

Hajkowicz, L. A. (1990), A global study of large scale travelling ionospheric disturbances (TID) following a step-like onset of auroral substorms in both hemispheres, *Planet. Space Sci.*, *39*, 913–923.

Hajkowicz, L. A. (1991), Auroral electrojet effect on the global occurrence pattern of large scale travelling ionospheric disturbances, *Planet. Space Sci.*, *39*, 1189–1196.

Hernandez, G., and R. G. Roble (1976), Direct measurements of nighttime thermospheric winds and temperatures: II. Geomagnetic storms, *J. Geophys. Res.*, *81*, 5173–5181.

Hernandez, G., and R. G. Roble (1978), Observations of large-scale thermospheric waves during geomagnetic storms, *J. Geophys. Res.*, *83*, 5531–5538.

Hernandez, G., and R. G. Roble (1979), On divergences of thermospheric meridional winds at midlatitudes, *Geophys. Res. Lett.*, *6*, 294–296.

Hernandez, G., R. G. Roble, E. C. Ridley, and J. H. Allen (1982), Thermospheric response observed over Fritz Peak, Colorado, during two large geomagnetic storms near solar cycle maximum, *J. Geophys. Res.*, *87*, 9181–9192.

Hocke, K., and K. Schlegel (1996), A review of atmospheric gravity waves and travelling ionospheric disturbances: 1982–1995, *Ann. Geophys.*, *14*, 917–940.

Maruyama, N., et al. (2007), Modeling storm-time electrodynamics of the low-latitude ionosphere–thermosphere system: Can long lasting disturbance electric fields be accounted for? *J. Atmos. Sol. Terr. Phys.*, *69*, 1182–1199.

Mendillo, M., X.-Q. He, and H. Rishbeth (1992), How the effects of winds and electric fields in *F*-2 layer storms vary with latitudes and longitude: a theoretical study *Planet. Space Sci.*, *40*, 595–606.

Meriwether, J. W. (2006), Studies of thermospheric dynamics with a FabryPerot interferometer network: A review, *J. Atmos. Sol. Terr. Phys.*, *68*, 1576–1589.

Meriwether, J. W., J. W. Moody, M. A. Biondi, and R. G. Roble (1986), Optical interferometric measurements of nighttime equatorial thermospheric winds at Arequipa, Peru, *J. Geophys. Res.*, 5547–5556.

Meriwether, J. W., M. A. Biondi, F. A. Herrero, C. G. Fesen, and D. C. Hallenback (1997), Optical interferometric studies of the nighttime equatorial thermosphere: Enhanced temperatures and zonal wind gradients, *J. Geophys. Res.*, *102*, 20,041–20,058.

Prolss, G. W. (1987), Storm-induced changes in the thermospheric composition at middle latitudes, *Planet. Space Sci.*, *35*, 807–811.

Prolss, G. W. (1997), Magnetic storm associated perturbations of the upper atmosphere, in *Magnetic Storms*, Geophys. Monogr.

Ser., vol. 98, edited by B. T. Tsurutani et al., pp. 227–241, AGU, Washington, D. C.

Rees, D. (1996), Observations and modeling of ionospheric and thermospheric disturbances during major geomagnetic storms: A review, *J. Atmos. Terr. Phys.*, *57*, 1433–1457.

Richards, P. G., P. L. Dyson, T. P. Davies, M. L. Parkinson, and A. J. Reeves (1998), Behavior of the ionosphere and thermosphere at a southern midlatitude station during magnetic storms in early March 1995, *J. Geophys. Res.*, *103*, 26,421–26,432.

Richmond, A. D., and S. Matsushita (1975), Thermospheric response to a magnetic substorm, *J. Geophys. Res.*, *80*, 2839–2850.

Rishbeth, H. (1989), F-region storms and thermospheric circulation, in *Electromagnetic Coupling in the Polar Clefts and Caps*, edited by P. E. Sandholt and A. Egeland, pp. 393–406, Kluwer Academic Publishers, Dordrecht, The Netherlands.

Roble, R. G., A. D. Richmond, W. L. Oliver, and R. M. Harper (1978), Ionospheric effects of the gravity wave launched by the September 18, 1974, sudden commencement, *J. Geophys. Res.*, *83*, 999–1009.

Straus, J. M., and M. Schulz (1976), Magnetospheric convection and upper atmospheric dynamics, *J. Geophys. Res.*, *81*, 5822–5832.

Tinsley, B. A., Y. Sahai, M. A. Biondi, and J. W. Meriwether Jr. (1988), Equatorial particle precipitation during magnetic storms and relationship to equatorial thermospheric heating, *J. Geophys. Res.*, *93*, 270–276.

Valladares, C. E., J. W. Meriwether, R. Sheehan, and M. A. Biondi (2002), Correlative study of neutral winds and scintillation drifts measured near the magnetic equator, *J. Geophys. Res.*, *107*(A7), 1112, doi:10.1029/2001JA000042.

J. W. Meriwether, Department of Physics and Astronomy, Clemson University, 208 Kinard Laboratory, Clemson, SC 29631, USA. (john.meriwether@ces.clemson.edu)

Disturbed O/N$_2$ Ratios and Their Transport to Middle and Low Latitudes

Geoff Crowley

Atmospheric and Space Technology Research Associates, San Antonio, Texas, USA

R. R. Meier

Department of Physics and Astronomy, George Mason University, Fairfax, Virginia, USA

The neutral atmospheric composition responds dramatically to geomagnetic storms. Much progress toward understanding the response came from *in situ* measurements on board the Atmospheric Explorer and Dynamics Explorer 2 satellites. These observations also provided significant databases for the Mass Spectrometer Incoherent Scatter empirical model. Strong evidence was found for upwelling of molecular species at high latitudes and downwelling at lower latitudes. Early authors connected these composition changes to ionospheric effects, including both positive and negative phases of geomagnetic storms. Models have confirmed some of their conclusions, but up until recently, the data have not yet been good enough to completely define the mechanisms of the drivers or evolution of the composition. Global imaging capabilities evolved from Dynamics Explorer 1 through the Polar and IMAGE satellites. However, the Global Ultraviolet Imager (GUVI) spectroscopic imager on the TIMED satellite has provided a revolutionary view of composition changes and their relationship to ionospheric storms. Even so, we still do not have much of the right data in the right places and the right times to unravel the combined effects of composition, winds, and electric fields on the ionosphere. In this short review, we explain what is known about disturbed O/N$_2$ ratios during storms and discuss how changes can occur at regions remote from the high latitude forcing. We also discuss their relationship with neutral winds and their effects on NmF_2. Finally, we discuss outstanding problems and required measurements for addressing them.

1. INTRODUCTION

The *F* region ionosphere is highly variable, especially during geomagnetic storms, which can produce large increases and decreases in the *F* region electron density. The mechanisms driving these ionospheric changes are still a very active topic of research. Changes in thermospheric composition, winds, and electrodynamics can all play a major role [e.g., *Buonsanto*, 1999; *Kintner*, 2007]. Figure 1 depicts a simplified view of the drivers of electron density. The figure reveals that understanding the thermosphere is key to understanding the ionosphere.

This brief review focuses on thermospheric composition, for which much remains to be learned. Most of the previous research on thermospheric composition has focused

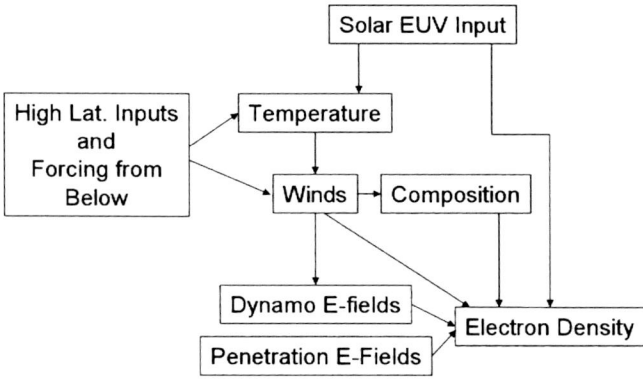

Figure 1. The drivers of ionospheric F region electron density imply the importance of measuring thermospheric parameters.

on heights around the ionospheric F region peak electron density, near 300–400 km. There were two reasons for this focus: (1) a desire to understand the role of composition in ionospheric storm processes and (2) the existence of much more satellite (and ground based) data from F region heights than from other altitudes.

Early measurements of thermospheric composition were obtained by rockets [e.g., *Meadows and Townsend*, 1958] and the Explorer and OGO 6 satellite carrying mass spectrometers [*Hedin et al.*, 1974]. Later, the ESRO and Atmospheric Explorer (AE) series of satellites [e.g., *Kayser and Potter*, 1976] made additional measurements with more comprehensive instrument packages. Variability was observed on a broad range of temporal and spatial scales for solar minimum conditions, including diurnal, seasonal, latitudinal, and storm responses. The AE measurements were obtained mainly in the F region, but many orbits dipped to altitudes between 130 and 170 km. These dipping orbits provided important information about the lower thermosphere, but due to fuel and safety concerns, there were insufficient measurements to build a comprehensive global picture of the lower thermosphere composition. The duty cycle on the AE satellite series was also quite low, so relatively few data were collected relative to present needs and capabilities.

The Dynamics Explorer (DE) mission provided *in situ* measurements at higher altitudes for solar maximum conditions [*Killeen and Roble*, 1988]. The AE mission had limited information about auroral forcing; however, the DE 2 *in situ* measurements were complemented by a high-altitude imager on DE 1 that often provided contextual auroral images for the low-altitude localized observations [*Frank and Craven*, 1988]. The *in situ* composition measurements from DE 2 revealed strong variations on daily and seasonal timescales. They also provided solar cycle complementarity to the AE data. Although the DE 2 *in situ* measurements were made at a higher altitude than AE, they were still near the F region ionospheric peak (h_mF_2), because at solar maximum, the F layer peak is at a much higher altitude than at solar minimum. Reviews of the early work were provided by *Mayr et al.* [1978] and *Prölss* [1980], while the DE work was included in review papers by *Breig* [1987], *Killeen et al.* [1991], and *Crowley* [1992].

Hedin et al. [1983, 1987, 1991] combined the AE and DE data sets (along with incoherent scatter radar data) as the basis for an empirical model of thermospheric composition and temperature variability. This model has been widely used to specify the atmospheric state for many different studies. The most recent update of the Mass Spectrometer Incoherent Scatter model is known as the NRLMSISE-00 [*Picone et al.*, 2002].

In this review, we consider the evidence for compositional influences on F region electron densities. Because the electron production is primarily by photoionization of atomic oxygen and the loss is proportional to N_2 (and to a lesser extent O_2), the electron density will be proportional to the local O/N_2 density ratio (when electrodynamic processes are small compared with photochemistry). We distinguish herein among the number density ratio at a given location, O/N_2 (typically measured with mass spectrometers and more recently with limb scanning spectroscopic imagers), and the column density ratio, designated by *Meier et al.* [2005] by $\Sigma O/N_2$ (measured with spectroscopic imagers). The latter quantity was defined by *Strickland et al.* [1995] as the column density of O above an altitude where the N_2 column density is 10^{17} cm^{-2}. This definition allows a nearly unique relationship between $\Sigma O/N_2$ and the ratio of the OI 135.6-nm emission to the N_2 Lyman–Birge–Hopfield (LBH) band emission. The relationship between the two ratios comes about because the column base of 10^{17} cm^{-2} is near the bottom of the photoelectron excited dayglow layer, at about 135 km. Further, the intensity ratio can be observed by spectroscopic imagers thereby providing regional or global views of $\Sigma O/N_2$. The column density ratio is an important diagnostic of thermospheric dynamical processes and a valuable parameter for comparison with models. However, it is only indirectly related to the electron density at any given altitude, and the exact relationship has not been quantified (except for the empirical correlation shown by *Strickland et al.* [2001, Plate 1], which may not be generally applicable).

2. COMPOSITION AND IONOSPHERIC STORMS

Geomagnetic storms can produce both increases and decreases in the F region electron density, depending on latitude, local time, season, and time elapsed since the storm onset. For example, Figure 2 shows the variation of the F

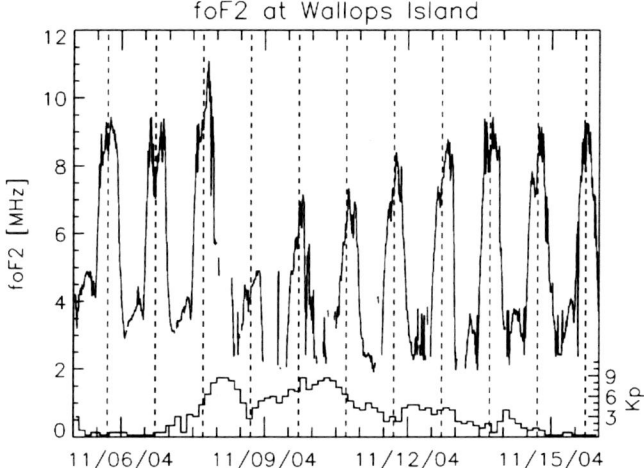

Figure 2. Variations in the *F* region critical frequency (f_oF_2) measured by the Wallops Island ionosonde during the November 2004 storm. Vertical lines indicate noon solar local time.

region critical frequency over several days at Wallops Island, Virginia, during geomagnetic activity in November 2004. The figure exhibits several classic features, including evening enhancement following the onset of significant geomagnetic activity on 7 November 2004, long-lived dayside depletion the next day (8 November 2004), and gradual recovery over several days to prestorm values. Composition changes are thought to play an important role both in long-lived increases and decreases of critical frequency and the corresponding electron density, called "positive" and "negative" ionospheric storms, respectively. In contrast, composition changes are not thought to be important for the evening enhancements, which are now generally called Storm-Enhanced Density features, which are regions of increased density often observed over North America [*Foster*, 1993]. There have been many reviews of ionospheric electron density changes during disturbed times [e.g., *Rodger et al.*, 1989; *Buonsanto*, 1999; *Kintner et al.*, 2007] describing storm-time variability and current theories regarding their driving mechanisms.

Seaton [1956] first suggested that storm-time electron density depletions during storms might be caused by thermospheric neutral composition changes. *Hays et al.* [1973] showed from theoretical considerations that composition changes could be produced by vertical winds driven by Joule heating at high latitudes. *Mayr and Volland* [1973] published the first global model of these processes. Their three-dimensional spectral model indicated that the mixing ratios of Ar, N_2, O_2, and O would be affected differently due to their different masses and scale heights. They were able to qualitatively explain many of the observed compositional variations, including diurnal, seasonal, solar cycle, and magnetic storm responses.

A picture of the compositional storm-time response in the upper thermosphere was further developed by *Prölss* [1980, 1981]. He extended the concept of upward winds in the lower thermosphere that carry molecular-rich air to higher altitudes. Strong transpolar winds driven by momentum coupling at *F* region heights then carry the uplifted parcels toward the nightside of the polar cap and out to middle latitudes, creating a "disturbance zone" in the midnight and early morning sector, as shown in Figure 3. The region of modified gas on the nightside corotates onto the dayside, where the composition recovers diffusively over several hours. The outflow of air from high latitudes creates a large-scale circulation system as horizontal winds converge equatorward of the auroral oval, resulting in downward winds. These downward winds carry atomic-oxygen-rich air parcels to lower altitudes, thereby enhancing the O/N_2 ratio at these latitudes. The circulation is completed by a return flow at low altitudes. To maintain continuity of the flow, much smaller velocities are required in the return flow than in the high-altitude flow because of the larger densities at lower altitudes.

The first general circulation model (GCM) simulation to comprehensively reproduce these characteristics was presented by *Crowley et al.* [1989a], who used the NCAR Thermosphere General Circulation Model [*Dickinson et al.*, 1984; *Roble and Ridley*, 1987] driven by Assimilative Mapping of Ionospheric Electrodynamics (AMIE) convection patterns to simulate the September 1984 storm period. The model confirmed the upwelling at high latitudes and showed for the first time that horizontal winds could transport the molecular rich gas well beyond the auroral zone, creating the "disturbance zone" envisaged by *Prölss* [1980]. Figure 4 shows how the disturbance zone develops over several hours, being created on the nightside and corotating onto the dayside. *Crowley et al.* [1989b] used various data sets to validate their model simulation, emphasizing the ionospheric *F* region effects of the composition changes, especially the electron density reduction corresponding to the negative storm response.

Fuller-Rowell et al. [1991, 1994, 1996] used the University College of London–GCM to study generic storm time variations in the thermosphere. They showed how the wind, composition, and electron density responses to storms are affected by the UT of the storm onset [*Fuller-Rowell et al.*, 1994] and how the response is influenced by season and solar cycle [*Fuller-Rowell et al.*, 1996]. They found that the picture developed by *Prölss* [1980] seemed to also be reproduced by their global GCM. A comprehensive description of thermospheric storm effects was provided by *Fuller-Rowell*

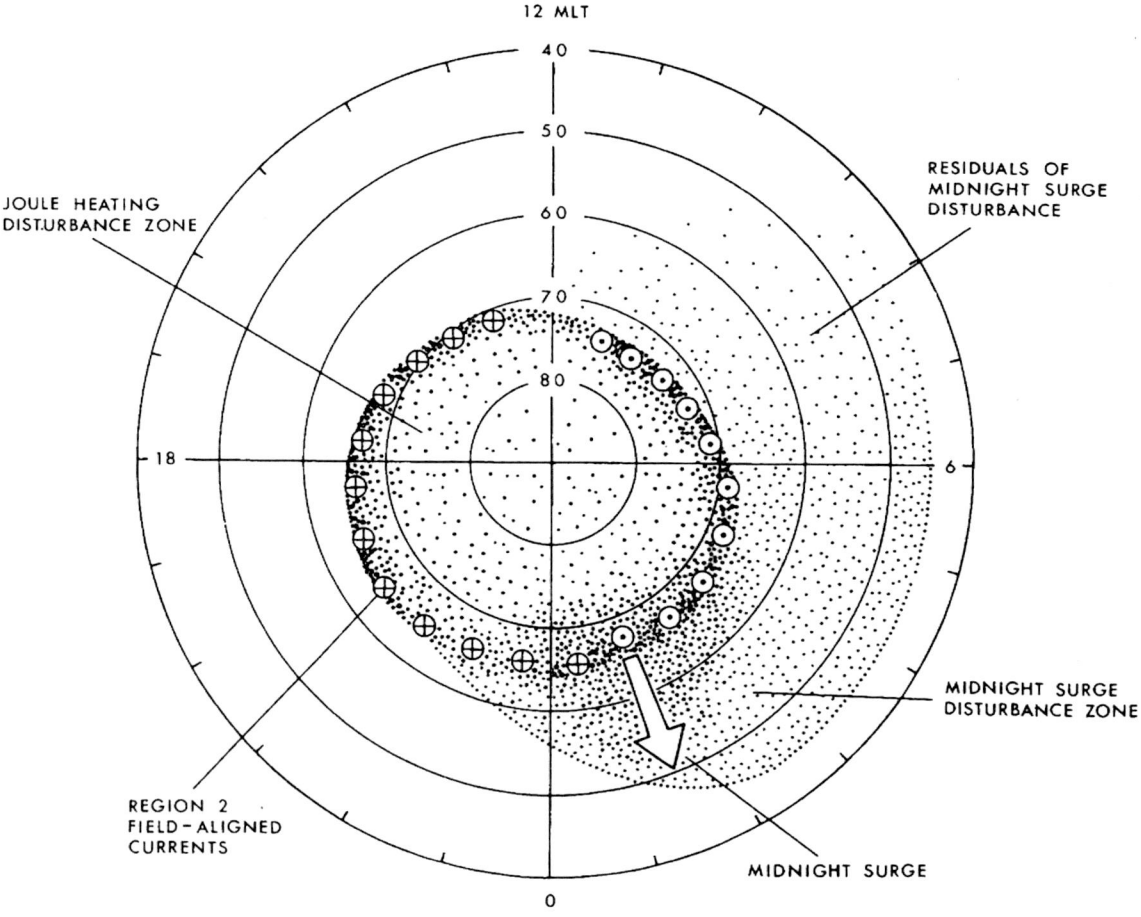

Figure 3. Extent of the composition disturbance zone during moderately disturbed conditions. The polar coordinates are invariant magnetic latitude and magnetic local time. The dotted area indicates the distribution of the composition perturbation, and the circles mark the position of region 2 Birkeland currents flowing into and out of the ionosphere. An arrow indicates the expansion of the composition disturbance zone toward middle latitudes, which occurs in the nighttime/early morning sector [after *Prölss*, 1981].

et al. [1997]. They reviewed much of the knowledge gleaned from observations and modeling studies over the previous 30–40 years.

In a series of papers, *Burns et al.* [1989, 1991, 1992, 1995, 2006] developed model postprocessors to help interpret the compositional response simulated by the NCAR model. They performed numerical experiments to simulate storm-like conditions using the NCAR model and showed that the major mechanisms for compositional change in the upper thermosphere are vertical advection, horizontal advection, and molecular diffusion [*Burns et al.*, 1991]. They discussed some of the specific features of composition effects including the recovery of the thermospheric composition after storms [*Burns et al.*, 1989] and the evening enhancement of O/N_2 in the winter hemisphere [*Burns et al.*, 1995a]. Figure 5a shows northern hemisphere N_2 mass mixing ratio on the $z = 2$ pressure level near 350 km, with winds superposed, shortly after the start of a storm [*Burns et al.*, 2006]. The perimeter latitude is at 20°. Figures 5b–5d show the corresponding compositional forcing terms: vertical advection, horizontal advection, and molecular diffusion, respectively. Vertical advection tends to carry molecular-rich air upward in the auroral zone, and this is counteracted in surrounding regions. Horizontal advection results from the horizontal winds and strong gradients in N_2 mass mixing ratio. Once the nitrogen-rich air has been lifted to these altitudes, it is advected toward the nightside midlatitudes, as shown here. Molecular diffusion then tends to return the composition profile to diffusive equilibrium in regions where it has been disturbed.

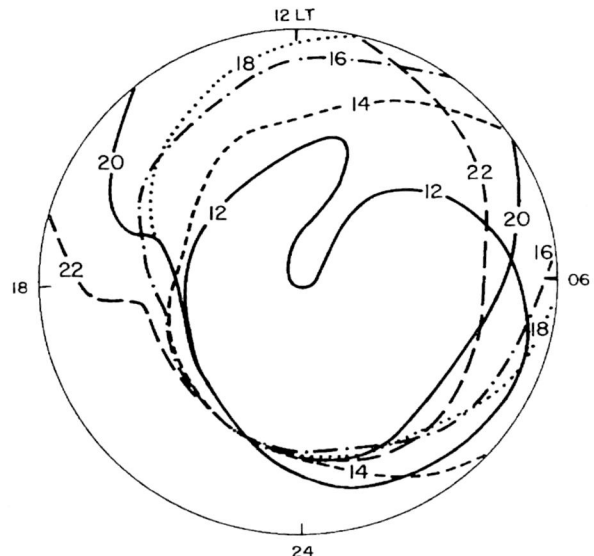

Figure 4. Simulated development of the disturbance zone in the Northern Hemisphere during the storm of September 1984 [after *Crowley et al.*, 1989a]. Polar plot shows development of contours where quiet and storm-time values of $O/(O_2 + N_2)$ are equal (i.e., zero difference) from 1200 to 2200 UT. Outer latitude is 30°. Simple corotation effects would move the pattern counterclockwise by 15° every hour.

3. GLOBAL AND REGIONAL OBSERVATIONS OF LARGE-SCALE COMPOSITION CHANGES

While a great deal has been learned from the available *in situ* observations, the data have been too sparse to confirm several of the fundamental storm effects simulated by the most advanced models that are now available. In particular, regional or global measurements of the neutral composition have been rare in the upper thermosphere and almost nonexistent in the lower thermosphere.

The state of affairs began to change in the 1980s, motivated at least in part by the remarkable global thermospheric and ionospheric images taken earlier from a lunar observing site during the Apollo 16 mission [*Carruthers and Page*, 1976]. Other missions carrying high-quality thermospheric airglow imagers began to emerge. The DE 1 satellite photometric imager revealed the first global images of composition changes during geomagnetic storms [*Frank and Craven*, 1988; *Craven et al.*, 1994]. A review of ultraviolet remote sensing was published by *Meier* [1991]. The UV airglow brightness was quantitatively related to the O/N_2 column density ratio by *Strickland et al.* [1995]. Plate 1a (reproduced from *Immel et al.* [2006]) shows sample DE 1 images from the storm of September 1981 revealing extended dayglow depletions on the dayside following a period of geomagnetic activity. The associated TIMEGCM simulation (Plate 1b) revealed that the depleted region had been formed on the nightside and corotated onto the dayside as predicted by *Prölss* [1980] and others. The ability to image the dayside thermosphere quantitatively provided a significant step forward in confirming the role of composition changes during the negative phase of ionospheric storms. *Strickland et al.* [2001] demonstrated that during the negative phase of a storm, F region electron densities are reduced in regions where the dayglow is depleted. Conversely, *Immel et al.* [2001] present cases where the dayglow and O/N_2 ratios are enhanced during the onset of a storm, in accord with the theoretical interpretation described in the previous section.

Further imaging capabilities were developed for the Polar and IMAGE missions [*Mende et al.*, 2000; *Zhang et al.*, 2003; *Immel et al.*, this volume]. Like the DE 1 images, these generally consisted of single-color (emission feature) views of almost an entire hemisphere, with large time gaps between images. In the case of DE 1, the images often were used to provide context for understanding the lower altitude DE 1 *in situ* data. Despite the quantitative results noted above, the view has persisted among many space scientists that thermospheric imaging only provides qualitative results. That perspective is now being dispelled by the much more detailed view of composition changes revealed by the GUVI instrument.

GUVI was one of four instruments launched on the TIMED satellite on 7 December 2001. *Christensen et al.* [2003] presented an overview of the GUVI instrument and operations. With the satellite in a 630-km circular orbit and a 74.1° inclination, GUVI provides a detailed multispectral view of a 2000-km-wide swath in a fixed local time every 100 minutes. The orbit precesses at a rate of 3° per day. In the imaging mode, GUVI returns data in five "colors": H (121.6 nm), O (130.4 nm), O (135.6 nm), N_2 (141.0–152.8 nm), and N_2 (167.2 – 181.2 nm). As noted above, the 135.6-nm and LBH dayside brightness ratio measured in the nadir can be used to obtain measurements of the column $\Sigma O/N_2$ ratio [*Strickland et al.*, 1995]. *Strickland et al.* [2004] reviewed the quiet time disk $\Sigma O/N_2$ distribution measured by GUVI for different seasons. *Zhang et al.* [2004] demonstrated good correspondence between the GUVI and IMAGE satellite observations of a storm on 1–4 October 2002.

The evolution of the $\Sigma O/N_2$ ratio during the November 2003 superstorm was discussed in detail by *Meier et al.* [2005]. Plate 2 (top) reproduces from their paper the GUVI-measured $\Sigma O/N_2$ during a 5-day period from day 322 to day 326 (18–22 November 2003). Note that time runs from right to left and the data from consecutive TIMED orbits are superposed on the appropriate geographic latitude and

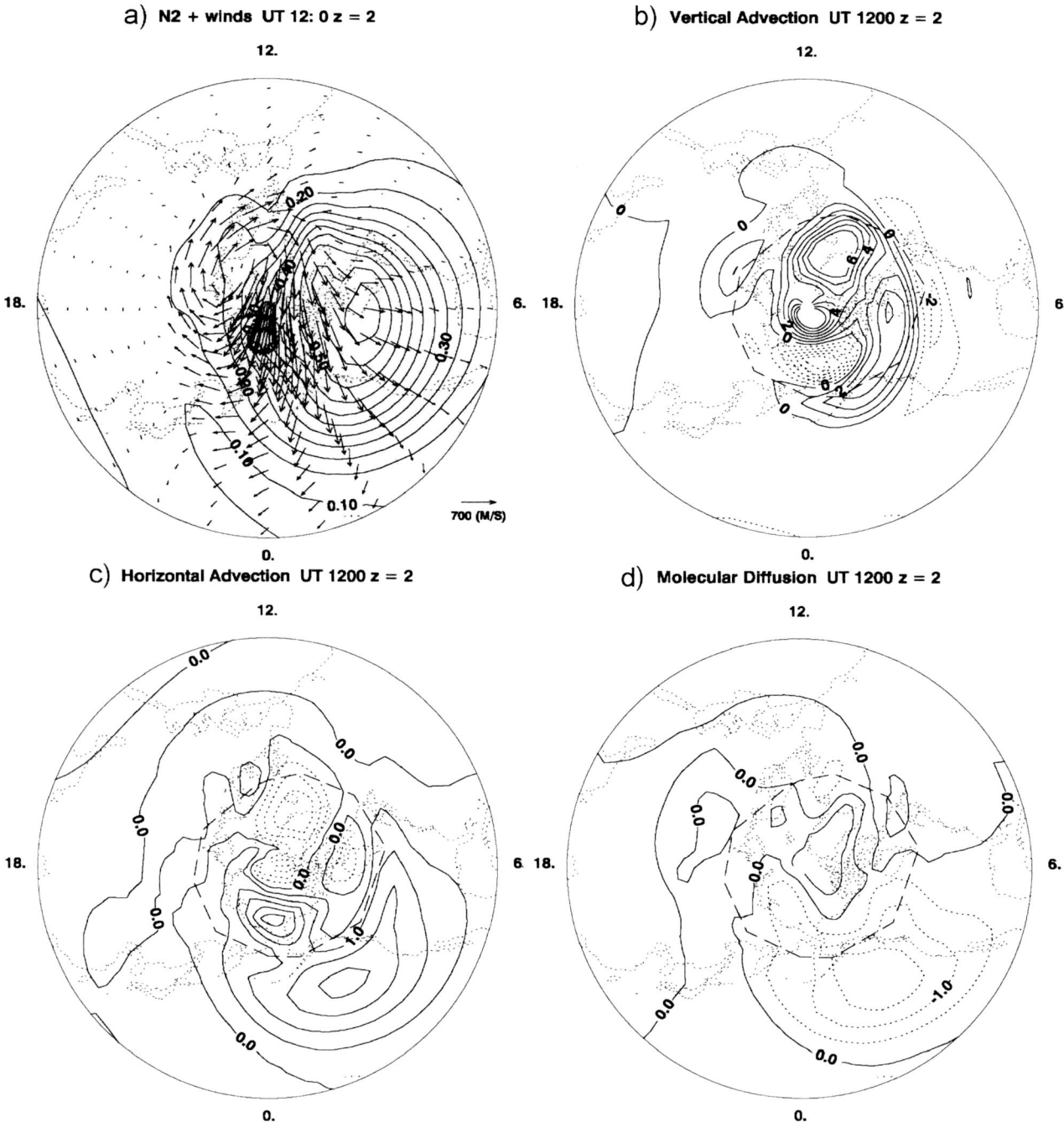

Figure 5. Evolution of N_2 mass mixing ratio and forcing terms from TIEGCM model [after *Burns et al.*, 2007]. Results are plotted for the z = 2 pressure surface near 350 km. The outer circle is at 20°.

longitude. The figure reveals the characteristic hemispheric asymmetry in $\Sigma O/N_2$ for November, with highest values in the northern midlatitudes. There is also a "scalloping" effect that recurs everyday due to the offset of the geographic and magnetic poles. There is a major perturbation on day 324 caused by a large geomagnetic storm. Plate 2 (middle and bottom) shows the results of two different simulations using the TIMEGCM model (which was developed by *Roble and*

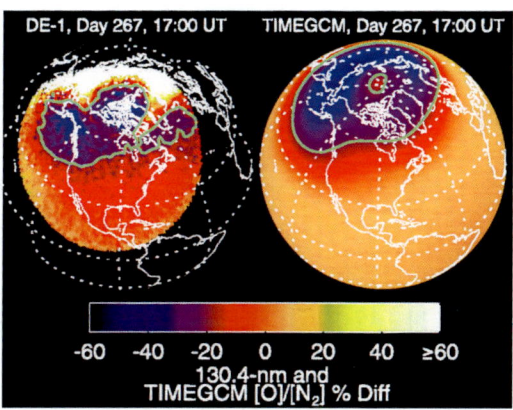

Plate 1. (a) Global images of depleted dayglow from DE 1 satellite revealed large regions of depleted $\Sigma O/N_2$ and (b) simulations of global $\Sigma O/N_2$ from TIMEGCM driven with realistic high-latitude inputs can often reproduce the observations reasonably well. In this case, AMIE was used to specify the polar cap potential distribution and auroral particle precipitation [*Immel et al.*, 2006].

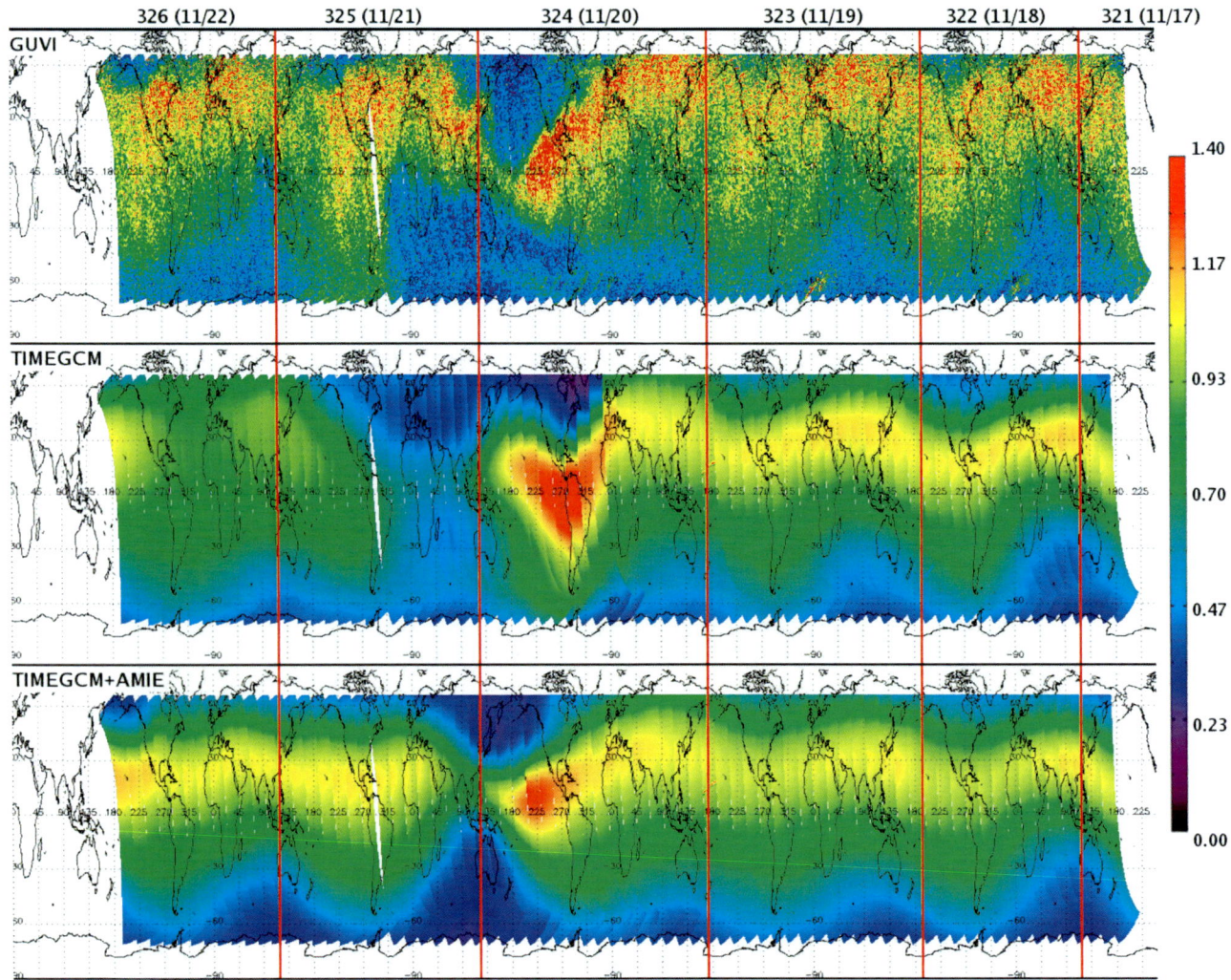

Plate 2. (top) GUVI observed significant changes in the column integrated O/N$_2$ ratio during the November 2003 storm. Five days of GUVI data are plotted as individual dayside orbits superposed on the continental outlines. Note that time runs from right to left in this format. (middle) Simulation using the TIMEGCM driven by climatological high-latitude drivers [after *Meier et al.*, 2005]. (bottom) Simulation using the TIMEGCM driven by AMIE.

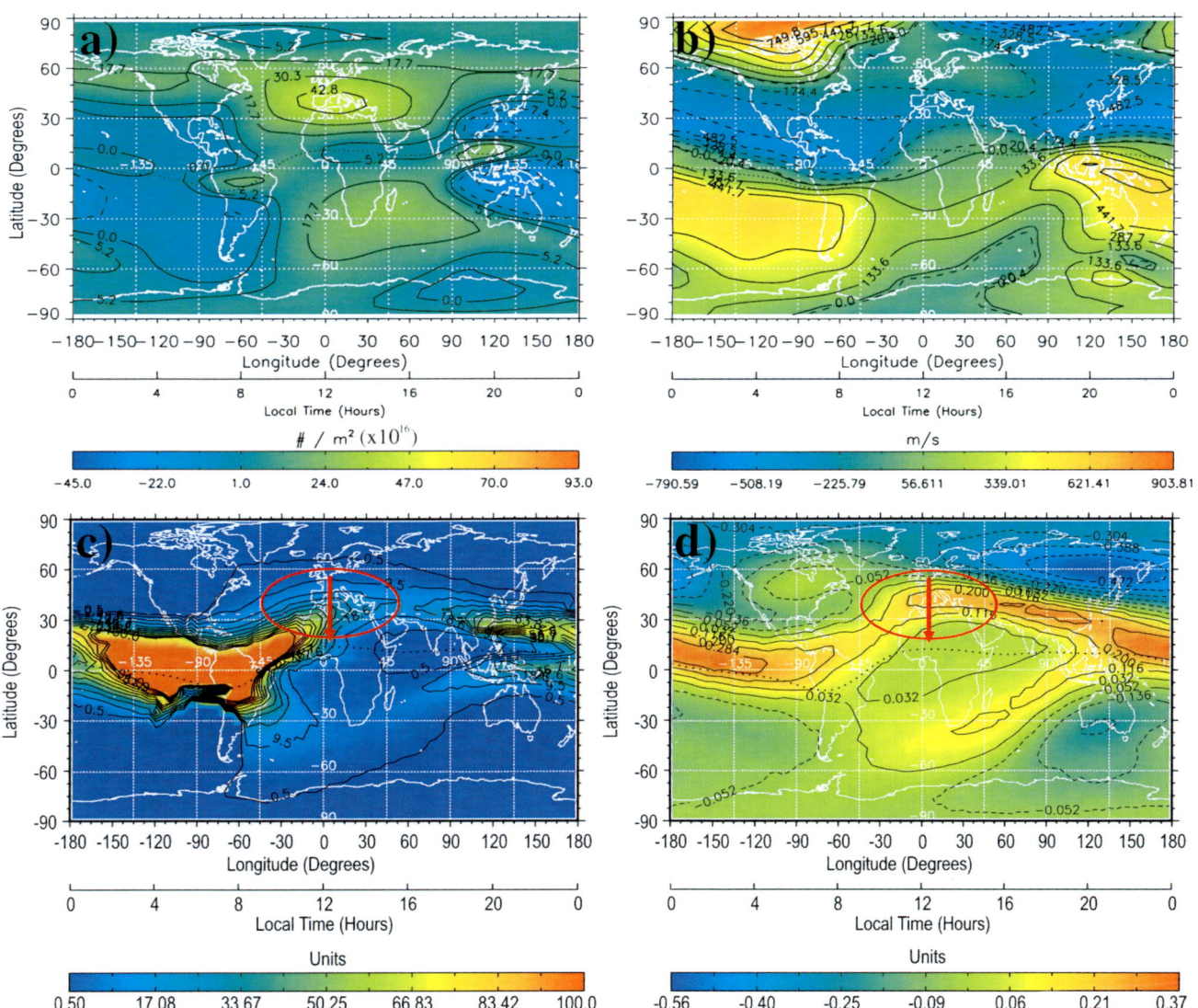

Plate 3. TIMEGCM predicts importance of winds and composition in the midlatitude electron density enhancement following the onset of the 20 November 2003 storm. Differences between the storm day at 1200 UT and the previous quiet day at 1200 UT are depicted for (a) modeled TEC, (b) equatorward meridional wind, (c) O/N_2 at the F region peak height (h_mF_2), and (d) $\Sigma O/N_2$. The red ellipse indicates the location of the TEC and NmF_2 increases. TIMED/GUVI orbits near 1130 LT on this day [after *Crowley et al.*, 2006].

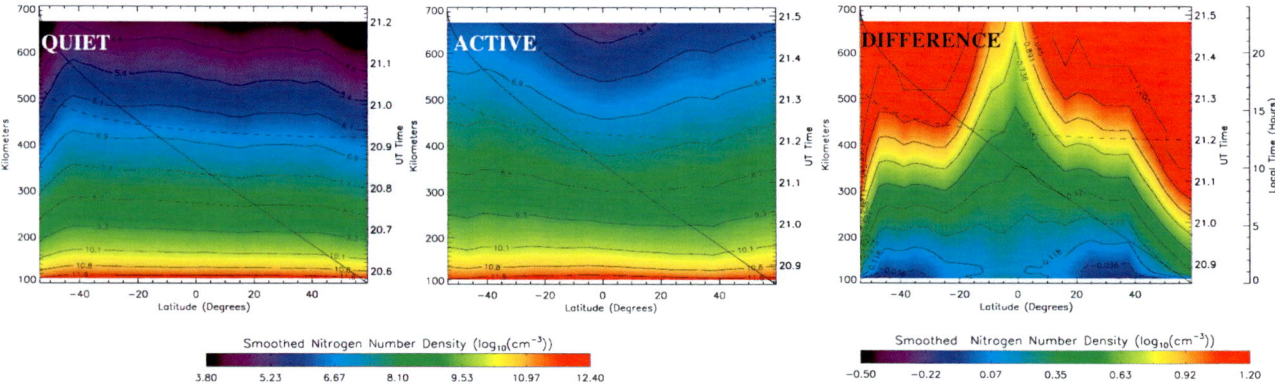

Plate 4. Latitudinal variation of vertical profiles for molecular nitrogen from single orbits of GUVI limb measurements: (a) quiet day orbit, (b) storm day orbit, and (c) difference.

Ridley [1994]). Plate 2 (middle) [after *Meier et al.*, 2005] was driven by the TIMEGCM default analytical high-latitude convection pattern [*Heelis et al.*, 1982] and aurora as described by *Roble and Ridley* [1987]. Plate 2 (bottom) shows a new simulation driven by convection patterns and auroral precipitation derived from the AMIE algorithm developed by *Richmond and Kamide* [1988] and *Richmond* [1992]. There is a significant improvement in the fidelity with which the model reproduces the northern hemisphere data when AMIE is used, rather than the default high-latitude inputs. In particular, Plate 2 (middle) implies two cycles of $\Sigma O/N_2$ depletion extending from high to low latitudes on day 325. However, the Plate 2 (bottom) and the GUVI data show only a single cycle, with recovery completed by about 1200 UT on day 325. In the southern hemisphere, both simulations recover too early on day 325. The most likely reason is poor representation of southern high-latitude forcing: the northern hemisphere AMIE pattern was used in both hemispheres. This claim will be investigated later in the next step of the analysis, in which we will produce separate southern hemisphere AMIE convection and particle fluxes and develop the ability to use them in the southern hemisphere of TIMEGCM. Other reasons for the quick recovery in the southern hemisphere, such as diffusion constants, NO cooling, or chemistry, would imply a similar problem should exist in the northern hemisphere, which we do not see.

The successful reproduction by the TIMEGCM of the gross features of $\Sigma O/N_2$ motivated *Crowley et al.* [2006] to investigate major increases in observed electron density and total electron content (TEC) in the midlatitude dayside ionosphere during the early phases of the same storm. The model predicted these TEC increases in the appropriate locations and within about 10%–20% of the observed values (Plate 3a). Their analysis showed that the observed Ne and TEC enhancement was due to neutral winds (Plate 3b) driving the ionosphere to higher altitudes where the O/N_2 ratio is larger (Plate 3c), recombination is slower, and the electron densities are enhanced. Reinforcing this effect, the model also suggested that there was an increase in the $\Sigma O/N_2$ ratio (Plate 3d) caused by the storm onset. The GUVI data added to the constraints imposed on the fidelity of the TIMEGCM output by confirming there was a $\Sigma O/N_2$ enhancement in the area (see Plate 2), thus lending credence to the argument that the model successfully accounted for the enhanced electron concentration. Unfortunately, no wind or electric field data were available to provide closure in our understanding of the event, although the model suggests that electrodynamic drift was not important in this case.

While we can learn a great deal about global or regional dynamics from downward-looking imagers, no altitudinal information is available from such line-of-sight observations. Vertical profiles of composition are fundamentally important for studying the thermospheric response to geomagnetic and solar forcing. While rockets had obtained occasional vertical profiles in the past, such profiles had never been available globally and systematically until the launch of the GUVI instrument on the TIMED satellite. Analysis of the GUVI limb brightness measurements has now provided altitude profiles of thermospheric composition and temperature from about 135 to 500 km [*Meier et al.*, 2005; *Stephan et al.*, 2008]. This is the first time that latitudinal variations in the entire thermospheric temperature and composition profiles have been available to test ideas that have been put forward previously on a (mainly) theoretical basis or have been suggested by limited *in situ* observations. *Emmert et al.* [2006] used satellite drag measurements to validate total mass densities constructed from GUVI limb composition data. The agreement between the two dramatically different approaches is good even though an earlier less accurate version of the GUVI limb data was used. The most recent version of the GUVI database shows even better consistency with the drag results (J. T. Emmert, unpublished data, 2007).

As an illustration of the compositional changes with altitude and latitude, we show in Plate 4 vertical profiles of molecular nitrogen from GUVI for the same November 2003 period as Plate 2. The latitudinal extent of the altitudinal changes offers dramatic new information about thermospheric responses to geomagnetic storms. The preliminary modeling of the data was reported by *Meier et al.* [2005]. Their work has been extended by G. Crowley, R. Meier, A. Christensen, L. Paxton, and D. Strickland (Observed latitudinal variation of thermospheric composition during a large storm, to be submitted to *Journal of Geophysical Research*, 2008), who reanalyzed the storm using the AMIE model to specify the inputs to the TIMEGCM.

4. CONCLUSIONS

Thermospheric compositional effects are important in the production of *F* region electron density enhancements and depletions during storms. The compositional variability is driven by both vertical and horizontal winds. Our ideas about compositional changes are still simple, and largely untested, and have focused on the *F* region because of their effects on the electron density there.

While there was some imaging capability on the DE 1, Polar, and IMAGE satellites, the recent GUVI data provide a revolutionary advance in our ability to measure thermospheric composition: they provide continuous coverage with high spatial resolution and low noise in five colors simultaneously, so the data can be unambiguously interpreted in terms of $\Sigma O/N_2$ ratios. In addition to column $\Sigma O/N_2$ ratios,

GUVI has provided the first limb retrievals of the composition as a function of altitude and latitude and has revealed new complexities in the global compositional response to storms.

Further work is needed to understand the variable composition of the thermosphere. The algorithms to analyze the GUVI data are still under development, and the data quality continues to be validated and improved. Once this is complete, the GUVI data will provide new insights into the storm response, which has not been completely documented or understood. There is still much work to be done on the effects of horizontal and vertical winds on composition. Other issues that need to be addressed include the effects of eddy diffusion and solar flares.

While compositional data may be available from GUVI, the other ionospheric drivers indicated in Figure 1a, such as winds, temperatures, and electric fields, are not generally measured, and therefore detailed scientific understanding of particular events is not usually possible. Global thermosphere–ionosphere models have made progress in simulating the compositional response, but the data required for validation of these models are lacking.

What data are needed? Our ability to measure and image the three-dimensional F region electron density is quite extraordinary, thanks to the growth of ground-based TEC and satellite occultation measurements and their ingestion into assimilation algorithms such as Ionospheric Data Assimilation-3D (see Mapping the time-varying distribution of high-altitude plasma during storms by *Bust* [this volume]). Supporting thermospheric data are much more sparse. GUVI provides compositional data for limited local times in any single orbit (although all dayside local times are covered as the orbit precesses over several days), and electric field measurements are available along DMSP satellite orbits at fixed local times. In addition, there are several ground-based radars that can provide local E field estimates. However, F region wind data are almost completely absent. A handful of ground-based Fabry–Perot interferometers provide winds, but only at night during good weather, and meridional winds can sometimes be deduced from incoherent scatter radars. Cross-track winds have recently been estimated from the CHAMP satellite [*Sutton et al.*, 2005], but require validation. With this varied assortment of measurements, we find that serendipity is not working well for Ionosphere–Thermosphere science: it is almost impossible to find cases where there is a comprehensive collocated set of measurements obtained simultaneously. One bright spot is the Air Force C/NOFS mission launched in April 2008. C/NOFS will measure *in situ* electron density, winds, and electric fields at low latitudes, hopefully while GUVI measures composition nearby. What is needed, however, is a satellite with comprehensive electron density, E field, wind, and $\Sigma O/N_2$ measurements (with high duty cycle) in a higher inclination orbit so that closure can finally be brought to our understanding of the higher-middle latitude ionospheric and thermospheric variations.

Acknowledgments. This work was partially supported by the TIMED/GUVI project through NASA grant NAG5-5001 to the Aerospace Corporation. G.C. was also supported by NASA grants NAG5-10059, NAG5-11055, and NAG5-13613 and by NASA Living with a Star grant NNG04GN04G. R.R.M. was also supported by the NASA Guest Investigator Program.

REFERENCES

Breig, E. L. (1987), Thermospheric ion and neutral composition and chemistry, *Rev. Geophys.*, *25*(3), 455–470.

Buonsanto, M. J. (1999), Ionospheric storms—A review, *Space Sci. Rev.*, *88*, 563.

Burns, A. G., T. L. Killeen, and R. G. Roble (1989), Processes responsible for the compositional structure of the thermosphere, *J. Geophys. Res.*, *94*(A4), 3670–3686.

Burns, A. G., T. L. Killeen, and R. G. Roble (1991), A theoretical study of thermospheric composition perturbations during a impulsive geomagnetic storm, *J. Geophys. Res.*, *96*(A8), 14,153–14,167.

Burns, A. G, T. L. Killeen, and R. G. Roble (1992), Thermospheric composition changes seen during a geomagnetic storm, *Adv. Space Res.*, *12*, 253.

Burns, A. G., T. L. Killeen, G. R. Carignan, and R. G. Roble (1995), Large enhancements in the O/N_2 ratio in the evening sector of the winter hemisphere during geomagnetic storms, *J. Geophys. Res.*, *100*, 14,661–14,671.

Burns, A. G., W. Wang, T. L. Killeen, S. C. Solomon, and M. Wiltberger (2006), Vertical variations in the N_2 mass mixing ratio during a thermospheric storm that have been simulated using a coupled magnetosphere–ionosphere–thermosphere model, *J. Geophys. Res.*, *111*, A11309, doi:10.1029/2006JA011746.

Bust, G. (2008), Mapping the time-varying distribution of high-altitude plasma during storms, this volume.

Carruthers, G. R., and T. Page (1976), Apollo 16 far ultraviolet imagery of the polar auroras, tropical airglow belts, and general airglow, *J. Geophys. Res.*, *81*, 483–496.

Christensen, A. B., et al. (2003), Initial observations with the Global Ultraviolet Imager (GUVI) in the NASA TIMED satellite mission, *J. Geophys Res.*, *108*(A12), 1451, doi:10.1029/2003JA009918.

Craven, J. D., A. C. Nicholas, L. A. Frank, D. J. Strickland, and T. J. Immel (1994), Variations in the FUV dayglow intense auroral activity, *Geophys. Res. Lett.*, *21*, 2793–2796.

Crowley, G. (1992), Dynamics of the Earth's thermosphere: A review, *U.S. National Report 1987–1990, Twentieth General Assembly, International Union of Geodesy and Geophysics, Vienna, Austria, 1991*, pp. 1143–1165.

Crowley, G., B. A. Emery, R. G. Roble, H. C. Carlson Jr., and D. J. Knipp (1989a), Thermospheric dynamics during September 18-19, 1984, 1, Model simulations, *J. Geophys. Res.*, *94*, 16,925–16,944.

Crowley, G., B. A. Emery, R. G. Roble, H. C. Carlson Jr., J. E. Salah, V. B. Wickwar, K. L. Miller, W. L. Oliver, R. G. Burnside, and F. A. Marcos (1989b), Thermospheric dynamics during September 1994 2, Validation of the NCAR Thermospheric General Circulation Model, *J. Geophys. Res.*, *94*, 16,945–16,960.

Crowley, G., et al. (2006), Global thermosphere–ionosphere response to onset of 20 November 2003 magnetic storm, *J. Geophys. Res.*, *111*, A10S18, doi:10.1029/2005JA011518.

Dickinson R. E., E. C. Ridley, and R. G. Roble (1984), Thermospheric general circulation with coupled dynamics and composition, *J. Atmos. Sci.*, *41*, 205–219.

Emmert, J. T., R. R. Meier, J. M. Picone, J. L. Lean, and A. B. Christensen (2006), Thermospheric density 2002–2004: TIMED/GUVI dayside limb observations and satellite drag, *J. Geophys. Res.*, *111*, A10S16, doi:10.1029/2005JA011495.

Foster, J. C. (1993), Storm time plasma transport at middle and high latitudes, *J. Geophys. Res.*, *98*(A2), 1675–1689.

Frank, L. A., and J. D. Craven (1988), Imaging results from Dynamics Explorer 1, *Rev. Geophys.*, *26*, 249–283.

Fuller-Rowell, T. J., D. Rees, H. Rishbeth, A. G. Burns, T. L. Killeen, and R. G. Roble (1991), Modelling of composition changes during F-region storms: A reassessment, *J. Atmos. Terr. Phys.*, *53*, 541–550.

Fuller-Rowell, T. J., M. V. Codrescu, R. J. Moffett, and S. Quegan (1994), Response of the thermosphere and ionosphere to geomagnetic storms, *J. Geophys. Res.*, *99*, 3893–3914.

Fuller-Rowell, T. J., M. V. Codrescu, H. Rishbeth, R. J. Moffett, and S. Quegan (1996), On the seasonal response of the thermosphere and ionosphere to geomagnetic storms, *J. Geophys. Res.*, *101*, 2343–2353.

Fuller-Rowell, T. J., M. V. Codrescu, R. G. Roble, and A. D. Richmond (1997), How does the thermosphere and ionosphere react to a geomagnetic storm?, in *Magnetic Storms, Geophys. Monogr. Ser.*, vol. 98, edited by B. T. Tsurutani, W. D. Gonzalez, Y. Kamide, and J. K. Arballo, p. 203, AGU, Washington, D. C.

Hays, P. B., R. A. Jones, and M. H. Rees (1973), Auroral heating and the composition of the neutral atmosphere, *Planet. Space Sci.*, *21*, 559–573.

Hedin, A. E. (1983), A revised thermospheric model based on mass spectrometer and incoherent scatter data: MSIS-83, *J. Geophys. Res.*, *88*, 10,170–10,188.

Hedin, A. E. (1987), MSIS-86 thermospheric model, *J. Geophys. Res.*, *92*, 4649–4662.

Hedin, A. E. (1991), Extension of the MSIS thermosphere model into the middle and lower atmosphere, *J. Geophys. Res.*, *96*, 1159–1172.

Hedin, A. E., H. G. Mayr, C. A. Reber, N. W. Spencer, and G. R. Carignan (1974), Empirical model of global thermospheric temperature and composition based on data from the OGO-6 quadrupole mass spectrometer, *J. Geophys. Res.*, *79*, 215–225.

Heelis, R. A., J. K. Lowell, and R. W. Spiro (1982), A model of the high latitude ionospheric convection pattern, *J. Geophys. Res.*, *87*, 6339–6345.

Immel, T. J., G. Crowley, J. D. Craven, and R. G. Roble (2001), Dayside enhancements of thermospheric O/N_2 following magnetic storm onset, *J. Geophys. Res.*, *106*, 15,471–15,488.

Immel, T. J., G. Crowley, C. L. Hackert, J. Craven, and R. G. Roble (2006), Effect of IMF B_Y on thermospheric composition at high and middle latitudes: 2. Data comparisons, *J. Geophys. Res.*, *111*, A10312, doi:10.1029/2005JA011372.

Immel, T. J., J. M. Forbes, R. S. Nerem, and E. K. Sutton, and G. Crowley (2008), Neutral composition and density effects in the October–November 2003 magnetic storms this storms, this volume.

Kayser, D. C., and W. E. Potter (1976), Molecular oxygen measurements at 200 km from AE-D near Winter solstice, 1975, *Geophys. Res. Lett.*, *3*, 455–458.

Killeen, T. L., and R. G. Roble (1988), Thermosphere dynamics: Contributions from the first 5 years of the Dynamics Explorer program, *Rev. Geophys.*, *26*, 329–367.

Killeen, T. L., F. G. McCormac, A. G. Burns, J. P. Thayer, R. M. Johnson, and R. J. Niciejewski (1991), On the dynamics and composition of the high-latitude thermosphere, *J. Atmos. Terr. Phys.*, *53*, 797.

Kintner, P., A. Coster, T. Fuller-Rowell, and A. J. Mannucci (2007), Overview of midlatitude ionospheric storms, *Eos Trans. AGU*, *88*(37), doi:10.1029/2007EO37002.

Mauersberger, K., D. C. Kayser, W. E. Potter, and A. O. Nier (1976), Seasonal variation of neutral thermospheric constituents in the Northern Hemisphere, *J. Geophys. Res.*, *81*, 7–11.

Mayr, H. G., and I. Harris (1977), Diurnal variations in the thermosphere, 2. Temperature, composition, and winds, *J. Geophys. Res.*, *82*, 2628–2640.

Mayr, H. G., and H. Volland (1973), Magnetic storm characteristics of the thermosphere, *J. Geophys. Res.*, *78*, 2251–2264.

Mayr, H. G., A. E. Hedin, C. A. Reber, and G. R. Carignan (1974), Global characteristics in the diurnal variations of the thermospheric temperature and composition, *J. Geophys. Res.*, *79*, 619–628.

Mayr, H. G., I. Harris, and N. W. Spencer (1978), Some properties of upper atmospheric dynamics, *Rev. Geophys. Space Phys.*, *16*, 539–565.

Meadows, E. B., and J. W. Townsend (1958), Diffusive separation in the winter nighttime arctic upper atmosphere, *Ann. Geophys.*, *14*, 80–93.

Meier, R. R. (1991), Ultraviolet spectroscopy and remote sensing of the upper atmosphere, *Space Sci. Rev.*, *58*, 1–185.

Meier, R. R., G. Crowley, D. J. Strickland, A. B. Christensen, L. J. Paxton, D. Morrison, and C. L. Hackert (2005), First look at the 20 November 2003 super storm with TIMED/GUVI: Comparisons with a thermospheric global circulation model, *J. Geophys. Res.*, *110*, A09S41, doi:10.1029/2004JA010990.

Mende, S. B., et al. (2000), Far ultraviolet imaging from the IMAGE spacecraft. 2. Wideband FUV imaging, *Space Sci. Rev.*, *91*, 271–285.

Picone, J. M., A. E. Hedin, D. P. Drob, and A. C. Aikin (2002), NRLMSISE-00 empirical model of the atmosphere: Statistical comparisons and scientific issues, *J. Geophys. Res*, *107*(A12), 1468, doi:10.1029/2002JA009430.

Prölss, G. W. (1980), Magnetic storm associated perturbations of the upper atmosphere: Recent results obtained by satellite-borne gas analyzers, *Rev. Geophys. Space Phys.*, *18*, 183.

Prölss, G. W. (1981), Latitudinal structures and extension of the polar atmospheric disturbance, *J. Geophys. Res.*, *86*, 2385–2396.

Prölss, G. W. (1997), Magnetic storm associated perturbations of the upper atmosphere, in *Magnetic Storms, Geophys. Monogr. Ser.*, vol. 98, edited by B. T. Tsurutani, W. D. Gonzalez, Y. Kamide, and J. K. Arballo, p. 227, AGU, Washington, D. C.

Roble, R. G., and E. C. Ridley (1987), An auroral model for the NCAR thermospheric general circulation model (TGCM), *Ann. Geophys., Ser. A*, *5*, 369.

Roble, R. G., and E. C. Ridley (1994), A thermosphere–ionosphere–mesosphere–electrodynamics general circulation model (TIME–GCM): Equinox solar cycle minimum simulations (30–500 km), *Geophys. Res. Lett.*, *21*(6), 417–420.

Richmond, A. D. (1992), Assimilative mapping of ionospheric electrodynamics, *Adv. Space Res.*, *12*, 59.

Richmond, A. D., and Y. Kamide (1988), Mapping electrodynamic features of the high latitude ionosphere from localized observations: Technique, *J. Geophys. Res.*, *93*, 5741–5759.

Seaton, M. J. (1956), A possible explanation of the drop in F-region critical densities accompanying major ionospheric storms, *J. Atmos. Terr. Phys.*, *8*, 122–124.

Stephan A. W., R. R. Meier, and L. J. Paxton (2008), Comparison of Global Ultraviolet Imager limb and disk observations of column O/N_2 during a geomagnetic storm, *J. Geophys. Res.*, *113*, A01301, doi:10.1029/2007JA012599.

Strickland, D. J., J. S. Evans, and L. J. Paxton (1995), Satellite remote sensing of thermospheric O/N_2 and solar EUV, 1. Theory, *J. Geophys. Res.*, *100*, 12,217–12,226

Strickland, D. J., R. E. Daniell, and J. D. Craven (2001), Negative ionospheric storm coincident with DE 1-observed thermospheric disturbance on October 14, 1981, *J. Geophys. Res.*, *106*, 21,049–21,062.

Strickland, D. J., R. R. Meier, R. L. Walterscheid, J. D. Craven, A. B. Christensen, L. J. Paxton, D. Morrison, and G. Crowley (2004), Quiet-time seasonal behavior of the thermosphere seen in the far ultraviolet dayglow, *J. Geophys. Res.*, *109*, A01302, doi:10.1029/2003JA010220.

Sutton, E. K., J. M. Forbes, and R. S. Nerem (2005), Global thermospheric neutral density and wind response to the severe 2003 geomagnetic storms from CHAMP accelerometer data, *J. Geophys. Res.*, *110*, A09S40, doi:10.1029/2004JA010985.

Zhang, Y., L. J. Paxton, D. Morrison, B. Wolven, H. Kil, C.-I. Meng, S. B. Mende, and T. J. Immel (2004), O/N_2 changes during 1–4 October 2002 storms: IMAGE SI-13 and TIMED/GUVI observations, *J. Geophys. Res.*, *109*, A10308, doi:10.1029/2004JA010441.

G. Crowley, Atmospheric and Space Technology Research Associates, 11118 Quail Pass, San Antonio, TX 78249, USA.

R. R. Meier, Department of Physics and Astronomy, George Mason University, 4400 University Drive, MS 3F3, Fairfax, VA 22030, USA.

Storm Time Energy Budgets of the Global Thermosphere

William J. Burke

Air Force Research Laboratory Space Vehicles Directorate, Hanscom Air Force Base, Massachusetts, USA

Space environment data acquired near L_1 and in low Earth orbit have allowed us to develop a deeper appreciation of the unity binding the dynamics of the magnetosphere-ionosphere-thermosphere system. This tutorial focuses on the main phase of large ($Dst < -200$ nT) magnetic storms when predictions of thermospheric density models appear "too little-too late" in comparison with neutral mass densities (ρ) inferred from accelerometers on the Gravity Recovery and Climate Experiment (GRACE) satellites. Empirically, we find that (1) Dst traces correlate well with orbit-averaged values of ρ during the main and early recovery phases, but not thereafter and (2) polar cap potentials and magnetospheric electric fields derived from L_1 measurements predict thermospheric responses in all storm phases with lead times of 4–6 h. These two results provide clues for tracking energy transport from the solar wind to the ring current via magnetospheric electric fields. The asymmetric ring current seems to act as a reservoir from which energy is extracted as a net particle and Poynting flux into the ionosphere. Ion-neutral collisions transfer energy to the thermosphere. As the ring current symmetrizes, coupling to the ionosphere-thermosphere diminishes. Radiative losses of energy cause the thermosphere to relax to prestorm levels in ~13 h. We demonstrate a new application of *Jacchia's* [1977] model to test this conjecture. With inputs from GRACE measurements, the model predicts thermospheric energy increases (ΔE_{th}) that are about a factor of four greater than the increase in ring current energy (ΔE_{RC}) predicted by the Dessler-Parker-Sckopke relation. A larger fraction of the thermosphere's storm time energy budget must enter directly into the polar cap from the solar wind rather than indirectly via the magnetosphere into the auroral oval.

1. INTRODUCTION

Let us begin with a "truth in lending" statement. Conveners of the Chapman Conference on *Mid-Latitude Ionospheric Dynamics and Disturbances* requested that I present a tutorial/review in the context of interplanetary-magnetosphere-ionosphere coupling. Lacking credible midlatitude experience and being a novice in thermospheric research, I demurred. They persisted. Our compromise was to smooth out the physics by addressing the subject on a global scale. My presentation at the Chapman Conference summarized practical steps for auditing energy flows from the Lagrange point (L_1) to the thermosphere found while analyzing data from the Advanced Composition Explorer (ACE) and Gravity Recovery and Climate Experiment (GRACE) spacecraft. It focused on discrepancies between storm time measurements and model predictions and the clues they provide

Midlatitude Ionospheric Dynamics and Disturbances
Geophysical Monograph Series 181
This paper is not subject to U.S. copyright. Published in 2008 by the American Geophysical Union.
10.1029/181GM21

about energy flows. Over the intervening months, supporting arguments were published [*Burke*, 2007; *Burke et al.*, 2007a, 2007b] and new ways to quantify energy budgets suggested themselves. Thus, it seems useful to summarize the empirical data in comparison with model predictions, then offer simple procedures to extract storm time energy budgets from neutral density measurements.

National responsibility for tracking and predicting the precise orbits of satellites and space debris lies with the U.S. Air Force. Considerable efforts underlie empirical and theoretical modeling in support of this responsibility. *Jacchia* [1964, 1970, 1977] integrated available information about neutral density and temperature profiles to develop models that predict thermospheric drag on satellites. Inputs include the solar declination, solar-cycle phase, daily, and 81-day averaged values of the $F_{10.7}$ flux and the *ap* magnetic index. Jacchia's work still lies at the heart of operational models, albeit augmented by more recent contributions such as the Mass Spectrometer and Incoherent Scatter (MSIS) [*Hedin et al.*, 1977], National Aeronautics and Space Administration Marshall Engineering Thermosphere (NASA/MET) [*Owens and Vaughn*, 2004], and Naval Research Laboratory Mass Spectrometer and Incoherent Scatter (NRLMSISE-00) [*Picone et al.*, 2002] models. Despite these efforts, even quiet-time position errors persisted at 10–15% levels through the 1990s [*Marcos et al.*, 2006]. The High Accuracy Satellite Drag Model (HASDM) [*Casali et al.*, 2002] assimilated drag measurements from 75 calibration satellites with perigees ranging from 200 to 800 km. HASDM data, coupled with new capabilities for monitoring UV emissions from the solar corona and chromosphere, further reduced quiet-time errors to the 5–8% range [*Bowman*, 2002; *Bowman et al.*, 2006]. Algorithms presently used to represent the disturbance-time drivers of models are based on the *ap* index. Recently available accelerometer data from the Challenging Minisatellite Payload for Geophysical Research and Application (CHAMP) and GRACE spacecraft allow inferences about satellite drag on unprecedented spatial and temporal scales.

Sutton et al. [2005] analyzed neutral density and wind measurements from the Spatial Triaxial Accelerometer for Research (STAR) sensor on the CHAMP satellite during the magnetic storm of late October 2003, focusing on mid to low latitudes. *Forbes et al.* [2005] found evidence in CHAMP STAR data for equatorward-propagating, long-wavelength-traveling atmospheric disturbances during the April 2000 storms. Of concern for this present analysis are the results of neutral density correlations with geophysical parameters. *Forbes et al.* [2005, Table 1] shows that the highest correlations occur on the night side with the magnitude of the interplanetary magnetic field (IMF). More recently, *Bruinsma et al.* [2006] analyzed simultaneous CHAMP and GRACE data acquired during the November 2003 storm to show that the high-latitude thermosphere responds to solar wind changes immediately; near equatorial responses occur ~4 h later. In comparing observations with storm time predictions, both studies found models wanting.

While the general features of magnetosphere-ionosphere-thermosphere (M-I-T) coupling are known, gaps remain in our understanding. During the main phase of a large magnetic storm on 6 April 2000, four DMSP spacecraft detected simultaneous and repeated episodes of intense field-aligned currents (FACs) [*Huang and Burke*, 2004]. The satellites' spatial distributions suggested that the FAC sheets extended at least 6 h in magnetic local time. Magnetic perturbations (δB_Y) with amplitudes >1300 nT indicated that the intensity of the FACs flowing into and out of the ionosphere exceeded >1 MA for each hour in magnetic local time extent. However, ground magnetic records showed no commensurate perturbations within local time sectors sampled by DMSP spacecraft. DMSP data showed that current-carrying electrons had average energies of ~0.5 keV. At magnetic latitudes of 55°–60°, these electrons stop at altitudes >180 km, where they support significant Pedersen but little Hall conductivity. Under such conditions, magnetic perturbations expected on the ground are weak [*Fukushima*, 1969]. *Huang and Burke* [2004] also demonstrated that the quantity $E_Y \delta B_Z / \mu_0$, constructed from DMSP drift meter and magnetometer data is the net (incident − reflected) Poynting flux into the ionosphere. Conservative estimates for rates of energy deposition into the night-side auroral oval during the large FAC episodes were >600 GW, exceeding the rates of particle and solar UV energy deposited in the global thermosphere [*Knipp et al.*, 2005]. Clearly, this and similar storm time energy depositions must affect the thermospheric budget profoundly. However, weak magnetic signatures on the ground used to drive specification/forecast models hide true energy deposition rates.

Wilson et al. [2006] suggested a new approach to quantify global thermospheric energy budgets. Storm time HASDM data showed that time-varying density changes observed between 200 and 800 km required thermal and gravitational energy histories in substantial agreement with Poynting fluxes into the high-latitude ionosphere predicted by the model of *Weimer* [2005]. This model combines statistically determined high-latitude electrostatic and magnetic Euler potential responses to interplanetary changes to estimate the total Poynting flux into the global ionosphere and thermosphere. As such, it does not rely on ionospheric conductance profiles that for a given geomagnetic state are subject to wide variability [*Hardy et al.*, 2008]. Note that most of the electromagnetic energy deposition occurs at auroral rather than polar cap latitudes [*Weimer*, 2005].

The remainder of this tutorial has three main sections. The first describes how neutral mass densities ρ are determined from accelerometer data and demonstrates that present models fail to capture their storm time dynamics. On the other hand, the *Dst* index and predicted magnetospheric electric fields show remarkable correlations with orbit-averaged densities $\bar{\rho}$. The second section shows that the J77 model can be adapted to estimate storm time energetics. Instead of predicting neutral density responses to time-varying external sources, we use $\bar{\rho}$ measured by GRACE to calculate exospheric temperatures T_∞ and global thermospheric energies E_{th} required by J77. This information can then provide realistic drivers for global circulation models that follow the transport of energy from auroral to mid and low latitudes during magnetic storms. The final section contains a critical summary and points out some questions for further investigation.

2. MEASUREMENTS AND MODELS

GRACE consists of two spacecraft that fly in tandem separated by ~220 km. They were launched in March 2002 into slightly elliptical polar orbits. Atmospheric densities are derived from the STAR sensor that monitors the electrostatic forces needed to maintain a proof mass (PM) at the center of a cage located within 2 mm of the spacecraft's center of mass [*Bruinsma et al.*, 2004]. Both the spacecraft and PM respond to gravity in the same way. Thus, changes in electrostatic forces that maintain the PM at its cage's center reflect responses to nongravitational forces such as atmospheric drag [*Bruinsma and Biancale*, 2003]. The acceleration from drag is given by

$$a_{\text{drag}} = \frac{C_D A_{SC}}{M_{SC}} \rho V^2 \qquad (1)$$

Here A_{SC} and M_{SC} represent the cross-sectional area and mass of the spacecraft. V is the spacecraft velocity in the rest frame of ambient neutrals. The drag coefficient C_D depends on the angle of flow to the spacecraft surface, the ratio of the temperatures of the satellite surface and the local atmosphere, and the ratio of the mean mass of atoms in the atmosphere to those on the satellite surface [*Bruinsma and Biancale*, 2003].

Ionospheric and thermospheric dynamics are mainly driven by two external sources: EUV and FUV radiation from the Sun and electromechanical energy carried in the solar wind. UV fluxes have immediate impacts on the Earth's day side but are subject to relatively slow day-to-day variations. Energy derived from solar wind sources first reaches the ionosphere at high magnetic latitudes on the day and night sides but fluctuates on more rapid timescales than UV. In both cases, internal transport and loss processes carry energy away from deposition sites. Magnetic storms are sustained episodes of strong solar-wind energy coupling to the global magnetosphere, ionosphere, and thermosphere. Sensors on the ACE satellite near L_1 provide nearly continuous information about the solar wind velocity (V_{sw}) and density (n_{sw}) as well as the IMF. To a first approximation, the dynamic pressure of the solar wind $P_{sw} \approx m_p n_{sw} V_{sw}^2$ specifies the size and shape of the magnetosphere [*Roelof and Sibeck*, 1993]. In the rest frame of the Earth, the convecting IMF carries an interplanetary electric field $\boldsymbol{E}_I = -\boldsymbol{V}_{sw} \times \boldsymbol{B}_I$ that imposes a potential across the polar ionosphere (Φ_{PC}) and magnetosphere. In the magnetosphere, this electric field energizes the particles that carry the storm time ring current. The different drift motions of energetic charged particles give rise to plasma pressure gradients that support the FACs that electrically couple the magnetosphere to the auroral ionosphere [*Vasyliunas*, 1970]. Magnetic and electric perturbations associated with FACs, regularly sampled by sensors on DMSP, are direct measures of the net electromagnetic (Poynting) energy flux into the ionosphere [*Huang and Burke*, 2004]. Some fraction of the Poynting flux drives winds and heats the thermosphere via ion-neutral collisions.

The black line trace in Plate 1 (top) provides an example of ρ(*t*), at 10-s resolution, acquired by GRACE during the magnetically disturbed period 7–10 November 2004. Superposed blue dots indicate orbit-averaged densities $\bar{\rho}(t)$. The orbital plane of GRACE was near the noon-midnight meridian. Following standard procedures, data were projected to a common altitude of 350 km via the MSIS model. The red line gives provisional –*Dst* (right scale).

Burke et al. [2007a, Figure 2] compares the same ρ(*t*) measurements with predictions of the MSIS model throughout this period. Sinusoidal structures seen in ρ(*t*) traces before and after the disturbance reflect day-to-night and latitudinal thermospheric variations. MSIS predictions replicated these features. In GRACE's frame of reference during the storm ρ(*t*) varied on timescales of several hours and a few minutes. The *ap*-based algorithm presently used to drive MSIS during magnetic storms captured neither feature. MSIS under predicted main-phase variations of $\bar{\rho}(t)$ by >100%. It was only in the late main phase that $\bar{\rho}(t)$ predictions and observations returned to agreement. Also seen in the storm time ρ(*t*) traces are upward and downward spikes. Through comparisons with simultaneous drift meter measurements from DMSP spacecraft, *Burke et al.* [2007a] showed that the spikes mark GRACE encounters with head and tail winds driven by antisunward plasma convection in the polar cap.

Plate 1 (top) indicates that –*Dst* variations were similar to those of $\bar{\rho}(t)$. The solid red line trace in Plate 1 (bottom) represents electric fields (E_{VS}) in the inner magnetosphere determined from ACE data. Methods used to derive these

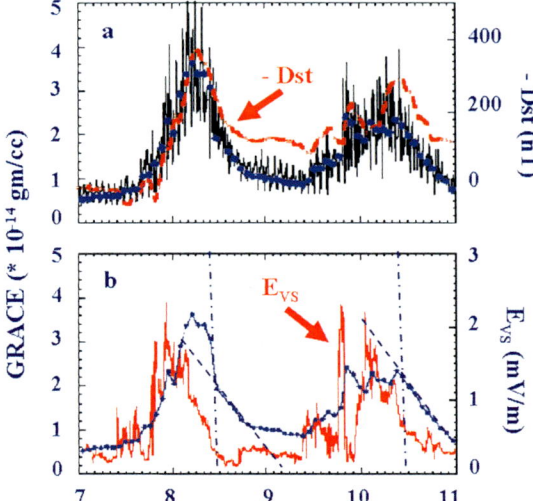

Plate 1. The black trace in (a) indicates mass densities measured by GRACE on 7–10 November 2004 at 10-s resolution. The red dashed line shows provisional $-Dst$ for the period. Blue dots in (a) and (b) indicate orbit-averaged mass densities. The red trace in (b) shows electric fields in the equatorial plane of the inner magnetosphere inferred from ACE measurements at L_1. Vertical lines mark responses to rapid downturns in the electric fields; slanted dashed lines indicate thermospheric relaxation rates after driving electric fields are removed.

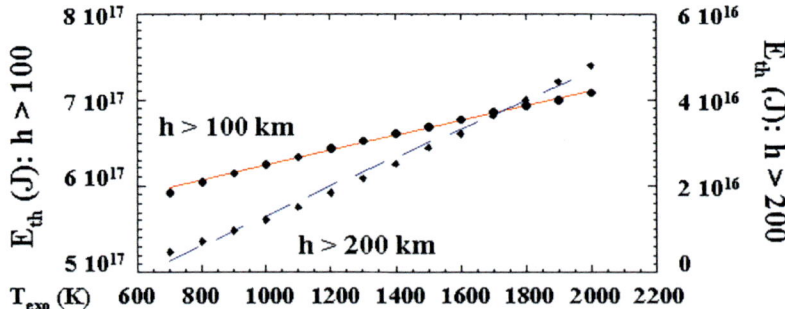

Plate 2. J77 total thermospheric energy E_{th} above 100 km (circles) and 200 km (diamonds) plotted as functions of exospheric temperatures. The straight lines represent linear regression fits to these "data."

quantities are summarized below. Vertical lines in Plate 1 (bottom) mark times when E_{VS} decreased significantly. Slanted dashed lines indicate rates at which $\bar{\rho}(t)$ relaxed after E_{VS} turned off. Data in Plate 1 support three empirical conclusions:

1. Variations in $-Dst$ were similar to those of $\bar{\rho}(t)$ throughout the main and early recovery phases of the magnetic disturbances. Both quantities attained maximum epochs at about the same time. However, they relaxed at different rates.

2. E_{VS} anticipated changes in $\bar{\rho}(t)$ with delay times of 4–6 h. With a fixed delay of 4 h the correlation coefficient between $\bar{\rho}(t)$ and E_{VS} is ~0.87.

3. In agreement with *Wilson et al.* [2006], $\bar{\rho}(t)$ returned to quiet values in ~13 h.

Before offering an explanation of these observations it is helpful to digress to describe our application of the Volland–Stern (V-S) model we used to estimate E_{VS} and the significance of Dst captured in the Dessler–Parker–Sckopke (D-P-S) [cf. review by *Carovillano and Siscoe*, 1973] and Burton–McPherron–Russell (B-M-R) [*Burton et al.*, 1975] relations.

2.1. Volland-Stern Electric Fields

V–S originated at Rice University in the late 1960s [*Kavanagh et al.*, 1968; *Chen and Wolf*, 1972]. *Volland* [1973], *Stern* [1975], and *Ejiri* [1978] developed it as a conceptual tool to estimate guiding-center trajectories of energetic particles. V-S operates in the equatorial plane of the magnetosphere and makes calculations in an inertial coordinate system. The magnetic field is represented as a rotating dipole with a superposed dawn-to-dusk electric field. *Burke* [2007] adapted the model to estimate magnetospheric electric fields as

$$\vec{E}(L,\phi) = -\frac{91\,(\text{kV})}{R_E L^2}\hat{r}$$
$$+ \frac{\Phi_{PC}\,(\text{kV})}{2R_E L_Y}\left(\frac{L}{L_Y}\right)^{\gamma-1}[\gamma\sin\phi\,\hat{r} + \cos\phi\,\hat{\phi}] \quad (2)$$

The first and second terms on the right side represent the corotation and cross-magnetosphere electric fields, respectively. Φ_{PC} is the potential drop across the polar cap, $2R_E L_Y$ is the width of the magnetosphere at the dawn–dusk meridian, γ is a shielding parameter [*Southwood and Kaye*, 1979] that is ~1 during the main phase of storms.

Following the suggestion of *Hill* [1984], *Siscoe et al.* [2002] postulated that a saturation term Φ_S contributes to the nonlinear Φ_{PC}

$$\Phi_{PC} = \frac{\Phi_E \Phi_S}{\Phi_E + \Phi_S} \quad (3)$$

Burke et al. [1999] showed that in the linear regime

$$\Phi_E\,(\text{kV}) \approx \Phi_0 + L_G V_{sw} B_T \sin^2\frac{\theta}{2} \quad (4)$$

The subscript E refers to interplanetary contributions; $\Phi_0 \approx 25$ kV is a residual potential, $B_T = \sqrt{(B_y^2 + B_z^2)}$, and $\theta = \cos^{-1}(B_Z/B_T)$ is the IMF clock angle in the Y–Z plane. If $V_{sw}B_T$ is expressed in kV/R_E (1 mV/m ≈ 6.4 kV/R_E), then the regression slope $L_G \approx 3.5 R_E$ represents the width of a "gate" in the solar wind through which geoeffective streamlines (equipotentials) must pass to reach the day-side magnetopause. Based on magnetohydrodynamic simulations, *Siscoe et al.* [2002] derived an expression for the saturation term:

$$\Phi_S\,(\text{kV}) = \frac{1600\sqrt[3]{P_{sw}\,(\text{nPa})}}{\Sigma_P\,(\text{mho})} \quad (5)$$

$\Sigma_P \approx 10$ mho is the effective Pedersen conductance of the polar ionosphere. *Ober et al.* [2003] used ACE data from the magnetic storm of 31 March 2001 to demonstrate that the Siscoe–Hill (S-H) model provides excellent upper-bound envelopes for DMSP measures of Φ_{PC}. Calculations below use $E_{VS} = \Phi_{PC}/2R_E L_Y$ with $L_Y \approx 14.4/\sqrt[6]{P_{sw}\,(\text{nPa})}$.

2.2. Dessler-Parker-Sckopke Relation

Unlike most magnetic indices, Dst was designed to reflect contributions of magnetospheric rather than ionospheric currents to perturbations observed at the Earth's surface [*Mayaud*, 1980]. D-P-S relates Dst to the total energy in the ring current E_{RC}. In this analysis, we use the formula [*Stern*, 2005]

$$E_{RC}\,(\text{joules}) \approx \frac{3}{2}\frac{|Dst^*|}{B_0\,(\text{nT})}E_M \approx 3.87\times10^{13}|Dst^*\,(\text{nT})| \quad (6)$$

B_0 is the equatorial magnetic field at the Earth's surface, and E_M is the energy of the Earth's magnetic field beyond its surface, ~8×10^{17} J. Dst^* includes solar wind pressure corrections for Chapman–Ferraro current contributions to the reported Dst.

2.3. Burton-McPherron-Russell Relation

B-M-R is an empirically based algorithm used to predict the storm time evolution of Dst using solar wind/IMF parameters

$$\frac{dDst^*}{dt} = -aE_y - \frac{Dst^*}{\tau} \quad (7)$$

where

$$Dst^* = Dst - b\sqrt{P_{sw}} - c \quad (8)$$

The terms E_y, τ, P_{sw}, and c represent the dawn–dusk component of the interplanetary electric field, the relaxation timescale, the dynamic pressure of the solar wind, and the quiet-time value of Dst, respectively. *Temerin and Li* [2002, 2006] conducted exhaustive least-square investigations of Dst to reveal three decay scales of ~1, 11, and 160 h. *Vasyliunas* [2006] demonstrated that D-P-S underlies the assumptions of B-M-R. Consistent with equation (6), *Burke et al.* [2007b] showed that the main-phase development of Dst highly correlates ($R > 0.95$) with $\int E_{VS}(t)dt$.

At the Chapman Conference, I argued that following the flow of energy was key to understanding evidence shown in Plate 1. Viewed in the light of D-P-S and B-M-R, the data suggested that during the main phase of storms, some fraction of the ring current's energy content diverts into the ionosphere-thermosphere in the forms of particle precipitation and Poynting flux. The latter contribution is dominant during storms [*Knipp et al.*, 2005]. In this scenario, the ring current acts as a reservoir from which most of the storm time energy budgets of the ionosphere and thermosphere are drawn. Deviations between traces of $\bar{\rho}$ and $-Dst$ during the recovery phase actually support the reservoir hypothesis. *Vasyliunas* [1970] showed that FACs electrically coupling the ring current to the ionosphere are proportional to the cross product of the magnetospheric pressure gradient and the gradient of the flux-tube volume, $j_\parallel \propto \nabla P_{RC} \times \nabla \int ds/B$. The flux tube gradient is mostly in the radially outward direction. During a storm's main phase, the ring current pressure gradient has both radial and azimuthal components. Several hours into recovery, the spatial distribution of ring current particles tends to become symmetric in azimuth. This leads to the weakening and cessation of the $j_\parallel$ that couples the ring current to the ionosphere [*Vasyliunas*, 1970]. Although the symmetric ring current still contains considerable energy, it is electrically decoupled from the ionosphere-thermosphere.

The reservoir hypothesis has other consequences that were not addressed in the Chapman Conference presentation. For example, if the ring current acts as a reservoir from which the thermosphere draws energy, then the increase in thermospheric energy ΔE_{th} should be less than, but of the same order as, the ring current energy increase ΔE_{RC}. The following section proposes a quantitative test of this consequence.

3. ENERGETICS OF THE J77 THERMOSPHERE

Plate 1 shows that while ρ varied rapidly, its orbit-averaged values changed slowly in response to external drivers. Despite strong local dynamics, global-scale changes appear adiabatic, in which case $\bar{\rho}$ never strays far from thermodynamic equilibrium. This perception suggests a certain utility in exploring consequences of treating J77 as though it represented the real thermosphere and using GRACE measurements of $\bar{\rho}$ as a diagnostic for drawing inferences about its energy requirements. The choice of J77 was dictated only by its availability; models such as J70 and MSIS would probably do as well.

J77 consists of a set of stable density and temperature profiles, each uniquely specified by an exospheric temperature T_∞. The modeled atmosphere consists of two diatomic (N_2 and O_2) and four monatomic (O, Ar, He, and H) species (σ). At altitudes ≤90 km, species are well mixed and maintain approximately their ground fractional densities. A mesopause temperature minimum of 188 K is assigned at $h = 90$ km. At higher altitudes all species are in states of diffusive equilibrium specified by temperature profiles $T(r)$ that approach asymptotic values of T_∞. For use in subsequent calculations, we define number density $n(r) = \sum_\sigma n_\sigma(r)$, mean mass $\bar{m}(r) = \frac{1}{n(r)}\sum_\sigma m_\sigma n_\sigma(r)$, mass density $\bar{\rho}(r) = \bar{m}(r)n(r)$, and heat capacity

$$C_V(r) = \frac{k_B A}{n(r)}\left\{\frac{5}{2}(n[N_2] + n[O_2]) + \frac{3}{2}(n[O] + n[Ar] + n[He] + n[H])\right\},$$

where $r = R_E + h$ is the distance from the Earth's center; k_B and A are Boltzmann's constant and Avogadro's number, respectively. *Jacchia* [1977] provides temperature and density profiles associated with a representative number of T_∞ values in tabular form. Assuming diffusive equilibrium, J77 temperature profiles and boundary conditions at $h = 90$ km are sufficient to calculate all of the quantities defined above.

Standard applications of J77 use empirical relationships between T_∞ and external drivers such as the $F_{10.7}$ flux, solar declination, and the magnitude of ap to estimate local mass densities that exert drag on space objects. However, at any given altitude, within J77 tables there exists a unique relationship between ρ and T_∞. Since orbit-averaged values of $\bar{\rho}$ and the altitude of GRACE are known one should be able to use them to determine T_∞. Least-square testing of J77 tables revealed quadratic relationships between T_∞ and $\rho(h)$ that can be represented in the form

$$T_\infty = \sum_{i=0}^{2} a_i(h)\rho^i(h). \tag{9}$$

In the 300- to 500-km altitude range sampled by GRACE and CHAMP during the period of interest, the coefficients $a_i(h)$ are well described by fifth-order polynomials of the form

$$a_i(h) = \sum_{j=0}^{5} b_{ij}h^j. \tag{10}$$

That is

$$\begin{pmatrix} a_0(h) \\ a_1(h) \\ a_2(h) \end{pmatrix}$$

$$= \begin{pmatrix} -28.10 & 2.69 & -2.03 \times 10^{-3} & 0 & 0 & 0 \\ -4.733 \times 10^{17} & 4.312 \times 10^{15} & -1.372 \times 10^{13} & 1.60 \times 10^{10} & 0 & 0 \\ 3.2695 \times 10^{32} & -4.620 \times 10^{30} & 2.618 \times 10^{28} & -7.456 \times 10^{25} & 1.071 \times 10^{23} & -6.237 \times 10^{19} \end{pmatrix}$$

$$\times \begin{pmatrix} 1 \\ h \\ h^2 \\ h^3 \\ h^4 \\ h^5 \end{pmatrix}$$

In calculating T_∞ and related polynomial coefficients, we required that $|R| > 0.999$. Also, ρ and h must be expressed in grams per cubic centimeter and kilometers, respectively.

The total energy density has two contributors, thermal $\eta_T(r) = C_V(r)n(r)T(r)/A$ and gravitation potential $\phi_G(r) = \rho(r)M_E G/r$ [*Wilson et al.*, 2006]. The symbols M_E and G represent the Earth's mass and the gravitational constant, respectively. To estimate the total energy content $E_{th} = H_T + \Phi_G$, it is necessary to integrate $\eta_T(r)$ and $\phi_G(r)$ over the volume of the thermosphere. Calculations performed in equations (11) and (12) ignore local time differences and limit the ranges of integration from selected altitudes $h_0 \geq 90$ km to 1000 km. In this case, the total thermal energy within this altitude range is

$$H_T = 4\pi \int_{R_E+h_0}^{R_E+1000} \eta(r)r^2 dr = \frac{4\pi}{A} \int_{R_E+h_0}^{R_E+1000} C_V(r)n(r)T(r)r^2 dr \quad (11)$$

Since r is usually represented in kilometers, it is necessary to transform $\bar{n}(r)$ into units of #/km³. Since we are only interested in changes in potential energy, it is useful to set the potential energy of the thermosphere to zero at the base of the integration range [*Wilson et al.*, 2006] and represent the total potential energy of thermospheric neutrals as

$$\Phi_G = 4\pi \int_{R_E+h_0}^{R_E+1000} [\phi(r) - \phi(r_0)]r^2 dr$$

$$= 4\pi M_E G \int_{R_E+h_0}^{R_E+1000} \rho(r)\left[\frac{1}{r} - \frac{1}{r_0}\right] r^2 dr \quad (12)$$

With available information, it is possible to obtain E_{th} by adding the integrands of $H_T(h)$ and $\Phi_G(h)$ to J77 tables for 700 K $\leq T_\infty \leq$ 2000 K at increments of 100 K. Ignoring local time variations of $\eta_T(r)$ and $\phi_G(r)$ may appear problematic. However, for polar orbiting satellites like CHAMP and GRACE, this concern is unwarranted. *Jacchia* [1977] showed that the maximum exospheric temperature is ~1.28 times its minimum value. Orbit averaged $T_\infty \approx \frac{1}{2}(T_{\infty max} + T_{\infty min}) \approx 1.14$. $T_{\infty min}$ is not very sensitive to the local time planes of measurements. During 2004 when the orbital planes of CHAMP and GRACE were separated by ~80°, calculated values of T_∞ and E_{th} differed by only 0.4%.

The final steps in this exploration of the J77 thermosphere are to integrate $4\pi\eta(r)r^2$ and $4\pi[\phi(r) - \phi(r_0)]r^2$ from base altitudes of 100 and 200 km to obtain E_{th} as a function of T_∞. The lower limit $h_0 = 100$ km includes E region contributions to the total thermospheric budget; setting $h_0 = 200$ km allows comparisons with *Wilson et al.* [2006].

Plate 2 shows plots of $E_{th}(h \geq 100$ km) and $E_{th}(h \geq 200$ km) as functions of T_∞ and linear regression fits to the J77 "data." In both cases, regression coefficients >0.99 were obtained for the fitted relations:

$$E_{th}(h \geq 100 \text{ km}) = 5.365 \times 10^{17} + 8.727 \times 10^{13} T_\infty \quad (13)$$

and

$$E_{th}(h \geq 100 \text{ km}) = -2.094 \times 10^{16} + 3.362 \times 10^{13} T_\infty \quad (14)$$

A T_∞ change of 100 K corresponds to a thermospheric energy gain/loss of ~8.7 × 10¹⁵ J. J77 places >60% of the energy change at altitudes between 100 and 200 km.

Inferences about energy budgets of the thermosphere and ring current were obtained from GRACE and *Dst* data streams via the following steps:

1. Determine the history of $\bar{\rho}(h,t)$ from GRACE accelerometer measurements.

2. Apply equations (9) and (10) to the $\bar{\rho}(h,t)$ data stream to obtain a time history of $T_\infty(t)$.

3. Use linear relations (13) and (14) to calculate $E_{th}(h_0 = 100$ km$)$ and $E_{th}(h_0 = 200$ km$)$.

4. Specify thermospheric power requirements from time derivatives of E_{th}.

5. Estimate history of E_{RC} from *Dst* and D-P-S relationship.

6. Specify magnetospheric power requirements from time derivatives of E_{RC}.

Data plotted in Figure 1a show orbit-averaged mass densities measured by GRACE between J.D. 150 (1 June) and J.D. 230 (17 August) 2004. While the mean altitude remained near 486.5 km, the orbital plane rotated from the noon–midnight to the dawn–dusk meridian over the course of the interval. Figure 1b is a trace of $T_\infty(t)$ obtained by an application of equation (9) to the $\bar{\rho}(t)$ stream. It shows T_∞ ranging between ~750 and 1200 K. Figure 1c shows daily (solid line) and 81-day averaged values of $F_{10.7}$ (dot–dash line). Figure 1d shows hourly averaged E_{VS} calculated with measurements of P_{sw} and the IMF from ACE. Two timescales of changes appear in the $\bar{\rho}(t)$ trace: (1) low-frequency variations that follow changes in $F_{10.7}$ with the 27 day solar-rotation period and (2) spikelike features that occur in response to changes in the interplanetary medium. Vertical lines superposed on Figure 1 highlight rapid increases/decreases in $\bar{\rho}(t)$ and T_∞, which correlate with turn on/off features of E_{VS}. The large structures observed between days 204 and 211 are discussed below.

Figure 2 shows the total energy content of the thermosphere and its power requirements for the same period. Again we find the solar rotation and spiked features in the E_{th} trace. Whereas the amplitude of solar rotation variations is on the order of 5×10^{15} J, the largest change in this data stream, $\sim 3 \times 10^{16}$ J, occurred on day 209 in response to interplanetary forcing. When the integration was limited to altitudes above 200 km, $\Delta E_{th} \approx 10^{16}$ J, a value that falls within the range of storms studied by *Wilson et al.* [2006]. Caution must be exercised in looking at the $P_{th}(t)$ trace, which represents the changes in energy input required by orbit-to-orbit changes in E_{th}. Clearly, UV radiation from the Sun is required to maintain the thermosphere against radiative losses [*Knipp et al.*, 2005]. Thus, the plot in Figure 2 (bottom) reflects changes up or down from this baseline. Spiked features in $\bar{\rho}(t)$ require substantial power changes above the baseline. On J.D. 210, a power surge of ~1500 GW was followed by a period of radiative losses at a rate of ~1000 GW.

Figure 3 shows traces of –*Dst* (nT), the ring current energy E_{RC} (J), and power requirements P_{RC} (GW) = dE_{RC}/dt. This example does not include P_{sw} corrections to *Dst* and sets $E_{RC} = 0$ for *Dst* ≥ 0. The –*Dst* trace shows that outside the J.D. 204–210 interval geomagnetic conditions were relatively quiet. The largest E_{RC} excursion of $\sim 7 \times 10^{15}$ J is substantially less than the $\sim 3 \times 10^{16}$ required by J77 to explain the largest $\bar{\rho}(t)$ increase observed by GRACE. Data traces plotted in Figure 4 allow easier comparisons of $E_{th}(t)$ with the development of E_{RC}, –*Dst*, and E_{VS} during the period of geomagnetic disturbance. Attention is directed to four aspects of the data in Figure 4:

1. The high density of E_{VS} points reflects its higher sampling resolution (4 min) in comparison with *Dst* and $\bar{\rho}$. Nonetheless, relatively fast E_{VS} excursions, such as that seen at the end of day 208, produced a noticeable increase in *Dst* and E_{RC}.

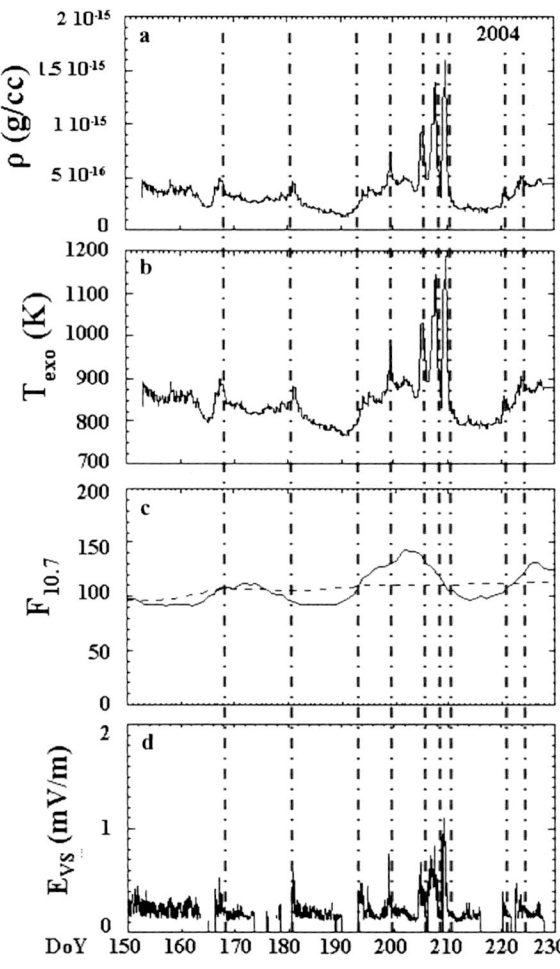

Figure 1. Traces of orbit-averaged mass densities from (a) GRACE, (b) exospheric temperatures inferred from J77, (c) $F_{10.7}$, and (d) magnetospheric electric fields determined from ACE observations between J.D. 150 and 230 in 2004. The vertical lines are guides for identifying coincidences between spikelike thermospheric and interplanetary driving.

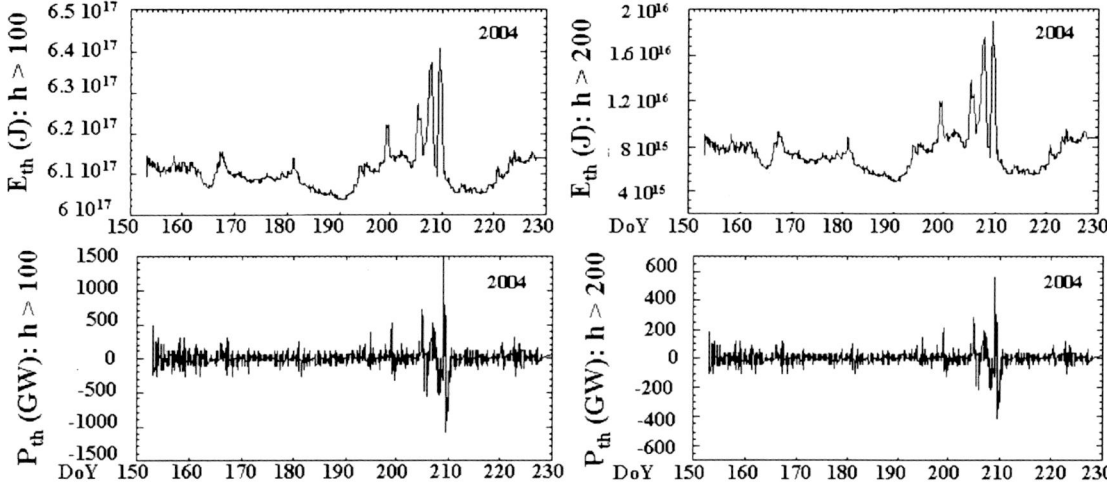

Figure 2. Total thermospheric energy E_{th} required by J77 for J.D. 150–230 in 2004 at altitudes above 100 (left) and 200 km (right). Corresponding thermospheric powers P_{th} in GW (bottom) were obtained by differentiating E_{th} with respect to time.

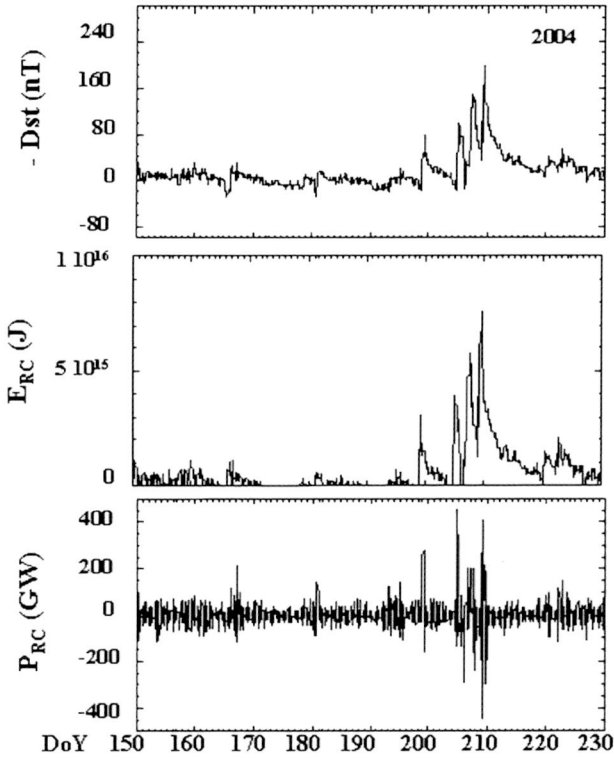

Figure 3. (top) Provisional $-Dst$ for J.D. 150–230 in 2004. (middle) Ring current energy E_{RC} was obtained using the Dessler-Parker-Sckopke theorem. (bottom) Corresponding ring current powers P_{RC} were obtained by differentiating E_{RC} with respect to time.

2. Despite obvious differences in the magnitudes of ΔE_{th} and ΔE_{RC}, they appear to scale proportionately to ring current intensifications.

3. The two vertical (dot–dash) lines mark rapid downturns as E_{VS} returned to background values. These downturns are accompanied by significant shifts in the slopes of E_{th} and E_{RC} (dashed lines). The slopes in the E_{th} trace at these times indicate the relaxation time constant is ~13 h. Recall that Plate 1 shows similar relaxation timescales during the magnetic storms of November 2004.

4. Recovery slopes for E_{RC} differ from each other and those of E_{th}. This reflects the slower loss rates for ring current ions via charge exchange than radiative losses as the thermosphere relaxes toward quiet conditions [*Mlynczak et al.*, 2005].

4. CONCLUSIONS AND RECOMMENDATIONS

During the main phases of the November 2004 magnetic storms accelerometers on GRACE measured increased atmospheric drag effects. Thermospheric densities inferred from equation (1) exceeded predictions of operational models by a factor of two or more [*Burke et al.*, 2007a]. However, the *Dst* index and E_{VS} followed or anticipated $\bar{\rho}(t)$ variations throughout the main and early recovery phases. Since DMSP spacecraft observed large energy depositions near the equatorward boundary of the auroral oval, *Burke et al.* [2007a] suggested that the ring current acts as a reservoir

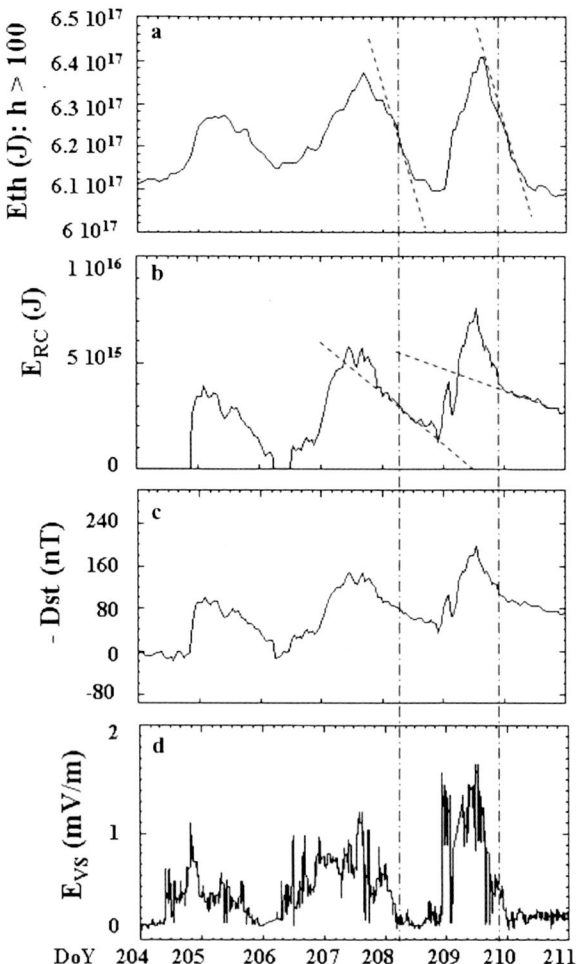

Figure 4. Traces of (a) E_{th}, (b) E_{RC}, (c) $-Dst$, and (d) E_{VS} for J.D. 204–211 in 2004. Vertical lines correspond to rapid decreases in E_{VS}. Slanted dashed lines indicate relaxation rates for E_{th} and E_{RC} after the driving electric field terminated.

and E_{th}. If we approximate the thermosphere's quiet-time baseline energy as 6.1×10^{17} J, then ΔE_{th} reached maxima of $\sim 1.8 \times 10^{16}$, 2.6×10^{16}, and 3.2×10^{16} J, respectively. Values of ΔE_{th} that are four times larger than E_{RC} maxima estimated using equation (6) are consistent with results reported by Knipp et al. [1998] and Lu et al. [1998]. As indicated above, if the simple ring current reservoir conjectures were correct, we anticipated that $E_{RC} > E_{th}$. However, the obvious correlation between variations of the two quantities (Figure 4) cautions that we consider the ring current's relationship with the ionosphere-thermosphere more carefully. Perhaps we should either regard the ring current as an energy conduit rather than a reservoir or both Dst and E_{th} as responding to the same electric field driver. Our reconsideration should begin with a critical review of assumptions about mechanisms by which energy reaches the thermosphere, as well as the accuracy of (1) GRACE estimates of $\bar{\rho}$, (2) J77's ρ and T profiles, and (3) the E_{RC} approximation encapsulated in equation (6). A few final comments appear useful.

4.1. Locus of Energy Transfer

The observed correlation between Dst and $\bar{\rho}(t)$ suggested that most of the energy transferred to the ionosphere-thermosphere occurs at auroral latitudes via particle precipitation and Poynting flux from the plasma sheet/ring current. However, the correlation may only point to a common cause whose dynamics are reflected in the histories of both parameters. The bottom trace in Figure 4 indicates that the interplanetary and magnetospheric electric fields act in just this way and suggests that some fraction of the thermosphere's energy budget was deposited in the polar cap and was communicated via collisions between anti-sunward drifting ions and ambient neutrals. The rate of energy transfer is $\Phi_{PC} \cdot I_P$, where I_p is the total Pedersen current across the polar cap. ACE data and equations (3)–(5) indicate that Φ_{PC} reached maxima of ~ 100, 150, and 200 kV during the three activations shown in Figure 4. Time profiles of Φ_{PC} resemble those of $E_{VS}(t)$. As an example, let us consider the event of day 209 whose main phase lasted ~ 8 h and assume that $I_p \approx 2$ MA. The rate of energy deposition would be ~ 400 GW and the total deposited energy 1.15×10^{16} J. We note that in modeling the dynamics of the July 2000 magnetic storm, Zhang et al. [2005] placed the largest energy deposition near the cusp–mantle part of the polar cap. Some fraction reaches the thermosphere in the form of anti-sunward neutral winds and by frictional heating. The only point to be made here is that the storm time energy entering the polar cap is of the same order as ΔE_{th} and must be considered in future investigations.

from which the ionosphere and thermosphere draw their storm time energy budgets. This paper outlined a method to test this hypothesis by using $\bar{\rho}(t)$ inputs from GRACE to calculate energy requirements of J77 during an 80-day period in the summer of 2004. The first law of thermodynamics stipulates that the total energy gained ΔE_{th} has two components, changes in internal energy ΔH_T and work against gravity $\Delta \Phi_G$. Equations (9) and (10) allowed calculations of $T_\infty(t)$, which were then transposed into the traces of $E_{th}(t)$ in Figures 2 and 4.

The period 22–28 July 2004 was marked by three disturbances that produced increases in $|Dst|$ of ~ 100, 150, and 200 nT, accompanied by proportionate increases in E_{RC}

4.2. Accuracy of GRACE Estimates

Calculating ρ from a_{drag} measurements is not a trivial exercise [*Bruinsma and Biancale*, 2003]. *Marcos* (private communication, 2007) compared GRACE measurements of ρ with those obtained from atmospheric drag exerted on spherical reference satellites. During 2004, he found that densities derived from GRACE accelerometer data were systematically too high by a factor of ~1.5. The nonlinear ρ–T_∞ relation (9) suggests that this overestimate may have a substantial impact on calculations of E_{th}. I recalculated Figure 4d with $\bar{\rho}(t)$ reduced by a factor of 1.5. Results (not shown) indicate that the baseline decreased to $E_{th} \approx 6.05 \times 10^{16}$ J and ΔE_{th} associated with *Dst* extrema were reduced by ~25%.

4.3. Accuracy of J77 ρ–T_∞ Profiles

If the method outlined here has utility, then the built-in physics and chemistry of the J77 model merits further investigation. The present exercise can be repeated using the ρ–T_∞ profiles of NRLMSISE-00 [*Picone et al.*, 2002] to specify the energy requirements of different models. However, ΔE_{th} ($h \geq 200$ km) excursions are similar in magnitude to the HASDM-based calculations of *Wilson et al.* [2006].

4.4. Accuracy of the D-P-S Estimate of E_{RC}

Sources of *Dst* and implications of D-P-S for understanding the energy of magnetospheric current systems are of fundamental concern. *Tsyganenko and Sitov* [2005] demonstrated that during the main phases of large storms, the tail and asymmetric ring currents are the dominant contributors to *Dst*. Only during recovery does the symmetric ring current dominate. These conclusions flow from an analysis of magnetic field and current distributions in storm time models [*Tsyganenko et al.*, 2003]. Perhaps the force balance $\nabla p = j \times B$ requirements of the model would allow better estimates of the total energy of magnetospheric particles referred to here as E_{RC}.

Acknowledgments. The presented work was supported by AFOSR Task 2311SDA3 and reflects many discussions with AFRL colleagues Cheryl Huang, Frank Marcos, Gordon Wilson, and John Wise. I am grateful to Louise Gentile of Boston College for editing the manuscript.

REFERENCES

Bowman, B. R. (2002), True satellite ballistic coefficient determination for HASDM, AIAA-2002-4887, *AAS/AIAA Astrodynamics Specialist Conference*, Monterey, Calif.

Bowman, B. R., W. K. Tobiska, and F. A. Marcos (2006), A new empirical thermospheric density model JB2006 using new solar indices, AIAA 2006-6166, *AIAA Astrodynamics Conference*, Keystone, Colo.

Bruinsma, S., and R. Biancale (2003), Total densities derived from accelerometer data, *J. Spacecr. Rockets*, *40*, 230.

Bruinsma, S., D. Tamagnan, and R. Biancale (2004), Atmospheric densities derived from CHAMP/STAR accelerometer observations, *Planet. Space Sci.*, *52*, 297.

Bruinsma, S., J. M. Forbes, R. S. Nerem, and X. Zhang (2006), Thermospheric density response to the 20–21 November 2003 solar and geomagnetic storm from CHAMP and GRACE accelerometer data, *J. Geophys. Res.*, *111*, A06303, doi:10.1029/2005JA011284.

Burke, W. J. (2007), Penetration electric fields: A Volland–Stern approach, *J. Atmos. Sol. Terr. Phys.*, *69*, 1114.

Burke, W. J., D. R. Weimer, and N. C. Maynard (1999), Geoeffective interplanetary scale sizes derived from regression analysis of polar cap potentials, *J. Geophys. Res.*, *104*, 9989.

Burke, W. J., C. Y. Huang, F. A. Marcos, and J. O. Wise (2007a), Interplanetary control of thermospheric densities during large magnetic storms, *J. Atmos. Sol. Terr. Phys.*, *69*, 279.

Burke, W. J., L. C. Gentile, and C. Y. Huang (2007b), Penetration electric fields driving main phase *Dst*, *J. Geophys. Res.*, *112*, A07208, doi:10.1029/2006JA012137.

Burton, R. K., R. L. McPherron, and C. T. Russell (1975), An empirical relationship between interplanetary conditions and *Dst*, *J. Geophys. Res.*, *80*, 4204.

Carovillano, R. L., and G. L. Siscoe (1973), Energy and momentum theorems in magnetospheric processes, *Rev. Geophys.*, *11*, 289.

Casali, S. J., W. N. Barker, and M. F. Storz (2002), Dynamic calibration atmosphere (DCA) tool for the high accuracy satellite drag model (HASDM), AIAA 2002-4888, *AAS/AIAA Astrodynamics Specialist Conference*, Monterey, Calif.

Chen, A. J., and R. A. Wolf (1972), Effect on the plasmasphere of a time-varying convection electric field, *Planet. Space Sci.*, *20*, 483.

Ejiri, M. (1978), Trajectory traces of charged particles in the magnetosphere, *J. Geophys. Res.*, *83*, 4798.

Forbes, J. M., G. Lu, S. Bruinsma, R. S. Nerem, and X. Zhang (2005), Thermosphere density variations due to the 15–24 April 2002 solar events from CHAMP/STAR accelerometer measurements, *J. Geophys. Res.*, *110*, A12S27, doi:10.1029/2004JA010856.

Fukushima, N. (1969), Equivalence in ground geomagnetic effects of Chapman–Vestine's and Birkeland–Alfvén's electric current systems for polar magnetic storms, *Rep. Ionos. Space Res. Jpn.*, *23*, 219.

Hardy, D. A., E. G. Holeman, W. J. Burke, L. C. Gentile, and K. H. Bounar (2008), Probability distributions of electron precipitation at high magnetic latitudes, *J. Geophys. Res.*, *113*, A06305, doi:10.1029/2007JA012746.

Hedin, A. E., C. A. Reber, G. P. Newton, N. W. Spencer, H. C. Brinton, H. G. Mayr, and W. E. Potter (1977), A global thermospheric model based on mass spectrometer and incoherent scatter data MSIS, 2, composition, *J. Geophys. Res.*, *82*, 2148.

Hill, T. W. (1984), Magnetic coupling between solar wind and magnetosphere: Regulated by ionospheric conductance, *Eos Trans. AGU, 65*, 1047.

Huang, C. Y., and W. J. Burke (2004), Transient sheets of field-aligned current observed by DMSP during the main phase of a magnetic superstorm, *J. Geophys. Res., 109*, A06303, doi:10.1029/2003JA010067.

Jacchia, L. G. (1964), Static diffusion models of the upper atmosphere with empirical temperature profiles, *SAO Special Report 170*.

Jacchia, L. G. (1970), Revised static models of the thermosphere and exosphere with empirical temperature profiles, *SAO Special Report 313*.

Jacchia, L. G. (1977), Thermospheric temperature, density and composition: A new model, *SAO Special Report 375*.

Kavanagh, L. D., Jr., J. W. Freeman Jr., and A. J. Chen (1968), Plasma flow in the magnetosphere, *J. Geophys. Res., 73*, 5511.

Knipp, D. J., et al. (1998), An overview of the early November 1993 geomagnetic storm, *J. Geophys. Res., 103*, 26,197.

Knipp, D. J., W. K. Tobiska, and B. A. Emery (2005), Direct and indirect thermospheric heating sources for solar cycles 21–23, *Sol. Phys., 224*, 495.

Lu, G., et al. (1998), Global energy deposition during the January 1997 magnetic cloud event, *J. Geophys. Res., 103*, 11,685.

Marcos, F. A., B. R. Bowman, and R. E. Sheehan (2006), Statistical evaluation of the Jacchia–Bowman 2006 thermospheric neutral density model, AIAA 2006-6166, *AIAA Astrodynamics Conference*, Keystone, Colo.

Mayaud, P. N. (1980), *Derivation, Meaning and Use of Geomagnetic Indices*, pp. 47–48, AGU, Washington, D. C.

Mlynczak, M. G., et al. (2005), Energy transport in the thermosphere during the solar storms of April 2002, *J. Geophys. Res., 110*, A12S25, doi:1029/2005JA011141.

Ober, D. M., N. C. Maynard, and W. J. Burke (2003), Testing the Hill model of transpolar potential saturation, *J. Geophys. Res., 108*(A12), 1467, doi:10.1029/2003JA010154.

Owens, J., and W. Vaughan (2004), Semi-empirical thermospheric modeling: The New NASA Marshall engineering thermosphere model—version 2.0 (MET-V2.0), *35th COSPAR Scientific Assembly*, 2708.

Picone, J. M., A. E. Hedin, D. P. Drob, and A. C. Aiken (2002), NRLMSISE-00 empirical model of the atmosphere: Statistical comparisons and scientific issues, *J. Geophys. Res., 107*(A12), 1468, doi:10.1029/2002JA009430.

Roelof, E. C., and D. G. Sibeck (1993), Magnetopause shape as a bivariate function of interplanetary magnetic field B_z and solar wind dynamic pressure, *J. Geophys. Res., 98*, 21,241.

Siscoe, G. L., G. M. Erickson, B. U. Ö. Sonnerup, N. C. Maynard, J. A. Schoendorf, K. D. Siebert, D. R. Weimer, W. W. White, and G. R. Wilson (2002), Hill model of transpolar potential saturation: Comparison with MHD simulation, *J. Geophys. Res., 107*(A6), 1075, doi:10.1029/2001JA000109.

Southwood, D. J., and S. M. Kaye (1979), Drift boundary approximations in simple magnetospheric convection models, *J. Geophys. Res., 84*, 5773.

Stern, D. P. (1975), The motion of a proton in the equatorial magnetosphere, *J. Geophys. Res., 80*, 595.

Stern, D. P. (2005), A historical introduction to the ring current, in *The Inner Magnetosphere: Physics and Modeling, Geophys. Monogr. Ser.*, vol. 155, p. 1, AGU, Washington, D. C., doi:10.1029/155GM01.

Sutton, E. K., J. M. Forbes, and R. S. Nerem (2005), Global thermospheric neutral density and wind response to the severe 2003 geomagnetic storms from CHAMP accelerometer data, *J. Geophys. Res., 110*, A09S40, doi:10.1029/2004JA010985.

Temerin, M., and X. Li (2002), A new model for the prediction of Dst on the basis of the solar wind, *J. Geophys. Res., 107*(A12), 1472, doi:10.1029/2001JA007532.

Temerin, M., and X. Li (2006), Dst model for 1995–2002, *J. Geophys. Res., 111*, A04221, doi:10.1029/2005JA011257.

Tsyganenko, N. A., and M. I. Sitnov (2005), Modeling the dynamics of the inner magnetosphere during strong geomagnetic storms, *J. Geophys. Res., 100*, A03208, doi:10.1029/2004JA010798.

Tsyganenko, N. A., H. J. Singer, and J. C. Kasper (2003), Storm-time distortion of the inner magnetosphere: How severe can it get?, *J. Geophys. Res., 108*(A5), 1209, doi:10.1029/2001JA009808.

Vasyliunas, V. M. (1970), Mathematical models of magnetospheric convection and its coupling to the ionosphere, in *Particles and Fields in the Magnetosphere*, edited by B. M. McCormac, pp. 60–71, Springer, Dordrecht.

Vasyliunas, V. M. (2006), Reinterpreting the Burton-McPherron-Russell equation for predicting Dst, *J. Geophys. Res., 111*, A07S04, doi:10.1029/2005JA011440.

Volland, H. (1973), A semiempirical model of large-scale magnetospheric electric fields, *J. Geophys. Res., 78*, 171.

Weimer, D. R. (2005), Improved ionospheric electrodynamic models and application to calculating Joule heating rates, *J. Geophys. Res., 110*, A05306, doi:10.1029/2005JA010884.

Wilson, G. R., D. R. Weimer, J. O. Wise, and F. A. Marcos (2006), Response of the thermosphere to Joule heating and particle precipitation, *J. Geophys. Res., 111*, A10314, doi:10.1029/2005JA011274.

Zhang, X. X., C. Wang, T. Chen, Y. L. Wang, A. Tan, T. S. Wu, G. A. Germany, and W. Wang (2005), Global patterns of Joule heating in the high-latitude ionosphere, *J. Geophys. Res., 110*, A12208, doi:10.1029/2005JA011222.

W. J. Burke, Air Force Research Laboratory Space Vehicles Directorate, Hanscom AFB, MA 01731, USA.

Sources of *F*-Region Height Changes During Geomagnetic Storms at Mid Latitudes

Mariangel Fedrizzi, T. J. Fuller-Rowell, and Naomi Maruyama

Cooperative Institute for Research in Environmental Sciences, University of Colorado, Boulder, Colorado, USA

Mihail Codrescu and Hargobind Khalsa

Space Weather Prediction Center, NOAA, Boulder, Colorado, USA

The increased high-latitude energy input in the thermosphere during geomagnetic storms, mainly resulting from Joule heating, causes the atmosphere to heat and expand. The heating at high latitudes drives a global wind surge that propagates from both polar regions to low latitudes and into the opposite hemisphere. Those winds are driven by pressure inequalities due to temperature differences between high latitudes and equatorial regions. To balance the divergence or convergence produced by large-scale horizontal wind systems, vertical motions of air must occur in the thermosphere. The vertical motion of the thermosphere due to the vertical wind velocity can be represented as the sum of the "divergence" and the "barometric" components. The divergence velocity (W_D) component of the vertical wind, so called because it arises from the divergence of the horizontal winds, represents the flow "across" pressure surfaces. Conversely, the convergent horizontal wind is associated with a downward divergence wind. The circulation is closed by a return flow in the lower thermosphere. The expansion of a fixed pressure level atmospheric parcel by the heating drives the second component of the vertical wind, the so-called barometric velocity (W_B). The barometric component represents the rise and fall of constant pressure levels due to thermal expansion or contraction. Barometric winds are therefore related to the thermal expansion of the atmosphere, whereas vertical divergence winds are associated with the conservation of mass relative to fixed pressure levels. The F_2 layer height can change during geomagnetic storms both from the change in horizontal winds pushing plasma parallel to the inclined magnetic field, and due to vertical winds from the thermal expansion of the neutral atmosphere. In this paper, numerical experiments are conducted using a global, three-dimensional, time-dependent, nonlinear coupled model of the thermosphere, ionosphere, plasmasphere, and electrodynamics to quantify the impact of the horizontal thermospheric wind and the thermal expansion on changes in the F_2 layer peak height (h_mF_2). The results demonstrate that height changes in the neutral atmosphere from thermal expansion are clearly reflected in the changes of h_mF_2. Comparisons between model results and mid-latitude ionosonde observations are carried out for the magnetic storm events on 31 March 2001 and 17 April 2002. The analysis of the horizontal thermospheric wind and thermal expansion's relative role during the 31 March 2001 event reveals that both processes contribute significantly to the F-region height changes. The relative importance of those physical mechanisms depends on the local time at the storm commencement, the spatial distribution of the energy input over the poles, and the storm development and recovery duration.

1. INTRODUCTION

Although it is well known that the F_2 layer height changes significantly during geomagnetic storms, the relative contribution of the physical mechanisms causing the change has not previously been quantified. In the past, much attention has focused on the effect of the horizontal wind on the height of the F_2 layer peak (h_mF_2). In this paper, attention is given to the effect of the vertical wind, in particular the barometric or thermal expansion component.

The identification, understanding, and quantification of each physical mechanism's role in the thermospheric–ionospheric (T–I) response during magnetic storms is only feasible when a physically based model that captures all the relevant processes is used. In this study, a global, three-dimensional, time-dependent, nonlinear coupled model of the thermosphere, ionosphere, plasmasphere, and electrodynamics (CTIPe) [*Fuller-Rowell et al.*, 1996; *Millward et al.*, 1996] is used to examine the role of the thermal expansion and the horizontal thermospheric wind in the T–I response during magnetically disturbed conditions. As part of this study, numerical experiments are conducted using CTIPe model to investigate the effect of thermal expansion on the F_2 layer peak height. Model results are compared to ionosonde observations during the 31 March 2001 and 17 April 2002 magnetic storm events. The contribution of the horizontal thermospheric wind and the thermal expansion is examined for two mid-latitude ionosonde stations during the 31 March 2001 storm.

In the next sections of this paper, the following aspects will be addressed: the thermospheric circulation and its effects during quiet and storm conditions are summarized in Section 2; a brief description of CTIPe is presented in Section 3; in Section 4, comparisons between CTIPe results and ionosonde observations during the 17 April 2002 magnetic storm (Section 4.1), a numerical experiment showing the effects of thermal expansion on h_mF_2 (Section 4.2), and the relative contribution of the horizontal thermospheric wind and the thermal expansion at two mid-latitude stations during the 31 March 2001 storm event (Section 4.3) are presented; the conclusions are stated in Section 5.

2. THERMOSPHERIC CIRCULATION

The dominant mechanism driving winds in the thermosphere is the diurnal variation in the absorption of solar UV radiation, which heats and expands the dayside thermosphere, setting up horizontal pressure gradients. The winds driven by those pressure gradients blow away from the hottest part of the thermosphere, which is in the afternoon sector, toward the coldest part in the early morning sector. They therefore blow across the polar regions and zonally around the Earth at low latitudes [*Rishbeth*, 1972]. Because of the atmosphere's much greater horizontal than vertical extent, the motions are constrained to be predominantly horizontal; vertical winds are typically only of the order of 1% the horizontal wind magnitude [*Richmond*, 1995].

Like the winds in the lower atmosphere, thermospheric winds are influenced by the Coriolis force due to the Earth's rotation. In addition, they are influenced by frictional forces due to the viscosity of the air and to collisions between the neutral air particles and the positive ions. At mid latitudes, the ions exert a drag on the air because their motion is strongly impeded by the Earth's magnetic field, so they cannot be freely blown along by the wind. The wind, however, can move the ions and electrons in the direction of the geomagnetic field. If the field lines are inclined, this ion motion has a vertical component that can affect the ion and electron concentration, mainly because the loss coefficient is very height-dependent. The effect of the wind depends on its orientation with respect to the geomagnetic field: poleward wind (which mainly occurs by day) causes downward drift and tends to reduce the ion concentration, while equatorward wind (which occurs mainly at night) causes upward drift and tends to increase the ion concentration. These effects, being dependent on the geometry of the geomagnetic field, vary with latitude and with magnetic declination [*Rishbeth*, 1972].

To balance the divergence or convergence produced by large-scale horizontal wind systems, vertical motions of air must occur in the thermosphere [*Rishbeth et al.*, 1969]. The vertical motion of the thermosphere due to the vertical wind velocity can be represented as the sum of the "barometric" and the "divergence" components $U_Z = W_B + W_D$, where $W_B = (\partial h/\partial t)_p$ and $W_D = -dp/\rho g dt$, and h represents real height in km, p is total pressure, ρ corresponds to the air density, and g is the gravitational acceleration [*Rishbeth and Müller-Wodarg*, 1999]. The barometric component represents the rise and fall of constant pressure levels due to thermal expansion or contraction. In this one-dimensional motion, the expansion or contraction causes only vertical up or down winds (with no horizontal flow), and the air does not move with respect to the pressure levels. The divergence wind, so called because it arises from the divergence of the horizontal wind, represents the flow "across" pressure surfaces. The real three-dimensional thermosphere contains both vertical component and horizontal motions. Neutral air is neither produced nor destroyed, so horizontal divergence winds must be accompanied by upward W_D and horizontal convergence winds by downward W_D. The circulation is completed by a return flow at lower levels in the thermosphere, with much smaller wind speeds corresponding to the higher gas density. According to simulations done by *Rishbeth and Müller-Wodarg* [1999], W_D is upward by day and downward at night, while W_B is upward in the morning when the thermosphere is heating up and expanding, and downward in the evening and night when the thermosphere is cooling and contracting.

Variations in solar or geomagnetic activity also cause the thermosphere to expand and contract. The atmospheric heating and expansion during increased solar activity raises the pressure levels to greater heights and thereby raises the altitude of ion production [*Rishbeth and Edwards*, 1989; *Clilverd et al.*, 2003]. Early drag research on artificial satellites established the fact that the temperature and density of the upper atmosphere increase during geomagnetic storms [*Jacchia et al.*, 1967, and references therein], modifying the dynamics of the quiet-time thermosphere and ionosphere. During magnetically disturbed periods, high-latitude convection electric fields and ionospheric currents heat and expand the polar thermosphere. The thermal expansion causes an apparent composition change at a fixed height that would be seen by a satellite, but in itself does not cause neutral composition changes on a pressure level. Pressure levels are important because they represent the levels of constant optical depth at a given solar zenith angle, so the level that ionization from solar radiation or particle precipitation remains fixed. The uneven expansion of the thermosphere due to the high-latitude heating produces pressure gradients that generate strong neutral winds flowing equatorward. It is the divergent effect of these winds that drives an upwelling of molecular rich thermospheric gas from lower altitudes, which enhances the molecular species in the upper thermosphere and increases the chemical loss rates of the ions [*Fuller-Rowell et al.*, 1997]. The disturbed thermospheric circulation modifies the neutral composition and moves the plasma along the magnetic field lines, changing rates of production and recombination of the ionized species at mid and low latitudes. The change in circulation also drives downwelling at low latitudes, which decreases the molecular species, and causes a slight positive ionospheric phase. The storm-time thermospheric winds under the influence of the geomagnetic field and the action of the Coriolis force produce polarization electric fields through the dynamo effect modifying the electrodynamics and the structure of the Earth's upper atmosphere [*Buonsanto*, 1999]. This paper will not focus on the generation or impact of electric fields at mid latitudes.

Excellent articles and reviews on the thermospheric circulation are given by *Garriott and Rishbeth* [1963], *Rishbeth et al.* [1969], *Rishbeth* [1972], *Richmond* [1995], *Titheridge* [1995], *Rishbeth* [1998], and *Rishbeth and Müller-Wodarg* [1999]. Reviews on the thermospheric circulation response to magnetic storms can be found in *Rishbeth* [1975], *Prölss* [1995], *Fuller-Rowell et al.* [1997], *Prölss* [1997], and *Buonsanto* [1999].

3. CTIPe MODEL

The global, three-dimensional, time-dependent, nonlinear CTIPe is a self-consistent model and solves the momentum, energy, and composition equations for the neutral and ionized atmosphere. The model has evolved over the past 25 years and it is currently a combination of four components. The first component is a global thermosphere model, which solves the momentum, energy, and composition equations for the neutral atmosphere, covering the altitude range of 80 km to about 500 or 600 km. The neutral winds, temperatures, and composition are solved self-consistently with the ionosphere, the plasma component of the upper atmosphere. The second component is a mid- and high-latitude ionospheric convection model, and the third component includes the plasmasphere and the low-latitude ionosphere. In this way, the ionosphere is divided in two parts: at high latitude it uses open flux tubes from 80 to 10,000 km, and at mid and low latitudes it assumes the closed flux tubes to allow for plasma to be transported between the hemispheres. Finally, as the fourth component, the electrodynamics at mid and low latitudes is solved using conductivities from the ionospheric models and neutral winds from the neutral atmosphere code. These parts are all solved self-consistently.

The CTIPe model requires a few external drives, such as the solar UV and EUV, which heats, ionizes, and dissociates the atmosphere. At high latitudes, it needs the magnetospheric sources—the magnetospheric convection electric field is imposed using the Weimer model and the auroral precipitation comes from an empirical model based on the TIROS/NOAA series of satellites. Finally, it requires tidal forcing from the lower atmosphere. The global atmosphere in CTIPe is divided into a series of elements in geographic latitude, longitude, and pressure. The latitude resolution is 2°, the longitude resolution is 18°, and each longitude slice sweeps through local time with a 1-min time step. In the vertical direction, the atmosphere is divided into 15 levels in logarithm of pressure from a lower boundary of 1 Pa at 80 km to more than 500 km altitude [*Fuller-Rowell et al.*, 2002]. The height of the peak in the F_2 layer tends to lie at a fixed pressure level in the atmosphere in the absence of winds. In CTIPe, the mean h_mF_2 corresponds approximately to the pressure level 12. A detailed description of CTIPe can be found in *Millward et al.* [2001], *Fuller-Rowell et al.* [2002], and *Maruyama et al.* [2007].

4. RESULTS

In this section, the validity of CTIPe model is demonstrated by comparing its simulation results with ionosonde observations, including two magnetic storm events differing in magnitude. The comparisons for the moderately intense 17 April 2002 storm are presented in Section 4.1. A numerical experiment showing the effects of thermal expansion on the F_2 layer peak height is described in Section 4.2. In Section 4.3, further comparisons between model and observations, and the relative impact of the horizontal thermospheric wind and the thermal expansion on h_mF_2 at two mid-latitude locations during the very intense 31 March 2001 storm event are presented.

4.1. Model and Observation Comparison During the 17 April 2002 Magnetic Storm

The April 2002 storm event was not one of the most severe magnetic storms that occurred during the Solar Cycle 23. Nevertheless, large coronal mass ejections (CMEs) and one of the largest solar flares of the current solar cycle impacted the Earth's space environment during the storm period. Despite its moderate intensity, the April 2002 magnetic storm was a long-lasting storm that started on April 17 and persisted for more than 5 days. A summary of the magnetic conditions for this period is presented in Figure 1. The north–south component of the interplanetary magnetic field (IMF Bz) is displayed on the top panel. In the middle panel, the

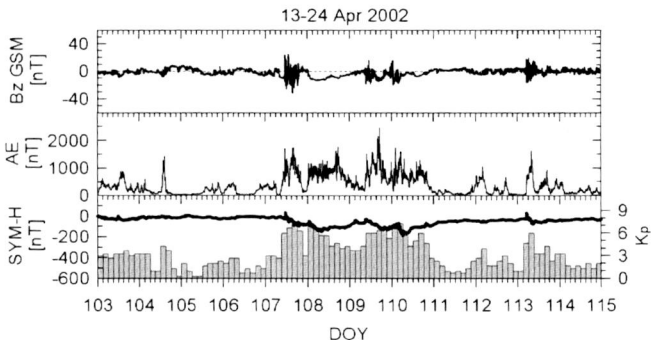

Figure 1. Interplanetary magnetic field (IMF) Bz component of the IMF in geocentric solar magnetospheric (GSM) coordinates provided by the Advanced Composition Explorer (ACE) spacecraft for 13–24 April 2002 (top), Quick-Look *AE* index (middle), and *SYM-H* and *Kp* indices for the same period (bottom). In the top panel, the IMF Bz propagation time from ACE to the subsolar magnetopause was taken into account.

auroral electrojet (*AE*) index shows that a significant amount of energy was injected at high latitudes during the storm period. In the bottom panel, the longitudinally symmetric (*SYM-H*) disturbance index initiated its negative incursion after 1200 UT on 17 April, reaching the minimum value of −185 nT at 0601 UT on 20 April. The *SYM-H* index has 1-min resolution and essentially follows the same variations as the disturbance storm-time (*Dst*) index [*Iyemori et al.*, 2008]. The planetary index *Kp* is presented in the bottom panel. Solar radio flux *F*10.7 cm ranges from 170 to 226 (10^{-22} J s^{-1} m^{-2} Hz^{-1}), according to Table 1.

To evaluate the realism of the physical processes in CTIPe model, its results are compared with observations from worldwide mid-latitude ionosonde stations during the period 13–24 April 2002. An example is presented in Figure 2 for Chilton (geographic coordinates, 51.5°N, 0.6°W; decl., −3.5°; dip, 66.2°). The electron density at the peak of the F_2 layer (N_mF_2) is shown in Figure 2a. The solid thick curve corresponds to CTIPe simulations. Results obtained from the empirical model International Reference Ionosphere (IRI-2001) are also shown for comparison purposes (dashed line). The agreement between model results and ionosonde observations is excellent for some days, but not so good for others. CTIPe captures the negative phase of the storm very well, but does not capture the occasional large positive phases, particularly during the extended period of reduced electron density between day of year (DOY) 108 and 110. The discrepancies between model and observations are possible due to the lack of penetration electric fields that were not included in this simulation. Comparisons for h_mF_2 are presented in Figure 2b. The overall agreement between model

Table 1. Solar Radio Flux 10.7 cm

YYYY-MM-DD	DOY[a]	Radio Flux[b] [10^{-22} J s^{-1} m^{-2} Hz^{-1}]
2000-03-30	089	256.8
2000-03-31	090	245.6
2002-04-13	103	226.0
2002-04-14	104	210.3
2002-04-15	105	203.3
2002-04-16	106	195.7
2002-04-17	107	193.5
2002-04-18	108	188.2
2002-04-19	109	179.7
2002-04-20	110	177.3
2002-04-21	111	173.4
2002-04-22	112	169.9
2002-04-23	113	175.3
2002-04-24	114	176.9

[a] DOY, day of year.
[b] Source: NGDC (2008).

and observations is very good. According to Galkin and Reinisch (private communication, 2008), the data gaps on DOY 108 and 110 occurred when the F_2 trace was uplifted beyond the ionogram top height. In 2002, Chilton digisonde upper sampling height was set to 650 km. Digisonde ionograms are inverted to vertical electron density profiles using the true height program NHPC [*Huang and Reinisch*, 1996]. Extensive validations were conducted by comparing digisonde profiles with several thousand incoherent scatter profiles showing excellent agreement, with h_mF_2 differences typically smaller than ±15 km [*Chen et al.*, 1994]. There are no special difficulties in applying the profile inversion technique for magnetic storm periods as long as f_oF_2 is not smaller than f_oF_1, and the virtual heights of the F_2 trace are not exceeding the upper sampling range of the ionosonde (B. W. Reinisch, private communication, 2008). The agreement between CTIPe model and ionosonde observations is sufficiently encouraging that the model can be used to unravel the physical mechanisms responsible for the height changes.

4.2. Thermal Expansion Effect on F-Region Height Changes

The hypothesis is that much of the height change experienced by the ionosphere during geomagnetic storms arises from the two physical processes described in Section 2—that is, the effect of the horizontal wind pushing plasma parallel to the inclined magnetic field, and the thermal expansion. The latter has not been highlighted in previous publications, so some theoretical explanation and model demonstration is required. The theory is that thermal expansion of the thermosphere will cause an upward neutral wind. This vertical wind will be experienced by the ions through collisions with the neutral atmosphere. Since the magnetic field is inclined at mid latitudes, this vertical wind will push the plasma parallel to the magnetic field and raise the height of the F_2 layer. The result is similar to the effect of the horizontal wind that has a component parallel to the magnetic field. The difference is that for a horizontal wind surge, the plasma is pushed out of equilibrium with its surrounding, so it continually wants to return to that equilibrium level. In contrast, as the plasma is moved with the thermal expansion it remains in equilibrium with its surroundings, so it does not attempt to return to its original height after the wind has abated. Therefore, the effects of thermal expansion integrate over the duration of the heating and cooling events. However, the question arises whether the collisions are sufficiently frequent that the plasma is carried with the neutral gas, or if it lags behind. There will be an accompanying horizontal movement in addition to the vertical motion of the plasma, but this effect is small on the scale sizes considered.

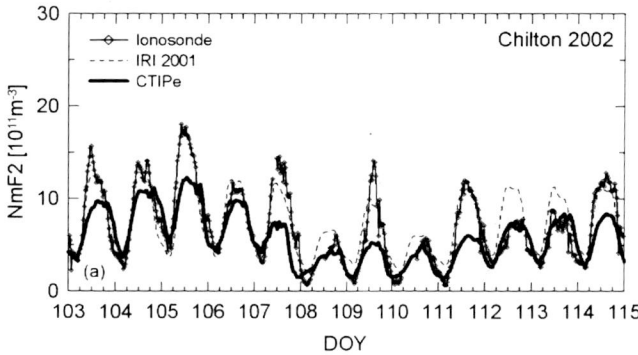

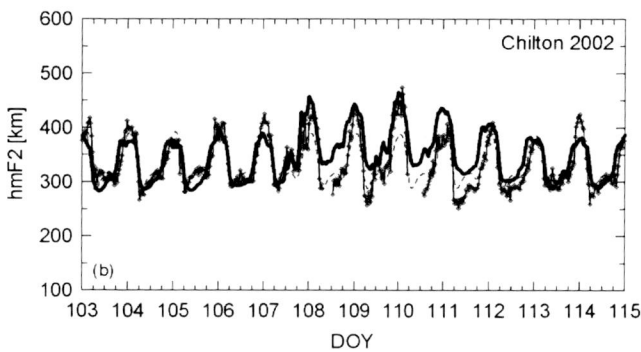

Figure 2. (a) F_2 layer peak electron density (N_mF_2) and (b) F_2 peak height (h_mF_2) at Chilton during the period 13–24 April 2002. CTIPe results are represented by the solid thick line. Ionosonde observations are shown by the solid line with diamonds. As an additional comparison, results from IRI-2001 (http://modelweb.gsfc.nasa.gov/models/iri.html) are represented by the dashed lines.

To test the effect of thermal expansion, a numerical experiment that simulates thermospheric heating without creating a change in the global wind pattern is conducted using CTIPe. This is accomplished by applying a constant increase in heating at all latitudes and longitudes, both day and night. This increased heating is a function of height but does not depend on local time and it is constant in UT. Because a fixed amount of heat is added to all CTIPe grid points, the neutral temperature increases uniformly around the globe, whereas the horizontal neutral wind changes are minimal. This enables the effect of thermal expansion on the F region height to be isolated from the other mechanisms.

Two simulations under steady-state conditions were done for this experiment, using exactly the same input parameters except for the additional globally averaged heat source as described above. Both simulations assumes $F10.7 = 194$ [10^{-22} J s^{-1} m^{-2} Hz^{-1}] on 10 April 2002 (DOY 100), and quiet geomagnetic conditions ($0 < K_p < 3^-$). Differences between neutral temperature resulting from both simulations at three different locations are presented in Figure 3. Port Stanley (geographic coordinates, 51.6°S, 57.9°W; decl., 3.5°; dip, −50.2°), Jicamarca (geographic coordinates, 12.0°S, 76.8°W; decl., 0.2°; dip, 1.0°), and Wallops Island (geographic coordinates, 37.9°N, 75.5°W; decl., −10.5°; dip, 66.0°) show the same increase in the neutral temperature at pressure level 12, indicating that the thermospheric heating was uniform all over the globe. Similar results were found at all geographic locations.

The uniform heating imposed in the numerical experiment was designed to minimize changes in the horizontal wind. This objective was reasonably well satisfied and it is demonstrated by the small change in the meridional wind velocity in the simulation with the additional heat source, according

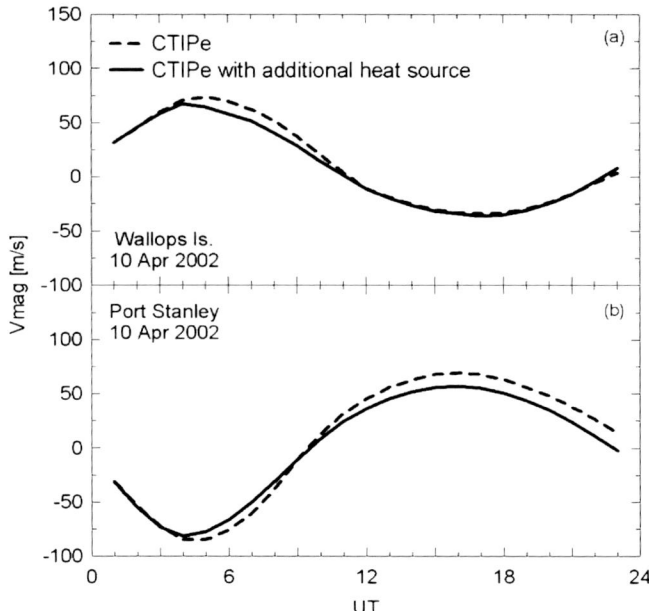

Figure 4. Meridional winds along the magnetic field line at (a) Wallops Island and (b) Port Stanley resulting from normal and increased heating CTIPe simulations.

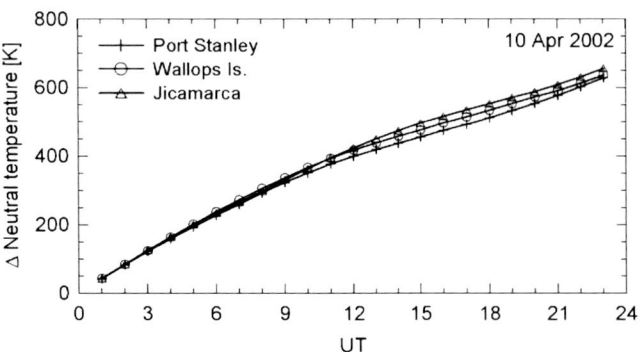

Figure 3. Differences between CTIPe neutral temperature resulting from normal and increased heating simulations at Port Stanley, Jicamarca, and Wallops Island.

to Figure 4. The dashed line represents the component of the meridional wind velocity along the magnetic field line during normal heating conditions, whereas the solid line corresponds to the wind during increased heating. Differences between meridional winds obtained from both simulations at each location were very small, and the change in h_mF_2 by the horizontal wind mechanism was negligible (see green line versus black solid thin line, in Plate 1). This effect is calculated using the method described by *Miller et al.* [1986], *Richards* [1991], and *Codrescu et al.* [1992], which derives meridional winds in the thermosphere from measurements of F_2 layer peak heights. The theory is based on the earlier "servo" model of *Rishbeth et al.* [1978].

Plate 1 shows the h_mF_2 obtained from the simulation with normal solar heating conditions (black thin line), and the h_mF_2 corrected for the meridional winds (green line). The height of pressure level 12 showed an immediate response to the increased thermospheric heating, gradually rising about 100 km in the period of 24 h. This uplift corresponds to the difference between heights of a constant pressure level obtained from both simulations (and it is represented by the orange area in Plate 1), which was added to the h_mF_2 corrected for the meridional wind effect (green curve, Plate 1). Thus, the orange curve in Plate 1 corresponds to the height increase in the pressure level 12 caused by the thermal expansion of

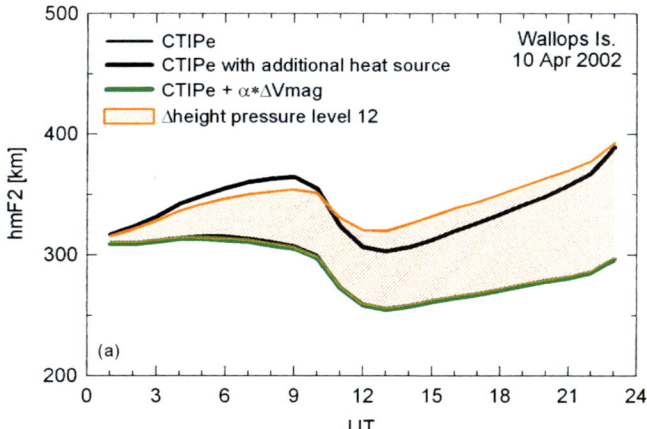

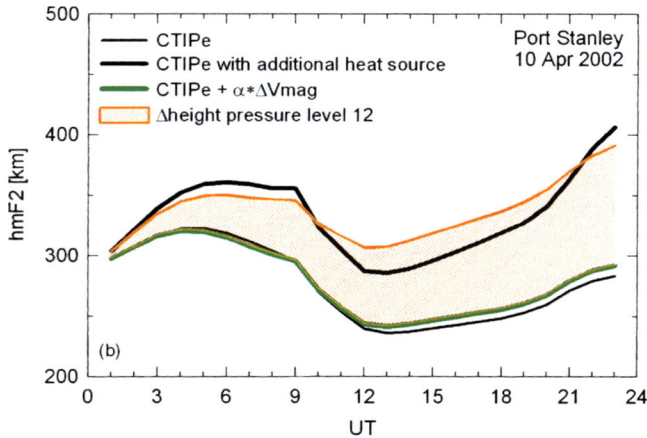

Plate 1. CTIPe h_mF_2 resulting from the simulation with normal (black thin line) and increased (black thick line) heating conditions at (a) Wallops Island and (b) Port Stanley. The h_mF_2 corrected for the meridional wind is represented by the green line, whereas the difference between heights of a constant pressure level obtained from both simulations is represented by the orange area.

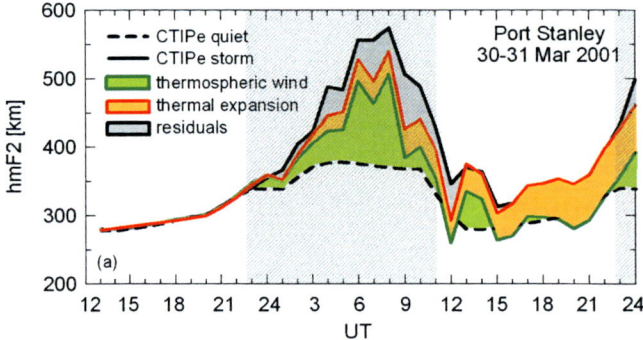
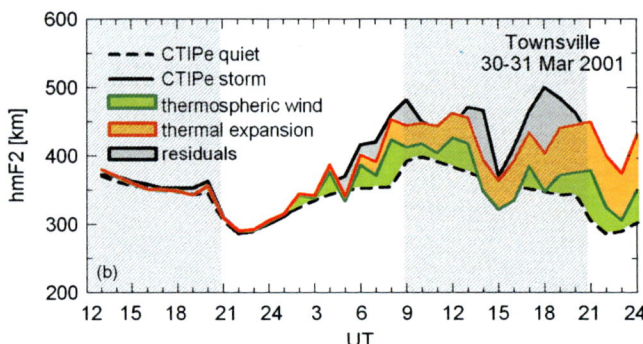

Plate 2. CTIPe h_mF_2 simulation results for quiet (black dashed line) and storm (black solid line) periods at (a) Port Stanley and (b) Townsville stations. The green area corresponds to the h_mF_2 changes due to the horizontal wind, and the orange area shows the thermal expansion contribution in uplifting h_mF_2. The gray area represents the residuals, which are a combination of effects due to uncertainties in the analysis and possible influence of electric fields at mid latitudes. The nighttime period at each location is represented by the shaded area. The storm commencement on 31 March 2001 occurred approximately at 01 UT.

the atmosphere. A similar response was observed in the h_mF_2, which uplifted about 100 km in 24 h, as illustrated by the black thick line in Plate 1. If the thermal expansion were the only mechanism responsible for h_mF_2 changes, then the orange and black thick lines would coincide. The differences between those lines are possibly caused by peak production height changes. At sunrise, the electron concentration near the F_2 peak increases at a rate that depends primarily on the production rate, and the height of the F_2 peak falls because of the rapid production of ionization in the lower F region, reaching a minimum before noon. Later in the day, N_mF_2 decreases and h_mF_2 rises as solar control weakens [*Rishbeth and Garriott*, 1969]. This behavior is shown in the results presented in Plate 1. During the day, h_mF_2 is lower than the height of pressure level 12 and, even though the thermal expansion uplifts h_mF_2, the ionization in the lower F region plays a role decreasing the F_2 layer peak height. In addition, the small difference between h_mF_2 and pressure level changes can arise from second-order effects resulting from temperature changes. This response was observed at both mid-latitude locations.

4.3. Relative Importance of Horizontal Winds and Thermal Expansion During the 31 March 2001 Magnetic Storm

The numerical experiment presented in the previous section demonstrated that thermal expansion is one possible mechanism responsible for the gradual uplift of h_mF_2 during heating events, such as geomagnetic storms. While the dominant heating source of the thermosphere is the solar EUV radiation during quiet periods, increased high-latitude convection electric fields and particle precipitation are the main heating drivers during magnetic storms. The heating and expansion of the polar thermosphere produce pressure gradients that generate strong neutral winds flowing equatorward, moving the plasma up along the geomagnetic field lines and raising the F_2 layer height to regions of reduced loss [e.g., *Fuller-Rowell et al.*, 1997; *Prölss*, 1997; *Buonsanto*, 1999]. The auroral zone heating is quickly spread to lower latitudes by those disturbed winds and also gravity waves, resulting in the thermal expansion of the neutral atmosphere [*Rishbeth*, 1975]. In this section, the relative contribution of the horizontal wind and the thermal expansion in modifying h_mF_2 during the 31 March 2001 magnetic storm event will be examined.

A summary of the magnetic conditions for 30–31 March 2001 is presented in Figure 5. On 31 March, the IMF Bz (top panel) exhibited two significant incursions to the south: the first one lasted about 4 h and the second one approximately 8 h. In the middle panel, the AE index shows the large amount of energy that was injected at high latitudes during the storm period. In the bottom panel, the $SYM\text{-}H$ index initiated its

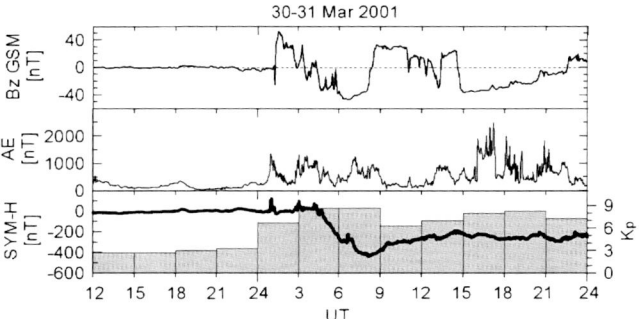

Figure 5. IMF Bz component of the IMF in GSM coordinates provided by the ACE spacecraft for 30–31 March 2001 (top), Quick-Look AE index (middle), and $SYM\text{-}H$ and Kp indices for the same period (bottom). In the top panel, the IMF Bz propagation time from ACE to the subsolar magnetopause was taken into account.

negative incursion around 0430 UT, reaching the minimum value of –437 nT at 0807 UT. Also, in the bottom panel, the Kp index shows the intense activity occurred on 31 March, reaching values of 9⁻ that lasted for about 6 h. Solar radio flux $F10.7$ cm for this high solar activity period is presented in Table 1.

The global T–I response to the 31 March 2001 magnetic storm was a consequence of the impact of a CME on Earth's magnetosphere, and a later substorm activity process during the recovery phase of the storm that caused the development and intensification of the electrojet activity. As a result, the observed T–I response to this event was shown to be dependent on the local time at the storm commencement and the magnetic conditions previous to and during the storm period. An example of this response is the fact that the equatorial anomaly development appeared to be more pronounced in the Australian/Asian sector in the first hours of the initial period of activity compared to the American sector during the second phase of the storm. Ionosonde measurements indicated that mechanisms such as prompt penetration and disturbance dynamo electric fields, thermospheric winds, and thermal expansion played an important role in the T–I behavior. Additional details and discussions on the response of the thermosphere and ionosphere to the 31 March 2001 storm, including observations and model simulations, are presented in the reports of *Fedrizzi et al.* [2005] and *Fuller-Rowell et al.* [2007].

As a further step in the 31 March 2001 magnetic storm investigation, the relative importance of the horizontal thermospheric wind and the thermal expansion on F_2 layer peak height changes during that storm event is presented in this study. The mid-latitude ionosonde stations of Port Stanley (geographic coordinates, 51.6°S, 57.9°W; decl., 3.5°; dip,

−50.2°) and Townsville (geographic coordinates, 19.6°S, 146.8°E; decl., 7.7°; dip, −48.9°) were chosen for the thermospheric wind and thermal expansion analysis, since those locations showed h_mF_2 uplifts possibly caused by both physical mechanisms on 31 March [*Fedrizzi et al.*, 2005]. As a first step, comparisons between h_mF_2 obtained from CTIPe simulations and ionosonde observations at Port Stanley and Townsville during both quiet (Figures 6a and 6c) and storm (Figures 6b and 6d) conditions were performed. Since the model was able to reproduce the observations with reasonable accuracy, it provided the confidence to use its results in the analysis. Adopting the method that derives meridional winds in the thermosphere from measurements of F_2 layer peak heights (referred in the previous section), the change in h_mF_2 due to the horizontal wind is calculated using the meridional component of the thermospheric wind velocity along the magnetic field line provided by CTIPe. Subsequently, the difference between storm- and quiet-time height of pressure level 12 was added to the h_mF_2 uplift due to the horizontal wind. The results are presented in Plate 2. The black dashed line represents CTIPe h_mF_2 for the quiet period, the green line corresponds to the h_mF_2 changes due to the horizontal wind, the red line shows the thermal expansion contribution in uplifting h_mF_2 (added on top of the green line), and the black solid line corresponds to the CTIPe h_mF_2 during the magnetic storm. The green and orange areas in Plate 2 illustrate, respectively, the relative contribution of the horizontal thermospheric wind and the thermal expansion mechanisms in modifying h_mF_2 during the storm period. The gray area represents the residuals, which are a combination of effects due to uncertainties in the analysis and possible influence of electric fields at mid latitudes.

During the first phase of the storm, the American sector was located in the nightside, while the Australian sector was located in the dayside. According to Plate 2, the h_mF_2 uplift due to the horizontal thermospheric wind at Port Stanley (Plate 2a) was larger than at Townsville (Plate 2b) in the first hours of the initial period of activity. A possible reason for this behavior is the superposition of diurnal quiet-time winds (poleward during the day, and equatorward at night) and storm-time winds. Later, during the second phase of the storm, the largest horizontal wind contribution in raising h_mF_2 is observed in Townsville, which was located in the nightside sector. The h_mF_2 uplift due to thermal expansion was larger at Port Stanley (in the nightside sector) during the first hours after the storm commencement and equally important at both stations in the second phase of the storm. The spatial distribution of the energy input over the poles is one possible reason for the differences in the thermal expansion response at those two locations during the first phase of the storm.

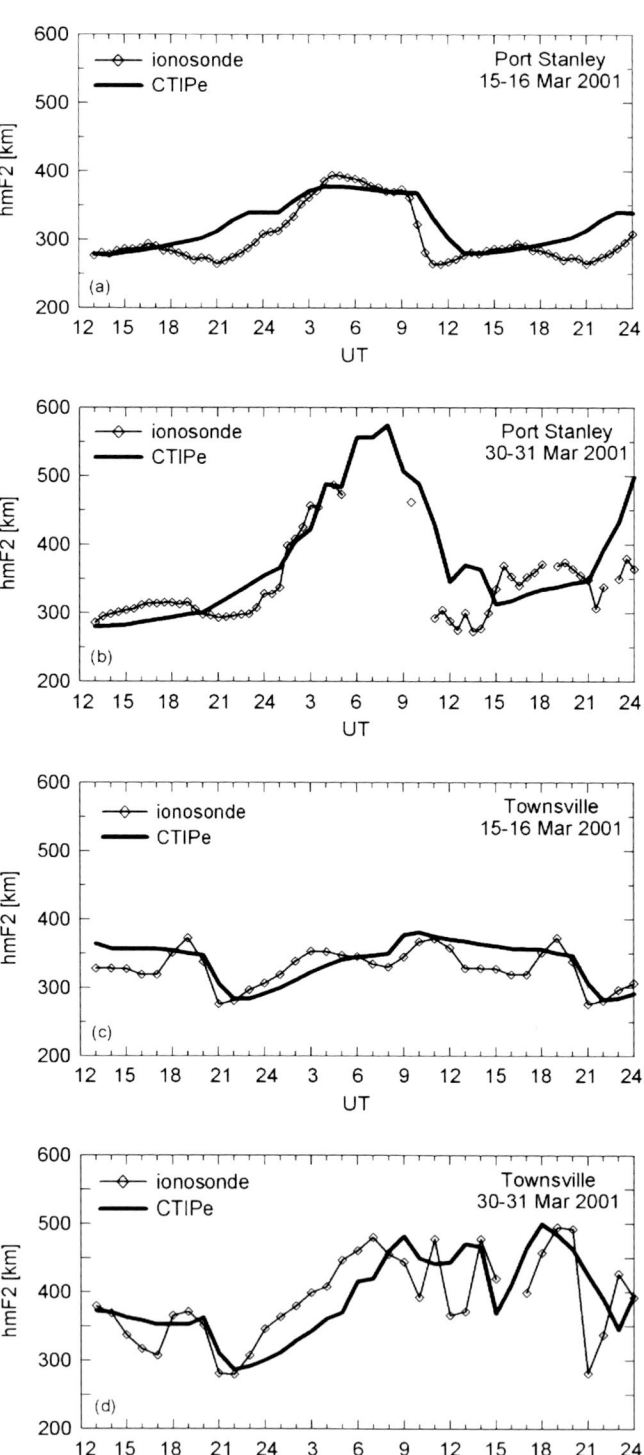

Figure 6. Comparison between CTIPe simulations (black thick line) and ionosonde observations (line with diamonds) at Port Stanley and Townsville during quiet (a, c) and storm (b, d) conditions.

5. CONCLUSIONS

Comparisons between CTIPe physically based model and ionosonde observations during two magnetic storm events showed good agreement for both quiet and storm periods, demonstrating that it is appropriate to use the model results in the analysis of the relative contribution of physical mechanisms responsible for the mid-latitude thermospheric–ionospheric response to magnetic storms.

The present work attempts to unravel the relative contribution of the thermal expansion and horizontal thermospheric wind mechanisms in modifying the F_2 layer peak height during geomagnetic storms. In particular, this study is focused on the effect of the barometric or thermal expansion component of the vertical wind in changing h_mF_2. Results from this study demonstrate that height changes in the neutral atmosphere are clearly reflected in the changes of h_mF_2, both in the transient and the approached equilibrium.

Analysis of the h_mF_2 response to the horizontal thermospheric wind and thermal expansion at two mid-latitude stations (located in opposite longitude sectors) during the 31 March 2001 magnetic storm revealed that during the first hours after the storm commencement the thermospheric wind was dominant in the nightside sector, agreeing with reports in the literature [e.g., *Fuller-Rowell et al.*, 1997; *Prölss*, 1997], and the thermal expansion was dominant in the longitude sector closest to energy input over the poles. A few hours after the storm commencement, the heating in the thermosphere was more evenly distributed, and equivalent integrated heating effects were indicated by the model at similar magnetic latitude locations. As long as there is high-latitude energy input, the thermosphere will expand and h_mF_2 will rise. Once the heating ceases, h_mF_2 will stop rising due to thermal expansion and will gradually fall as the thermosphere cools, returning to its original altitude. In summary, both the horizontal thermospheric wind and the thermal expansion mechanisms account for most of the h_mF_2 response during disturbed periods. Their relative importance will depend on the local time at the storm commencement, the spatial distribution of the energy input over the poles, and the storm development and recovery duration.

Acknowledgments. We thank UMLCAR DIDBase, NOAA/NGDC Solar Terrestrial Physics Division, IPS Radio Space Services, NASA/GSFC Coordinated Data Analysis Web, and World Data Center for Geomagnetism at Kyoto for providing data, as well as NASA/GSFC/SPDF/ Modelweb for IRI-2001 online computation availability. We also thank Dr. Bodo W. Reinisch and Dr. Ivan Galkin for consultations regarding storm-time NHPC performance and Chilton digisonde data revision. This material is based on work supported by the National Science Foundation CEDAR Postdoctoral Research under Grant No. ATM-0524144 "Physical Interpretation of TEC Response During Intense Geomagnetic Storms Using Data Assimilation and Physically-Based Models."

REFERENCES

Buonsanto, M. J. (1999), Ionospheric storms—A review, *Space Sci. Rev.*, *88*, 563–601.

Chen, C. F., B. W. Reinisch, J. L. Scali, X. Huang, R. R. Gamache, M. J. Buonsanto, and B. D. Ward (1994), The accuracy of ionogram-derived N(h) profiles, *Adv. Space Res.*, *14*, 43–46.

Clilverd, M. A., T. Ulich, and M. J. Jarvis (2003), Residual solar cycle influence on trends in ionospheric F_2-layer peak height, *J. Geophys. Res.*, *107*(A12), 1450, doi:10.1029/2003JA009838.

Codrescu, M. V., R. G. Roble, and J. M. Forbes (1992), Interactive ionosphere modeling: a comparison between TIGCM and ionosonde data, *J. Geophys. Res.*, *97*, 8591–8600.

Fedrizzi, M., E. R. de Paula, I. J. Kantor, I. S. Batista, R. B. Langley, and A. Komjathy (2005), Study of the March 31, 2001 magnetic storm effects on the ionosphere using GPS data, *Adv. Space Res.*, *36*, 534–545.

Fuller-Rowell, T. J., D. Rees, S. Quegan, R. J. Moffett, M. V. Codrescu, and G. H. Millward (1996), A coupled thermosphere-ionosphere model (CTIM) in *Handbook of Ionospheric Models*, STEP Report, edited by R. W. Schunk, pp. 217–238.

Fuller-Rowell, T. J., M. V. Codrescu, R. G. Roble, and A. D. Richmond (1997), How does the thermosphere and ionosphere react to a geomagnetic storm?, in *Magnetic Storms*, Geophys. Monogr. Ser., vol. 98, edited by B. T. Tsurutani et al., pp. 203–225, AGU, Washington, D. C.

Fuller-Rowell, T. J., G. H. Millward, A. D. Richmond, and M. V. Codrescu (2002), Storm-time changes in the upper atmosphere at low latitudes, *J. Atmos. Sol. Terr. Phys.*, *64*, 1383–1391.

Fuller-Rowell, T., M. Codrescu, N. Maruyama, M. Fedrizzi, E. Araujo-Pradere, S. Sazykin, and G. Bust (2007), Observed and modeled thermosphere and ionosphere response to superstorms, *Radio Sci.*, *42*, RS4S90, doi:10.1029/2005RS003392.

Garriott, O. K., and H. Risbeth (1963), Effects of temperature changes on the electron density profile in the F_2 layer, *Planet. Space Sci.*, *11*, 587–590.

Huang, X., and B. W. Reinisch (1996), Vertical electron density profiles from the digisonde network, *Adv. Space Res.*, *18*, 121–129.

Iyemori, T., T. Araki, T. Kamei, and M. Takeda (2008), *Mid-latitude Geomagnetic Indices "ASY" and "SYM" for 1999 (Provisional)*, http://swdcwww.kugi.kyoto-u.ac.jp/aeasy/asy.pddf, Accessed 22 April 2008.

Jacchia, L. G., J. Slowey, and F. Verniani (1967), Geomagnetic perturbations and upper-atmosphere heating, *J. Geophys. Res.*, *72*, 1423–1434.

Maruyama, N., et al. (2007), Modeling storm-time electrodynamics of the low-latitude ionosphere–thermosphere system: Can long lasting disturbance electric fields be accounted for?, *J. Atmos. Sol. Terr. Phys.*, *69*, 1182–1199.

Miller, K. L., Torr, D. G., and Richards, P. G. (1986), Meridional winds in the thermosphere derived from measurement of F_2-layer height, *J. Geophys. Res.*, *91*, 4531–4535.

Millward, G. H., R. J. Moffett, S. Quegan, and T. J. Fuller-Rowell (1996), A coupled thermosphere–ionosphere–plasmasphere model (CTIP), in *Handbook of Ionospheric Models*, STEP Report, edited by R. W. Schunk, pp. 239–279.

Millward, G. H., I. C. F. Müller-Wodarg, A. D. Aylward, T. J. Fuller-Rowell, A. D. Richmond, and R. J. Moffett (2001), An investigation into the influence of tidal forcing on F region equatorial vertical ion drift using a global ionosphere–thermosphere model with coupled electrodynamics, *J. Geophys. Res.*, *106*, 24,733–24,744.

Prölss, G. W. (1995), Ionospheric F-region storms, in *Handbook of Atmospheric Electrodynamics*, vol. II, edited by H. Volland, pp. 195–248, CRC Press, Boca Raton, Fla.

Prölss, G. W. (1997), Magnetic storm associated perturbations of the upper atmosphere, in *Magnetic Storms, Geophys. Monogr. Ser.*, vol. 98, edited by B. T. Tsurutani et al., pp. 227–241, AGU, Washington, D. C.

Richards, P. G. (1991), An improved algorithm for determining neutral winds from the height of the F_2 peak electron density, *J. Geophys. Res.*, *96*, 17,839–17,846.

Richmond, A. D. (1995), Ionospheric electrodynamics, in *Handbook of Atmospheric Electrodynamics*, vol. II, edited by H. Volland, pp. 249–290, CRC Press, Boca Raton, Fla.

Rishbeth, H. (1972), Thermospheric winds and the F-layer: A review, *J. Atmos. Terr. Phys.*, *34*, 1–47.

Rishbeth, H. (1975), F-region storms and thermospheric circulation, *J. Atmos. Terr. Phys.*, *37*, 1055–1064.

Rishbeth, H. (1998), How the thermospheric circulation affects the ionospheric F_2-layer, *J. Atmos. Sol. Terr. Phys.*, *60*, 1385–1402.

Rishbeth, H., and R. Edwards (1989), The isobaric F_2-layer, *J. Atmos. Terr. Phys.*, *51*, 321–338.

Rishbeth, H., and O. K. Garriott (1969), *Introduction to Ionospheric Physics*, Academic, New York.

Rishbeth, H., and I. C. F. Müller-Wodarg (1999), Vertical circulation and thermospheric composition: A modelling study, *Ann. Geophys.*, *17*, 794–805.

Rishbeth, H., R. J. Moffett, and G. J. Bailey (1969), Continuity of air motion in the mid-latitude thermosphere, *J. Atmos. Terr. Phys.*, *31*, 1035–1047.

Rishbeth, H., Ganguly, S., and Walker, J. C. G. (1978), Field-aligned and field-perpendicular velocities in the ionospheric F_2-layer, *J. Atmos. Terr. Phys.*, *40*, 767–784.

Titheridge, J. E. (1995), Winds in the ionosphere—A review, *J. Atmos. Terr. Phys.*, *57*, 1681–1714.

M. Codrescu and H. Khalsa, Space Weather Prediction Center, NOAA, W/NP9 325 Broadway, Boulder, CO 80305, USA.

M. Fedrizzi, T. J. Fuller-Rowell, and N. Maruyama, Cooperative Institute for Research in Environmental Sciences, University of Colorado, Space Weather Prediction Center, NOAA, 325 Broadway, Boulder, CO 80305, USA. (mariangel.fedrizzi@noaa.gov)

Neutral Composition and Density Effects in the October-November 2003 Magnetic Storms

T. J. Immel

Space Sciences Laboratory, University of California, Berkeley, California, USA

Geoff Crowley

Atmospheric and Space Technology Research Associates, San Antonio, Texas, USA

J. M. Forbes, R. S. Nerem, and E. K. Sutton

Department of Aerospace Engineering, University of Colorado, Boulder, Colorado, USA

The thermospheric effects of magnetic storm activity occurring in October and November of 2003 are studied using a combination of in situ and remote imaging instruments. The CHAMP (Challenging Minisatellite Payload) satellite near 400 km can detect the slight satellite drag force from which the in situ neutral density is then determined. Global images from the Far Ultraviolet Spectrographic Imager onboard the NASA-IMAGE satellite provide a remote measure of the disturbance in thermospheric composition at lower altitudes (130–200 km). Observed from high altitudes, variations in atomic oxygen emissions at 135.6 nm closely follow the ratio of the column integrated abundances of O and N_2. In this work, the relationship between density and composition perturbations is investigated. High-latitude Joule heating is known to produce global-scale reductions of O density relative to that of N_2, so when observed in daytime these areas are expected to exhibit both enhanced temperatures/densities and reduced 135.6-nm brightness. In this work we find that, in addition to effects that can be attributed to direct Joule heating, some large variations in temperature/density at 400 km can more likely be attributed to adiabatic compression and expansion associated with (1) traveling atmospheric disturbances launched by impulsive auroral inputs and (2) convergence and downwelling in the global-scale storm-time thermospheric circulation driven by those same auroral inputs. These are discussed along with other possible explanations for the departure of the present findings from recent work describing a straightforward relationship between thermospheric temperature/density and composition perturbation.

1. INTRODUCTION

The geomagnetic storm is a global disturbance in the geomagnetic field, usually indicated by large departures of the horizontal magnetic field near Earth's magnetic dip equator from normal levels. Geomagnetic storms occurring late in 2003 were remarkable for their prompt occurrence following a set of powerful solar coronal mass ejections (CME) directed toward Earth [*Gopalswamy et al.*, 2005], and their extreme and well-documented effects on geospace [*Tsurutani et al.*, 2005; *Sutton et al.*, 2005; *Yizengaw et al.*, 2005]. Of particular interest here are the effects of auroral Joule heating on the upper atmosphere that accompany these geomagnetic storms. Such effects extend from high to low latitudes in the form of traveling neutral density waves, enhanced neutral winds, and neutral composition disturbances.

Several processes are thought to contribute, by varying degree, to enhancements in ionospheric electron densities during magnetic storms, or positive ionospheric storms. Enhanced equatorward neutral winds and the imposition of expanded magnetospheric convection electric fields can both have the effect of lifting plasma to higher altitudes, thereby reducing recombination and/or enhancing peak plasma densities [cf. *Crowley et al.*, 2006]. Reductions of N_2 relative to atomic oxygen (O) may also contribute to middle latitude and afternoon enhancements in F-layer densities during storms. Large-scale reductions in ionospheric plasma densities, which also occur in the 3- to 24-h span following periods of enhanced geomagnetic activity, are termed negative ionospheric storms and are primarily attributed to changes in thermospheric composition (reductions in O relative to N_2). Regional-scale reductions of plasma densities such as the ionospheric trough and the evacuation of plasma at the magnetic equator during geomagnetic storms are other well-known "negative" effects driven by electric fields.

It is the neutral composition component of the ionospheric storm that can be observed in global imaging of Earth in the far ultraviolet (FUV). The increase in temperatures and ion convection at high latitudes has the combined effect of enhancing the molecular composition throughout the atmospheric column via upward vertical winds, and enhancing horizontal neutral winds that carry these effects away from the auroral zone and to middle latitudes [*Prölss and Roemer*, 1987; *Crowley and McCrea*, 1988; *Crowley and Meier*, 2008]. Usually ahead of this transport, associated with downwelling and the closed global circulation of neutral gas, is a region of enhanced atomic oxygen column density (cf. Figure 2 of *Meier et al.* [2005]). The enhancement of N_2 densities relative to O in the ionosphere enhances the loss rates of ionospheric O^+, reducing peak ionospheric densities and total electron content. The converse is true where O densities are enhanced relative to N_2. Thus, during magnetic storms, thermospheric composition disturbances contribute to negative ionospheric storm effects at high-to-middle latitudes [*Prölss and Craven*, 1998; *Strickland et al.*, 2001], and positive effects at middle-to-low latitudes [*Förster et al.*, 1999; *Goncharenko et al.*, 2007].

The absolute neutral density at thermospheric altitudes depends mainly on the scale height of the thermosphere, and thus the thermospheric temperature. With the initial auroral and Joule heating at high latitudes occurring in the lower thermosphere, it takes a few minutes for the thermospheric column to reach hydrostatic equilibrium, and thus for scale height to be a meaningful parameter [*Rishbeth et al.*, 1987]. This is the time required for the pressure increase to be communicated upward at the sound speed by the vertical propagation of acoustic waves. These pressure enhancements also propagate horizontally in large-scale traveling atmospheric disturbances (TADs) with a group speed of 600–800 m s^{-1} that depends on the background temperature and increases with altitude. The first indication of the onset of a thermospheric storm at middle and low latitudes is the arrival of a traveling thermospheric disturbance, usually detected by its effects on the ionosphere [cf. *Crowley and McCrea*, 1988; *Sastri et al.*, 2000; *Immel et al.*, 2001; *Nicolls and Kelley*, 2005].

This study compares the simultaneous observations of thermospheric density variations near 400-km altitude with global observations of the column-integrated thermospheric composition effects through several days of the magnetic storm of 28 October–1 November 2003. The density effects are measured by Challenging Minisatellite Payload (CHAMP) through the use of a sensitive onboard accelerometer that measures changes in the atmospheric drag on the satellite [e.g., *Marcos et al.*, 1977; *Marcos and Forbes*, 1985; *Crowley et al.*, 1995; *Schoendorf et al.*, 1996]. Its high-inclination circular orbit processes in local time, and was near the noon sector at 400-km altitude during this magnetic storm. *Liu and Lühr* [2005] showed that the density enhancements for this event were much larger on the dayside than the nightside, even in the absence of solar flare-related density enhancements that were also present during this storm.

The NASA IMAGE mission provided global FUV images on the dayside of the planet during the storm at 135.6 nm, the wavelength of a bright emission of O I [*Meier et al.*, 1991]. Deviations in brightness from quiet time levels are determined to quantify the changes in column integrated O/N_2 ($\Sigma O/N_2$) from baseline quiet-time levels in a manner similar to that developed by *Immel et al.* [2000] for Dynamics Explorer 1 FUV images. Relative, quiet-time levels of O I 135.6-nm brightness are determined from images obtained during quiet times occurring earlier in October. When dayside imaging operations are underway, the CHAMP satellite passes through or near the field of view (FOV) of the imager every 100 min. The relationship of thermospheric density and composition variations during a magnetic storm can therefore be examined. The immediate question is whether the two characteristics vary mutually or independently. The results of *Meier et al.* [2005], using neutral densities and O/N_2 from limb retrievals using the Global Ultraviolet Imager (GUVI) on board the NASA TIMED spacecraft [*Christensen et al.*, 2003], suggest that regions of storm-related composition perturbations should contain concomitant disturbances in temperature/density. In particular, large-scale temperature enhancements of 500 K in the upper thermosphere were observed in regions of reduced $\Sigma O/N_2$, a composition effect known to be driven by Joule heating. Although long suspected, the finding of Meier et al. provided the most significant evidence for this relationship. This work will further investigate the relationship between composition and temperature disturbances in the thermosphere.

A National Center for Atmospheric Research thermosphere–ionosphere–mesosphere electrodynamics general circulation model (TIME-GCM) [*Roble and Ridley*, 1994] run is performed for validation and comparison. Predicted in situ neutral densities and column integrated O/N_2 (down to the altitude at which N_2 concentration reaches 10^{17} cm^{-3}) are retrieved although interpolation to the location of CHAMP, which provided samples at ~2° intervals of latitude. The focus is to determine whether the model agrees with (1) the observations of thermospheric neutral density perturbations by CHAMP, (2) the composition disturbance effects observed by IMAGE, and (3) whether the modeled neutral density and composition fields vary relative to one another in a manner that matches the observations. Good agreement would support the use of the GCM as a laboratory for these effects to understand the physical connection between upper atmospheric density, composition, and temperature. These runs are performed using the *Weimer* [1996] empirical model to predict cross-cap potentials that drive the *Heelis et al.* [1982] potential distribution model that is native to the TIME-GCM, and solar EUV inputs driven by solar 10.7-cm radio flux in the same manner as that described by *Meier et al.* [2005].

2. OBSERVATIONS: MAGNETIC ACTIVITY, NEUTRAL DENSITY MEASUREMENTS, AND FUV IMAGING

A summary of neutral density effects measured by CHAMP is shown with pertinent inputs (*Dst* and solar X-ray flux) in Figure 1a. A similar summary from the TIME-GCM provided at the locations of the CHAMP satellite (as noted above) is shown in Figure 1b, along with a running 1-h average of the 1-min Auroral Electrojet (*AE*) index. To produce these summaries, the mean neutral density measured (or modeled) equatorward of ±40° is determined on the dayside,

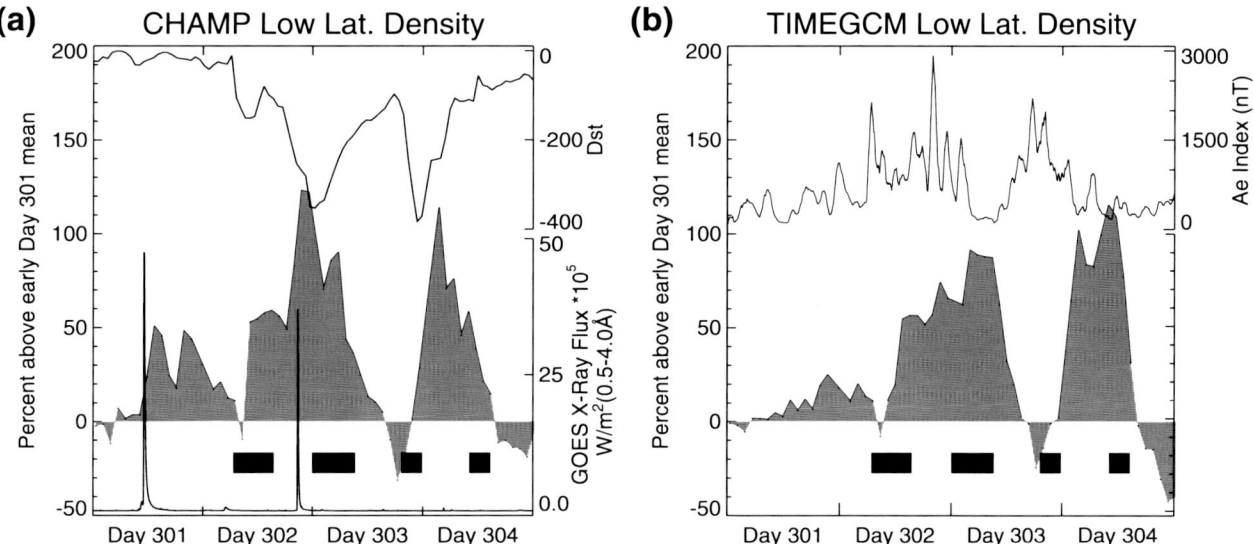

Figure 1. Initial comparison of TIME-GCM (thermosphere–ionosphere–mesosphere electrodynamics general circulation model) and CHAMP (Challenging Minisatellite Payload) densities, with magnetic and solar parameters. (a) Percent changes in low- to middle-latitude CHAMP neutral density (shaded line), shown with *Dst* (top) and solar X-ray flux (bottom). (b) Percent changes in low-middle latitude TIME-GCM neutral density, with 1-h running average of *AE*. Heavy bars show times of far ultraviolet (FUV) imaging shown in this report. The GOES X-ray instrument briefly reached its detection limit on day 301.

producing one value per orbit. The percent density difference is shown, calculated by using the minimum mean density observed in the first six orbits of the CHAMP satellite on day 301, before the occurrence of the first X-class flare. Notably, the density drops well below this value at later times in the 4-day interval, an effect observed both by CHAMP and in the TIME-GCM simulation. This suggests a significant reduction in temperatures late in the storm due possibly to enhanced NO densities and associated cooling rates that are expected after strong auroral precipitation events [*Mlynczak et al.*, 2005]. Nightside density differences are not determined, since the focus here is on comparisons with dayglow.

Three main enhancements in density are observed in the CHAMP data (Figure 1a). The first peak on day 301, driven mainly by direct effects of the first major solar flare, produces a >50% enhancement in density, an effect discussed in detail by *Sutton et al.* [2006]. The second peak at the end of day 302 is the greatest overall density perturbation, where high-latitude heating effects caused by the arrival of the first flare-related CME combine with flare-related heating effects of the second major flare, with remarkable results. The third density peak on day 304 is driven solely by the geomagnetic storm effects associated with the arrival of the CME associated with the day 302 flare that occurred ~24 h before this final peak in storm density.

Comparing the CHAMP and TIME-GCM densities (the latter in Figure 1b), one finds good correspondence between the overall time histories of neutral density variation. The main difference is that the TIME-GCM reports a smaller proportional response to solar EUV enhancements than that due to high-latitude heating effects. The reader may note the absence of flare-induced peaks on days 301 and 302 in Figure 1b that are clearly present in Figure 1a. This likely originates in an underestimate of the EUV heating because the TIME-GCM only used daily 10.7-cm solar flux values to characterize the solar inputs, and cannot therefore capture the extraordinary, short-term thermospheric heating that comes from solar flares. That said, densities from CHAMP clearly show that it is the geomagnetic forcing, rather than the solar flare inputs, which produces the larger overall change in thermospheric densities. This larger forcing component is also the main source of longer-lived composition perturbations, an effect not produced by the more evenly distributed solar flare heating. Modification of the TIME-GCM solar and high-latitude inputs could be made in order to tune the model to more closely reproduce the solar forcing, or to moderate the storm-time response on day 304, where the modeled density enhancement outlasts the observed effect. The goal, though, is not to force an exact match of data and model, but to produce a realistic, self-consistent history of behavior of the thermosphere during the storm to assist in understanding the complex relationships between density and composition. This initial survey of the modeled and measured density responses indicates that TIME-GCM provides that realistic history.

Imaging of the terrestrial 135.6-nm emission was performed using the IMAGE-FUV Spectrographic Imager [*Mende et al.*, 2000] with 2-min cadence. For this analysis, images are combined into 10-min average images, improving the ratio of instrumental signal-to-noise while increasing the effective sensitivity of the instrument to approximately 70 counts/kilorayleigh. The eccentric polar orbit had a near-equatorial apogee at 6.8 Re altitude and a period of 14 h. This imaging scenario afforded good views of auroral activity for a few hours of each orbit, with observations otherwise dominated by full-Earth dayglow imaging from the prenoon sector. FUV observations from three main storm periods will be shown in Plates 1, 2, and 3, where these periods are indicated by the heavy bar common to both Figures 1a and 1b.

The CHAMP satellite is in an 87° inclination near-circular orbit near ~410 km at the onset of the magnetic activity [*Reigber et al.*, 2002]. In the retrieval of thermospheric neutral density, horizontal winds originating in auroral forcing at high latitudes can affect the density retrieval, introducing an approximately 8% error per 100 m s^{-1} wind speed [*Sutton et al.*, 2005]. During a major magnetic storm, winds exceeding 300 m s^{-1} can be produced at high latitudes over regions larger than the CHAMP density sample spacing, thus introducing errors on the order of 25%. However, as will be shown, density variations are observed to be much larger than this potential source of error, and which are sometimes seen at low latitudes in subsequent orbits, from which one may imply the propagation of density effects in TADs.

To quantify thermospheric disturbances, baseline values representative of lower levels of activity must be determined for comparison to storm-time data. Quiet-time airglow brightnesses are determined by using the method of *Immel et al.* [2000], whereas CHAMP quiescent values are determined from the observations on day 301, where averages are now determined as a function of geographic latitude, using data from all of the 15 orbits occurring on that day, in a manner similar to that used by *Liu and Lühr* [2005]. In determining the average on day 301, the flare-time density enhancement is reduced in magnitude by fitting the daily values of middle-low latitude mean density (described in the previous section) with a cubic polynomial function, and using the higher-order terms thus determined to correct the instantaneous density observations. The main effect of this correction is simply to produce a lower baseline density than what would be calculated had the flare effects not been thus treated.

For comparison to the data, density and composition perturbations predicted by the TIME-GCM at the location of

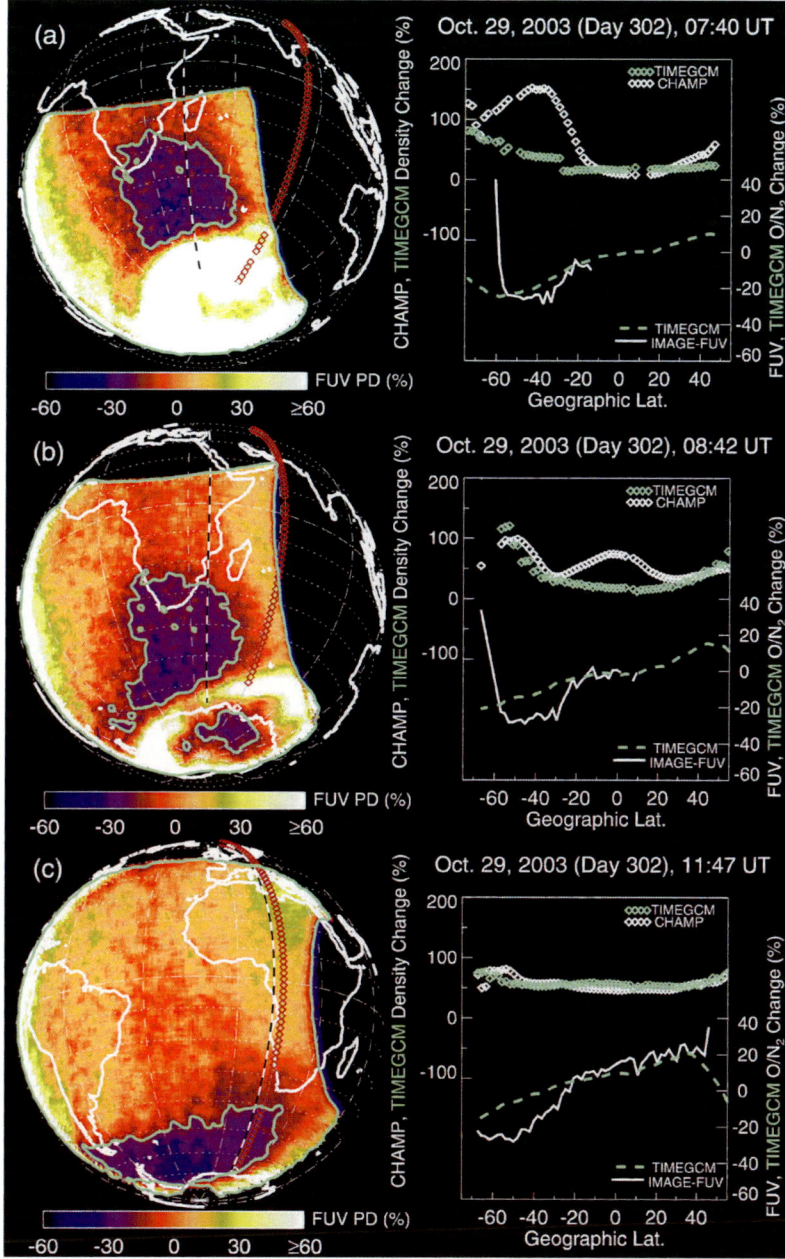

Plate 1. IMAGE FUV and CHAMP comparisons after magnetic storm onset, 29 October 2003. Plots 2a, 2b, and 2c show the 135.6-nm differences from quiet time background, with the central point at 1100 local solar time and the latitude of the IMAGE spacecraft. Overlaid are the CHAMP locations (red diamonds) and the sampling track for the FUV percent differences (dashed line). These tracks are not always coincident for times when CHAMP data come from mainly outside the imaging field of view early in the IMAGE orbit, with better coincidence later in the IMAGE orbit. The 10-min average images are shown at 0740, 0842, and 1147 UT. Each image is accompanied by a plot comparing the CHAMP density departures to those predicted by the TIME-GCM, and (below) a comparison of the TIME-GCM $\Sigma O/N_2$ perturbation to the IMAGE FUV brightness percent difference.

CHAMP are determined. Once the densities predicted by the TIME-GCM at CHAMP are determined, those values are compared to a day 301 baseline that is calculated in a manner similar to how the CHAMP data are treated (described above), including a fit to, and correction for, the late day density enhancements in the model (see Figure 1b). Finally, storm-time $\Sigma O/N_2$ perturbations predicted by the TIME-GCM at the location (latitude and longitude only) of the CHAMP observations are determined, using values from the day 301 model run at comparable UT as a baseline. The resultant average density is not a purely quiet-time baseline, as is determined for the FUV observations, but given the extreme level activity and composition disturbances seen on days 302–304, it serves as a moderately quiet baseline that contains background seasonal and longitudinal variability for the best determination of departures from these effects during the magnetic storm.

3. STORM ONSET, TAD, AND INITIAL $\Sigma O/N_2$ STATE

Extreme magnetic activity began suddenly at 0612 UT on 29 October 2003 (day 302) when the 1-min resolution *AE* index increased from a moderate level about 400 nT to more than 3000 nT in 1 min, and thereafter exceeded 2000 nT for most of the following hour (the 1-h running average shown in Figure 1 only reflects this latter fact). Data from successive orbits of CHAMP just after the onset of magnetic activity are compared to FUV images at 0740 and 0842 UT. The obvious signature of an activity-induced density enhancement is observed in successive orbits, first with a peak around 40°S (Plate 1a), and then at the equator (Plate 1b), propagating with an apparent resultant speed of 800 m s^{-1}. Such a substantial speed could be expected at the CHAMP satellite altitude for a large TAD generated in the sudden onset of a magnetic storm, whereas propagation at lower altitudes will be somewhat slower [*Hines*, 1974]. The FUV observations at the same latitude show a significant reduction in O versus N$_2$. Although the CHAMP satellite does not pass directly through this area, the density wave accompanying a TAD is not expected to have a greatly preferred longitude of propagation on the dayside [cf. *Balthazor and Moffett*, 1999], so it is assumed to have passed through this region as well.

The area of reduced $\Sigma O/N_2$ initially seen by IMAGE at 0740 UT on day 302 is related to activity occurring during the previous day, when the *AE* index reached values exceeding 1000 nT and maintained levels indicating moderate activity all day. This area of reduced $\Sigma O/N_2$ persists during the three CHAMP orbits shown, with reductions greater than 20% extending from the morning terminator to noon local time by 1147 UT. The continued presence of $\Sigma O/N_2$ reductions in the morning sector indicates replenishment of disturbed parcels of gas. The CHAMP pass through this region shows a general trend of greater densities (and therefore temperatures) toward the geographic pole, in a region where IMAGE and the TIME-GCM both show reduction of $\Sigma O/N_2$. This is in general agreement with the findings of *Meier et al.* [2005], although the large temperature variations implied from the large TAD effects in Plates 1a and 1b are never reflected in the $\Sigma O/N_2$ variability at middle to low latitudes on the dayside.

4. THERMOSPHERIC STORM EFFECTS, CONTINUED AURORAL FORCING

Magnetic activity increases again at 1700 UT on day 302 reaching a minimum in *Dst* (−353 nT) at the end of the day. FUV images and density data/model parameters are shown for >4 h of the following recovery period in Plate 2, beginning at 0012 UT. At this time, there is a global scale enhancement in neutral density, with CHAMP reporting >100% enhancements at all latitudes. This is driven in part by the heating effects of the second X-class flare that occurred at ~2100 UT on day 302, similar to the enhancement that occurred on day 301, described by *Sutton et al.* [2006]. However, these effects are obviously combined with extreme magnetic storm effects. The overall density enhancement drops over the 4.5-h period of observations, whereas the FUV imagery demonstrates the extreme redistribution of molecular constituents, with large reductions of $\Sigma O/N_2$ in the southern hemisphere and a sharp transition to enhancements of $\Sigma O/N_2$ in the northern hemisphere. Contamination of the FUV signal by auroral emissions is evident about −50° latitude.

The successive CHAMP orbits at 0012 and 0145 (Plates 2a and 2b) reveal a trend of increasing density toward both poles punctuated by the direct observation of the signature of Joule heating in the vicinity of the auroral brightness peaks (compare white diamonds (density) to white line (FUV)). In the two CHAMP passes at 0318 and 0444 UT, the trend in neutral density changes, and as auroral activity drops to low levels, a minimum in density emerges in the southern hemisphere, with a maximum in the northern hemisphere (Plates 2c and 2d). Comparing the CHAMP and FUV signatures (again, white diamonds and line), the zone of reduced FUV and thus $\Sigma O/N_2$ now corresponds to a minimum in density, whereas the peak in brightness contains the peak in density as well. This is in contrast to the behavior described by *Meier et al.* [2005].

During this period the TIME-GCM does not consistently show a correspondence between density and composition, but suggests at first a disconnect between the two processes, predicting only moderate latitudinal variation in density at CHAMP altitudes accompanied by extreme changes in the

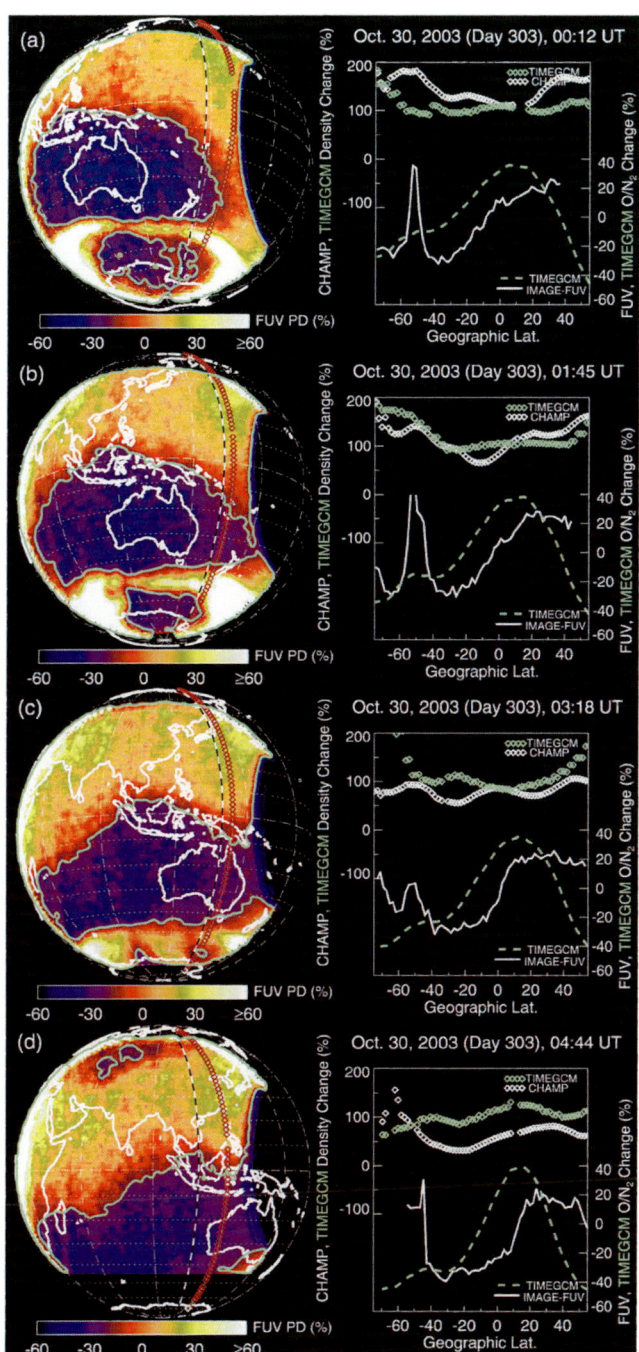

Plate 2. Same as Plate 3, but for FUV imaging at 0012, 0145, 0318, and 0444 UT on day 303 (30 October 2003).

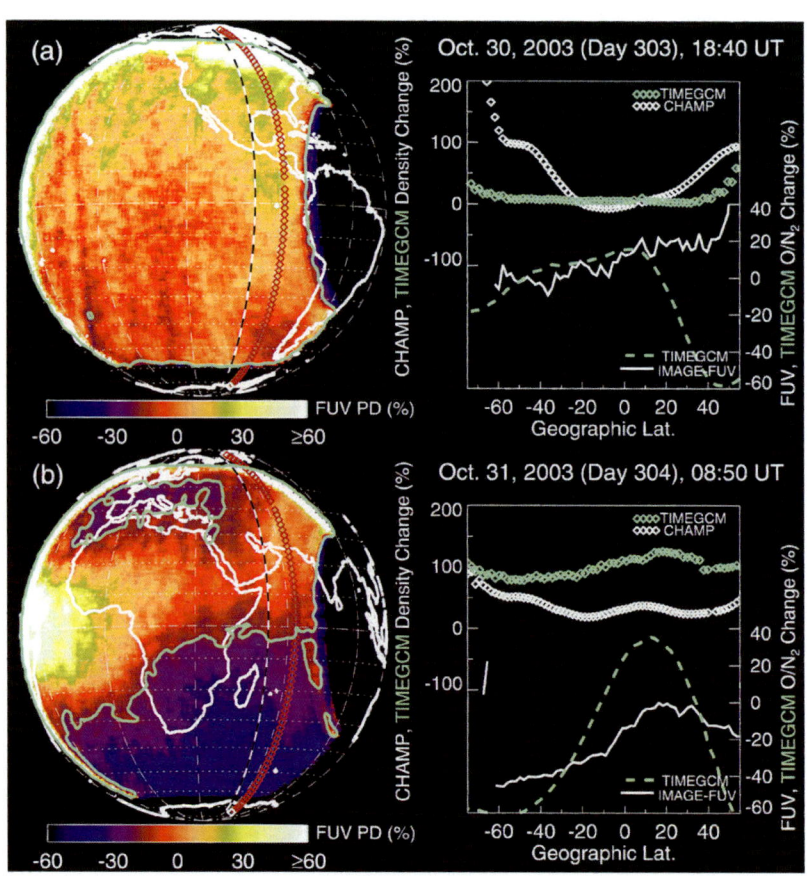

Plate 3. Same as Plate 2, but for FUV imaging at 1840 UT on day 303 (30 October 2003) and 0850 UT on day 304 (31 October 2003).

underlying $\Sigma O/N_2$ (Plates 2a and 2b). As magnetic activity subsides, the model settles into a state remarkably similar to that which is observed, where the peak in density is found at the peak in $\Sigma O/N_2$ (compare green diamonds, line). Although the modeled $\Sigma O/N_2$ distribution is more symmetric about the equator than that shown by the strongly asymmetric FUV distribution, the relationship between density and $\Sigma O/N_2$ is the same. Although the $\Sigma O/N_2$ reduction zone is clearly a result of Joule heating, other processes are in effect that produce more significant temperature variations on the dayside. The convergence of neutral winds and subsequent downwelling of atmospheric species is a likely source of some portion of this enhancement.

5. STORM PEAK AND RECOVERY

At 1840 UT on day 303, about 2 h into the second great decrease in *Dst*, FUV images were obtained from high altitude in the next orbit of IMAGE. The 10-min composite image (Plate 3a) shows a remarkably undisturbed view of the thermosphere, where the only indications of the extreme levels of heating currently occurring in the auroral zones is the auroral contamination of the dayglow signature over Canada, and the moderate enhancement in $\Sigma O/N_2$ north of the equator, with a concomitant moderate reduction in the southern hemisphere. Density measurements by CHAMP, however, show the remarkable effects of Joule heating, and a density enhancement at 40°S that is likely the leading edge of a TAD. At this point in the storm, the TIME-GCM does not yet show large perturbations in density or composition except at high northern latitudes, in disagreement with the IMAGE observations.

The effect of this major geomagnetic storm on composition is clearly observed in the FUV image taken ~14 h later during the next IMAGE orbit at 0850 UT (Plate 3b), showing again the strong hemispheric asymmetry in the $\Sigma O/N_2$ perturbation, and a particularly strong enhancement in $\Sigma O/N_2$ at the equator over the Atlantic. The neutral density variations along the CHAMP orbit show a trend toward higher density at higher latitudes that, unlike the pass accompanying the previous image, is not likely associated with a TAD given the low level of magnetic activity at this time and its gradual variation over several thousands of kilometers. Finally, another peak in density appears at ~10° latitude, near the location of the peak in FUV brightness above baseline values. This pair of images exhibits possibly three particular effects on density, first a major TAD beginning to carry large density perturbations to middle latitudes (Plate 3a), and then both density enhancements in the zone of heated gas, and a neighboring density enhancement at low latitudes resulting from wind convergence (Plate 3b).

TIME-GCM shows a significant variation in density along the CHAMP track at 0850 UT on day 304, reporting mainly 100% or greater increases in density over baseline values. The underlying $\Sigma O/N_2$ variation in the model is very large, exceeding the IMAGE-FUV observations of reduced $\Sigma O/N_2$ at middle and high latitudes, while predicting a very large peak in $\Sigma O/N_2$ near the equator. Clearly, the TIME-GCM run has overestimated the auroral forcing effects. That said, it predicts a peak in density not in regions of reduced $\Sigma O/N_2$, but rather in close proximity to the peak in $\Sigma O/N_2$ at low latitudes, corresponding to the finding of the CHAMP-IMAGE comparison. Apart from the data–model discrepancies at high latitudes, the TIME-GCM clearly predicts a peak in density at low latitudes that is not attributable to direct Joule heating of the gas.

6. DISCUSSION AND CONCLUSION

The main finding of this study is that during magnetic storms, large, column-integrated composition disturbances can develop steadily whereas the thermospheric density varies tremendously. Some of the neutral density variation CHAMP observes at high latitudes is due to upwelling of heavier air in the auroral oval (cf. Plates 1a, 2a, and 2b) and possibly due to neutral winds along the track of the satellite (the aforementioned errors discussed by *Sutton et al.* [2005]). However, at times where auroral activity subsides (cf. Plates 1c and 2d), the relationship of composition and density is quite variable, sometimes showing greater neutral densities/temperatures in regions of depleted $\Sigma O/N_2$, and at other times showing a peak in neutral densities colocated with a peak in enhanced $\Sigma O/N_2$. *Meier et al.* [2005] showed a remarkable case where large-scale reductions in $\Sigma O/N_2$ seemed to be directly related to enhanced exospheric temperatures. In this work we show that this relationship does not always carry over to the storm effects measured at ~400 km. Comparing the results of the present study to the work by Meier et al., who made data–model comparisons at only slightly earlier local times on 20 November 2003 (3 weeks later in the year), one may expect the two studies to have similar conclusions.

There are several possible explanations for the largely different findings. It is clear that TADs can carry large variations in density across the dayside, producing changes in density on the order of 100% that last less than 1 h at any one middle-to-low latitude location. The retrieval of exospheric temperature performed by *Meier et al.* [2005, after the work of *Meier and Picone* [1994]] is weighted significantly by emissions at altitudes below 400 km, where TAD effects on density/temperature are relatively small compared to that which CHAMP might observe. TADs result in only small changes in column integrated O/N_2 and thus produce minimal changes in overall FUV brightness [*Immet et al.*, 2001]. This fact alone clearly suggests are times where the temperature at 400 km and the composition below will not be closely related.

Another important consideration is the effect of vertical winds on temperature and composition. The large-scale enhancement of $\Sigma O/N_2$ in the northern hemisphere is seen several times here to be accompanied by a peak in density, seen in both Plates 2d and 3b. In each of these cases, the model and data show a similar effect, even if they disagree on the latitude of the peak. The convergence of storm-time winds will lead to downwelling of neutral constituents, and because the high-altitude parcels of air consist mainly of O, will enhance the column integrated ratio of O/N_2, thus leading to the 135.6-nm brightness enhancements. A result of the downwelling/subsidence is an enhancement in temperature through adiabatic heating. This occurs regardless of whether the parcels advected into the region of subsidence had previously been heated by ionospheric currents. These storm-time wind effects build up over hours of time and so, like the reductions in $\Sigma O/N_2$ driven by heating, are strongly influenced by conditions in the nighttime ionosphere. The outstanding question is, does a limb-imaging spectrograph such as GUVI find temperature enhancements due to both adiabatic and Joule heating, and if not, why not? Further comparisons of nadir and limb FUV measurements are required to address this question.

The TIME-GCM runs for this work have been analyzed for comparison directly to the two sources of data. The comparisons to the two data sets at hand, global 135.6-nm observations and satellite drag at ~400 km, have been performed. A continuation of this study would include a thorough investigation of the particular model fields that contribute to temperature changes, including NO densities and vertical winds with solar forcing and Joule heating, to assign importance to each of the driving terms under all of the conditions occurring during this remarkable storm event. The recent development of global FUV imaging techniques that simultaneously retrieves $\Sigma O/N_2$ and temperatures near the photoelectron peak (~150 km) through determination of the broadening of N_2 emission bands promises a new avenue for understanding the global connections between thermospheric composition, density, and temperature during magnetic storms.

Acknowledgments. IMAGE FUV analysis is supported by NASA through Southwest Research Institute subcontract number 83820 at the University of California, Berkeley, contract NAS5-96020, and by NSF grant ATM-0640362. The CHAMP density analysis at the University of Colorado was supported under NSF grant ATM-0208482 as part of the National Space Weather Program and NASA grant NNG04GN20H as part of the Graduate Student Researchers Program. GC is supported by NSF grant ATM0332307, and NASA grants NG04GN04G and NNX07AB66G to SwRI.

REFERENCES

Balthazor, R. L., and R. J. Moffett (1999), Morphology of large-scale traveling atmospheric disturbances in the polar thermosphere, *J. Geophys. Res.*, *104*, 15–24, doi:10.1029/1998JA900039.

Christensen, A. B., et al. (2003), Initial observations with the Global Ultraviolet Imager (GUVI) in the NASA TIMED satellite mission, *J. Geophys. Res.*, *108*(A12), 1451, doi:10.1029/2003JA009918.

Crowley, G., and I. W. McCrea (1988), A synoptic study of TIDs observed in the United Kingdom during the first WAGS campaign, October 10–18, 1985, *Radio Sci.*, *23*, 905–917.

Crowley, G., and R. R. Meier (2008), Disturbed O/N_2 ratios and their transport to middle and low latitudes, this volume.

Crowley, G., J. Schoendorf, R. G. Roble, and F. A. Marcos (1995), Satellite observations of neutral density cells in the lower thermosphere at high latitudes, in *The Upper Mesosphere and Lower Thermosphere: A Reivew of Experiment and Theory, Geophys. Monogr. Ser.*, vol. 87, edited by R. M. Johnson and T. L. Killeen, pp. 339–348, AGU, Washington, D. C.

Crowley, G., et al. (2006), Global thermosphere–ionosphere response to onset of 20 November 2003 magnetic storm, *J. Geophys. Res.*, *111*, A10S18, doi:10.1029/2005JA011518.

Förster, M., A. A. Namgaladze, and R. Y. Yurik (1999), Thermospheric composition changes deduced from geomagnetic storm modeling, *Geophys. Res. Lett.*, *16*, 2625–2628.

Goncharenko, L. P., J. C. Foster, A. J. Coster, C. Huang, N. Aponte, and L. J. Paxton (2007), Observations of a positive storm phase on September 10, 2005, *J. Atmos. Sol. Terr. Phys.*, *69*, 1253–1272, doi:10.1016/j.jastp.2006.09.011.

Gopalswamy, N., S. Yashiro, Y. Liu, G. Michalek, A. Vourlidas, M. L. Kaiser, and R. A. Howard (2005), Coronal mass ejections and other extreme characteristics of the 2003 October–November solar eruptions, *J. Geophys. Res.*, *110*, A09S15, doi:10.1029/2004JA010958.

Heelis, R. A., J. K. Lowell, and R. W. Spiro (1982), A model of the high latitude ionospheric convection pattern, *J. Geophys. Res.*, *87*, 6339–6345.

Hines, C. O. (1974), Observed ionospheric waves considered as gravity or hydromagnetic waves, *J. Atmos. Sol. Terr. Phys.*, *36*, 1205–1216.

Immel, T. J., J. D. Craven, and A. C. Nicholas (2000), The DE-1 auroral imager's response to the FUV dayglow for thermospheric studies, *J. Atmos. Sol. Terr. Phys.*, *62*, 47–64.

Immel, T. J., G. Crowley, J. D. Craven, and R. G. Roble (2001), Dayside enhancements of thermospheric O/N_2 following magnetic storm onset, *J. Geophys. Res.*, *106*, 15,471–15,488.

Liu, H., and H. Lühr (2005), Strong disturbance of the upper thermospheric density due to magnetic storms: CHAMP observations, *J. Geophys. Res.*, *110*, A09S29, doi:10.1029/2004JA010908.

Marcos, F. A., and J. M. Forbes (1985), Thermospheric winds from the satellite electrostatic triaxial accelerometer system, *J. Geophys. Res.*, *90*, 6543–6552.

Marcos, F. A., H. B. Garrett, K. S. W. Champion, and J. M. Forbes (1977), Density variations in the lower thermosphere from analysis of the AE-C accelerometer measurements, *Planet. Space Sci.*, *25*, 499–507, doi:10.1016/0032-0633(77)90082-4.

Meier, R. R. (1991), Ultraviolet spectroscopy and remote sensing of the upper atmosphere, *Space Sci. Rev.*, *58*, 1–185.

Meier, R. R., and J. M. Picone (1994), Retrieval of absolute thermospheric concentrations from the far UV dayglow: An application of discrete inverse theory, *J. Geophys. Res.*, *99*, 6307–6320.

Meier, R. R., G. Crowley, D. J. Strickland, A. B. Christensen, L. J. Paxton, D. Morrison, and C. L. Hackert (2005), First look at the 20 November 2003 superstorm with TIMED/GUVI: Comparisons with a thermospheric global circulation model, *J. Geophys. Res.*, *110*, A09S41, doi:10.1029/2004JA010990.

Mende, S. B., et al. (2000), Far ultraviolet imaging from the IMAGE spacecraft: 3. Spectral imaging of Lyman-α and OI 135.6 nm, *Space Sci. Rev.*, *91*, 287–318.

Mlynczak, M. G., et al. (2005), Energy transport in the thermosphere during the solar storms of April 2002, *J. Geophys. Res.*, *110*, A12S25, doi:10.1029/2005JA011141.

Nicolls, M. J., and M. C. Kelley (2005), Strong evidence for gravity wave seeding of an ionospheric plasma instability, *Geophys. Res. Lett.*, *32*, L05108, doi:10.1029/2004GL020737.

Prölss, G. W., and J. D. Craven (1998), Perturbations of the FUV dayglow and ionospheric storm effects, *Adv. Space Res.*, *22*(1).

Prölss, G. W., and M. Roemer (1987), Thermospheric storms, *Adv. Space Res.*, *7*(10), 223–235.

Reigber, C., et al. (2002), A high-quality global gravity field model from CHAMP GPS tracking data and accelerometry (EIGEN-1S), *Geophys. Res. Lett.*, *29*, 37–1.

Rishbeth, H., T. J. Fuller-Rowell, and D. Rees (1987), Diffusive equilibrium and vertical motion in the thermosphere during a severe magnetic storm: A computational study, *Planet. Space Sci.*, *35*, 1157–1165, doi:10.1016/0032-0633(87)90022-5.

Roble, R. G., and E. C. Ridley (1994), A thermosphere–ionosphere–mesosphere–electrodynamics general circulation model (TIME-GCM): Equinox solar cycle minimum simulations (300–500 km), *Geophys. Res. Lett.*, *22*, 417–420.

Sastri, J. H., N. Jyoti, V. V. Somayajulu, H. Chandra, and C. V. Devasia (2000), Ionospheric storm of early November 1993 in the Indian equatorial region, *J. Geophys. Res.*, *105*, 18,443–18,456, doi:10.1029/1999JA000372.

Schoendorf, J., G. Crowley, R. G. Roble, and F. A. Marcos (1996), Neutral density cells in the high latitude thermosphere: 1. Solar maximum cell morphology and data analysis, *J. Atmos. Terr. Phys.*, *58*, 1751–1768.

Strickland, D. J., R. E. Daniell, and J. D. Craven (2001), Negative ionospheric storm coincident with DE 1-observed thermospheric disturbance on October 14, 1981, *J. Geophys. Res.*, *106*, 21,049–21,062.

Sutton, E. K., J. M. Forbes, and R. S. Nerem (2005), Global thermospheric neutral density and wind response to the severe 2003 geomagnetic storms from CHAMP accelerometer data, *J. Geophys. Res.*, *110*, A09S40, doi:10.1029/2004JA010985.

Sutton, E. K., J. M. Forbes, R. S. Nerem, and T. N. Woods (2006), Neutral density response to the solar flares of October and November, 2003, *Geophys. Res. Lett.*, *33*, L22101, doi:10.1029/2006GL027737.

Tsurutani, B. T., et al. (2005), The October 28, 2003 extreme EUV solar flare and resultant extreme ionospheric effects: Comparison to other Halloween events and the Bastille day event, *Geophys. Res. Lett.*, *32*, L03S09, doi:10.1029/2004GL021475.

Weimer, D. R. (1996), A flexible, IMF dependent model of high-latitude electric potentials having "space weather" applications, *Geophys. Res. Lett.*, *23*, 2549–2552, doi:10.1029/96GL02255.

Yizengaw, E., H. Wei, M. B. Moldwin, D. Galvan, L. Mandrake, A. Mannucci, and X. Pi (2005), The correlation between mid-latitude trough and the plasmapause, *Geophys. Res. Lett.*, *32*, L10102, doi:10.1029/2005GL022954.

G. Crowley, Atmospheric and Space Technology Research Associates, 12703 Spectrum Drive, Suite 101, San Antonio, TX 78249, USA.

J. M. Forbes, R. S. Nerem, and E. K. Sutton, Department of Aerospace Engineering, University of Colorado, Boulder, CO 80309-0431, USA.

T. J. Immel, Space Sciences Laboratory, University of California, Berkeley, 7 Gauss Way, Berkeley, CA 94720, USA. (immel@ssl.berkeley.edu)

Optical and Radio Observations and AMIE/TIEGCM Modeling of Nighttime Traveling Ionospheric Disturbances at Midlatitudes During Geomagnetic Storms

K. Shiokawa,[1] T. Tsugawa,[1] Y. Otsuka,[1] T. Ogawa,[1] G. Lu,[2] A. Saito,[3] and M. Yamamoto[4]

This paper summarizes our recent observations of storm-time traveling ionospheric disturbances (TIDs) over Japan. The storm-time TIDs, which are often referred as large-scale TIDs or traveling atmospheric disturbances, are generated at high-latitude auroral zone and propagate predominantly equatorward at midlatitudes as atmospheric waves in the thermosphere. The imaging measurements by airglow imagers and GPS networks give reliable wave parameters (wavelength, phase velocity, and amplitude) of TIDs. Comparison of imaging observations of TIDs between the northern and southern hemispheres revealed their nonconjugacy. Combinations of these imaging observations with vertical sounding measurements by ionosondes and radars provide a comprehensive description of TID structures at midlatitudes. Comparison between these observations and the assimilative mapping of the ionospheric electrodynamics technique with the thermosphere–ionosphere electrodynamics general circulation model results in several insights into the generation and propagation of TIDs at the high-latitude source region as well as limitations of the current global model.

1. INTRODUCTION

Storm-time traveling ionospheric disturbances (TIDs) are distinct ionospheric features at midlatitudes during geomagnetic storms. The auroral disturbances at high latitudes cause atmospheric waves in the thermosphere in various spatial scales [e.g., *Hooke*, 1968; *Francis*, 1975; *Oyama et al.*, 2001]. These atmospheric waves, which are often referred as traveling atmospheric disturbances (TADs), propagate equatorward and are recognized as TIDs by ionospheric sounding techniques. The TADs/TIDs with larger spatial scales propagate more effectively to middle and low latitudes. The TIDs with a horizontal scale of more than 1000 km and a phase speed of 400–1000 m/s are defined as large-scale TIDs (LSTIDs) by *Hunsucker* [1982]. These storm-time TIDs/TADs are different from the so-called medium-scale TIDs (MSTIDs), which are observed even during geomagnetically quiet time and are more likely caused by ionospheric plasma instabilities rather than neutral waves [*Miller and Kelley*, 1997; *Shiokawa et al.*, 2003a, *Otsuka et al.*, 2004].

Hocke and Schlegel [1996] has reviewed the TID/TAD observations reported in 1982–1995. TIDs/TADs have been measured during magnetically disturbed periods by various ground-based remote-sensing techniques of the thermosphere and the ionosphere, such as by ionosondes [e.g., *Maeda and Handa*, 1980; *Ogawa and Kumagai*, 1985; *Hajkowicz and*

---

[1]Solar–Terrestrial Environment Laboratory, Nagoya University, Toyokawa, Japan.
[2]High Altitude Observatory, National Center for Atmospheric Research, Boulder, Colorado, USA.
[3]Faculty of Science, Kyoto University, Kyoto, Japan.
[4]Research Institute for Sustainable Humanosphere, Kyoto University, Uji, Japan.

Midlatitude Ionospheric Dynamics and Disturbances
Geophysical Monograph Series 181
Copyright 2008 by the American Geophysical Union.
10.1029/181GM24

Hunsucker, 1987; Hajkowicz, 1990; Prölss, 1993; Lee et al., 2002, 2004], high-frequency (HF) radars [e.g., Bristow et al., 1994, 1996], incoherent scatter radars [e.g., Rice et al., 1988; Reddy et al., 1990; Pi et al., 2000; Oyama et al., 2001; Nicolls et al., 2004], total electron content (TEC) measurement by GPS networks [e.g., Ho et al., 1996, 1998; Afraimovich et al., 2000a, 2000b; Nicolls et al., 2004; Tsugawa et al., 2003, 2004, 2006], and airglow imagers [Pi et al., 2000; Shiokawa et al., 2002, 2003b, 2005]. The imaging of ionospheric disturbances by GPS networks and airglow imagers can give reliable wave parameters (wavelength, phase velocity, and amplitude) of TIDs and their changes during the propagation [Tsugawa et al., 2003, 2004]. By combining such imaging measurements with vertical sounding techniques and neutral wind measurements, one can obtain detailed descriptions of the vertical and horizontal structure of storm-time TADs.

An important question on storm-time TIDs/TADs is their generation/propagation mechanisms in the thermosphere. Global modeling of the thermosphere has given various insights into this question [e.g., Richmond, 1978; Millward et al., 1993; Fuller-Rowell et al., 1994; Fujiwara et al., 1996; Balthazor and Moffett, 1999; Fujiwara and Miyoshi, 2006]. Recent modeling efforts have connected more realistic data-driven high-latitude energy input derived using the assimilative mapping of the ionospheric electrodynamics (AMIE) technique with the thermosphere–ionosphere electrodynamics general circulation model (TIEGCM) to describe storm-time thermospheric disturbances and TID generations [Emery et al., 1996, 1999; Buonsanto et al., 1997; Lu et al., 2001]. Such a realistic modeling effort makes it possible to compare the modeling results with observations for TIDs/TADs during particular storms [Lee et al., 2004; Shiokawa et al., 2007].

In this paper, as an article of the AGU Monograph issue on midlatitude ionospheric disturbances and dynamics, we review our recent observations of storm-time TIDs/TADs over Japan, on the basis of the results published by Shiokawa et al. [2002, 2003b, 2005, 2007]. These observations are combinations of imaging techniques by airglow imagers and GPS receiver network and vertical sounding techniques by ionosondes, an incoherent scatter radar, and a Fabry–Perot interferometer (FPI). Comparison with AMIE/TIEGCM global modeling was also carried out to investigate generation and propagation of the observed TIDs/TADs.

2. IMAGING OBSERVATIONS AND COMPARISON WITH VERTICAL SOUNDINGS

Figure 1 shows an example of clear TID observed in Japan during the moderate geomagnetic storm of 15 September 1999, as reported by Shiokawa et al. [2002]. As shown in Figure 1a, the enhancement was seen in the all-sky images at Rikubetsu, Japan [43.5°N, 143.8°E, geomagnetic latitude (magnetic latitude) = 34.7°N]. Figure 1b is the north–south cross section of these all-sky images (keogram), showing a clear southward propagation of the TID. From these airglow measurements, the horizontal phase velocity of TID is estimated to be ~450 m/s. It is interesting to note that the airglow is brighter in the west than in the east, showing a longitudinal inhomogeneity of the TID.

Plate 1 is two-dimensional map of TEC variations measured by the GPS Earth Observation Network (GEONET) over Japan during this TID event. A clear TID is seen in these maps, propagating equatorward with a velocity of ~430 m/s. The meridional scale of the TID was more than 1000 km.

The vertical sounding data of the ionosphere in Figures 1c and 1d show that this TID is accompanied by the decrease in F layer virtual height and increase in f_oF_2, which is proportional to the F layer peak electron density. These features clearly propagate from Wakkanai (northern edge of Japan near Rikubetsu) to Okinawa (southern edge of Japan).

On the basis of these observations, Shiokawa et al. [2002] drew a schematic picture of the TID, as caused by enhancement of poleward neutral wind [see Shiokawa et al., 2002, Figure 9]. The poleward wind pushes the F layer down along the magnetic field line, which has a finite angle to the horizontal plane at midlatitudes. The F layer height decrease causes enhancement of 630.0-nm airglow intensity by increasing reaction of ionospheric O^+ ions with O_2-rich atmosphere. As expected from various model calculations, the poleward wind enhancement is limited in the thermospheric altitudes and not in the lower thermosphere [e.g., Richmond, 1978; Millward et al., 1993]. This vertical shear of meridional wind causes temporal accumulation of plasma in the lower thermosphere, resulting in the increase in f_oF_2 and the F layer peak electron density. The TEC enhancement may be caused by supply of plasmas from higher altitudes. According to Shiokawa et al. [2002], the poleward wind enhancement during this event was estimated to be ~200 m/s by using a model calculation by Bailey and Balan [1996].

3. CONJUGACY OF STORM-TIME TIDS

The imaging measurement makes it possible to investigate geomagnetic conjugacy of wave phenomena between the northern and southern hemispheres. If electrodynamic coupling occurs between the two hemispheres connected by geomagnetic field line, one can expect conjugacy of the wave structure. Such example was presented by Otsuka et al. [2004] for the quiet-time MSTIDs, for which mirror images of waves are obtained at the conjugate hemispheres.

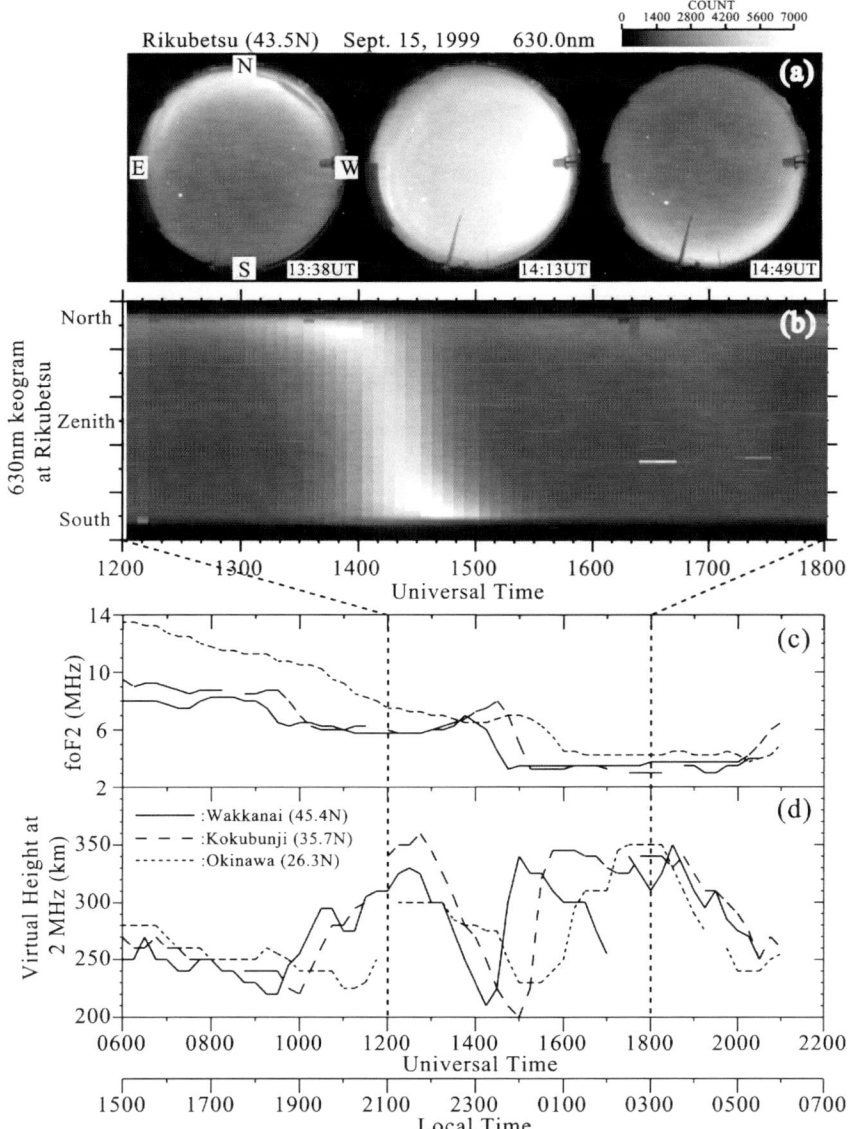

Figure 1. (a) Three representative all-sky airglow images at 630.0 nm obtained at Rikubetsu, (b) north–south cross sections (keogram) of all-sky airglow images at Rikubetsu at 1200–1800 UT, and (c) and (d) variations in f_oF_2 and virtual height at 2 MHz obtained at three ionosonde stations in Japan at 0600–2100 UT on 15 September 1999. The TID is characterized by southward-propagating airglow enhancement, f_oF_2 increases and decreases in the virtual height. Partly from Shiokawa et al. [2002].

Although the storm-time TIDs are basically disturbances in the neutral atmosphere, the neutral variations may cause inhomogeneity of ionospheric conductivity, which can couple with the horizontal ionospheric current to produce the polarization electric field that can propagate to the other hemisphere to cause conjugate signatures.

Figure 2 shows simultaneous observations of an equatorward-moving TID by 630.0-nm airglow imagers at geomagnetic conjugate points at Shigaraki, Japan (34.8°N, 136.1°E), and Renner Springs, Australia (18.3°S, 133.8°E), during the F region Radio and Optical measurement of Nighttime TID 3 (FRONT3) campaign [Shiokawa et al., 2005]. This event was observed during geomagnetically disturbed period with a Dst value of −40 nT. Figures 2a and 2b are north–south cross sections (keograms) of airglow images along a longitude of 136.1°E, whereas Figures 2c–2e are intensities at

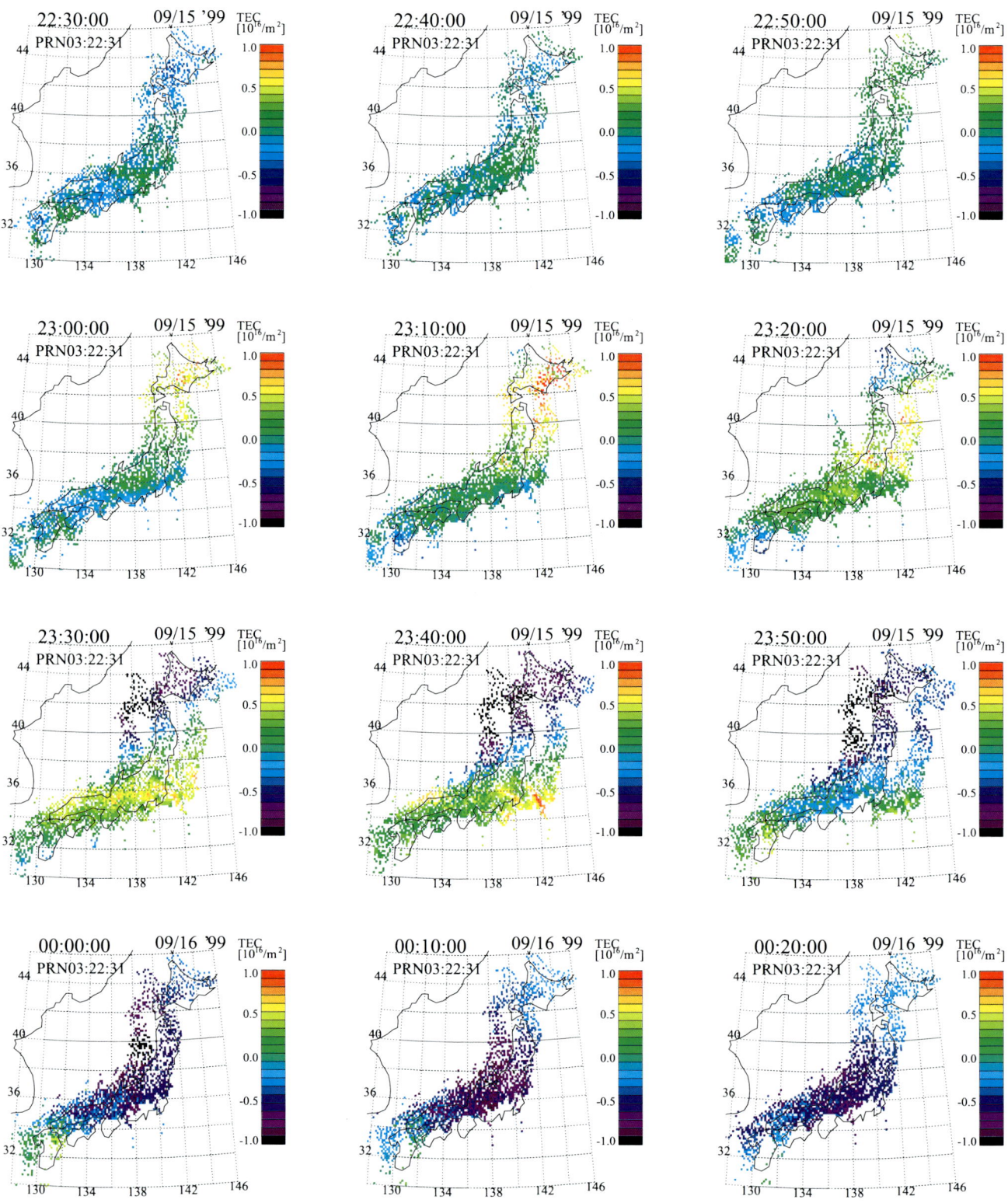

Plate 1. Two-dimensional maps of the total electron content (TEC) variations obtained by more than 1000 GPS receivers in Japan during the passage of the traveling ionospheric disturbance (TID) of September 15, 1999. A running average of TEC over 1 h was subtracted from the raw TEC values. The TEC values are mapped at an altitude of 250 km. [From *Shiokawa et al.*, 2002.]

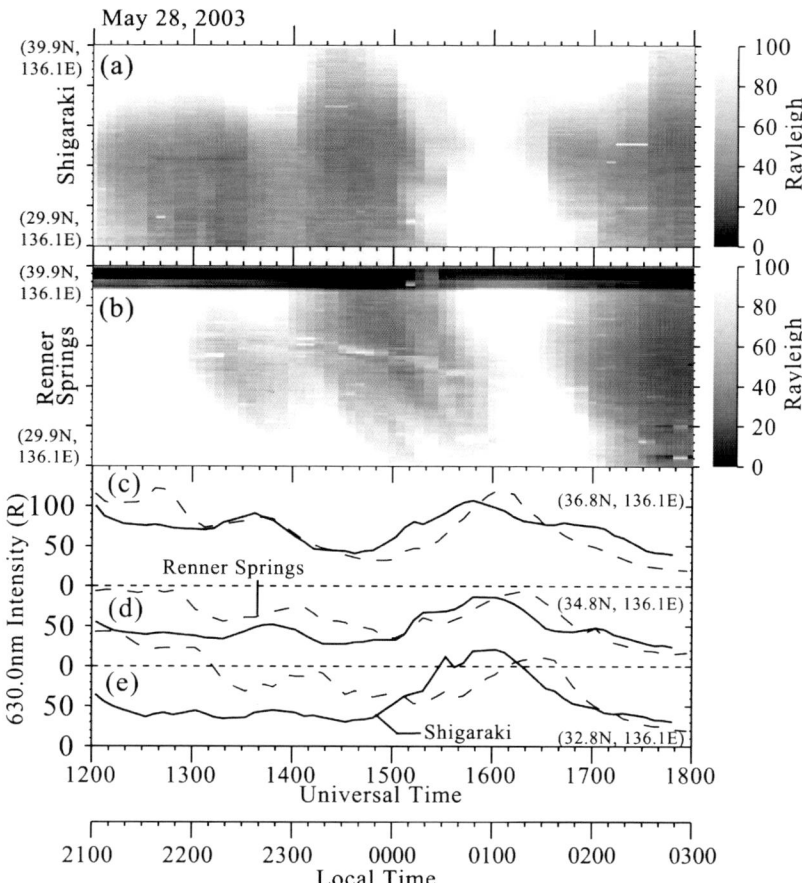

Figure 2. North–south keograms of 630-nm airglow images obtained at (a) Shigaraki, (b) Renner Springs, and (c–e) airglow intensities at Shigaraki (solid curves) and Renner Springs (dashed curves) at three conjugate points in the images, obtained on 28 May 2003. The airglow data at Renner Springs were converted to the conjugate northern hemisphere (same area as that of Shigaraki). The TID is identified at 1500–1700 UT. From *Shiokawa et al.* [2005].

three latitudes in the keograms. The data of Renner Springs are shown at the conjugate northern hemisphere by mapping the images to the northern hemisphere along the geomagnetic field line. An equatorward-propagating TID was observed at 1500–1700 UT at both hemispheres. However, the phase of the TID at Renner Springs clearly delays by ~20 min from that of Shigaraki, showing that the storm-time TIDs do not have geomagnetic conjugacy, although the TID was nearly simultaneously observed at both hemispheres for this case.

Tsugawa et al. [2006] further confirmed the nonconjugacy of LSTIDs statistically using GPS network data over Japan and Australia. They identified 20 (21) events of LSTIDs in Japan (Australia) when Kp is larger than 5– and 15 (10) when Kp is smaller than 4+ in 2002. In these events, the number of simultaneous occurrences of LSTIDs within ±1 hour in both hemispheres was only 5 for $Kp \geq 5-$ and 0 for $Kp \leq 4+$. Moreover, for these five simultaneous events, the crossing times at 30° geomagnetic latitude were different by several tens of minutes. These results indicate that the LSTIDs observed simultaneously in both hemispheres were not connected electromagnetically through the geomagnetic field but were generated by atmospheric gravity waves propagating to the equator independently in the two hemispheres.

4. COMPARISON WITH AMIE/TIEGCM

It has been considered that the TIDs during geomagnetic storms are caused at high latitudes in association with auroral disturbances. To investigate generation of TIDs, modeling approach is essentially needed. Global and local models of the thermosphere and ionosphere have successfully generated TIDs, as cited in the Introduction. On the other hand, comparison of these models with observations has not sufficiently been done yet.

Plate 2 shows LSTIDs generated in the AMIE/TIEGCMs for the severe geomagnetic storm of 31 March 2001 [*Shiokawa et al.*, 2007]. Plates 2a and 2b show AU/AL indices and joule heating rate obtained by AMIE, while Plate 2c shows a latitude–time plot of thermospheric temperature at an altitude of 500 km and a longitude of 135.0° (Japanese longitude) in the TIEGCM. The variation of the thermosphere and the ionosphere during this storm was modeled by the TIEGCM code [*Richmond et al.*, 1992] on a global scale. The code uses the high-latitude energy input calculated from AMIE. The AMIE estimates auroral precipitation and plasma convection in the high-latitude ionosphere quantitatively, by using data from ground magnetometers, SuperDARN radars, and Defense Meteorological Satellite Program and NOAA satellites [*Richmond and Kamide*, 1988; *Lu et al.*, 1998].

In Plate 2, two TADs are generated at 1200 UT and 1400 UT, as indicated by the oblique red dashed lines. They are generated in association with two large substorm activities that can be identified as two peaks in the AL index and Joule heating rate. On the other hand, the intense substorm after 1600 UT basically cause strong heating of high- and middle-latitude thermosphere that gradually expands to lower latitudes until 2100 UT. Some TAD-like signatures propagating equatorward (e.g., at ~1900 UT) are embedded on this global reconfiguration of the thermosphere. These TAD-like signatures also seem to be generated by the impulsive activities seen in the AL index and Joule heating. A high-temperature bulge suddenly appears at 1500 UT in the low-latitude region of the southern hemisphere.

During this geomagnetic storm, an intense TID was observed in Japan by airglow imagers, the middle and upper (MU) atmosphere radar, ionosondes, the GEONET GPS network, and an FPI. *Shiokawa et al.* [2007] made a detailed comparison of these observations with the AMIE/TIEGCM modeling. Their result is summarized in Plate 3. The observed thermospheric wind shown in Plate 3b (crosses and dashed line) varies significantly in that night probably due to the storm effect. Note that typical meridional wind in the thermosphere smoothly changes from 0 to 50–100 m/s southward from evening to midnight because of the diurnal tide.

The two TADs identified in the AMIE/TIEGCM modeling are characterized by southward wind in Plate 3a (blue area) and Plate 3b (solid line), as shown by the two red arrows. The rarefaction (decrease) of this southward wind speed for the second TAD occurs at 1600 UT in the model. On the other hand, the MU radar wind (crosses) and FPI wind (dashed line) in Plate 3b show sudden decrease in southward wind speed at ~1700 UT. This timing difference can be explained by the difference in phase speed of the TAD (370–640 m/s in observation and ~1100 m/s in the model). This wind speed decrease causes significant decrease in F layer height at 1600–1800 UT and subsequent F layer density decrease due to recombination both in the model and observations, as shown in Plates 3c and 3d, respectively. Signatures of the first TAD seen in the model are not clear in the FPI and MU radar data, but are seen in the virtual height data obtained by an ionosonde at Kokubunji, Japan (shown in Plate 3g), as a temporal decrease of the 2-MHz virtual height at 1500 UT.

5. SUMMARY AND DISCUSSION

The two-dimensional images of TADs/TIDs have been recently obtained by airglow imagers and GPS networks. These images clearly reveal propagating features of TADs/TIDs. Comparison of these imaging observations in the northern and southern hemispheres indicates that the storm-time TADs/TIDs are essentially nonconjugate phenomena propagating independently between the two hemispheres. The nonconjugacy of the TADs can be also recognized in the TIEGCM result in Plate 2c. Combinations of these imaging observations with vertical sounding measurements give a comprehensive description of TAD/TID structures at midlatitudes.

One of the major targets of storm-time TID research is the connection between the high-latitude source region and the mid- and low-latitude observations. Comparison with the global model is essential for such purpose, as shown in the examples in Plates 2 and 3. When TIDs are observed at midlatitudes, one can check geomagnetic indices, such as AU/AL indices, and global auroral images, to make a one-to-one correspondence between the high-latitude energy input and the TID/TAD. However, the result of TIEGCM in Plate 2 indicates such correspondence is not always kept, as shown in the schematic pictures of Figure 3.

In Figure 3a, impulsive heating of the high-latitude thermosphere can cause TID/TAD that propagates to lower latitudes. The TID/TAD is characterized by enhancements of temperature and equatorward wind speed and subsequent decreases as rarefaction waves [e.g., *Richmond et al.*, 1978]. Such correspondence is seen at 1200 UT and 1400 UT in Plate 2, as shown by the oblique red lines. The rarefaction (poleward wind) is easier to be observed by ground instruments as decrease in F layer virtual height and increases in 630-nm airglow intensity, f_oF_2, and TEC. On the other hand, continuous heating of the high-latitude thermosphere in Figure 3b may not cause TID/TAD but cause global reconfiguration (heating) of the thermosphere. Such example is seen in Plate 2 after 1600 UT. In addition, interaction of different waves in the high-latitude region, such as in the polar cap, may cause TID/TAD that propagates lower latitudes. In that

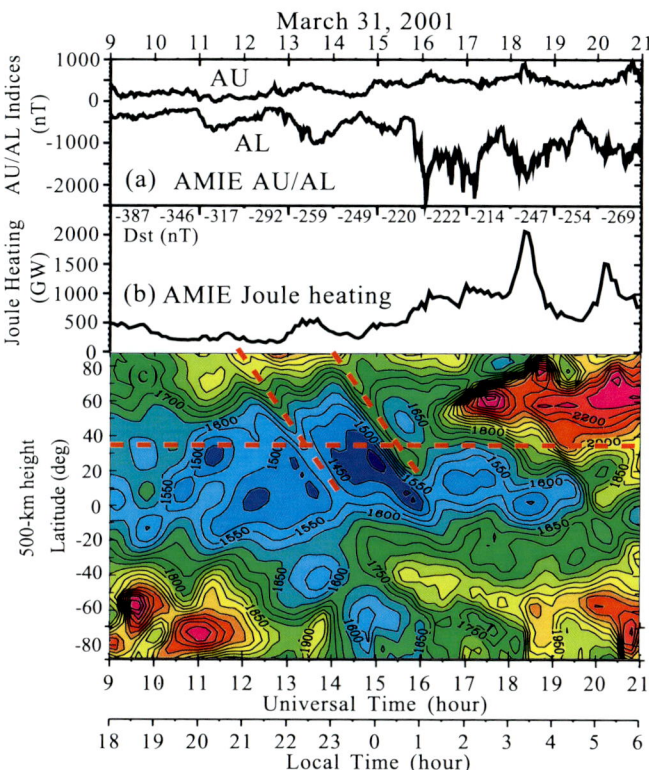

Plate 2. (a) AU/AL indices calculated from 192 ground magnetometers used for the AMIE procedure. (b) The northern hemisphere Joule heating rate. (c) Latitude–time contours of thermospheric temperatures at an altitude of 500 km calculated by the AMIE/TIEGCM at a longitude of 135.0°E for the magnetic storm of March 31, 2001. The horizontal red dashed line in Plate 2c indicates the latitude of Shigaraki (34.8°N). The two oblique red dashed lines in Plate 2c indicate southward propagation of two TADs. [From *Shiokawa et al.*, 2007.]

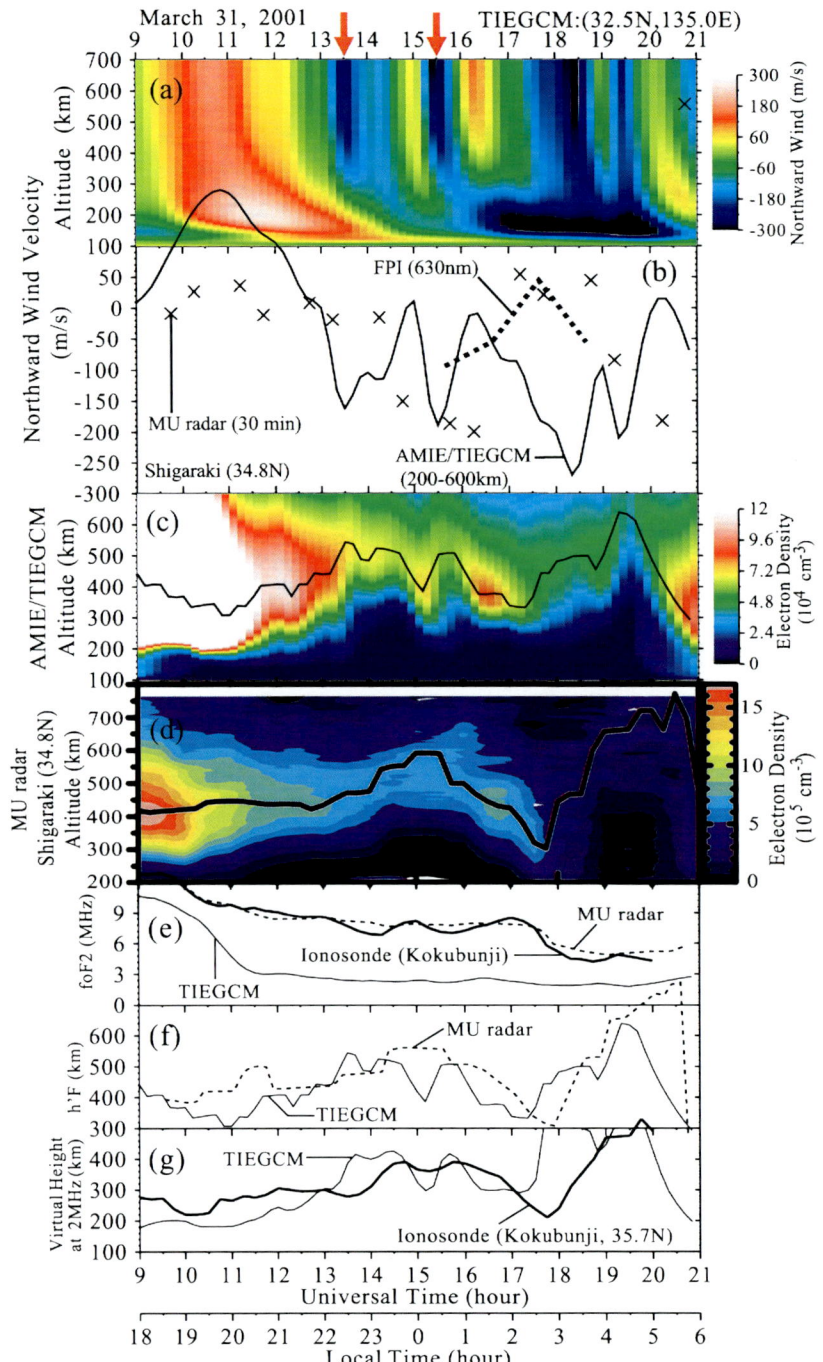

Plate 3. (a) Height profile of northward wind speed calculated by the AMIE/TIEGCM. (b) Northward wind speeds averaged over altitudes of 200–600 km by the AMIE/TIEGCM (thin solid curves), measured by the MU radar (crosses), and measured by FPI through the Doppler shift of the 630-nm airglow line (dashed curve). (c) Electron density profile calculated by the AMIE/TIEGCM. (d) Electron density profile measured by the MU radar. (e) f_oF_2 measured by the MU radar at Shigaraki and by the ionosonde at Kokubunji, Japan, and calculated by the AMIE/TIEGCM. (f) $h'F$ measured by the MU radar and calculated by the AMIE/TIEGCM. (g) Virtual height at 2 MHz measured by the ionosonde at Kokubunji and calculated by the AMIE/TIEGCM. The MU radar and FPI data were obtained at Shigaraki (34.8°N, 136.1°E). The AMIE/TIEGCM data are sampled at (32.5°N, 135.0°E). The solid curves in Plates 3c and 3d indicate the peak altitude of the F-layer electron density. The two red arrows at the top of the figure indicate two TADs identified in the model. The MU radar was operated with a cycle time of 1.5 h (1 and 0.5 h for ionospheric and mesospheric measurements, respectively). [From *Shiokawa et al.*, 2007.]

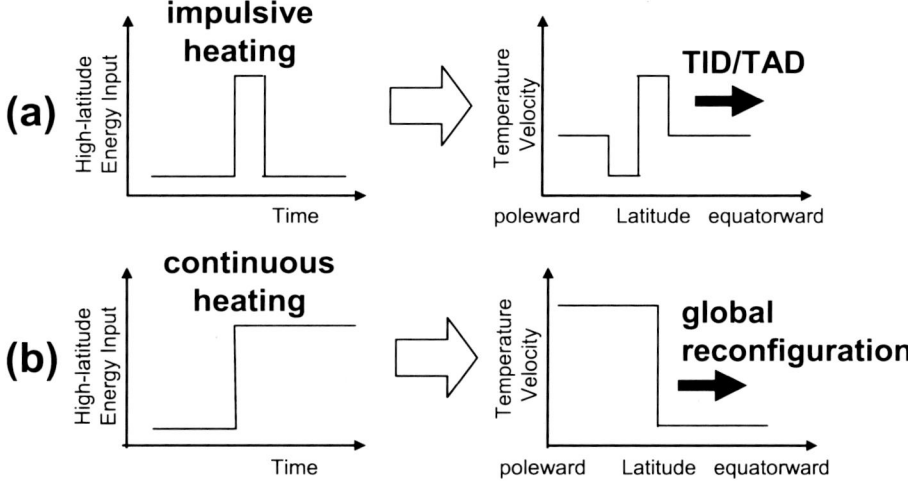

Figure 3. Schematic figures on the generation of TIDs/TADs from energy input in the high-latitude auroral zone: (a) impulsive heating causes impulsive TIDs/TADs at midlatitudes and (b) continuous heating causes global reconfiguration of the midlatitude thermosphere, which may not be identified as TIDs/TADs at middle latitudes.

case, the TID/TAD observed at midlatitudes may not have corresponding heating of the high-latitude region.

As discussed above, comparison of global model with observations gives various insights into the generation and propagation of storm-time TIDs. Such comparison also reveals the current limits of global models. For example, the absolute value of electron density in TIEGCM in Plates 3c and 3e is several factors smaller than that observed by the MU radar and the ionosonde in Plates 3d and 3e, suggesting that more precise information on the supply of O^+ ions from the plasmasphere (which determines the absolute density at midlatitudes) is needed in the model. The TAD phase speed in the model (~1100 m/s) in Plate 2 is much faster than the observed speed (370–640 m/s) from the imaging measurements, suggesting more precise information of the amount of high-latitude energy input (which controls the phase speed of TADs) is needed in the model [e.g., *Codrescu et al.*, 1997; *Lu et al.*, 2001]. For such comparison between models and observations, data-based estimation of high-latitude energy input, such as that by AMIE, would be essentially needed.

Acknowledgments. We thank Y. Katoh, M. Satoh, and T. Katoh of the Solar–Terrestrial Environment Laboratory, Nagoya University, for their kind support of the development and operation of the FPI. The observation at Shigaraki was carried out in collaboration with the Research Institute for Sustainable Humanosphere (RISH), Kyoto University. The GPS data from GEONET are provided by the Geographical Survey Institute, Japan. The MU radar at Shigaraki belongs to and is operated by RISH. The ionograms were supplied through WDC-C2 for ionosphere, National Institute of Information and Communication Technology, Tokyo. The work at HAO/NCAR was partly supported by the NASA Sun–Earth Connection Guest Investigator Program. This work was supported by a Grant-in-Aid for Scientific Research (11440145, 13573006 and, 13136201 and Priority Area 764) and Dynamics of the Sun–Earth–Life Interactive System (G-4, the 21st Century COE Program) of the Ministry of Education, Culture, Sports, Science and Technology of Japan.

REFERENCES

Afraimovich, E. L., E. A. Kosogorov, L. A. Leonovich, K. S. Palamartchouk, N. P. Perevalova, and O. M. Pirog (2000a), Determining parameters of large-scale traveling ionospheric disturbances of auroral origin using GPS-arrays, *J. Atmos. Sol. Terr. Phys.*, 62(7), 553–565.

Afraimovich, E. L., E. A. Kosogorov, L. A. Leonovich, K. S. Palamartchouk, N. P. Perevalova, and O. M. Pirog (2000b), Observation of large-scale traveling ionospheric disturbances of auroral origin by global GPS networks, *Earth Planets Space*, 52(10), 669–674.

Bailey, G. J., and N. Balan (1996), A low-latitude ionosphere–plasmasphere model, in *STEP Handbook*, edited by R. W. Schunk, p. 173, Utah State University, Logan.

Balthazor, R. L., and R. J. Moffett (1999), Morphology of large-scale traveling atmospheric disturbances in the polar thermosphere, *J. Geophys. Res.*, 104, 15–24.

Bristow, W. A., R. A. Greenwald, and J. C. Samson (1994), Identification of high-latitude acoustic gravity wave sources using the Goose Bay HF radar, *J. Geophys. Res.*, 99(A1), 319–331.

Bristow, W. A., R. A. Greenwald, and J. P. Villain (1996), On the seasonal dependence of medium-scale atmospheric gravity waves in the upper atmosphere at high latitudes, *J. Geophys. Res.*, *101*(A7), 15,685–15,699.

Buonsanto, M. J., M. Codrescu, B. A. Emery, C. G. Fesen, T. J. Fuller-Rowell, D. J. Melendez-Alvira, and D. P. Sipler (1997), Comparison of models and measurements at Millstone Hill during the January 24–26, 1993, minor storm interval, *J. Geophys. Res.*, *102*, 7267–7277.

Codrescu, M. V., T. J. Fuller-Rowell, and I. S. Kutiev (1997), Modeling the F layer during specific geomagnetic storms, *J. Geophys. Res.*, *102*(A7), 14,315–14,329.

Emery, B. A., et al. (1996), Assimilative mapping of ionospheric electrodynamics in the thermosphere–ionosphere general circulation model comparisons with global ionospheric and thermospheric observations during the GEM/SUNDIAL period of March 28–29, 1992, *J. Geophys. Res.*, *101*, 26,681–26,696.

Emery, B. A., C. Lathuillere, P. G. Richards, R. G. Roble, M. J. Buonsanto, D. J. Knipp, P. Wilkinson, D. P. Sipler, and R. Niciejewski (1999), Time dependent thermospheric neutral response to the 2–11 November 1993 storm period, *J. Atmos. Sol. Terr. Phys.*, *61*, 329–350.

Francis, S. H. (1975), Global propagation of atmospheric gravity waves: A review, *J. Atmos. Terr. Phys.*, *37*, 1011–1054.

Fujiwara, H., and Y. Miyoshi (2006), Characteristics of the large-scale traveling atmospheric disturbances during geomagnetically quiet and disturbed periods simulated by a whole atmosphere general circulation model, *Geophys. Res. Lett.*, *33*, L20108, doi:10.1029/2006GL027103.

Fujiwara, H., S. Maeda, H. Fukunishi, T. J. Fuller-Rowell, and D. S. Evans (1996), Global variations of thermospheric winds and temperatures caused by substorm energy injection, *J. Geophys. Res.*, *101*, 225–239.

Fuller-Rowell, T. J., M. V. Codrescu, R. J. Moffett, and S. Quegan (1994), Response of the thermosphere and ionosphere to geomagnetic storms, *J. Geophys. Res.*, *99*, 3893–3914.

Hajkowicz, L. (1990), A global study of large scale travelling ionospheric disturbances (TIDS) following a step-like onset of auroral substorms in both hemispheres, *Planet. Space Sci.*, *38*(7), 913–923.

Hajkowicz, L. A., and R. D. Hunsucker (1987), A simultaneous observation of large-scale periodic TIDs in both hemispheres following an onset of auroral disturbances, *Planet. Space Sci.*, *35*, 785–791.

Ho, C. M., A. J. Mannucci, U. J. Lindqwister, X. Pi, and B. T. Tsurutani (1996), Global ionosphere perturbations monitored by the worldwide GPS network, *Geophys. Res. Lett.*, *23*, 3219–3222.

Ho, C. M., A. J. Mannucci, L. Sparks, X. Pi, U. L. Lindqwister, B. D. Wilson, B. A. Iijima, and M. J. Reyes (1998), Ionospheric total electron content perturbations monitored by the GPS global network during two northern hemisphere winter storms, *J. Geophys. Res.*, *103*, 26,409–26,420.

Hocke, K., and K. Schlegel (1996), A review of atmospheric gravity waves and traveling ionospheric disturbances: 1982–1995, *Ann. Geophys.*, *14*, 917–940.

Hooke, W. H. (1968), Ionospheric irregularities produced by internal atmospheric gravity waves, *J. Atmos. Terr. Phys.*, *30*, 795–823.

Hunsucker, R. D. (1982), Atmospheric gravity waves generated in the high-latitude ionosphere: A review, *Rev. Geophys. Space Phys.*, *20*, 293–315.

Lee, C.-C., J.-Y. Liu, B. W. Reinisch, Y.-P. Lee, and L. Liu (2002), The propagation of traveling atmospheric disturbances observed during the April 6–7, 2000 ionospheric storm, *Geophys. Res. Lett.*, *29*(5),1068, doi:10.1029/2001GL013516.

Lee, C.-C., J.-Y. Liu, M.-Q. Chen, S.-Y. Su, H.-C. Yeh, and K. Nozaki (2004), Observation and model comparisons of the traveling atmospheric disturbances over the Western Pacific region during the 6–7 April 2000 magnetic storm, *J. Geophys. Res.*, *109*, A09309, doi:10.1029/2003JA010267.

Lu, G., et al. (1998), Global energy deposition during the January 1997 magnetic cloud event, *J. Geophys. Res.*, *103*(A6), 11,685–11,694.

Lu, G., A. D. Richmond, R. G. Roble, and B. A. Emery (2001), Coexistence of ionospheric positive and negative storm phases under northern winter conditions: A case study, *J. Geophys. Res.*, *106*, 24,493–24,504.

Maeda, S., and S. Handa (1980), Transmission of large-scale TIDs in the ionospheric F2-region, *J. Atmos. Terr. Phys.*, *42*, 853–859.

Miller, C. A., and M. C. Kelley (1997), Horizontal plasma flow at midlatitudes: More mechanisms and the interpretation of observations, *J. Geophys. Res.*, *102*(A6), 11,549–11,555, doi:10.1029/97JA03842.

Millward, G. H., R. J. Moffett, S. Quegan, and T. J. Fuller-Rowell (1993), Effects of an atmospheric gravity wave on the midlatitude ionospheric F layer, *J. Geophys. Res.*, *98*, 19,173–19,179.

Nicolls, M. J., M. C. Kelley, A. J. Coster, S. A. González, and J. J. Makela (2004), Imaging the structure of a large-scale TID using ISR and TEC data, *Geophys. Res. Lett.*, *31*, L09812, doi:10.1029/2004GL019797.

Ogawa, T., and H. Kumagai (1985), Deep depletions of total electron content associated with severe mid-latitude gigahertz scintillations during geomagnetic storms, *J. Geophys. Res.*, *90*, 6652–6656.

Otsuka, Y., K. Shiokawa, T. Ogawa, and P. Wilkinson (2004), Geomagnetic conjugate observations of medium-scale traveling ionospheric disturbances at midlatitude using all-sky airglow imagers, *Geophys. Res. Lett.*, *31*, L15803, doi:10.1029/2004GL020262.

Oyama, S., M. Ishii, Y. Murayama, H. Shinagawa, S. C. Buchert, R. Fujii, and W. Kofman (2001), Generation of atmospheric gravity waves associated with auroral activity in the polar F region, *J. Geophys. Res.*, *106*(A9), 18,543–18,554.

Pi, X., M. Mendillo, W. J. Hughes, M. J. Buonsanto, D. P. Sipler, J. Kelly, Q. Zhou, G. Lu, and T. J. Hughes (2000), Dynamical effects of geomagnetic storms and substorms in the middle-latitude ionosphere: An observational campaign, *J. Geophys. Res.*, *105*, 7403–7417.

Prölss, G. W. (1993), Common origin of positive ionospheric storms at middle latitudes and the geomagnetic activity effect at low latitudes, *J. Geophys. Res.*, *98*, 5981–5991.

Reddy, C. A., S. Fukao, T. Takami, M. Yamamoto, T. Tsuda, T. Nakamura, and S. Kato (1990), A MU radar-based study of mid-latitude F region response to a geomagnetic disturbance, *J. Geophys. Res.*, *95*, 21,077–21,094.

Rice, D. D., R. D. Hunsucker, L. J. Lanzerotti, G. Crowley, P. J. S. Williams, J. D. Craven, and L. Frank (1988), An observation of atmospheric gravity wave cause and effect during the October 1985 WAGS campaign, *Radio Sci.*, *23*, 919–930.

Richmond, A. D. (1978), Gravity wave generation, propagation, and dissipation in the thermosphere, *J. Geophys. Res.*, *83*, 4131–4145.

Richmond, A. D., and Y. Kamide (1988), Mapping electrodynamic features of the high-latitude ionosphere from localized observations: Technique, *J. Geophys. Res.*, *93*, 5741–5759.

Richmond, A. D., E. C. Ridley, and R. G. Roble (1992), A thermosphere/ionosphere general circulation model with coupled electrodynamics, *Geophys. Res. Lett.*, *19*, 601–604.

Shiokawa, K., Y. Otsuka, T. Ogawa, N. Balan, K. Igarashi, A. J. Ridley, D. J. Knipp, A. Saito, and K. Yumoto (2002), A large-scale traveling ionospheric disturbance during the magnetic storm of September 15, 1999, *J. Geophys. Res.*, *107*(A6), 1088, doi:10.1029/2001JA000245.

Shiokawa, K., Y. Otsuka, C. Ihara, T. Ogawa, and F. J. Rich (2003a), Ground and satellite observations of nighttime medium-scale traveling ionospheric disturbance at midlatitude, *J. Geophys. Res.*, *108*(A4), 1145, doi:10.1029/2002JA009639.

Shiokawa, K., et al. (2003b), Thermospheric wind during a storm-time large-scale traveling ionospheric disturbance, *J. Geophys. Res.*, *108*(A12), 1423, doi:10.1029/2003JA010001.

Shiokawa, K., Y. Otsuka, T. Tsugawa, T. Ogawa, A. Saito, K. Ohshima, M. Kubota, T. Maruyama, T. Nakamura, M. Yamamoto, and P. Wilkinson (2005), Geomagnetic conjugate observation of nighttime medium-scale and large-scale traveling ionospheric disturbances: FRONT3 campaign, *J. Geophys. Res.*, *110*, A05303, doi:10.1029/2004JA010845.

Shiokawa, K., G. Lu, Y. Otsuka, T. Ogawa, M. Yamamoto, N. Nishitani, and N. Sato (2007), Ground observation and AMIE–TIEGCM modeling of a storm-time traveling ionospheric disturbance, *J. Geophys. Res.*, *112*, A05308, doi:10.1029/2006JA011772.

Tsugawa, T., A. Saito, Y. Otsuka, and M. Yamamoto (2003), Damping of large-scale traveling ionospheric disturbances detected with GPS networks during the geomagnetic storm, *J. Geophys. Res.*, *108*(A3), 1127, doi:10.1029/2002JA009433.

Tsugawa, T., A. Saito, and Y. Otsuka (2004), A statistical study of large-scale traveling ionospheric disturbances using the GPS network in Japan, *J. Geophys. Res.*, *109*, A06302, doi:10.1029/2003JA010302.

Tsugawa, T., K. Shiokawa, Y. Otsuka, T. Ogawa, A. Saito, and M. Nishioka (2006), Geomagnetic conjugate observations of large-scale traveling ionospheric disturbances using GPS networks in Japan and Australia, *J. Geophys. Res.*, *111*, A02302, doi:10.1029/2005JA011300.

G. Lu, High Altitude Observatory, National Center for Atmospheric Research, 3450 Mitchell Lane, Boulder, CO 30301, USA.

T. Ogawa, Y. Otsuka, K. Shiokawa, Solar–Terrestrial Environment Laboratory, Nagoya University, Toyokawa 442-8507, Japan. (shiokawa@stelab.nagoya-u.ac.jp)

A. Saito, Graduate School of Science, Kyoto University, Kitashirakawa-Oiwakecho, Sakyo-ku, Kyoto 606-8502, Japan.

T. Tsugawa, National Institute of Information and Communications Technology, Koganei 184-8795, Japan. (tsugawa@nict.go.jp)

M. Yamamoto, Research Institute for Sustainable Humanosphere, Kyoto University, Gokanosho, Uji, Kyoto 611-0011, Japan.

A Digest of Electrodynamic Coupling and Layer Instabilities in the Nighttime Midlatitude Ionosphere

Roland T. Tsunoda

Center for Geospace Studies, SRI International, Menlo Park, California, USA

The basic elements of electrodynamic coupling and positive feedback between a sporadic E (E_s) layer and the F layer of the nighttime midlatitude ionosphere are briefly described. This coupled system is interesting because each layer, in isolation, can be unstable to perturbations in vertical displacement (i.e., the so-called E_s layer and Perkins instabilities). The two processes are similar generically and can be referred to as layer instabilities. Moreover, both have maximum growth rates when frontal perturbations are aligned northwest to southeast in azimuth (northern hemisphere), which leads to interactive behavior with positive feedback because the polarization electric fields that arise from both processes are in phase and additive. Consequently, the growth rates of both instabilities are enhanced by their mutual presence; the largest enhancement, however, occurs in the structuring of the F layer. Some of the interesting aspects of coupled-system behavior are described and discussed.

1. INTRODUCTION

There is considerable interest in the development of structure in plasma density (N) in the nighttime F region ionosphere. From a space–weather perspective, large-scale gradients in N (∇N) can cause refractive errors in navigation and tracking, while smaller-scale scintillation-producing irregularities in N (ΔN) can cause disruptions in communications. Because hope of mitigating these effects is small, short-term forecasts based on an understanding of the underlying physics appears to be the best alternative approach. Toward this end, we attempt to capsulate the physics relevant to the midlatitude ionosphere.

There is little doubt that the most intense ΔN in the nighttime midlatitude F layer arises from an "interchange process." That is, ΔN is produced by the transport of plasma in directions aligned with a background ∇N. Local depletions (enhancements) in N appear when there is local transport of lower (higher) N into regions of higher (lower) N. The growth rate of ΔN is simply proportional to V/L, where V is the local transport velocity along ∇N and L is a measure of the steepness of ∇N. The spatial scale of ΔN is related to that of V. The term "interchange" comes from descriptions in which the plasma, incompressible and "frozen" on geomagnetic field ($\vec{B}$) lines, is transported in the plane transverse to $\vec{B}$ by an electric field ($\vec{E}$). This description is appropriate for F region plasma because the ion-neutral collision frequency (ν_{in}) is much smaller than the ion gyro-frequency (Ω_i) and plasma transport is dictated by the requirement, $\nabla \times \vec{E} = 0$.

The source for interchange is most commonly a polarization $\vec{E}$ ($\vec{E}_p$), which must arise from a need to maintain $\nabla \cdot \vec{J} = 0$, where $\vec{J}$ is the field line-integrated current transverse to $\vec{B}$. In the midlatitude ionosphere (unlike that over the dip equator), $\vec{B}$ lines have sizeable inclination angles (I), measured from the horizontal plane. Consequently, excess electrical charges of opposite sign that arise in regions of upward and downward ∇N on a $\vec{B}$ line would be neutralized by electrons moving freely along that $\vec{B}$ line. Hence, the first midlatitude problem was to determine how an $\vec{E}_p$ could develop in a horizontally stratified F layer, that is, without a horizontal component of ∇N. A breakthrough occurred when *Perkins* [1973] showed that the midlatitude F layer is susceptible to another kind of plasma instability that differs from the usual form of the interchange instability. The nature of the underlying physics, described in terms of a $\vec{J} \times \vec{B}$ force [e.g., *Perkins*, 1973; *Kelley and Miller*, 1997; *Cosgrove et al.*, 2004; *Zhou*, 2006], remained difficult to comprehend, especially in terms of the role played by $\vec{E}_p$.

Until recently, F region plasma structure has been attributed to the Perkins instability, operating alone, despite the fact that its growth rate (γ_p) seems discouragingly small. That is, thermal fluctuations would not be amplified to a level large enough to explain observed F layer altitude modulations of 50–100 km [e.g., *Behnke*, 1979; *Kelley et al.*, 2000] or $\vec{E}_p$ as large as 17–22 mV/m [e.g., *Behnke*, 1979; *Swartz et al.*, 2002]. However, the seed is likely to be nonthermal and with an amplitude much larger than that of thermal noise, such as an atmospheric gravity wave (AGW). In this case, γ_p could prove to be large enough [*Huang et al.*, 1994]. *Cosgrove* [2007], for example, invoked a seed perturbation of ±5 km in altitude that led to sizeable F region structure in 2–3 h. The main reason for not discarding the Perkins instability, however, has been the belief that it alone could account for the peculiar northwest-to-southeast alignment of the frontal F region structures that are detected in the northern hemisphere [e.g., *Behnke*, 1979; *Kelley and Miller*, 1997; *Shiokawa et al.*, 2003].

There is now mounting evidence that γ_p may be significantly enhanced by an $\vec{E}_p$, which develops in the presence of a sporadic E (E_s) layer and maps to the F region. The first notion that there should be interactive behavior between E_s and F layer processes [*Tsunoda and Cosgrove*, 2001] was prompted by the discovery that a large $\vec{E}_p$ can develop at midlatitudes from the polarization of a Hall current [*Haldoupis et al.*, 1996] and a demonstration that $\vec{E}_p$ can develop under commonly occurring conditions [*Tsunoda*, 1998]. Given the apparent vigor of this process, *Tsunoda and Cosgrove* [2001] suggested that $\vec{E}_p$ from an F region source could map to the E region and be amplified through positive feedback by the process just mentioned. (Earlier, evidence of correlated occurrences of E_s and an altitude-modulated F layer [e.g., *Bowman*, 1960a] had been interpreted in terms of AGWs and ion-drag effects along $\vec{B}$ [*Bowman*, 1960b; 1968], but not in terms of electrodynamic coupling.)

Interest in coupled interactive behavior between the E_s and F layer grew with the discovery that the E_s layer is itself unstable in the presence of the very same vertical shear in the neutral wind ($\vec{U}$) that formed the E_s layer [*Cosgrove and Tsunoda*, 2002b]. Most intriguing was the finding that this E_s layer (E_sL) instability has a growth rate (γ_E) that maximizes when frontal E_s perturbations are aligned northwest to southeast in azimuth, a behavior that is similar to that for γ_p. Given that unstable frontal perturbations in both E_s and F layers have similar azimuthal alignments, it was not surprising to find that the electrically coupled system is more unstable than either of the instabilities operating alone [*Cosgrove and Tsunoda*, 2004].

A coherent picture has now begun to emerge. A first step was to understand how the Perkins instability functions in terms of $\vec{E}_p$ [*Tsunoda*, 2006]. With both instabilities described explicitly in terms of the coupling parameter $\vec{E}_p$, it is evident that both the Perkins and E_sL instabilities can be classified generically as layer instabilities [*Tsunoda*, 2006; *Cosgrove*, 2007]. That is, the whole layer, not just a ∇N region (as in the case of the interchange instability), is dynamically unstable to perturbations in local layer displacement. In following sections, we describe and discuss the elements of interactive behavior between these two processes.

2. ELEMENTS OF COUPLED INTERACTIVE BEHAVIOR

2.1. Hall Polarization

The terms Hall and Pedersen polarization [*Tsunoda and Cosgrove*, 2001] are used to indicate the nature of the primary current, that is, a Hall or a Pedersen current. Current continuity is always sustained by a Pedersen current, driven by an $\vec{E}_p$. A sketch of Hall polarization and the development of the $\vec{E}_p$ (that can lead to the E_sL instability) is shown in Figure 1 (left). We assume presence of a geographically extended E_s layer, which contains altitude perturbations in the form of two highly elongated bands. Projection of those bands onto a plane transverse to $\vec{B}$ is shown in Figure 1. We neglect presence of a background E layer. The axes are as follows: x is magnetic east, y is along $\vec{B}$, and z is in the magnetic meridian plane. The high (low) band is assumed to be located slightly above (below) a shear node for the zonal $\vec{U}$ (U_x); U_x is westward (eastward) in the high (low) band, as indicated by the arrows. We assume the meridional component ($U_{y'}$) is zero, which means that the background current is zero. The expression for the $\vec{E}_p$ that sets up to maintain a continuous current in the direction ($\vec{k}_E$) normal to the bands (see *Tsunoda et al.* [2004] for details), is given by

$$E_p = -\frac{J_x C_\alpha + J_z S_\alpha}{\Sigma_P} = -U_x B \left[\frac{\Sigma_H}{\Sigma_P} C_\alpha + S_\alpha\right]$$

$$\approx -\frac{\Sigma_H}{\Sigma_P} C_\alpha U_x B$$

(1)

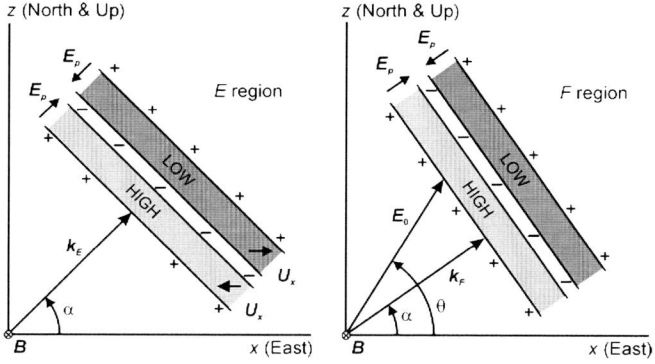

Figure 1. Sketches (in planes transverse to $\vec{B}$) show frontal perturbations that are displaced in altitude from their equilibrium altitudes. In the E region (left), the E_s bands are displaced relative to the zonal wind shear node; in the F region (right), the F layer bands are displaced relative to the equilibrium altitude where the ambient F layer is balanced by vertical component of forces, which could include $\vec{E}$, $\vec{U}$, and $\vec{g}$. Vector quantities are shown by bold and italicized fonts.

where $C_\alpha \equiv \cos\alpha$, $S_\alpha \equiv \sin\alpha$, J_x (J_z) is the field line-integrated Hall (Pedersen) current, Σ_H is the field line-integrated Hall conductivity of the E_s band, and Σ_P is the sum of the field line-integrated Pedersen conductivities of the E_s and F layers. It is clear from equation (1) that $|\vec{E}_p|$ could become larger than $|U_x B|$ for intermediate values of α and in cases where $\Sigma_H > \Sigma_P$.

Besides electrical loading (by Σ_H / Σ_P) [e.g., *Shalimov et al.*, 1998; *Hysell and Burcham*, 2000], $|\vec{E}_p|$ is affected by any restrictions in the flow of the secondary Hall current that is driven by $\vec{E}_p$ [e.g., *Shalimov et al.*, 1998; *Tsunoda*, 1998]. Studies have shown that closure of this secondary current is possible through a Pedersen current in the F region [*Shalimov et al.*, 1998; *Hysell and Burcham*, 2000] or through E_s bands above and below a wind shear node [*Cosgrove and Tsunoda*, 2001, 2002a]. *Yokoyama et al.* [2004] performed three-dimensional computer simulations and showed that both closure paths are reasonable. *Yokoyama et al.* [2005] also showed that theoretical estimates of secondary Hall currents are consistent with rocketborne magnetometer measurements.

2.2. Pedersen Polarization

A sketch of Pedersen polarization in the F layer, which leads to the Perkins instability, is shown in Figure 1 (right). As with *Perkins* [1973], we assume that the F layer, with a constant total electron content, is describable in terms of v_{in} at the average layer altitude ($\langle v_{in} \rangle$). We assume presence of an applied $\vec{E}$ ($\vec{E}_0$) as shown. The expression for the $\vec{E}_p$ that sets up to maintain a continuous current in the direction ($\vec{k}_F$) normal to the bands (see *Tsunoda* [2006] for details), is given by

$$E_p = \frac{E_0 C_{\theta-\alpha}(\Sigma^F_{P0} - \Sigma^F_P)}{\Sigma^F_P} = E_0 C_{\theta-\alpha}\left[\frac{\langle v_{in}\rangle_0}{\langle v_{in}\rangle} - 1\right] \quad (2)$$

where $C_{\theta-\alpha} \equiv \cos(\theta-\alpha)$. Here, the change in current is produced by a change in Σ^F_P, which is produced by a change in $\langle v_{in}\rangle$. Hence, $\vec{E}_p$ is determined by a change in Σ^F_P, which results from a vertical displacement in altitude away from equilibrium. Notice from equation (2) that large values of $\vec{E}_p$ are difficult to realize.

2.3. Preferred Azimuthal Orientation

The basis for interactive behavior boils down to the fact that the vertical component of ion transport, in the presence of $\vec{E}_p$, is in the same direction in both E and F regions. That is, displacement is in the Pedersen direction in the E region and the Hall direction in the F region. Hence, vertical transport comes from the meridional component of $\vec{E}_p$ in the E region and from the zonal component of $\vec{E}_p$ in the F region. This means that a northeastward (southwestward) $\vec{E}_p$ produces upward (downward) transport in both E and F regions (northern hemisphere) [*Tsunoda*, 2006]. As can be seen from Figure 1, such a favorably aligned $\vec{E}_p$ arises from the polarization of frontal structures that are aligned northwest to southeast in both E and F regions. The expressions for the vertical (upward) ion velocity ($V_{z'}$) in the E and F regions (see *Tsunoda* [2006] for details) are given by

$$V^E_{z'} \approx \frac{C_I}{\rho_i}\left[U_x + \frac{E_p S_\alpha}{B}\right] \quad (3)$$

where ρ_i is v_{in} normalized by the ion gyrofrequency, and

$$V^F_{z'} = \frac{E_0 S_{\theta-\alpha} S_\alpha C_I}{B}\left[\frac{\langle v_{in}\rangle_0}{\langle v_{in}\rangle} - 1\right] = \frac{S_{\theta-\alpha} S_\alpha C_I}{C_{\theta-\alpha}}\frac{E_p}{B}. \quad (4)$$

The first term in equation (3) leads to the formation of E_s in the presence of a zonal wind shear. Again, it should be apparent from equations (3) and (4) that a northeastward $\vec{E}_p$ produces upward drift in both E_s and F layers. Notice that while $\vec{E}_p$ from Hall polarization in equation (1) can be large, $V^E_{z'}$ from equation (3) is small compared with $V^F_{z'}$ from equation (4). Note also that the F layer plasma is transported in the $\vec{E}_p \times \vec{B}$ direction, which produces altitude modulation along the bottomside of the F layer. This modulation has been referred to as medium-scale traveling ionospheric disturbances (MSTIDs) [e.g., *Mendillo et al.*, 1997; *Taylor et al.*, 1998; *Garcia et al.*, 2000; *Shiokawa et al.*, 2003].

3. SOURCES OF INSTABILITY

3.1. Sporadic E Layer Instability

Wind shear theory is well established as the primary source for E_s [e.g., *Whitehead*, 1970]. However, previous derivations of this theory have not allowed for the development of $\vec{E}_p$; the likely reason is that Hall polarization was not expected in the midlatitude ionosphere (see discussion of *Tsunoda et al.* [2004]). Inclusion of $\vec{E}_p$ from Hall polarization led to the discovery of the E_sL instability [*Cosgrove and Tsunoda*, 2002b]. Descriptions of the physics underlying this instability are given by *Cosgrove and Tsunoda* [2003], *Tsunoda et al.* [2004], and *Tsunoda* [2006]. The growth rate of the E_sL instability (γ_E) is essentially the derivative of equation (3) with respect to altitude (z') [*Tsunoda et al.*, 2004] and is given by

$$\gamma_E \approx \frac{U_x C_I}{H_U \rho_i} \left[\frac{\Sigma_H^{E_s}}{\Sigma_P} S_\alpha C_\alpha - 1 \right] \quad (5)$$

where

$$H_U \equiv \left[\frac{1}{U_x} \frac{\partial U_x}{\partial z'} \right]^{-1}. \quad (6)$$

This growth rate for a perturbation in vertical displacement is identical with that of *Cosgrove and Tsunoda* [2002b, equation (11)] if we set Σ_P^F to zero (unloaded case) and ignore the last term in equation (5). The last term, which is responsible for the formation of E_s, acts to create an instability threshold. For example, if $S_\alpha C_\alpha \approx 0.5$ (i.e., $\alpha \approx 45°$), $\gamma_E = 0$ when $\Sigma_H^{E_s} \approx 2\Sigma_P$. A similar threshold was obtained by *Cosgrove and Tsunoda* [2004], whose results for an E_sL instability loaded by a passive F layer show that $\gamma_E \approx 0$, when they assumed $\Sigma_H^{E_s} \approx \Sigma_P$. (Earlier, *Cosgrove and Tsunoda* [2003] showed in computer simulations that altitude modulation of the E_s layer is quite small when loaded by the presence of a typical F layer; they used $\Sigma_H^{E_s}/(\Sigma_P^E + \Sigma_P^F) \approx 2.5$.)

Cosgrove and Tsunoda [2002b] used a rotational wind shear and showed that inclusion of a meridional $\vec{U}$ affects Σ_H^E, which increases γ_E slightly. An important feature of equation (5) is the azimuthal dependence of γ_E (through $S_\alpha C_\alpha$), which has a maximum (γ_E^{max}), where $\alpha = 45°$. We, therefore, find that the alignment of the E_s bands, as shown in Figure 1, is not fortuitous. We can see from equations (1) and (3) that this dependence originates from the control of the direction of $\vec{E}_p$ by the frontal alignment and from the dependence of vertical transport on the meridional component of $\vec{E}_p$.

While γ_E provides the basis for frontal alignment and $\vec{E}_p$ generation, *Cosgrove and Tsunoda* [2003] showed how nonlinear processes control the temporal evolution of the E_s layer. Perhaps the most interesting result was the finding that a slightly raised portion of the E_s sheet is transported horizontally by the northward component of $\vec{U}$ (U_y), relative to a lowered (or unlifted) portion of the E_s sheet. When the higher and lower E_s sections overlapped, $\vec{E}_p$ was drastically reduced, which led to convergence and coalescence of the two E_s sections. Upon merging back into a single layer, the E_sL instability was rejuvenated and the cycle repeated (see *Cosgrove and Tsunoda* [2003] or *Tsunoda et al.* [2004]). This process of repeatedly folding an E_s band onto itself is intriguing because N must increase progressively, which should lead to large values. An E_s layer would then be transformed into a train of frontal bands containing enhanced N.

3.2. Perkins Instability

Before describing this instability, we can see from equation (2) that F layer bands, such as sketched in Figure 1, are unstable to the development of $\vec{E}_p$. That is, $\vec{E}_p = 0$ when $\langle v_{in} \rangle = \langle v_{in} \rangle_0$ (i.e., no displacement); $\vec{E}_p$ increases (in northeastward direction) as upward displacement increases (i.e., $\langle v_{in} \rangle$ decreases). Hence, the rate of displacement increases as displacement increases, which is an unstable situation. The only fundamental difference between the E_sL and Perkins instabilities is that the latter has an equilibrium altitude that depends on $\langle v_{in} \rangle$ and the strength of the eastward component of $\vec{E}_0$, as follows,

$$\frac{E_0 C_\theta C_I}{B} = \frac{g S_I^2}{\langle v_{in} \rangle_0}. \quad (7)$$

For example, if we lift an F layer band, a northeastward $\vec{E}_p$ appears. At that new altitude, $\langle v_{in} \rangle$, the eastward component of $\vec{E}_p$, given by equation (2) multiplied by C_α, would add to the left side of equation (7). But at $\langle v_{in} \rangle$, the right side of equation (7) would become $gS_I^2/\langle v_{in} \rangle$. Clearly, if the left side is larger (smaller) than the right side, the lifted F layer would continue to rise (fall), which would be an unstable (stable) situation. *Tsunoda* [2006] showed that marginal stability occurs when $\vec{E}_0$ is in the $\vec{k}_F$ direction ($\theta = \alpha$), normal to the phase front. He then showed that instability (stability) occurs when $\theta > \alpha$ ($\theta < \alpha$). In essence, the normal component of $\vec{E}_0$ determines the strength of $\vec{E}_p$, while a tangential component of $\vec{E}_0$ acts to reduce the influence of gravity without reducing $\vec{E}_p$. We prefer this description over the use of northward and eastward components of $\vec{E}_0$ [e.g., *Perkins*, 1973; *Cosgrove*, 2007] because the former allows separation of polarization and equilibrium effects, whereas the latter mixes the two (see *Tsunoda* [2006] for details.)

Tsunoda [2006] showed that the γ_P derived by *Perkins* [1973] is essentially an instability in vertical displacement, which is obtained by differentiating equation (4) with altitude (z') and is given by

$$\gamma_P \approx \frac{E_0 S_{\theta-\alpha} S_\alpha C_l}{BH} \quad (8)$$

The maximum γ_P is evident from equation (8) and given by

$$\gamma_P^{\max} = \frac{E_0 C_l}{BH} \sin^2 \frac{\theta}{2}. \quad (9)$$

These expressions are for an unloaded F layer, which is reasonable because the conductivity of the ambient nighttime E layer is small. Numerical simulations have shown that the azimuthal alignment of frontal perturbations appears to preserve the directional property of equation (9) [e.g., *Kelley and Miller*, 1997; *Zhou et al.*, 2006].

3.3. Linear Aspects of Coupling

As expected from the above discussions, *Cosgrove and Tsunoda* [2004] found that the growth rate of the fastest growing mode, when the E_s and F layers are electrically coupled, is significantly larger than either γ_E or γ_P. Because their derivation is not easily interpretable in physical terms, *Tsunoda* [2006] presented a simplified derivation of the $E_s L$ and Perkins instabilities, for the coupled case. To summarize that derivation, the $\vec{E}_p$ generated in the E_s and F layers are added and used in equations (3) and (4) to modify the expressions for $V_{z'}^E$ and $V_{z'}^F$. The coupled growth rates, γ_E^C and γ_P^C, are given by

$$\gamma_E^c \approx \gamma_E + \frac{E_0 C_{\theta-\alpha} S_\alpha C_l}{\rho_i BH} \quad (10)$$

where γ_E is given by equation (5) and

$$\gamma_P^c \approx \gamma_P + \frac{\Sigma_H^{E_s}}{\Sigma_P} \frac{U_x C_\alpha^2 C_l}{H_U} \quad (11)$$

where γ_P is given by equation (8).

Using a different approach, *Cosgrove et al.* [2004] derived coupled growth rates from a circuit model. The two roots for growth rates (λ_{EFCL}) are given by

$$\lambda_{EFCL} = \frac{\Sigma_P^E \gamma_E^0 + \Sigma_P^F \gamma_P}{\Sigma_P^F + \Sigma_P^E} \approx \frac{\Sigma_P^E}{\Sigma_P^F} \gamma_E^0 + \gamma_P \quad (12)$$

for the positive root, and $\lambda_{EFCL} = 0$ for the negative root. (Both γ_E^0 and γ_P are for unloaded cases.) We see that equation (12) is similar to equation (11).

We can compare these formulas with the more comprehensive results obtained by *Cosgrove and Tsunoda* [2004], which are reproduced in Figure 2, where various growth rates are plotted as a function of the effective eastward $\vec{E}$ (i.e., $E'_{y''}$ in the figure). The growth rate of the fastest growing mode is shown by the solid line, γ_E by the dashed line and γ_P by the dotted line. Our result for γ_E in equation (5) should correspond with the γ_E curve in Figure 2, where $E'_{y''} = 0$. Both indicate that γ_E is very small for the case considered by *Cosgrove and Tsunoda* [2004], that is, $\Sigma_H \approx \Sigma_P$, where Σ_H and Σ_P are the total field line-integrated Hall and Pedersen conductivities, respectively. We see from equation (5) that the quantities used in the square brackets essentially cancel for this case. The γ_P curve is a constant because $U_{y'}$ in the F region was adjusted to keep the F layer at a constant altitude, which cancels the variation of the $E'_{y'}$ in Figure 2. According to *Cosgrove and Tsunoda* [2004], the increase in γ_E away from $E'_{y''} = 0$ occurs because $E'_{y''}$ drives a meridional current through the E_s layer, which interacts with the modulation in $\Sigma_H^{E_s}$ that is produced by the tilts associated with the assumed sinusoidal perturbation of the E_s layer. Both a tilt and a nonzero $E'_{y''}$ do not exist in our simple model (Figure 1), but the finding that γ_E can be larger than γ_P, even under F layer-loaded conditions, is interesting.

The solid-line curve in Figure 2 displays a local maximum at $E'_{y''} = 0$, which is the condition for "spatial resonance," that is, when the F layer plasma drifts together with the E_s layer in the magnetic meridian plane. In this example, we find that the coupled growth rate is larger than γ_p, even when $\gamma_E \approx 0$. To explain this behavior, we have noted earlier that $\gamma_E \approx 0$ occurs when the bracketed quantity in equation (5) is zero. This means that $V_{z'}^E = 0$, but $E_p^F = -U_x B/S_\alpha$, which is not zero. Hence, we have a contribution to $V_{z'}^F$ from E_p^E through E_p in equation (4), even when $\gamma_E \approx 0$. The contribution to γ_P^c, through the second term in equation (11), is obtained by simply setting $\Sigma_H^{E_s} = \Sigma_P$. In fact, this coupling term continues to contribute to γ_P^c even when $\gamma_E < 0$ ($\Sigma_H^{E_s} < \Sigma_P$). We also see that the coupling term in equation (11) is larger than that in

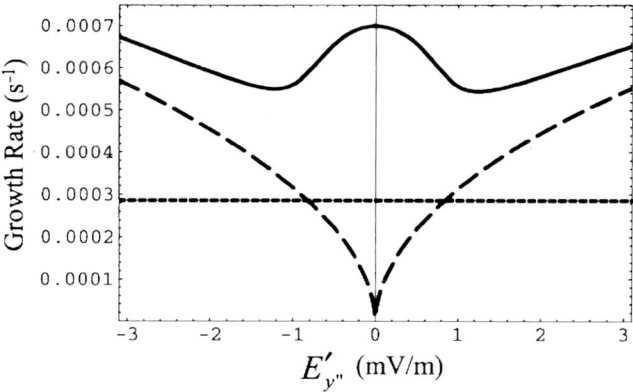

Figure 2. Plots of the growth rate of the fastest growing mode in the coupled case (solid line), together with those for the loaded $E_s L$ instability (long dashed line) and the Perkins instability (short dashed line). From *Cosgrove and Tsunoda* [2004].

equation (10), which means that the local maximum in Figure 2 is probably related to γ_p^c, not to γ_E^c. As $|E'_{y''}|$ increases, the interaction time between the two processes decreases, which leads to the decrease seen in Figure 2. The mode of the fastest growth rate, therefore, changes near the inflection points in the solid-line curve, as indicated by the fact that this curve tracks the dashed curve for large $|E'_{y''}|$. These results are important because they show that an E_s layer acts to enhance the structuring of the F layer by the Perkins instability and that it does not behave simply as a passive load [*Klevans and Imel*, 1978], even when $\gamma_E^c \leq 0$.

It is interesting to note that the strength of the coupling term in equation (11) depends on U_x/Σ_p, which arises from E_p^E in equation (1). We see from equation (3) that $V_{z'}^E > 0$ as long as $E_p^E + E_p^F > U_x B/S_\alpha$. As the E_s band is lifted in altitude, E_p^E increases with U_x. In other words, γ_p^c increases with time during the interaction between E_s and F layers. In turn, $V_{z'}^F$ increases as E_p^E increases, and as the F layer band is lifted further in altitude, Σ_p will decrease, and the strength of the coupling term in equation (11) would increase as Σ_p decreases. Hence, we have reason to believe that the growth rate of F region plasma structure would increase with time, beyond that indicated in Figure 2, because of the presence of this positive feedback [*Tsunoda*, 2006]. Thus, this process, besides the seeding of the uncoupled Perkins instability by AGWs, provides another interpretation for occurrences of 20-mV/m electric fields and 50- to 100-km modulations of the F layer.

3.4. Nonlinear Aspects of Coupling

Numerical simulations of the coupled E_s–F layer system [*Cosgrove*, 2007] have verified some of the behavior predicted from linear analysis, while providing further insight into its temporal development. Unfortunately, because of space restrictions, we are only able to summarize some of the key findings. First, by applying a horizontal seed perturbation of 200 km to the F layer, *Cosgrove* [2007] demonstrated that the electrodynamic coupling effects involving E_s are not confined to spatial scales of only a few tens of kilometers. Second, under spatially resonant conditions ($E'_{y''} = 0$ in Figure 2) and $\Sigma_H^{Es}/\Sigma_P^F = 1$, when $\gamma_E \approx 0$, the simulations showed that the E_sL instability was still active, which seems to be consistent with the above linear analysis. The process, however, became quickly nonlinear, displaying the cyclical overlapping of E_s bands (i.e., "wave breaking") followed by a coalescing of the overlapping bands; this pattern is similar to the nonlinear behavior of the uncoupled E_sL instability [*Cosgrove and Tsunoda*, 2003]. The coalescing of bands led to an increase in Σ_H^{Es} and $\vec{E}_p$. Modulation growth appeared to still be controlled by the Perkins instability, but with a shortened growth time. But in the case where $\Sigma_H^{Es}/\Sigma_P^F \geq 1$, modulation growth appeared to be controlled by the E_sL instability. These results also verify the behavior suggested by equation (11).

Third, the behavior found under nonresonant conditions ($E'_{y''} \neq 0$ in Figure 2) is essentially that produced by a meridional Hall current driven by $E'_{y''}$ and the enhanced Σ_H^{Es} produced by E_sL instability, as part of the wave breaking process described above. *Cosgrove* [2007] found that $\vec{E}_p$ could be enhanced by a factor of five. Hence, we have a situation in which $\vec{E}_p$ (associated with E_s bands) increases, and the interaction time between that $\vec{E}_p$ and the F layer decreases, with increasing $E'_{y''}$. Consequently, F layer modulation could be large under resonant conditions, but tends to be small under nonresonant conditions, although $\vec{E}_p$ tends to be larger under nonresonant conditions. Modulation amplitude evidently depends more on interaction time than on $|E'_{y''}|$.

4. DISCUSSION

In this paper, we have attempted to describe, in physical terms, why coupled interactive behavior should be expected between layer instabilities of the E_s and the F layer in the nighttime midlatitude ionosphere. Moreover, the feedback should be positive, given that the $\vec{E}_p$ generated by both processes are more or less in phase and hence additive. Indications are that the coupled response could become "explosive" in the sense that both γ_E and γ_p^c depend inversely on Σ_p, and Σ_p^F can decrease by a factor of four [e.g., *Makela and Kelley*, 2003] or more. If loading of the E_sL instability is reduced progressively in this manner [*Tsunoda*, 2006], it may be responsible for some of the more spectacular effects that have been observed in the nighttime F layer [e.g., *Behnke*, 1979; *Fukao et al.*, 1991; *Kelley et al.*, 2000; *Swartz et al.*, 2002]. Once structuring is initiated by the layer instabilities, secondary processes such as interchange instabilities should appear and assist with the production of ΔN, as described in the Introduction.

The results also indicate that plasma structuring from a coupled response would be most noticeable in the F region. Spatial scales typically associated with MSTIDs (e.g., 200 km) could very well be initiated in the F region, but it would be enhanced by the coupled response of the system [*Cosgrove*, 2007] in the presence of an underlying E_s layer. While this digest has focused on conveying the basic physics underlying the nature of the coupled system, it should be evident that nonlinear response plays an important role in the temporal development of the plasma structure [e.g., *Cosgrove and Tsunoda*, 2003; *Cosgrove*, 2007]. Most notable of these effects are the cyclical "folding and compression" of E_s bands, which should increase Σ_H^{Es} and $\vec{E}_p$, and the relative importance of longer interaction times when $|E'_{y''}|$ is small, versus larger $\vec{E}_p$, because $|E'_{y''}|$ is large but with shorter available interaction times. Other effects remain to be investigated.

It is worth mentioning that *Haldoupis et al.* [2003] have proposed that Hall polarization operates on a different model of E_s, which could be responsible for altitude modulation of the F layer. Their model differs from ours in several respects. Their E_s patch is not at a wind shear node and is assumed to drift westward with $\vec{U}$. They argue that a prevailing westward $\vec{U}$ produces vertical ion convergence and hence E_s. While true, the rate of ion convergence is smaller by an order of magnitude than that produced by wind shear [*Whitehead*, 1970], and *Larsen* [2002] has shown that the large $\vec{U}$ in the altitude range of interest is usually associated with a rotational wind shear. Hence, if E_s is present, it is more likely to be associated with wind shear. Their evidence that E_s patches must be drifting westward comes from azimuth scan radar backscatter from field-aligned irregularities. The middle and upper atmosphere (MU) radar has also found from azimuth scan data that these echo regions appear to drift westward. However, from a two-radar experiment, which included the MU radar, *Yamamoto et al.* [1997] showed that these regions appeared to have fronts that were aligned northwest to southeast and drifted southwestward. We suggest that the apparent westward drift is more likely the trace speed of the frontal structures just described. We can also add that uplifted frontal structures of the kind sketched in Figure 1 is associated with a large northeastward $\vec{E}_p$, which would produce large negative Doppler velocities in observations with a northward-looking radar. We, therefore, suggest that their evidence is actually consistent with the frontal model we have described. While their model seems theoretically plausible [*Shalimov and Haldoupis*, 2005], it would not seem to be representative of most observations.

In closing, we note that we have discussed only the case of perfect mapping of $\vec{E}_p$ between E_s and F layers. We envision that the E_sL instability can operate vigorously at smaller spatial scales, where the related $\vec{E}_p$ would not map to the F layer and be shorted out. This topic, however, is beyond the scope of this paper.

Acknowledgments. The author thanks Russell Cosgrove for useful discussions. Research was supported by the National Science Foundation under grant ATM-0333138.

REFERENCES

Behnke, R. (1979), *F* layer height bands in the nocturnal ionosphere over Arecibo, *J. Geophys. Res.*, *84*, 974–978.

Bowman, G. G. (1960a), Some aspects of sporadic *E* at midlatitudes, *Planet. Space Sci.*, *2*, 195.

Bowman, G. G. (1960b), Further studies of 'spread *F*' at Brisbane—II. Interpretation, *Planet. Space Sci.*, *2*, 150.

Bowman, G. G. (1968), Movements of ionospheric irregularities and gravity waves, *J. Atmos. Terr. Phys.*, *30*, 721.

Cosgrove, R., and R. T. Tsunoda (2001), Polarization electric fields sustained by closed-current dynamo structures in midlatitude sporadic *E*, *Geophys. Res. Lett.*, *28*, 1455.

Cosgrove, R., and R. T. Tsunoda (2002a), Wind-shear-driven, closed-current dynamos in midlatitude sporadic *E*, *Geophys. Res. Lett.*, *29*, 7.

Cosgrove, R., and R. T. Tsunoda (2002b), A direction-dependent instability of sporadic-*E* layers in the nighttime midlatitude ionosphere, *Geophys. Res. Lett.*, *29*(18), 11.

Cosgrove, R. B. (2007), Generation of mesoscale *F* layer structure and electric fields by the combined Perkins and E_s layer instabilities, in simulations, *Ann. Geophys.*, *25*, 1579.

Cosgrove, R. B., and R. T. Tsunoda (2003), Simulation of the nonlinear evolution of the sporadic-*E* layer instability in the nighttime midlatitude ionosphere, *J. Geophys. Res.*, *108*, A1283, doi:10.1029/2002JA009728.

Cosgrove, R. B., and R. T. Tsunoda (2004), Instability of the *E–F* coupled nighttime midlatitude ionosphere, *J. Geophys. Res.*, *109*, A04305, doi:10.1029/2003JA010243.

Cosgrove, R. B., R. T. Tsunoda, S. Fukao, and M. Yamamoto (2004), Coupling of the Perkins instability and the sporadic *E* layer instability derived from physical arguments, *J. Geophys. Res.*, *109*, A06301, doi:10.1029/2003JA010295.

Fukao, S., M. C. Kelley, T. Shirakawa, T. Takami, M. Yamamoto, T. Tsuda, and S. Kato (1991), Turbulent upwelling of the midlatitude ionosphere 1. Observational results by the MU radar, *J. Geophys. Res.*, *96*, 3725–3746.

Garcia, F. J., M. C. Kelley, J. J. Makela, and C.-S. Huang (2000), Airglow observations of mesoscale low-velocity traveling ionospheric disturbances at midlatitudes, *J. Geophys. Res.*, *105*(18), 407.

Goodwin, G. L., and R. N. Summers (1970), E_s-layer characteristics determined from spaced ionosondes, *Planet Space Sci.*, *18*, 1417.

Haldoupis, C., K. Schlegel, and D. T. Farley (1996), An explanation for type 1 radar echoes from the midlatitude *E* region ionosphere, *Geophys. Res. Lett.*, *23*, 97.

Haldoupis, C., M. C. Kelley, G. C. Hussey, and S. Shalimov (2003), Role of unstable sporadic-*E* layers in the generation of midlatitude spread *F*, *J. Geophys. Res.*, *108*, A1446, doi:10.1029/2003JA009956.

Huang, C. S., C. A. Miller, and M. C. Kelley (1994), Basic properties and gravity wave initiation of the midlatitude *F* region instability, *Radio Sci.*, *29*, 395.

Hysell, D. L., and J. D. Burcham (2000), The 30 MHz radar interferometer studies of midlatitude *E* region irregularities, *J. Geophys. Res.*, *105*, 12,797.

Kelley, M. C., and C. A. Miller (1997), Electrodynamics of midlatitude spread *F* 3. Electrohydrodynamic waves? A new look at the role of electric field in thermospheric wave dynamics, *J. Geophys. Res.*, *102*, 11,539–11,547.

Kelley, M. C., J. J. Makela, A. Saito, N. Aponte, M. Sulzer, and S. A. Gonzalez (2000), On the electrical structure of airglow depletion/height layer bands over Arecibo, *Geophys. Res. Lett.*, *27*, 2837.

Klevans, E. H., and G. Imel (1978), E region coupling effects on the Perkins spread F instability, *J. Geophys. Res.*, *83*, 199–202.

Larsen, M. F. (2002), Winds and shears in the mesosphere and lower thermosphere: Results from four decades of chemical release wind measurements, *J. Geophys. Res.*, *107*, 28.

Makela, J. J., and M. C. Kelley (2003), Using the 630.0 nm nightglow emission as a surrogate for the ionospheric Pedersen conductivity, *J. Geophys. Res.*, *108*, A1253, doi:10.1029/2003JA009894.

Mendillo, M. M., J. Baumgardner, D. Nottingham, J. Aarons, B. Reinisch, J. Scali, M. C. Kelley (1997), Investigations of thermospheric–ionospheric dynamics with 6300-A images from the Arecibo observatory, *J. Geophys. Res.*, *102*, 7331–7343.

Perkins, F. (1973), Spread F and ionospheric currents, *J. Geophys. Res.*, *78*, 218.

Shalimov, S., and C. Haldoupis (2005), E-region wind-driven electrical coupling of patchy sporadic E and spread F at midlatitude, *Ann. Geophys.*, *23*, 2095.

Shalimov, S., C. Haldoupis, and K. Schlegel (1998), Large polarization electric fields associated with midlatitude sporadic E, *J. Geophys. Res.*, *103*, 11,617.

Shiokawa, K., C. Ihara, Y. Otsuka, and T. Ogawa (2003), Statistical study of nighttime medium-scale traveling ionospheric disturbances using midlatitude airglow images, *J. Geophys. Res.*, *108*, A1052, doi:10.1029/2002JA009491.

Swartz, W. E., S. C. Collins, M. C. Kelley, J. J. Makela, E. Kudeki, S. Franke, J. Urbina, N. Aponte, S. Gonzalez, M. P. Sulzer, and J. S. Friedman (2002), First observations of an F-region turbulent upwelling coincident with severe E-region plasma and neutral atmosphere perturbations, *J. Atmos. Sol. Terr. Phys.*, *64*, 1545.

Taylor, M. J., J.-M. Jahn, S. Fukao, and A. Saito (1998), Possible evidence of gravity wave coupling into the midlatitude F region ionosphere during the SEEK campaign, *Geophys. Res. Lett.*, *25*, 1801.

Tsunoda, R. T. (1998), On polarized frontal structures, type-1 and quasi-periodic echoes in midlatitude sporadic E, *Geophys. Res. Lett.*, *25*, 2641.

Tsunoda, R. T. (2006), On the coupling of layer instabilities in the nighttime midlatitude ionosphere, *J. Geophys. Res.*, *111*, A11304, doi:10.1029/2006JA011630.

Tsunoda, R. T., and R. B. Cosgrove (2001), Coupled electrodynamics in the nighttime midlatitude ionosphere, *Geophys. Res. Lett.*, *28*, 4171."

Tsunoda, R. T., R. B. Cosgrove, and T. Ogawa (2004), The azimuth-dependent E_s-layer instability: A missing link found, *J. Geophys. Res.*, *109*, A12303, doi:10.1029/2004JA010597.

Whitehead, J. D. (1970), Report on the production and prediction of sporadic E, *Rev. Geophys. Space Phys.*, *8*, 65.

Yamamoto, M., F. Kumura, S. Fukao, R. T. Tsunoda, K. Igarashi, and T. Ogawa (1997), Preliminary results from joint measurements of E-region field-aligned irregularities using the MU radar and the frequency-agile radar, *J. Atmos. Solar Terr. Phys.*, *59*, 1655.

Yokoyama, T., M. Yamamoto, S. Fukao, and R. B. Cosgrove (2004), Three-dimensional simulation on generation of polarization electric field in the midlatitude E-region ionosphere, *J. Geophys. Res.*, *109*, A01309, doi:10.1029/2003JA01238.

Yokoyama, T., M. Yamamoto, S. Fukao, T. Takahashi, and M. Tanaka (2005), Numerical simulation of mid-latitude ionospheric E region based on SEEK and SEEK-2 observations, *Ann. Geophys.*, *23*, 2377.

Zhou, Q., and J. D. Mathews (2006), On the physical explanation of the Perkins instability, *J. Geophys. Res.*, *111*, A12309, doi:10.1029/2006JA011696.

Zhou, Q.-N., J. D. Mathews, C. A. Miller, and I. Sekar (2006), The evolution of nighttime midlatitude mesoscale F region structures: A case study utilizing numerical solution of the Perkins instability equations, *Planet. Space Sci.*, *54*, 710.

R. T. Tsunoda, Center for Geospace Studies, SRI International, Menlo Park, CA 94025, USA.

Irregularities Within Subauroral Polarization Stream-Related Troughs and GPS Radio Interference at Midlatitudes

Evgeny Mishin

Institute for Scientific Research, Boston College, Chestnut Hill, Massachusetts, USA

Natan Blaunstein

Department of Communication Systems Engineering, Ben Gurion University of the Negev, Beer-Sheva, Israel

We report on the Defense Meteorological Satellite Program satellite observations of irregular plasma density troughs related with UHF radio scintillations at subauroral latitudes during the magnetic storms of September 1999 and 2001. The troughs were associated with either structured subauroral polarization streams or elevated electron temperature. The observed irregularity spectrum within the range of 0.75–7.5 km is well-approximated by a power law with the spectral index $-5/3 \geq -p_n \geq -2$.

1. INTRODUCTION

It is well-known that ionospheric irregularities cause phase and amplitude scintillations of radio signals, interfering communication and navigation [e.g., *Kintner and Ledvina*, 2005]. Irregularities intensify during magnetically perturbed periods. It is commonly believed that midlatitude scintillation events occur inside the storm time auroral zone as it expands toward midlatitudes. However, *Foster and Rich* [1998] reported on observations of strong irregularities equatorward of the auroral zone during the 21 March 1990 magnetic storm. Additionally, strong GPS phase and amplitude scintillations were observed well-equatorward of the expanded auroral oval by *Basu et al.* [2001] and *Ledvina et al.* [2002] during the magnetic storms of 22–23 September 1999 and 25–26 September 2001, respectively. *Basu et al.* [2001] noted that the Defense Meteorological Satellite Program (DMSP) satellites F13 and F14 had flown near the subauroral scintillation region and observed structured convection flows.

The dynamics of the storm time subauroral ionosphere is driven by the ring current (RC)–plasmasphere interaction, resulting in enhanced flows of the westward convection, named subauroral polarization streams (SAPS) [e.g., *Foster and Burke*, 2002]. The SAPS region is often irregular [*Erickson et al.*, 2002; *Mishin et al.*, 2002; 2003; *Foster et al.*, 2004]. In particular, strong SAPS wave structures (SAPSWS) and irregular plasma density troughs are associated with fresh injections of RC ions [*Mishin et al.*, 2004; *Mishin and Burke*, 2005]. From Plate 1 [*Basu et al.*, 2001], it follows that the DMSP F13 satellite encountered a developing SAPSWS event. Thus, a search for SAPSWS-associated density irregularities appears to be a logical next step. This paper presents the results of the DMSP F13 and F14 satellite observations of irregular plasma density troughs colocated with the regions of either structured SAPS or elevated electron temperature nearby the subauroral scintillation regions [*Basu et al.*, 2001; *Ledvina et al.*, 2002].

2. OBSERVATIONS

2.1. Instrumentation

DMSP satellites are three-axis stabilized spacecrafts that fly in circular, sun-synchronous polar (inclination 98.7°) orbits at an altitude of ~840 km. The geographic local times of the orbits are either near the 1800–0600 (F13) or 2100–0900 (F14, F15) meridians. Each satellite carries a suite of sensors to measure (1) fluxes of precipitating electrons and ions in the energy range between 30 eV and 30 keV (SSJ4), (2) the densities, temperatures, and drift motions of ionospheric ions and electrons (SSIES), and (3) perturbations of the Earth magnetic field (SSM). The SSJ4 sensors consist of four detectors: one high (1–30 keV) energy detector and one low (30–1000 eV) energy detector for the each of the particle types [*Hardy et al.*, 1984]. Nineteen-point ion and electron spectra are returned once per second. SSIES consists of (1) an ion drift meter to measure the horizontal (V_H) and vertical (V_V) cross-track components of plasma drift within the range of ±3000 m/s and a one-bit resolution of 12 m/s for ambient ion densities greater than 5×10^3 cm^{-3}, (2) a retarding potential analyzer to measure ion temperatures (T_i), composition, and the in-track component of plasma drift ($V_\parallel$), (3) an ion trap to measure the total ion density (n_i), and (4) a spherical Langmuir probe to measure the density (n_e) and temperature (T_e) of ambient electrons [*Rich and Hairston*, 1994]. The plasma drift and density measurements are sampled at the rates of 6 and 24 Hz, respectively.

2.2. Storm Time Subauroral Density Irregularities

From *Basu et al.* [2001, Figure 2], it follows that the DMSP F13 and F14 satellites practically intersected the region of scintillations near 2236:00 and 0058:15 UT on 22 and 23 September 1999, respectively. The former (latter) occurred at the expansion (recovery) phase of the magnetic storm. Figure 1 shows snapshots of plasma observations from DMSP F13 (left) and F14 (right). Vertical dashed lines indicate the equatorward boundaries of auroral electron precipitation. Shown in Figure 1a are the horizontal V_H (positive westward) and vertical V_V (positive upward) components of the convection velocity (in meters per second) and the electron temperature $\frac{1}{3}T_e$ (in kelvin). Note that the short-scale velocity oscillations in the SAPSWS region faded away during the recovery phase. Such behavior is typical of SAPSWS, which develop within 10–20 min [*Mishin and Mishin*, 2007] and decay within ~1 hour after (storm time) substorm expansion onsets [*Mishin and Burke*, 2005]. Figure 1c presents the variations in the plasma density along the satellite tracks. Figure 1b shows the corresponding waveforms obtained applying 0.1- to 9.5-Hz band-pass elliptic filter. Its upper fre-

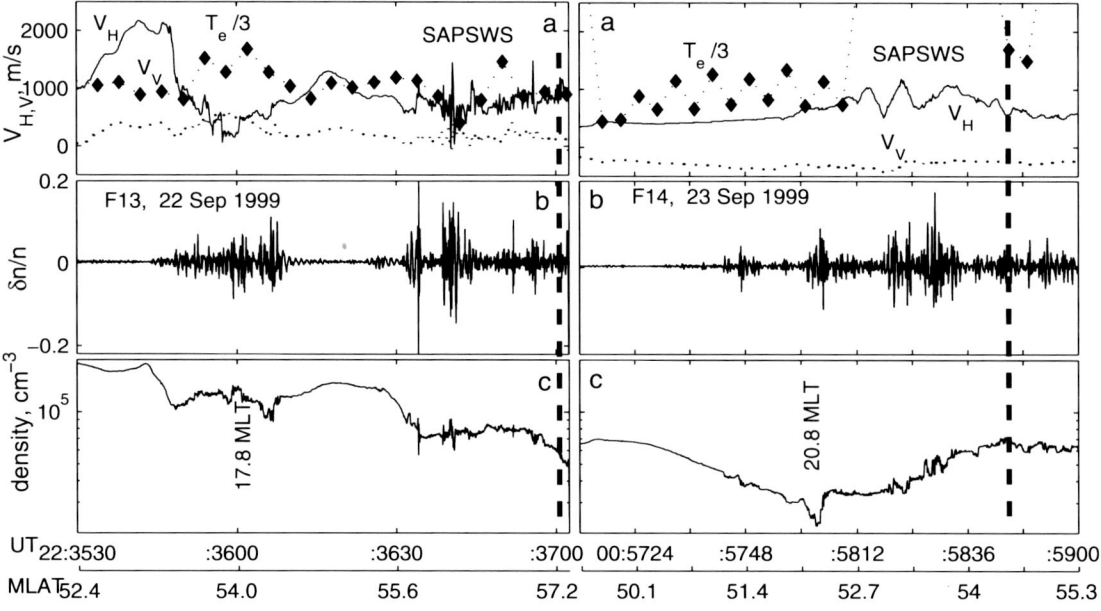

Figure 1. SAPS events observed by DMSP F13 (left) and F14 (right) in September 1999 near Boston, Mass. (a) Westward V_H and vertical V_V components of the convection velocity (in meters per second) and $\frac{1}{3}T_e$ (in kelvin). (b) Waveforms of relative density variations in the frequency range of 0.1–9.5 Hz. (c) The density variations along the satellite tracks. Vertical dashed lines indicate the equatorward boundaries of the auroral zone.

quency cutoff is defined by the 24-Hz sampling rate. One can see that enhanced density irregularities were present near the region of radio scintillations.

Figure 2 shows the DMSP F14 satellite pass near Ithaca, New York, right at the beginning of the recovery phase of the 25–26 September 2001 magnetic storm and in the middle of *Ledvina et al.*'s [2002] scintillation event near 53.2 magnetic latitude (MLAT). Color spectrograms present directional differential number fluxes of downcoming electrons

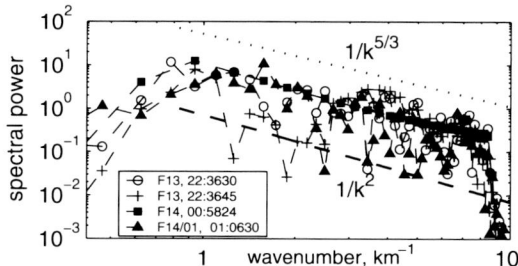

Figure 3. Power spectral densities of plasma irregularities as a function of spatial wave number during the periods indicated.

and ions (Figures 2a and 2b). Figure 2c shows the horizontal (mainly westward) and vertical components of the convection velocity (in meters per second) and $\frac{1}{7}T_e$ (in kelvin). Note gaps in the temperature data from F14 in the regions of strong irregularities and the bad quality of the ion temperature data (not shown) in both events. The ion density variation and its waveform obtained by applying the same 0.1- to 9.5-Hz filter are displayed in Figures 2d and 2e, respectively. The vertical dashed line indicates the equatorward boundary of the electron plasma sheet. Two density troughs equatorward of the auroral zone are evident. One coincides with a structured SAPS (the remains of SAPSWS) near 56.5 MLAT, while that at the equatorward side colocates with the elevated electron temperature. These features are typical of the SAPSWS events associated with storm time RC injections [*Mishin et al.*, 2003, 2004; *Mishin and Burke*, 2005].

As the subauroral regions of strongest irregularities and GPS scintillations in both events virtually colocate, we suggest that they are causally related. Furthermore, the waveforms in Figures 1 and 2 appear alike. Figure 3 shows the power spectral density of relative density irregularities as a function of spatial wave number. We emphasize that spatial and temporal variations cannot be separated in data from a single spacecraft, and thus, "frequencies" ("wavelengths") mean apparent frequencies (scale lengths). Given the satellite speed $v_s = 7.5$ km/s, apparent frequencies of 1–10 Hz correspond to wavelengths $\lambda_s = 2\pi/k_s = v_s/f = 7.5$–0.75 km along the satellite track, implying the satellite crossing time be much shorter than the wave period. On the average, the spectra in the range of $10 > k \geq 1$ km^{-1} are well-represented by a power law $(\delta n_k/n_0)^2 \propto k^{-p_n}$ with the spectral index $5/3 \lesssim p_n \lesssim 2$. Note that the upper bound is due to the applied band-pass filter.

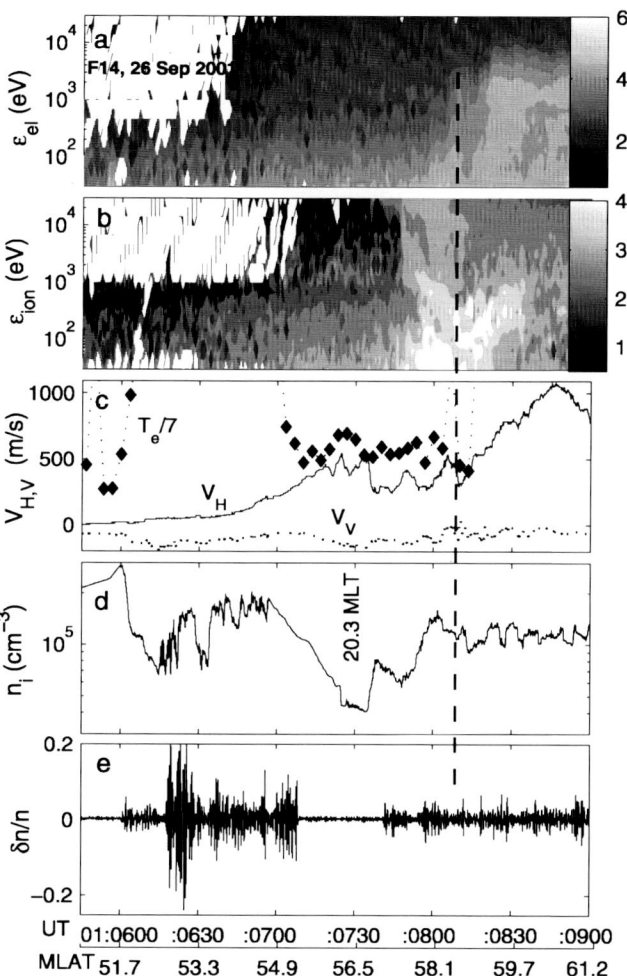

Figure 2. A SAPS event observed by DMSP F14 at ~ 01:06–08:00 UT on 26 September 2001 near Ithaca, New York: (a) electron and (b) ion energy–time spectrograms of downcoming directional differential number fluxes (in cubic centimeters per second per steradian per electron volt); (c) westward V_H and vertical V_V components of the convection velocity (meters per second) and $\frac{1}{7}T_e$ (in kelvin); (d) plasma density variation; (e) waveform of relative density variations in the frequency range of 0.1–9.5 Hz. The vertical dashed line indicates the equatorward boundary of the electron plasma sheet.

3. DISCUSSION

The spatial structure of the above density troughs is similar to that of *Foster and Rich* [1998, Figure 9] in particular and is typical of the SAPSWS events in general [*Mishin et al.*, 2004; *Mishin and Burke*, 2005]. Namely, the perturbed density region comprises two distinctive parts, one of which embedded within the region of structured convection flows (SAPSWS) and another located well-equatorward, in the region of a locally elevated electron temperature. High-T_e-related density troughs are consequent to thermal electrons vibrationally exciting thermospheric molecules and thereby increasing rates of charge exchange and subsequent recombination of O_2^+ and NO^+ [*Mishin et al.*, 2004]. In turn, elevated T_e in the subauroral topside ionosphere derive from heat generated in the RC/plasmasphere overlap region. The main source of the electron heating near the RC's inner edge is the lower hybrid drift instability driven by the diamagnetic current [*Mishin and Burke*, 2005].

The spectra of (magnetic field-aligned) density irregularities in Figure 3 are close to that of *Foster and Rich* [1998, Figure 9]. Their generation mechanism(s) has not yet been established. The gradient drift instability (GDI) requires that the dot product $[E \times B] \cdot \nabla n$ be positive [e.g., *Kadomtsev*, 1965]. Numerical simulations of GDI [e.g., *Keskinen*, 1984; *Guzdar et al.*, 1998] show that saturated density irregularities obey a power law spectrum with the spectral index of the order of 2, i.e., close to that in Figure 3. In addition, the temperature GDI can develop in the region where $\nabla n \cdot \nabla T_e < 0$ [e.g., *Kadomtsev*, 1965; *Hudson and Kelley*, 1976]. When $\nabla n \rightarrow 0$, meridional fields $E_x > 20$ mV/m or $V_H > 500$ m/s can generate ~10-km-scale irregularities via the ion frictional heating instability [*Keskinen et al.*, 2004]. This can trigger the positive ionospheric feedback instability inside the depleted region, thereby enhancing SAPS (Alfvén) wave structures and further strengthen the "seed" irregularities [*Streltsov and Mishin*, 2003].

For a longitudinally extended ($|\partial n/\partial x| >> |\partial n/\partial y| \rightarrow 0$) density trough, GDI implies the presence of the eastward/westward electric field on its poleward/equatorward wall. Hereafter, we define the axes x and y to be directed northward and eastward, respectively. The presence of the eastward electric field ($E_y > 0$) in Figure 1a (left) is manifested by the upward vertical velocity. Thus, whenever $E_y \cdot \partial n/\partial x$ exceeds the instability threshold GDI contributes to the source of short-scale irregularities, which is consistent with the modulation of their amplitude near 2236:00 UT by tens of kilometers length-scale density depletions. On the other hand, irregularities in the SAPSWS region are likely because of the concerted effort of the GDI, ion heating, and ionospheric feedback instabilities.

The downward velocity in Figures 1a (right) and 2a indicates the westward electric field. Thus, weak irregularities on the equatorward wall near 0106:00 UT could be because of the combined action of GDI and temperature GDI. However, the strongest irregularities develop on the poleward wall in both cases. Field-aligned currents of either sign could make this system unstable [*Kadomtsev*, 1965; *Ossakow and Chaturvedi*, 1979]. However, virtually no such currents were present. Additional experimental investigations and theoretical efforts are needed to address this problem, which is beyond our observation-based study.

4. CONCLUSION

The DMSP F13 and 14 satellites, while flying near the regions of intense radio signal scintillations, detected small-scale plasma density and electromagnetic oscillations. The enhanced fluctuations appeared at the poleward edges of irregular density troughs, embedded within structured SAPS. Twenty-four-hertz Langmuir probe data show that during the scintillation intervals the power spectra of plasma density irregularities in the range $8 \geq k \geq 1$ km^{-1} or meridional wavelengths ≃0.5–4 km are well-represented by a power law $(\delta n_k/n_0)^2 \propto k^{-p_n}$ with the spectral index $5/3 \lesssim p_n \leq 2$.

Acknowledgments. We thank Fred Rich of AFRL for help with DMSP data processing, Paul Kintner of Cornell University, and Sunanda Basu of BU for discussions. This research was supported in part by AFRL contract FA8718-04-C-0055 with Boston College.

REFERENCES

Basu, S., et al. (2001), Ionospheric effects of major magnetic storms during the International Space Weather Period of September and October 1999: GPS observations, VHF/UHF scintillations and in situ density structures at middle and equatorial latitudes, *J. Geophys. Res., 106*, 30,389–30,413.

Erickson, P. J., J. C. Foster, and J. M. Holt (2002), Inferred electric field variability in the polarization jet from Millstone Hill E region coherent scatter observations, *Radio Sci., 37*(2), 1027, doi:10.1029/2000RS002531.

Foster, J. C., and W. J. Burke (2002), SAPS: A new categorization for subauroral electric fields, *Eos Trans. AGU, 83*(36), 393.

Foster, J. C., and F. J. Rich (1998), Prompt midlatitude electric field effects during severe magnetic storms, *J. Geophys. Res., 103*, 26,367–26,372.

Foster, J. C., P. J. Erickson, F. D. Lind, and W. Rideout (2004), Millstone Hill coherent-scatter radar observations of electric field variability in the sub-auroral polarization stream, *Geophys. Res. Lett., 31*, L21803, doi:10.1029/2004GL021271.

Guzdar, P. N., N. A. Gondarenko, P. K. Chaturvedi, and S. Basu (1998), Three-dimensional nonlinear simulations of the gradient drift instability in the high-latitude ionosphere, *Radio Sci., 33*, 1901–1913.

Hardy, D., L. Schmidt, M. Gussenhoven, F. Marshall, H. Yeh, T. Shumaker, A. Huber, and J. Pantazis (1984), Precipitating electron and ion detectors (SSJ/4) for block 5D/Flights 4–10 DMSP satellites: Calibration and data presentation, *AFGL Tech. Rep. AFGL-TR-84-0317*, Hanscom Air Force Base, Mass.

Hudson, M. K., and M. C. Kelley (1976), The temperature gradient drift instability at the equatorward edge of the ionospheric plasma trough, *J. Geophys. Res.*, *81*, 3913–3918.

Kadomtsev, B. (1965), *Plasma Turbulence*, Academic Press, New York.

Keskinen, M. J. (1984), Nonlinear theory of the **E** × **B** instability with an inhomogeneous electric field, *J. Geophys. Res.*, *89*, 3913–3920.

Keskinen, M. J., S. Basu, and S. Basu (2004), Midlatitude subauroral small scale structure during a magnetic storm, *Geophys. Res. Lett.*, *31*, L09811, doi:10.1029/2003GL019368.

Kintner, P., and B. Ledvina (2005), The ionosphere, radio navigation, and global navigation satellite systems, *Adv. Space Res.*, *35*, 788.

Ledvina, B. M., J. J. Makela, and P. M. Kintner (2002), First observations of intense GPS L1 amplitude scintillations at midlatitude, *Geophys. Res. Lett.*, *29*(14), 1659, doi:10.1029/2002GL014770.

Mishin, E. V., and W. J. Burke (2005), Stormtime coupling of the ring current, plasmasphere and topside ionosphere: Electromagnetic and plasma disturbances, *J. Geophys. Res.*, *110*, A07209, doi:10.1029/2005JA011021.

Mishin, E., and V. Mishin (2007), Prompt response of SAPS to stormtime substorms, *J. Atmos. Sol. Terr. Phys.*, *69*, 1233.

Mishin, E. V., J. C. Foster, A. P. Potekhin, F. J. Rich, K. Schlegel, K. Yumoto, V. I. Taran, J. M. Ruohoniemi, and R. Friedel (2002), Global ULF disturbances during a stormtime substorm on 25 September 1998, *J. Geophys. Res.*, *107*(A12), 1486, doi:10.1029/2002JA009302.

Mishin, E. V., W. J. Burke, C. Y. Huang, and F. J. Rich (2003), Electromagnetic wave structures within subauroral polarization streams, *J. Geophys. Res.*, *108*(A8), 1309, doi:10.1029/2002JA009793.

Mishin, E. V., W. J. Burke, and A. A. Viggiano (2004), Stormtime subauroral density troughs: Ion-molecule kinetics effects, *J. Geophys. Res.*, *109*, A10301, doi:10.1029/2004JA010438.

Ossakow, S. L., and P. K. Chaturvedi (1979), Current convective instability in the diffuse aurora, *Geophys. Res. Lett.*, *6*, 332–334.

Rich, F. J., and M. Hairston (1994), Large-scale convection patterns observed by DMSP, *J. Geophys. Res.*, *99*, 3827–3844.

Streltsov, A. V., and E. V. Mishin (2003), Numerical modeling of localized electromagnetic waves in the nightside subauroral zone, *J. Geophys. Res.*, *108*(A8), 1332, doi:10.1029/2003JA009858.

N. Sh. Blaunstein, Department of Communication Systems Engineering, Ben Gurion University of the Negev, Beer-Sheva, Israel.

E. V. Mishin, Institute for Scientific Research, Boston College, 402 Saint Clements Hall, 140 Commonwealth Avenue, Chestnut Hill, MA 02467-3862, USA. (evgenii.mishin@hanscom.af.mil)

DEMETER Satellite Observations of Plasma Irregularities in the Topside Ionosphere at Low, Middle, and Sub-Auroral Latitudes and Their Dependence on Magnetic Storms

Robert F. Pfaff Jr.,[1] Carmen Liebrecht,[1] Jean-Jacques Berthelier,[2] Michel Malingre,[2] Michel Parrot,[3] and Jean-Pierre Lebreton[4]

Observations of plasma density structure and electric field irregularities gathered with probes on the Detection of Electro-Magnetic Emissions Transmitted from Earthquake Regions (DEMETER) satellite are presented to characterize the topside ionosphere at low, middle, and sub-auroral latitudes. Data from successive DEMETER orbits during magnetic storms illustrate how low, mid-, and sub-auroral latitude plasma density, plasma density structures, and electric field irregularities correlate and evolve with changes in *Dst*. The observations reveal: (1) electric field irregularities associated with density depletions at mid latitudes are similar to those that characterize equatorial spread *F* at low latitudes; (2) large, well-defined density depletions that extend to mid-latitude display zonal widths and interdepletion spacings that are similar to spread *F* density depletion widths and spacings observed at the equator; (3) in some cases, ULF/ELF magnetic field irregularities are observed in association with the electric field irregularities particularly on the walls of the plasma density structures and may be related to finely structured spatial currents and/or Alfvén waves; (4) precisely during the main phase of severe geomagnetic storms, increased ambient plasma densities and broad regions of irregularities are observed near 710 km, initially at storm commencement near the magnetic equator and then extending to mid- and sub-auroral latitudes within the ~8-h period corresponding to the negative *Dst* excursions; and (5) intense, broadband electric and magnetic field irregularities are often observed at sub-auroral latitudes and are typically associated with the trough region and its poleward plasma density gradient. The observations provide a general framework showing how low, mid-, and sub-auroral latitude plasma density structuring and associated electric field irregularities respond to geomagnetic storms.

[1] NASA Goddard Space Flight Center, Greenbelt, Maryland, USA.
[2] CETP, St. Maur, France.
[3] CNRS, Orleans, France.
[4] ESA, Noordwijk, Netherlands.

Midlatitude Ionospheric Dynamics and Disturbances
Geophysical Monograph Series 181
Copyright 2008 by the American Geophysical Union.
10.1029/181GM27

1. INTRODUCTION

The Earth's ionosphere at low latitudes is known to become highly unstable at night, creating broad spectra of plasma density irregularities that range in scales from more than 100 km to less than 1 m. At low latitudes, such irregularities at *F* region altitudes (>250 km) are collectively called equatorial spread *F* (ESF). Although this term is often invoked to describe the mid-latitude extension of this phenomenon, the characteristics and generation mechanisms of mid-latitude irregularities are, by comparison, not well established. Furthermore, another class of F-region ionospheric irregularities are those observed at upper mid latitudes or "sub-auroral" latitudes, which are associated with the plasmaspheric trough and which have distinct generation mechanisms compared to those generally associated with spread F.

Over the last several decades, ground-based and *in situ* observations, together with modeling and theoretical advances, have considerably advanced our understanding of ESF [see, e.g., *Kelley*, 1989; *Hysell*, 2000; *Hysell and Kudeki*, 2004]. However, fundamental knowledge concerning mid-latitude "spread F-like" irregularities and density structures, including their conditions for growth, spectral distributions, and cross-scale coupling mechanisms, still eludes our understanding, despite growing experimental evidence of such structures and irregularities [e.g., *Fukao et al.*, 1991; *Saito et al.*, 1995; *Mendillo et al.*, 1997; *Swartz et al.*, 2000; *Kelley et al.*, 2000a, 2000b, 2002; *Makela and Kelley*, 2005]. In particular, the extension of ESF to mid latitudes, particularly during magnetic storms, is neither well characterized nor understood. Indeed, mid-latitude plasma structuring at F region altitudes may be due to a variety of causes, including ones distinct from those associated with conventional ESF. At somewhat higher latitudes, recent observations of plasma density and electric field structures and irregularities in the sub-auroral ionosphere, including their relation to magnetic storms, have been reported using data from the DMSP (Defense Meteorological Satellite Program) satellites [e.g., *Mishin et al.*, 2002, 2003; *Mishin and Burke*, 2005] and incoherent scatter and backscatter radars [e.g., *Foster and Rich*, 1998; *Erickson et al.*, 2002; *Foster et al.*, 2004; *Greenwald et al.*, 2006; *Oksavik et al.*, 2006]. Despite advances in theory and modeling to help explain these observations [e.g., *Streltsov and Mishin*, 2003; *Keskinen et al.*, 2004], their large-scale conditions for growth and possible association with irregularities at mid latitudes during storm periods have not been established.

Observations gathered on the French polar-orbiting Detection of Electro-Magnetic Emissions Transmitted from Earthquake Regions (DEMETER) satellite are well suited to advance our understanding of topside plasma gradients and irregularities as they provide direct comparisons of mid-latitude and sub-auroral irregularities with those at low latitudes. Furthermore, because of its near-continuous measurements at the same local time and its dedicated plasma wave instrumentation, the DEMETER satellite provides an opportunity to study in detail the evolution of plasma density structures and electric field irregularities during magnetic storms. Such observations at low, mid, and sub-auroral latitudes gathered by instruments on the DEMETER satellite are the subject of this paper.

Before proceeding, we define the terms used herein to describe the latitudinal structure of the ionosphere. In general, we refer to "low" latitudes as those below 15° (magnetic), "mid" latitudes as those between roughly 15° and 45° (magnetic), and "sub-auroral" latitudes, as those between 45° and 60° (magnetic). During storm periods, the equatorward edge of the sub-auroral region, as defined by a distinct trough in the ambient plasma density, may be found at still lower latitudes, and, similarly, the poleward edge of both the mid-latitude and sub-auroral regions may extend to higher latitudes when magnetic activity is weak.

2. BRIEF OVERVIEW OF THE DEMETER SATELLITE

The French Space Agency (CNES) developed the DEMETER microsatellite to search for electromagnetic signals associated with earthquakes as the primary objective [*Cussac et al.*, 2006]. A secondary objective of the DEMETER mission is to carry out ionospheric research. The satellite was launched on 29 June 2004 into a polar, circular orbit near 720-km altitude that is Sun-synchronous, providing data at or near 1030 LT and 2230 LT. Operating at a near 100% duty cycle at latitudes below about 65°, this satellite includes measurements of vector wave electric and magnetic fields, plasma density, temperature, and velocity, and energetic particles. Details may be found in *Cussac et al.* [2006] and on two DEMETER web sites: http://smsc.cnes.fr/DEMETER/GP_satellite.htm, and http://demeter.cnrs-orleans.fr/. Below, we provide a few comments on those DEMETER instruments whose data will be presented in this paper.

2.1. Electric Field Instrument

The electric field instrument includes four rigid 4-m booms with 8-cm-diameter spherical sensors with embedded preamps. Sphere biasing is included to maintain optimum coupling of the probe to the plasma, as described by *Berthelier et al.* [2006a]. The booms are oriented in a tetrahedron yielding three orthogonal measurements and thus providing the vector DC and AC electric field. Because of the moderate length of the booms, differences in the surface potentials of the various electrodes, and effects of the sheath around the booms and the spacecraft, it is difficult to measure low-amplitude DC electric fields on a nonspinning spacecraft such as DEMETER. DC electric field measurements at mid and low latitudes with the desired accuracy of better than ~1 mV/m are thus not available. AC electric fields, however, are available and of excellent quality. One boom pair creates a baseline of 8 m that is in the zonal direction and is always perpendicular to the orbit plane. The VLF electric field measurements reported herein were gathered with this double probe.

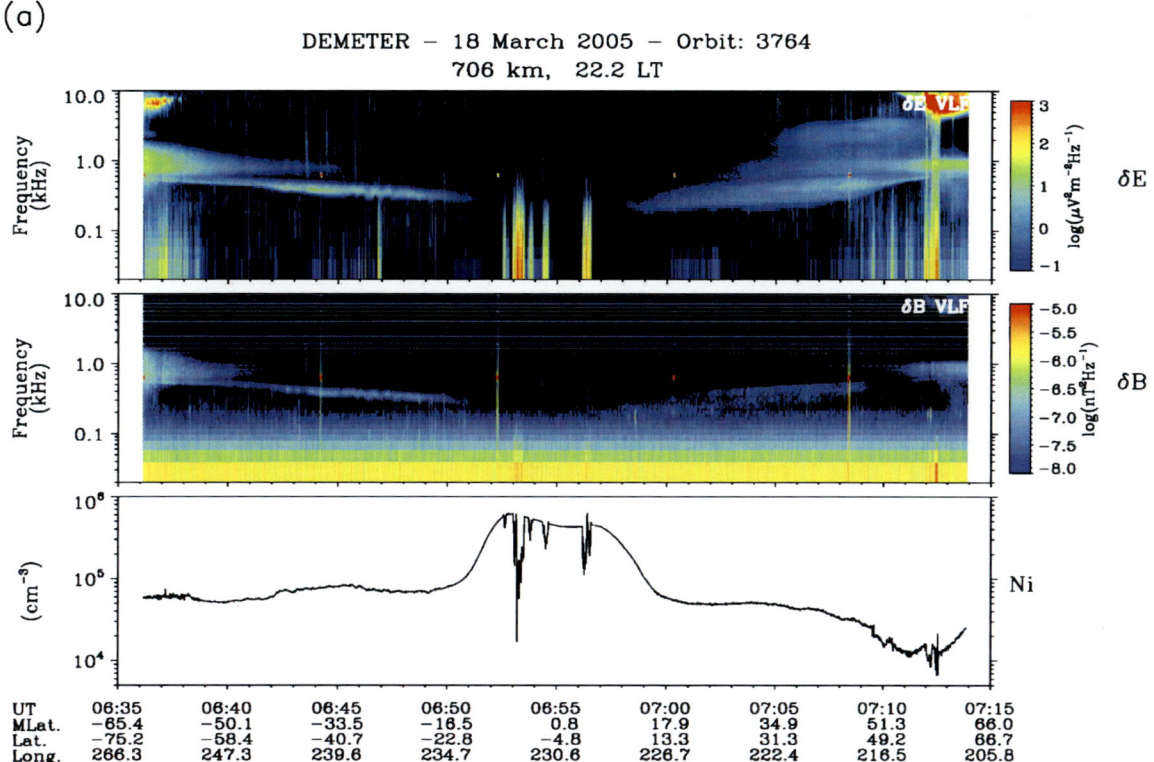

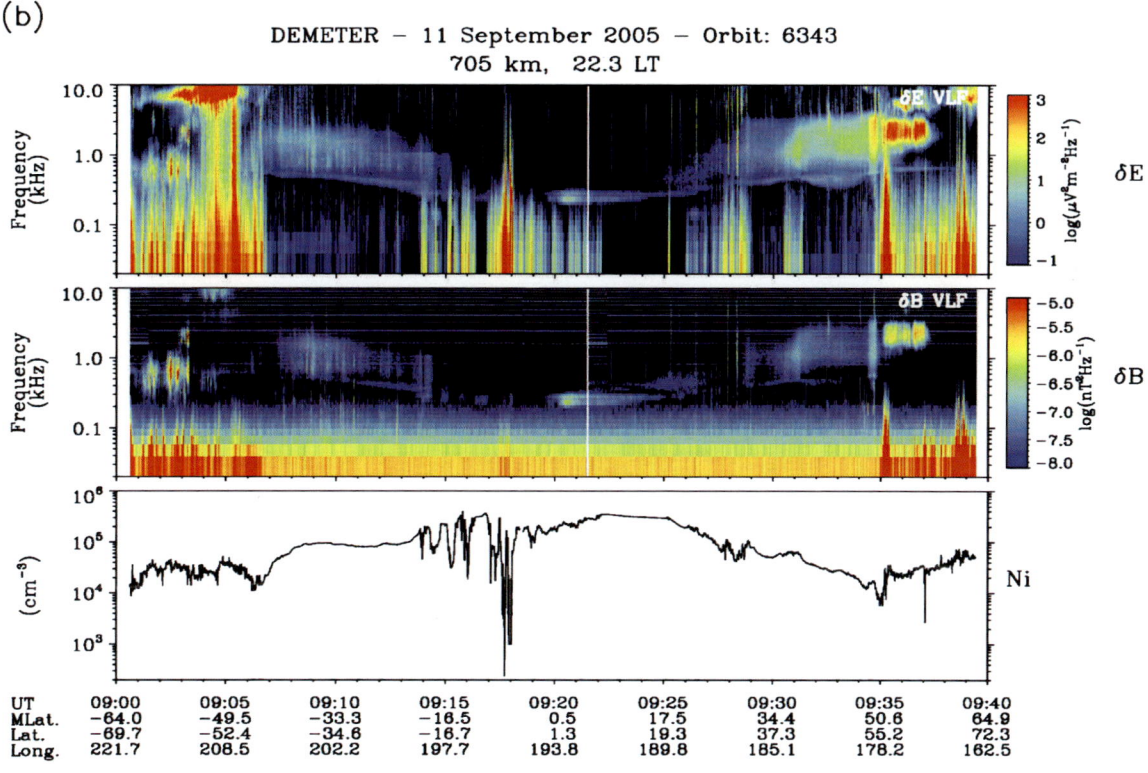

Plate 1. (a, b, c, d) Examples of nightside DEMETER satellite data within times corresponding to approximately -65° to +65° magnetic latitude (MLAT) for different ionospheric conditions. The top and middle panels show VLF electric field and magnetic field spectrograms, respectively, and the bottom panel shows the plasma number density.

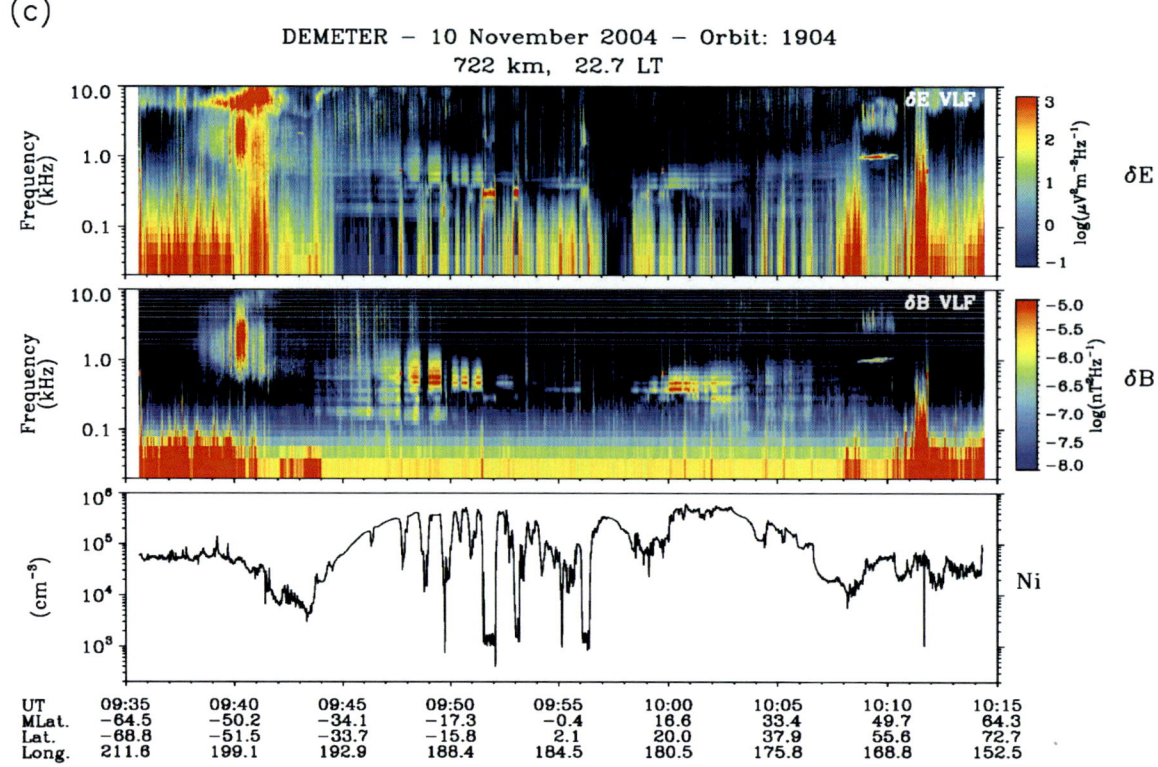

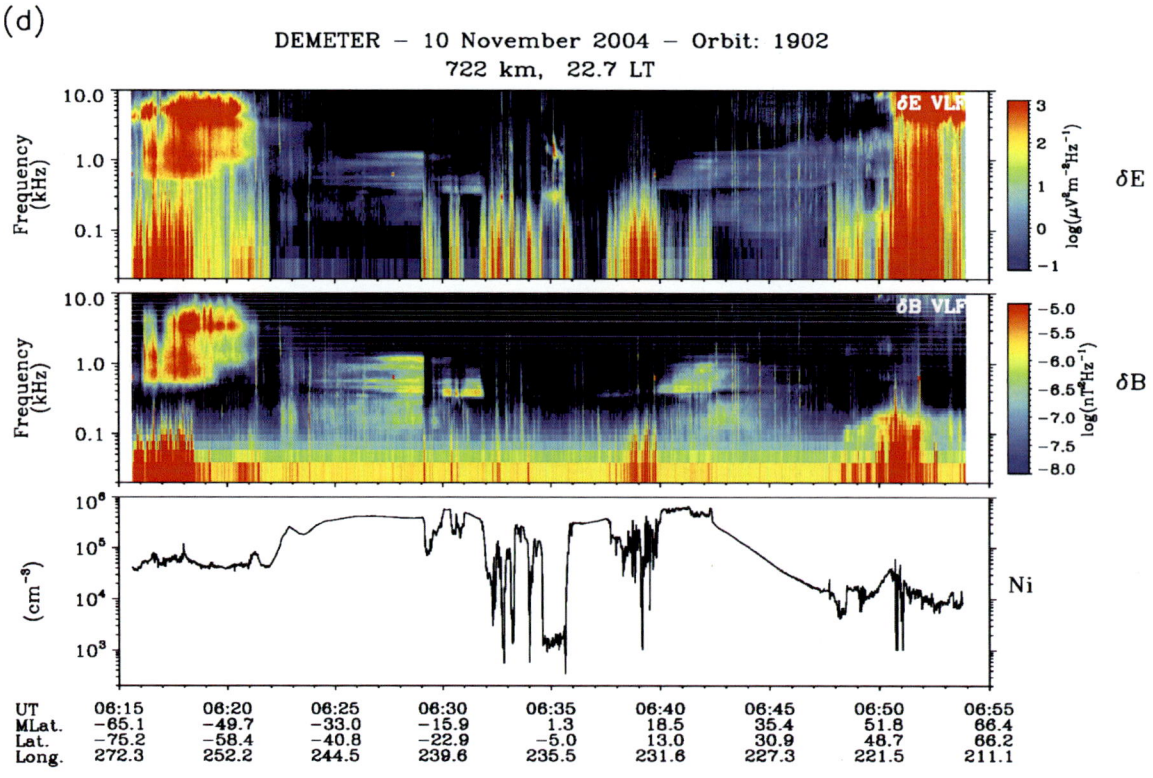

Plate 1. (continued)

2.2. AC Magnetic Fields

AC or wave magnetic field components are gathered using a three-axis search coil instrument in the 15-Hz to 17.4-kHz range. During burst mode, the three waveforms in the 15-Hz to 1-kHz range are recorded. Details of this instrument may be found in *Parrot et al.* [2006].

2.3. Plasma Density

Plasma density and temperature measurements are gathered using a swept Langmuir probe that consists of a sphere and a cylinder. Details of the Langmuir probe instrument may be found in *Lebreton et al.* [2006].

3. REPRESENTATIVE DEMETER OBSERVATIONS

3.1. Observations of Irregularities at Low, Mid, and Sub-Auroral Latitudes

Examples of relevant DEMETER observations gathered on four representative orbits near 710 km, 2230 LT, and between roughly ±65° magnetic latitude (MLAT) are shown in Plates 1a, 1b, 1c, and 1d. The first and second panels show VLF electric field and magnetic field wave spectrograms. The third panel shows the plasma number density measured by the Langmuir probe. Data from each orbit are described below in sequence.

A well-defined ESF event is shown in the DEMETER observations of 18 March 2005 in Plate 1a that reveal distinct plasma density depletions near 0652–0657 UT with broadband electric field irregularities observed precisely coincident with the depletions. Faint ELF hiss is observed in both the electric field and magnetic field wave spectrograms with evidence for a low frequency cutoff near 300 Hz that increases slightly with increasing latitude. This wave emission propagates in the whistler mode and has a well-defined low-frequency cutoff at the $L = 0$ frequency that depends on the magnetic field strength and ion composition [e.g., *Smith and Brice*, 1964]. A small patch of sub-auroral irregularities are observed in the electric field data near 60° MLAT, particularly in the northern hemisphere. These DEMETER data correspond to a relatively quiet period ($Kp = 1+$) and are presented for contrast with the following three examples.

Considerably more intense electric field irregularities were observed on 11 September 2005 and are presented in Plate 1b. Near the center of the panels at 0918 UT, spread F irregularities are associated with density depletions that extend below $10^3/cm^3$, an order of magnitude lower than those shown in Plate 1a. (Note the change in scale in the density panel between Plates 1a and 1b.) These "deep" plasma density depletions have associated strong electric field irregularities whose amplitudes are more intense than those in Plate 1a and whose spectra extend to higher frequencies in the satellite frame, corresponding to shorter spatial scales. Nearby density depletions on either side of the largest depletions are more modest and have corresponding spectra that are both less intense and do not extend as high in frequency. Density structures near 0928 and 0931 UT (near 25° and 35° MLAT) also display associated electric field irregularities. Although these mid-latitude irregularities do not correspond to well-defined plasma density depletions, they are nevertheless associated with distinct density gradients and large-scale plasma density structures.

The data in Plate 1b also show intense electric field irregularities associated with the poleward edge of the trough region near ±50° MLAT. Correlative data for these events reveal latitudinally confined poleward-directed DC electric fields (not shown) and enhanced temperatures (not shown), indicative of sub-auroral ion drifts (SAID) events [e.g., *Anderson et al.*, 1993]. Note that these poleward DC electric fields are of sufficient amplitude to be confidently measured with the DEMETER instrumentation, despite the measurement limitations discussed above. These sub-auroral irregularities in Plate 1b appear to extend into the auroral zone at higher latitudes, although a region of distinctly enhanced waves are present at 0935 UT that coincides with the local, poleward plasma gradient. Note also the intense magnetic field component of these irregularities, which may be due to static current structures or Alfvén waves [e.g., *Knudsen et al.*, 1990]. These waves may in fact represent higher-frequency extensions of the electromagnetic structures reported by *Mishin et al.* [2003] within sub-auroral polarization streams.

A third representative orbit, gathered during a strong magnetic storm that occurred on 10 November 2004, is shown in Plate 1c in which multiple density depletions are present. Note that during part of the orbit, the depletions display a quasi-periodic spacing of ~60 s, corresponding to ~150-km spacing in the east–west direction, since the zonal component of the satellite velocity was about 2.5 km/s. In this work, we assume that the proper motion of the density structures was sufficiently small that they could be considered stationary compared to the satellite zonal velocity component. This observed spacing is typical of that of the zonal separation of successive depletion regions that characterize ESF [e.g., *Kelley*, 1989]. Furthermore, the data suggest that the spacecraft "sliced" across a family of depletions as it journeyed between −30° and the equator (i.e., between 0946 and 0955 UT) that appear to represent a collection of interhemispherical, elongated, narrow-width density depletions that extend across the equator and well into mid latitudes.

We further note that the most well-defined density depletions have characteristic depletion widths of 10–20 s in the spacecraft frame, corresponding to zonal widths of 25–50 km, again estimated using the zonal component of the satellite velocity. In one case, near 0952 UT in Plate 1c, the zonal depletion width corresponded to ~90 km.

Associated with these distinct, well-defined narrow density depletions are corresponding electric field irregularity spectra as also noted in Plates 1a and 1b. Detailed inspection reveals that, in some cases, the spectra are enhanced along the walls or steep gradients of the depletions, although it is difficult to discern in this presentation. Analysis of the irregularity spectra further show that their largest amplitudes consistently correspond to the lowest frequencies in the satellite frame, or, in general, the longest wavelengths, displaying a power law-like spectral distribution, regardless of whether they are observed at low or mid latitudes. As this is typical of ESF irregularities [e.g., *Kelley*, 1989], we conclude that these mid-latitude irregularities are likely generated by similar processes to those at the equator where the magnetic field lines are more nearly horizontal, and furthermore, that their associated density structures are simply extensions of ESF depletions that are sufficiently long that they extend to mid latitudes. We shall return to this topic in section 4.

Before we leave the discussion of Plate 1c, note in the magnetic wave spectrogram that the background ELF hiss appears to sharply turn off in the depletion regions where low-frequency, broadband irregularities are present in the electric field spectrogram (e.g., between 0947 and 0952 UT). We speculate that the density component of the electrostatic irregularities appears to "scatter" the ELF hiss via a process not unlike the scattering of radiowave signals by density irregularities in space. This phenomenon may be related to whistler-mode scattering from small-scale density inhomogeneities, as discussed, for example, by *Trakhtengerts and Hayakawa* [1993].

DEMETER data gathered during the same magnetic storm on 10 November 2004 are shown in Plate 1d for an orbit that preceded the data shown in Plate 1c by 3 h. This orbit also depicts an extended, highly structured plasma density profile with significant depletions and associated strong electric field irregularities. The extended density depletion near 0635 UT at first resembles a spread F depletion, but may, instead, be evidence that the F peak had risen above the spacecraft altitude since the magnetic activity was greatly elevated (*Dst* was ~−300 nT at this time, as shown later on). We return to this point below. Note also in Plate 1d that, again, the ambient ELF hiss (e.g., at 0629 UT) is sharply "interrupted," presumably scattered by the density component of the irregularities, as is particularly evident in the magnetic wave spectrogram at this time.

Finally, we remark on the small but significant ULF/ELF magnetic field component associated with the broadband electric field irregularities that are also strongest at the lowest frequencies. This is particularly evident at 0639 UT in Plate 1d. Faint low-frequency magnetic field emissions are also present along the poleward edge of some of the depletions. We interpret these magnetic fluctuations as spatial currents created via $\nabla \times \mathbf{B} = \mu_0 \mathbf{J}$, where $\mathbf{B}$ and $\mathbf{J}$ are the magnetic field and current vectors and μ_0 is the permeability constant, as the spacecraft traversed the structures. Careful inspection of the data in Plate 1c, and even very faintly in the data in Plates 1a and 1b, provides additional evidence that these small but distinct magnetic field signatures are a common feature of ionospheric F region irregularities at low and mid latitudes. Indeed, we speculate that these ULF magnetic field signatures may be extensions of the DC magnetic field structures associated with spread F depletions observed in the CHAMP satellite data discussed by *Lühr et al.* [2002] and *Stolle et al.* [2006]. Alternatively, such magnetic field perturbations may be due to Alfvén waves, as recently discussed by *Potellette et al.* [2007] using DEMETER data. We will further explore the characteristics and origins of these ULF magnetic field signatures in a future work.

3.2. Correlations With Magnetic Storms

The near-continuous data gathered by the DEMETER satellite at latitudes below 65° and at the same local time enables a straightforward examination of how these ionospheric parameters change with respect to geomagnetic activity. We first illustrate this in Plate 2 by showing a sequence of DEMETER passes gathered on successive orbits during the magnetic storm of 10 November 2004. Here, data from consecutive nightside DEMETER passes are shown that correspond to orbits 1900–1907. (Unfortunately, most of the data from orbit 1901 are not available.) The data have been arranged such that the usual time scale along the abscissa is replaced with the magnetic latitude between ±65° in which the data have been mapped. For each orbit, a pair of data panels are shown that include the VLF electric field spectrogram and the plasma number density. As for the data shown in Plates 1a–1d, the plasma density data are provided from the ion saturation portion of the Langmuir probe sweeps. At the very lowest densities ($<10^3$ cm^{-3}), these data are not well determined and hence, for the extended depletion regions near the magnetic equator for orbits 1903 and 1905 (described below), we have substituted the ion density data from the onboard retarding potential analyzer [*Berthelier et al.*, 2006b, 2008].

As shown in Plate 2, the occurrence of both irregularities and density depletions/structures increases dramatically

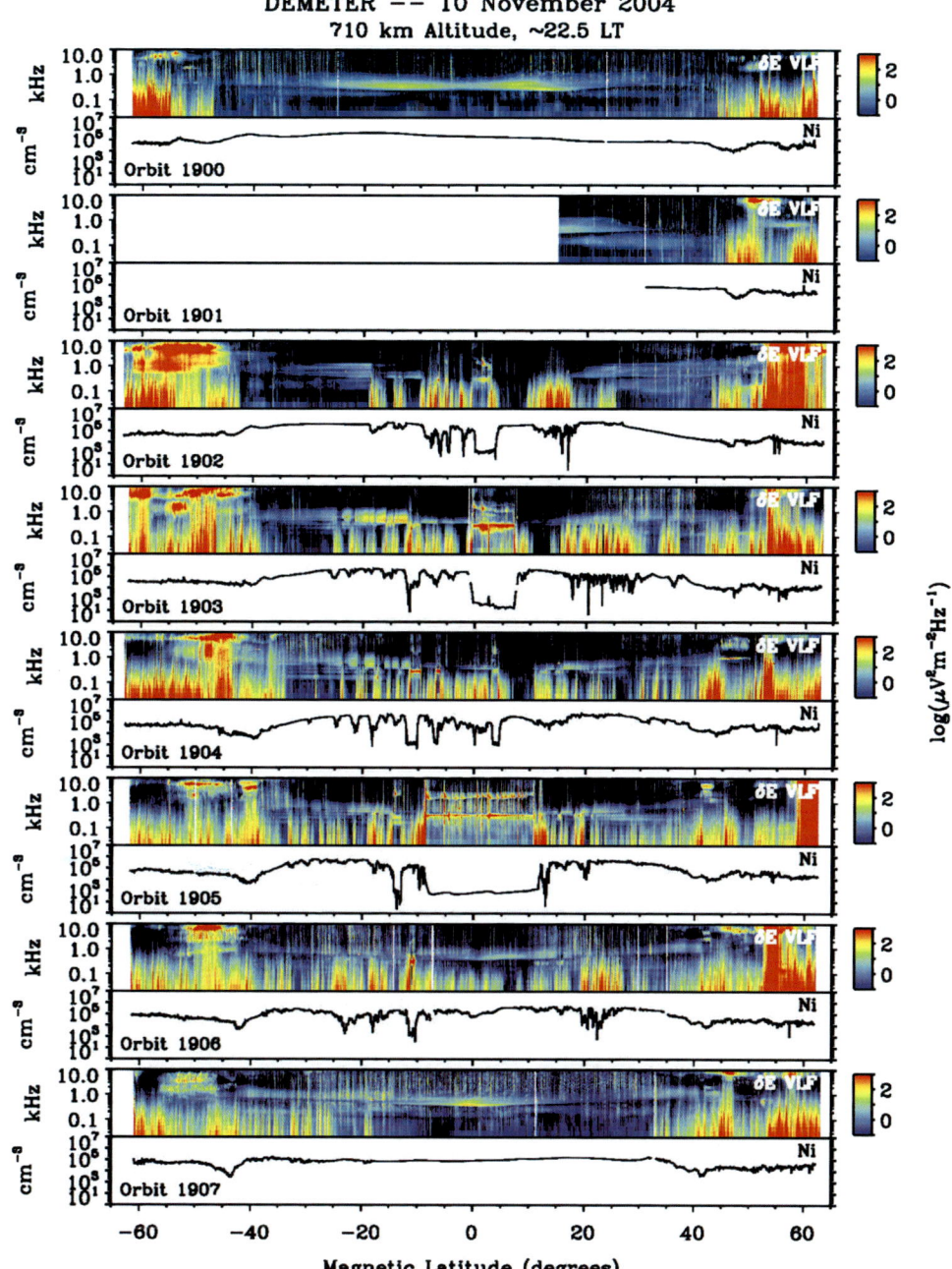

Plate 2. DEMETER satellite data for eight consecutive nighttime orbits during the magnetic storm of 10 November 2004, displayed from −65° to +65° MLAT. A pair of panels is shown for each orbit, which include the VLF electric field spectrogram on the top and the plasma number density at the bottom.

between orbits 1902 and 1906, which correspond to the period of the main phase of the magnetic storm. (*Dst* for this event will be shown in Plate 3.) Furthermore, the sub-auroral irregularity amplitudes are also increased and move equatorward during the storm period, such that the entire mid-latitude ionosphere at 710 km appears to be unstable. In this regard, it is important to note that the ionosphere had also been disrupted by a previous storm on 8 November 2004, for which sub-auroral structuring and associated irregularities were initially excited. The sub-auroral features were thus already active when the second storm of 10 November 2004 occurred, as will become much clearer in Plate 3.

The extended regions of very low plasma density observed near the magnetic equator for orbits 1902, 1903, and 1905 are suggestive of regions where the ionosphere rose above the height of the DEMETER satellite, as discussed earlier with respect to orbit 1902 shown in Plate 1d. The main argument for this interpretation is that the long duration of these depletions in the satellite reference frame corresponds to zonal distances that are on the order of several hundred kilometers, which is considerably larger than that of typical spread *F* depletion widths. For example, for orbit 1905, the large depletion centered on the magnetic equator corresponds to a width of ~850 km, given the zonal component of the satellite velocity of 2.4 km/s. The rise of the *F* peak during magnetic storms to altitudes above 840 km has been reported by *Greenspan et al.* [1991] using DMSP data and to altitudes above 600 km by *Su et al.* [2003] using ROCSAT data. Furthermore, *Basu et al.* [2007] not only reported extended density depletions recorded in DMSP data at 840 km centered on the magnetic equator during the intense magnetic storm of 30 October 2003, but also showed concurrent equatorial ionosonde data gathered beneath the satellite that clearly demonstrated that the bottomside of the ionosphere had abruptly risen to altitudes of greater than 1000 km within the same observing period. Other interpretations of such extended depletion regions are also possible, including that they are the consequences of the bifurcation of the depletions, as discussed by *Berthelier et al.* [2008], or that they simply represent extremely large "bubbles," as discussed by *Kit et al.* [2006]. It is interesting to note that *Berthelier et al.* [2008] interpret the VLF wave emissions observed within these extended regions of very low plasma density (same data as shown here for DEMETER orbits 1903 and 1905 in Figure 2) as due to lightning-induced plasma instabilities at the lower hybrid frequency [see also *Malingre et al.*, 2008], whose energy clearly originates in the troposphere below.

Another representation of these same observations are presented in Plate 3, which shows 5 days of consecutive DEMETER passes (7–11 November 2004) in which an extended, "double-peaked" geomagnetic storm occurred, as revealed in the *Dst* data in the uppermost panel. The second panel shows integrated electric field irregularity power within the band of 10–100 Hz and the third panel shows the plasma density. In each case, the data along the orbit have been converted to geomagnetic latitude and the amplitude of the wave power and plasma density have been color-coded. In other words, the data gathered on 10 November along the eight orbits shown in Plate 2 have been "turned on their side" and are now represented as condensed data within the 7–11 November presentation in Plate 3. Again, the irregularities and plasma density have been mapped to a uniform scale of magnetic latitude to enable a ready comparison of successive orbits as a function of time.

Notice that the plasma density is highly variable at mid latitudes during this period but appears to increase overall in association with the large, negative *Dst* excursions. A more definitive correlation with *Dst* is provided by the integrated 10- to 100-Hz electric field irregularity power shown in the second panel of Plate 3. Here, the irregularity activity clearly increases at precisely the times of the large *Dst* excursions, displaying simultaneous activity in both the northern and southern hemispheres (at least to the extent that this can be measured along a single satellite orbit). There is also a small, but significant response in the irregularity data during the last few orbits of 9 November 2004 (see just after 2000 UT), which correlates with a small negative excursion of the *Dst*.

The most intense electric field irregularities in Plate 3 are those at higher latitudes, for example, between roughly 50° and 65° MLAT, and are also clearly associated with the storm. These higher latitude irregularities, however, appear as soon as magnetic activity commences, as seen in the *Dst* signature, and hence both precede and continue to exist for times beyond those of the appearance of the low and mid-latitude irregularities discussed above. In fact, the irregularities at the highest latitudes shown here may be generated by auroral zone processes that have simply extended equatorward as a result of the onset of magnetic activity. There is, however, a clear "ribbon-like" component of irregularities slightly equatorward yet within the sub-auroral region that are coincident with depletions in the plasma density or mid-latitude trough. This latitudinally narrow band of irregularities displays similar modulation with latitude in both hemispheres simultaneously. Notice in particular that both the plasma density trough and the narrow band of irregularities are well defined and "track" each other throughout the storm period of 8–12 November, including not only the times of the large negative *Dst* excursions but also during the recovery periods after the main phase of each storm has subsided. We note that part of the diurnal variation of the trough and its associated irregularities may be due to the asymmetry of the geographic and geomagnetic poles that could account

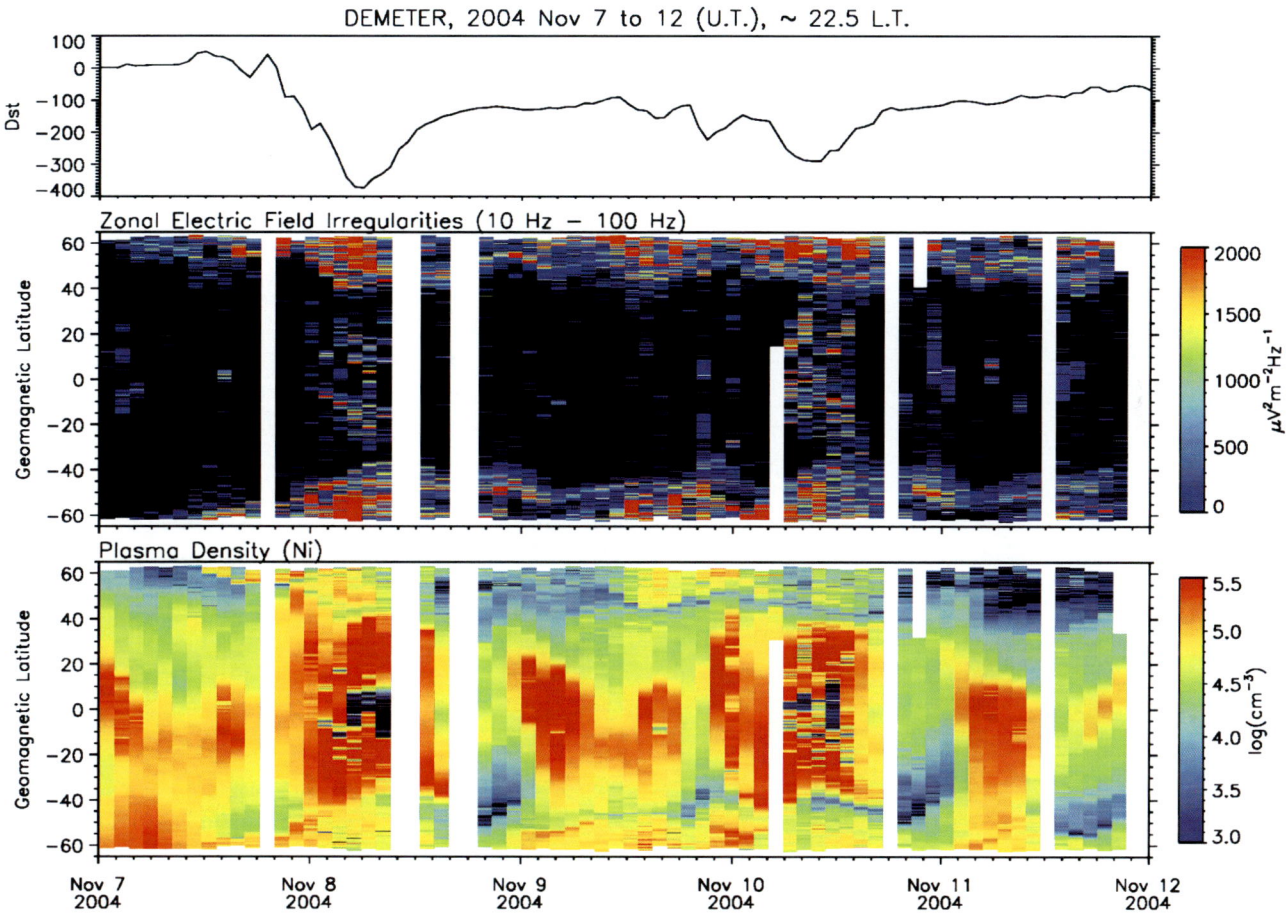

Plate 3. DEMETER satellite nighttime data (near 2230 LT) for a 5-day period from 7 to 11 November 2004 during a double peaked magnetic storm. The top panel shows the *Dst*, whereas the middle and bottom panels show the integrated electric field irregularity power from 10 to 100 Hz and the plasma number density along each orbit. The satellite data have been converted to color scales and mapped to bins of MLAT between –65° and +65°.

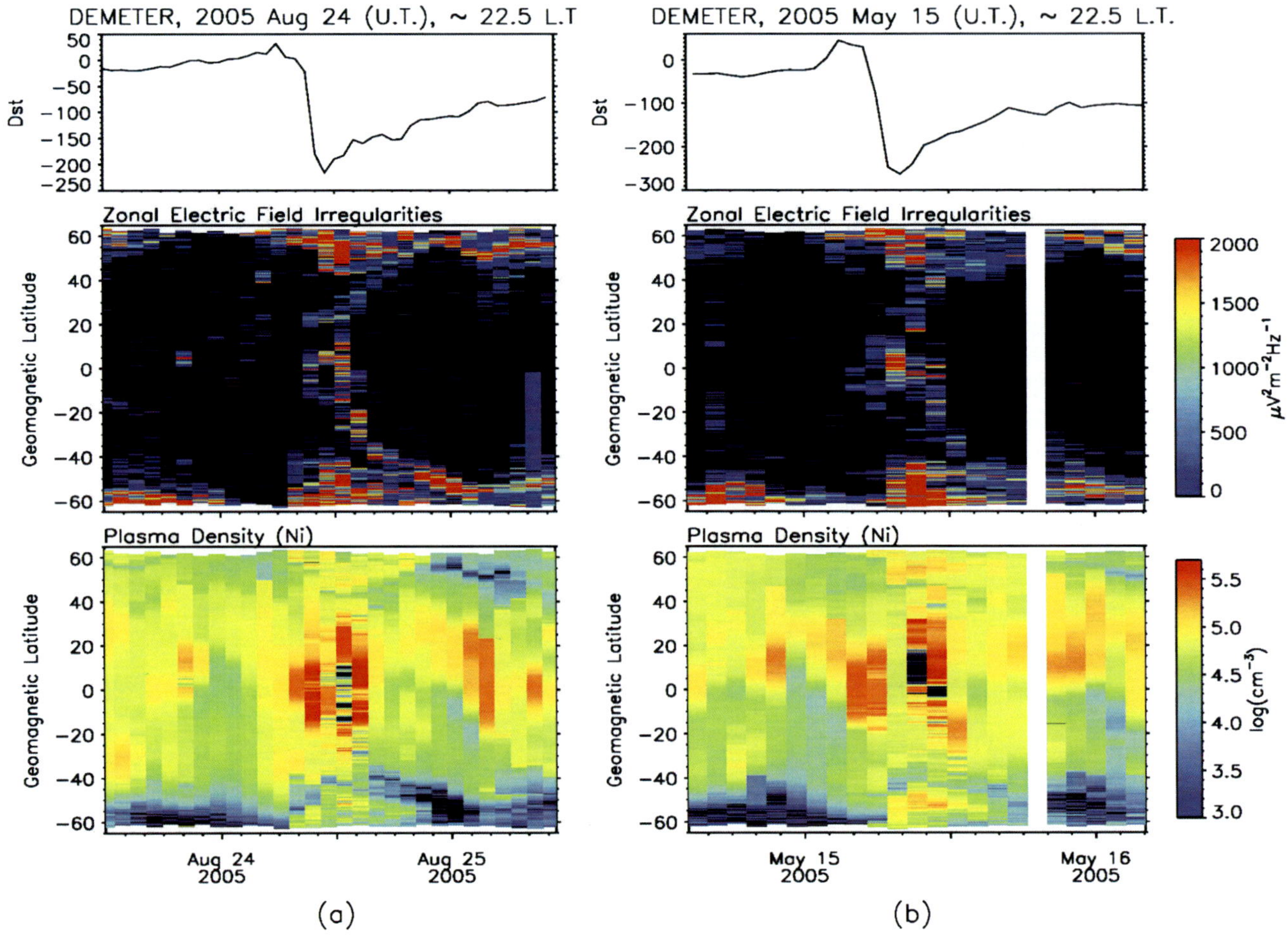

Plate 4. Same as Plate 3 except the data are shown for (a) the storm period of 24 August 2005 and (b) for the storm period of 15 May 2005.

for a 3–5° offset in magnetic latitude as a function of longitude, when observed by a Sun-synchronous, polar orbiting satellite (P. Anderson, personal communication, 2008).

Two additional examples of groups of data from successive DEMETER orbits and gathered during magnetic storms are displayed in the same format as that of Plate 3. These are shown in Plates 4a and 4b for the 2-day period of 24–25 August 2005 and 15–16 May 2005, respectively. In each case, a distinct, isolated magnetic storm occurred, as shown in the *Dst* data in the uppermost panels. In both Plates 4a and 4b, the irregularities appeared first at the magnetic equator, coincidentally (or perhaps just before) the onset of the negative *Dst* excursion, and then proceeded to "fill out" the mid-latitude ionosphere in both the northern and southern hemispheres. Again, the low and mid-latitude irregularities are present only within the narrow time interval (~6–8 h) associated with the negative *Dst* signature and its initial recovery. The mid-latitude irregularities eventually appear to "merge" with the sub-auroral irregularities that are clearly associated with distinct depletions or troughs in the plasma density, in a similar fashion as in the data shown in Plate 3.

Finally, although the low and mid-latitude nighttime plasma density is inherently variable for a number of reasons, notice in both examples in Plates 4a and 4b that the plasma density increases somewhat just before the onset of irregularities and furthermore that this increase initially occurred at the equator and neighboring latitudes. This observation is consistent with the hypothesis that the ionosphere was "rising" in response to the storm onset such that the DEMETER satellite encountered enhanced ionospheric density at low latitudes at 710 km that would normally (i.e., during a nonstorm period) be present at lower altitudes. This uplifting of the *F* region and associated topside plasma depletions associated with geomagnetic storms was observed and discussed by *Foster and Rich* [1998]. Other explanations for the increased plasma density are also possible, including increased transport along the magnetic field, perhaps driven by neutral winds or other processes. We return to this hypothesis in the next section.

4. DISCUSSION

The DEMETER data present a unique opportunity to study and understand how ionospheric irregularities near 710 km, particularly at low, middle, and sub-auroral latitudes, form and evolve during magnetic storms. Through an examination of the electric field spectra in association with the density depletions and in particular with steep density gradients, we have concluded that a significant fraction of the observed mid-latitude irregularities at 710 km have similar characteristics to those normally associated with ESF and hence are likely generated by similar local instability processes. How these irregularities and plasma structures are distributed, in a general sense, among the low, mid, and sub-auroral latitudes and with respect to a global-scale, externally driven "perturbation" such as that of a magnetic storm, is not established and is the focus of the discussion here.

Magnetic storm effects are revealed through detailed correlations between the low, middle, and sub-auroral latitude irregularities and plasma density structures with *Dst* and other indicators of magnetic activity. The fact that the entire low and mid-latitude topside ionosphere is shown to essentially "snap on" and be unstable in a relatively short period (~1 h) supports a large-scale process, perhaps governed by penetration electric fields and other global drivers [e.g., *Burke*, 2007], as the energy source for the large-scale plasma structures and associated irregularities, as opposed to a more confined, local energy source such as neutral winds propagating from one region of the globe to another, perhaps pushing plasma up field lines in each hemisphere. We emphasize that the intermediate and short-scale irregularities themselves, however, very likely result from local generation mechanisms associated with the larger-scale density structures and available sources of free energy.

The evolution of the irregularity patterns observed in successive DEMETER orbits are consistent with the hypothesis of rising, unstable, interhemispherical depletions driven by eastward electric fields, from which the following chain of events is proposed to explain, at least qualitatively, the observations. From the vantage point of the DEMETER satellite, which samples the nightside (2230 LT) ionosphere at 710 km on successive orbits spaced ~100 min apart, the observations first indicate an increase in plasma density at the equator and low latitudes that occurs near the onset of the storm activity as measured by *Dst*, consistent with a rise in the *F*-region ionospheric density from below. In fact, there may be unstable, interhemispherical depletion regions below the satellite at this time whose irregularities are not observed by DEMETER because they have not yet risen to its altitude of 710 km. Next, the fastest upward-moving unstable depletion region is encountered by DEMETER, and its associated irregularities are first observed at the magnetic equator since this corresponds to the apex of the flux-tube-shaped depletion. Data from successive orbits suggest that the depletion regions then extend to altitudes higher than DEMETER, such that their associated local irregularities fill out the mid-latitude ionosphere at increasingly higher magnetic latitudes with time. The fact that these irregularities appear within both hemispheres simultaneously, or at least within the same satellite orbit, supports this proposed scenario. As the storm progresses, the equatorial *F* region ionosphere may continue to rise to altitudes such that the bottomside

F region ledge surpasses the altitude of the DEMETER satellite, thus explaining the large density depletions centered at approximately the magnetic equator that are observed toward the latter portion of the large negative *Dst* excursion in the examples presented here. Eventually, as the magnetic storm (and *Dst* excursion) subsides, the irregularities are no longer observed at low and mid latitudes. At first glance, this suggests that the eastward electric field diminished quickly, as opposed to gradually subsiding, since there is no evidence of the depletions and their associated irregularities transitioning to progressively lower latitudes in subsequent orbits as the negative *Dst* excursion relaxes during the recovery phase. However, since the depletions and associated irregularities encounter the 2230 LT sampling region of the DEMETER satellite presumably from the west (i.e., earlier local times) as well as from below, it may be that irregularities are still being generated as the magnetic storm subsides, but they are simply not present at the necessary local time or high altitude to be encountered by the DEMETER observing platform.

A consequence of this proposed scenario is that the apex of the unstable interhemispherical depletions would rise to altitudes much greater than 1000 km, indeed to altitudes of 2500 km or more for flux tubes with footprints near ±30–35° MLAT. This conclusion is consistent with that reported by *Mendillo et al.* [1997], who suggested that airglow depletion patterns observed at the Arecibo Observatory in Puerto Rico (30° MLAT) during a magnetically active period represented optical signatures of ESF that had risen to altitudes of 2500 km at the equator in order to intersect the flux tube that extended to this mid-latitude observing station. Similarly, *Kelley et al.* [2002] reported airglow observations of ESF that extended north of Hawaii and hence reached altitudes of more than 1500 km at the magnetic equator.

The presence of irregularities at still higher "mid" latitudes (>40°) at 710 km could be due to unstable interhemispherical depletions that continue to rise to still higher altitudes. Although this is possible for some, perhaps very rare, events, it seems more likely—and is supported by the observations presented here—that sub-auroral irregularities move equatorward during the storms in association with the equatorward motion of the plasmaspheric trough, ultimately merging with the mid-latitude spread *F* irregularity signatures.

The patterns and evolution of sub-auroral irregularities and their associated plasma density troughs with the magnetic storm activity are revealed by the DEMETER observations at 2230 LT presented here. The observations build on a number of recent studies on the sub-auroral ionospheric response to enhanced plasma drifts during magnetic storms observed by satellite and radars, as discussed by numerous authors cited in section 1. The DEMETER observations demonstrate that the trough continues to exist after the main phase of the storm subsides, moving poleward during the recovery phase (see, in particular, Plates 4a and 4b), while maintaining a well-defined latitudinal band of associated electric field irregularities. Indeed, these irregularities appear to be governed by the poleward edge of the trough, supporting a gradient drift-type instability to account for their generation since the DC electric field may be expected to be directed positive along the poleward gradient of the trough during enhanced SAID events typically associated with such phenomena. Other possible generation mechanisms may also be at work and will be explored in subsequent studies.

Acknowledgment. This work was supported by a grant (to R.F.P.) from NASA's Living With a Star Targeted Research and Technology program.

REFERENCES

Anderson, P. C., W. B. Hanson, R. A. Heelis, J. D. Craven, D. N. Baker, and L. A. Frank (1993), A proposed production model of rapid subauroral ion drifts and their relation to substorm evolution, *J. Geophys. Res.*, *98*, 6069–6078.

Basu, S., S. Basu, F. J. Rich, K. M. Groves, E. MacKenzie, C. Coker, Y. Sahai, P. R. Fagundes, and F. Becker-Guedes (2007), Response of the equatorial ionosphere at dusk to penetration electric fields during intense magnetic storms, *J. Geophys. Res.*, *112*, A08308, doi:10.1029/2006JA012192.

Berthelier, J. J., et al. (2006a), ICE, the electric field experiment on DEMETER, *Planet Space Sci.*, *54*, 456–471.

Berthelier, J. J., M. Godefroy, F. Leblanc, E. Seran, D. Peschard, P. Gilbert, and J. Artru (2006b), IAP, the thermal plasma analyzer on DEMETER, *Planet Space Sci.*, *54*, 487–501.

Berthelier, J. J., M. Malingre, R. Pfaff, E. Seran, R. Pottelette, J. Jasperse, J.-P. Lebreton, and M. Parrot (2008), Lightning-induced plasma turbulence and ion heating in equatorial ionospheric depletions, *Nat. Geosci.*, doi:10.1038.

Burke, W. J. (2007), Penetrating electric fields: A Volland–Stern approach, *J. Atmos. Sol. Terr. Phys.*, *69*, 1114–1126.

Cussac, T., M-A. Clair, P. Ultre-Guerard, F. Buisson, G. Lassalle-Balier, M. Ledu, C. Elisabelar, X. Passot, and N. Rey (2006), The DEMETER microsatellite and ground segment, *Planet Space Sci.*, *54*, 413–427.

Erickson, P. J., J. C. Foster, and J. M. Holt (2002), Inferred electric field variability in the polarization jet from Millstone Hill E region coherent scatter observations, *Radio Sci.*, *37*(2), 1027, doi:10.1029/2000RS002531.

Foster, J. C., and F. J. Rich (1998), Prompt midlatitude electric field effects during severe geomagnetic storms, *J. Geophys. Res.*, *103*, 26,367–26,372.

Foster, J. C., P. J. Erickson, F. D. Lind, and W. Rideout (2004), Millstone Hill coherent-scatter radar observations of electric field variability in the sub-auroral polarization stream, *Geophys. Res. Lett.*, *31*, L21803, doi:10.1029/2004GL021271.

Fukao, S., M. C. Kelley, T. Shirakawa, T. Takami, M. Yamamoto, T. Tsuda, and S. Kato (1991), Turbulent upwelling of the midlatitude ionosphere, 1. Observational results by the MU radar, *J. Geophys. Res.*, *96*, 3725–3746.

Greenspan, M. E., C. E. Rasmussen, W. J. Burke, and M. A. Abdu (1991), Equatorial density depletions observed at 840 km during the great magnetic storm of March 1989, *J. Geophys. Res.*, *96*(A8), 13,931–13,942.

Greenwald, R. A., K., Oksavik, P. J. Erickson, F. D. Lind, J. M. Ruohoniemi, J. B. H. Baker, and J. W. Gjerloev (2006), Identification of the temperature gradient instability as the source of decameter-scale ionospheric irregularities on plasmapause field lines, *Geophys. Res. Lett.*, *33*, L18105, doi:10.1029/2006GL026581.

Hysell, D. L. (2002), An overview and synthesis of plasma irregularities in equatorial spread-F, *J. Atmos. Sol. Terr. Phys.*, *62*, 1027.

Hysell, D. L., and E. Kudeki (2004), Collisional shear instability in the equatorial F region ionosphere, *J. Geophys. Res.*, *109*, A11301, doi:10.1029/2004JA010636.

Kelley, M. C. (1989), *The Earth's Ionosphere*, Academic, New York.

Kelley, M. C., F. Garcia, J. Makela, T. Fan, E. Mak, C. Sia, and D. Alcocer (2000a), Highly structured tropical airglow and TEC signatures during strong geomagnetic activity, *Geophys. Res. Lett.*, *27*, 465–468.

Kelley, M. C., J. J. Makela, W. E. Swartz, S. C. Collins, S. Thonnard, N. Aponte, and C. A. Tepley (2000b), Caribbean Ionosphere Campaign, Year One: Airglow and plasma observations during two intense mid-latitude spread-F events, *Geophys. Res. Lett.*, *27*, 2825–2828.

Kelley, M. C., J. J. Makela, B. M. Ledvina, and P. M. Kintner (2002), Observations of equatorial spread-F from Haleakala, Hawaii, *Geophys. Res. Lett.*, *29*(20), 2003, doi:10.1029/2002GL015509.

Keskinen, M. J., S. Basu, and S. Basu (2004), Midlatitude sub-auroral ionospheric small scale structure during a magnetic storm, *Geophys. Res. Lett.*, *31*, L09811, doi:10.1029/2003GL019368.

Kil, H., L. Paxton, S.-Y. Su, Y. Zhang, and H. Yeh (2006), Characteristics of the storm-induced big bubbles (SIBBs), *J. Geophys. Res.*, *111*, A10308, doi:10.1029/2006JA011743.

Knudsen, D. J., M. C. Kelley, G. D. Earle, J. F. Vickrey, and M. Boehm (1990), Distinguishing Alfvén waves from quasi-static field structures associated with the discrete aurora: Sounding rocket and Hilat satellite measurements, *Geophys. Res. Lett.*, *17*, 921–924.

Lebreton, J.-P., et al. (2006), The ISL Langmuir probe experiment and its data processing onboard DEMETER: Scientific objectives, description and first results, *Planet. Space Sci.*, *54*, 472–486.

Lühr, H., S. Maus, M. Rother, and D. Cooke (2002), First in-situ observation of night-time F region currents with the CHAMP satellite, *Geophys. Res. Lett.*, *29*(10), 1489, doi:10.1029/2001GL013845.

Makela, J. J., and M. C. Kelley (2005), Two-dimensional imaging of the development phase of plasma instabilities in the Earth's ionosphere, *IEEE Trans. Plasma Sci.*, *33*, 502.

Malingre, M., J.-J. Berthelier, R. Pfaff, J. Jasperse, and M. Parrot (2008), Lightning-induced lower hybrid turbulence and trapped ELF electromagnetic waves observed in deep equatorial plasma density depletions during intense magnetic storms, *J. Geophys. Res.*, in press.

Mendillo, M., J. Baumgardner, D. Nottingham, J. Aarons, B. Reinisch, J. Scali, and M. Kelley (1997), Investigations of thermospheric-ionospheric dynamics with 6300-Å images from the Arecibo Observatory, *J. Geophys. Res.*, *102*, 7331–7343.

Mishin, E. V., and W. J. Burke (2005), Stormtime coupling of the ring current, plasmasphere, and topside ionosphere: Electromagnetic and plasma disturbances, *J. Geophys. Res.*, *110*, A07209, doi:10.1029/2005JA011021.

Mishin, E. V., J. C. Foster, A. P. Potekhin, F. J. Rich, K. Schlegel, K. Yumoto, V. I. Taran, J. M. Ruohoniemi, and R. Friedel (2002), Global ULF disturbances during a stormtime substorm on 25 September 1998, *J. Geophys. Res.*, *107*(A12), 1486, doi:10.1029/2002JA009302.

Mishin, E. V., W. J. Burke, C. Y. Huang, and F. J. Rich (2003), Electromagnetic wave structures within subauroral polarization streams, *J. Geophys. Res.*, *108*(A8), 1309, doi:10.1029/2002JA009793.

Oksavik, K., R. A. Greenwald, J. M. Ruohoniemi, M. R. Hairston, L. J. Paxton, J. B. H. Baker, J. W. Gjerloev, and R. J. Barnes (2006), First observations of the temporal/spatial variation of the sub-auroral polarization stream from the SuperDARN Wallops HF radar, *Geophys. Res. Lett.*, *33*, L12104, doi:10.1029/2006GL026256.

Parrot, M., et al. (2006), The magnetic field experiment IMSC and its data processing onboard DEMETER: Scientific objectives, description and first results, *Planet. Space Sci.*, *54*, 441–455.

Pottelette, R., M. Malingre, J.-J. Berthelier, E. Seran, and M. Parrot (2007), Filamentary Alfvénic structures excited at the edges of equatorial plasma bubbles, *Ann. Geophys.*, *25*, 2159–2165.

Saito, A, T. Iyemori, M. Sugiura, N. C. Maynard, T. L. Aggson, L. H. Brace, M. Takeda, and M. Yamamoto (1995), Conjugate occurrence of the electric field fluctuations in the nighttime mid-latitude ionosphere, *J. Geophys. Res.*, *100*, 21,439–21,451.

Smith, R. L., and N. Brice (1964), Propagation in multicomponent plasmas, *J. Geophys. Res.*, *69*, 5029–5040.

Stolle, C., H. Lühr, M. Rother, and G. Balasis (2006), Magnetic signatures of equatorial spread F as observed by the CHAMP satellite, *J. Geophys. Res.*, *111*, A02304, doi:10.1029/2005JA011184.

Streltsov, A. V., and E. V. Mishin (2003), Numerical modeling of localized electromagnetic waves in the nightside subauroral zone, *J. Geophys. Res.*, *108*(A8), 1332, doi:10.1029/2003JA009858.

Su, S.-Y., H. C. Yeh, C. K. Chao, and R. A. Heelis (2002), Observation of a large density dropout across the magnetic field at 600 km altitude during the 6–7 April 2000 magnetic storm, *J. Geophys. Res.*, *107*(A11), 1404, doi:10.1029/2001JA007552.

Swartz, W. E., M. C. Kelley, J. J. Makela, S. C. Collins, E. Kudeki, S. Franke, J. Urbina, N. Aponte, M. P. Sulzer, and S. A. Gonzalez (2000), Coherent and incoherent scatter radar observations during intense mid-latitude spread F, *Geophys. Res. Lett.*, *27*, 2829–2832.

Trakhtengerts, V. Y., and M. Hayakawa (1993), A wave–wave interaction in whistler frequency range in space plasma, *J. Geophys. Res.*, *98*, 19,205–19,217.

J.-J. Berthelier and M. Malingre, CETP, 4 avenue de Neptune, F-94100 St. Maur, France.

J.-P. Lebreton, ESA, Keplerlaan 1, NL-2200 AG Noordwijk, Netherlands.

C. Liebrecht and R. F. Pfaff Jr., NASA Goddard Space Flight Center, Mail Code 612.3, Greenbelt, MD 20771, USA. (Robert.F.Pfaff@nasa.gov)

M. Parrot, CNRS, 3A Avenue de la Recherche, F-45071 Orleans CEDEX 2, France.

Optical and Radio Observations of Structure in the Midlatitude Ionosphere: Midlatitude Ionospheric Dynamics and Disturbances

Jonathan J. Makela

Department of Electrical and Computer Engineering, University of Illinois at Urbana–Champaign, Urbana, Illinois, USA

We present a reanalysis of optical and radio observations of nighttime ionospheric structure at midlatitudes during heightened geomagnetic activity ($Kp > 4$). This is done using several examples of structures that have been studied in past publications. The reanalysis is performed in light of recent advances in our understanding of these structures and in search of common properties. It is found that these structures are characterized by enhancements in the background total electron content, development of instabilities on the edges of the structure due to the generalized $\mathbf{E} \times \mathbf{B}$ instability, and a counterclockwise rotation aligning the structures significantly off the magnetic meridian. The instabilities generated on the gradients occur over several decades of scale sizes, and there is evidence that associated irregularities can occur at the magnetic equator along field lines connected to the structures' edges. We discuss the possibilities of the structures being equatorial or local in origin and suggest that future experiments are required with expanded coverage of the ionosphere.

1. INTRODUCTION

Prior to the late 1990s, examples of irregularities in the midlatitude ionosphere were few and far between. *Basu et al.* [1981] presented Arecibo incoherent scatter radar (ISR) data showing kilometer-scale irregularities that they attributed to the generalized $\mathbf{E} \times \mathbf{B}$ instability, although the origin of the required gradient in electron density was unknown. *Behnke* [1979] presented observations of "height layered bands" observed using the Arecibo ISR. These bands were the location of some of the largest electric fields ever reported equatorward of the auroral zone. Behnke interpreted these in terms of the Perkins instability [*Perkins*, 1973], an interpretation that remains as the de facto explanation. A new dimension to the observations was gained during the 10-day run of the Arecibo ISR in 1993 when modern optical instrumentation was used to make the first images of two-dimensional ionospheric structure at midlatitudes. This experiment resulted in important new observations comparing mesoscale depletions in the 630.0-nm nighttime emission with Arecibo ISR-derived electric fields [e.g., *Mendillo et al.*, 1997; *Miller et al.*, 1997].

Since these initial midlatitude observations in the early 1990s, several long-duration experiments have been undertaken. In fact, almost a complete solar cycle of near-continuous nighttime data has been collected from the Arecibo Observatory located in Puerto Rico through the efforts of several groups. Similarly robust data sets have also been collected in the Japanese and Pacific sectors. These optical observations have been augmented by data obtained from collocated radio equipment, ranging from dual- and single-frequency GPS receivers providing total electron content (TEC) and information on scintillations, to large radars providing information on the state of the ionospheric plasma.

Through these various experiments, beginning with the 10-day run, a consistent picture of nighttime midlatitude structure has arisen. The most common type of nighttime structures observed at midlatitudes is typically referred to as medium-scale traveling ionospheric disturbances (MSTIDs). During quiet geomagnetic conditions ($Kp < 4$), the characteristics of MSTIDs are fairly well understood [e.g., *Garcia et al.*, 2000a; *Shiokawa et al.*, 2003]. The typical MSTID in the northern hemisphere is aligned from the northwest to southeast (approximately 20° west of the magnetic meridian), propagates to the southwest at a velocity of approximately 100 m/s, and has a wavelength (parallel to **k**) of 100–300 km. They appear as alternating bands of bright and dark regions of airglow in the 630.0-nm emission, which are coincident with low-level variations (0.5–2.5 TECU; 1 TECU = 10^{16} el/m^2) in the TEC [*Ogawa et al.*, 2002] and a radially outward electric field [*Saito et al.*, 1995; *Kelley et al.*, 2000]. This large-scale electric field maps to both hemispheres, meaning that structures are present in both conjugate hemispheres [*Otsuka et al.*, 2004].

However, several unresolved mysteries pertaining to these structures and their development still remain, especially during periods of enhanced geomagnetic activity. Most reports in the literature of such structure have been case studies examining a single event. Here, we synthesize these individual events in light of conclusions that were drawn from each event to come to a more complete understanding of the general properties of these structures. We also report on recent observations suggesting that coupling effects along the magnetic field lines must be considered to fully understand these structures. Finally, we suggest future experiments that should be carried out to come to a more complete understanding of these vexing midlatitude disturbances.

2. DATA PRESENTATION

The data to be synthesized here were collected by a variety of instruments during four distinct events, as summarized in Table 1. Here, we reanalyze the available data to understand the common trends in terms of background geophysical conditions, the ionospheric state (electron content), as well as the electrodynamics and development of instabilities associated with the structures. The primary data that are common to all of these events are images of the 630.0-nm emission, characterized by complex two-dimensional structure.

The background geophysical characteristics (Kp and Dst) during the four events collected here are presented in Table 1. There have been several studies relating these terms to the development of irregularities in the postsunset equatorial ionosphere [see *Martinis et al.*, 2005, and references therein], and so we choose them here to search for a possible relationship between the equatorial and midlatitude structures. Kp values are reported in the second column for the 3-h interval closest to local sunset. The hourly Dst values corresponding to 1800–2000 LT are given in the third column. The Dst values for the various events range from a slightly disturbed level of approximately −30 nT during the 17–18 February 1999 event, to more severely disturbed levels closer to −100 nT during the other three events. These data present a consistent picture in which geomagnetic activity is high and the ionosphere/thermosphere system is likely prone to the effects of disturbance electric fields (both prompt penetrating and disturbance dynamo).

In contrast to the low-level variations in electron density of 0.5–2.5 TECU observed to be coincident with MSTIDs, extreme midlatitude TEC gradients develop during these pe-

Table 1. Parameters and References for the Four Storm Cases of Midlatitude Structure

Date and Location	Kp[a]	Dst[b] (nT)	Available Instruments	Reference
22–23 Nov 1997, Puerto Rico	7+	−44, −72, −89	Imaging	*Garcia et al.* [2000b]
			Dual-frequency GPS	*Makela et al.* [2001]
17–18 Feb 1998, Puerto Rico	7−	−92, −95, −100	Imaging	*Kelley et al.* [2000]
			Dual-frequency GPS	*Swartz et al.* [2000]
			ISR	*Swartz et al.* [2000]
			Coherent scatter radar (local ionosphere)	*Swartz et al.* [2000]
17–18 Feb 1999, Puerto Rico	5−	−29, −27, −25	Imaging	*Kelley et al.* [2000]
			Dual-frequency GPS	
			ISR	*Kelley et al.* [2000]
10–11 Sep 2005, Hawaii	8−	−45, −57, −101	Imaging	*Makela et al.* [in review]
			Coherent scatter radar (magnetic equator)	
			Dual-frequency GPS	

[a] Corresponding to local sunset.
[b] The three values given correspond to Dst at 1800, 1900, and 2000 LT, respectively.

riods of heightened geomagnetic activity. For example, data from 22–23 November 1997, reproduced in Figure 1, show a TEC environment consisting of large-scale spatiotemporal gradients coincident with a complicated airglow signature in the 630.0-nm emission. The gradients of 20 TECU over the course of one hour show just how disturbed the nighttime midlatitude ionosphere can become. We further note that in each of the cases studied here, the gradients in TEC seem to point toward enhanced levels of TEC corresponding to the bright regions seen in the airglow data, rather than depleted electron densities corresponding to dark bands. That is, the TEC levels within the dark bands are roughly equivalent to quiet-time TEC levels, while the TEC within the bright regions is enhanced by a factor of 3–6.

Turning now to a description of the electrodynamics of the structures, we note that each event is characterized by an apparent surge of a dark region from the southeast shortly after local sunset. In the case of the 22–23 November 1997 data, the velocity of this surge was approximately 480 m/s, as estimated from the imaging data [*Garcia et al.*, 2000b]. For the 17–18 February 1999 event, ISR data are available, and measurements show that the surge was coincident with an enhanced meridional drift velocity, northward and upward perpendicular to **B** [*Kelley et al.*, 2000]. Similar data exist for the 17–18 February 1998 event, in which a poleward surge of the structure was less prevalent. *Swartz et al.* [2000] show that there were enhanced velocities northward and upward perpendicular to **B** (360 m/s line of sight) associated with the passage of the structure through the radar beams.

Although the structures during three of the four events listed in Table 1 are observed to drift eastward early in the evening (all but the 22–23 November 1997 event), by local midnight the structures switch directions and drift in the southwestward direction typical of MSTIDs. During this period, the structures rotate counterclockwise, becoming aligned significantly off the magnetic meridian.

The gradients in electron density observed using GPS and ISR coincident with the storm-enhanced structure seen in the images serve as the breeding ground for the local development of instabilities. *Kelley et al.* [2004] show that the western edge of the dark structure observed on 17–18 February 1998 became unstable to the generalized **E** × **B** instability, while the gradient on the other side remained stable. This is also apparent for the structure shown in Figures 1 in which structuring is only apparent on one side of the dark regions. Structuring on the western edge of a dark region is consistent with the generalized **E** × **B** instability being driven by an eastward neutral wind or poleward electric field. Note that a

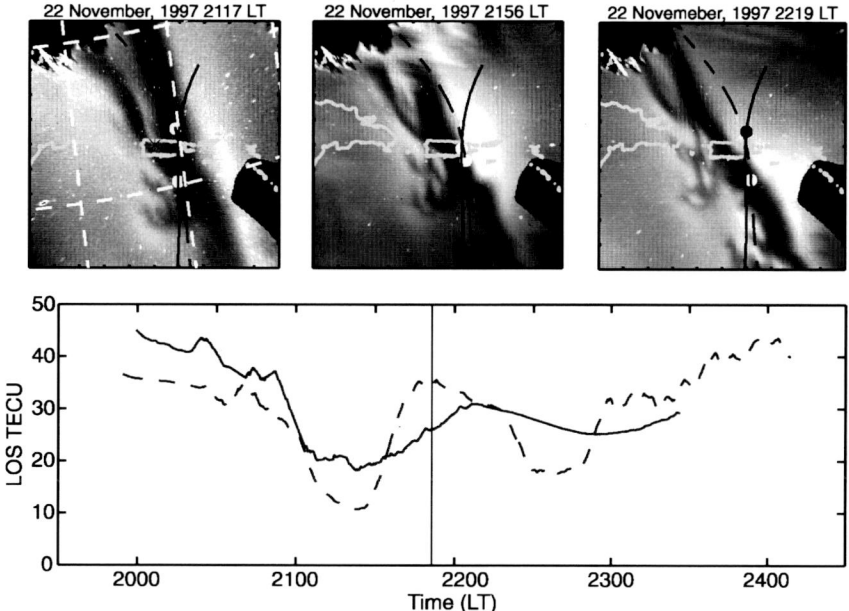

Figure 1. Observations of the 630.0-nm emission obtained from an imager located at the Arecibo Observatory on 22–23 November 1997. The bottom panel shows the TEC derived from receivers located at Isabela, PR (solid), and St. Croix, VI (dashed) [*Makela et al.*, 2001]. Lines of constant magnetic latitude and longitude are reproduced on the left image. Reproduced by permission of the American Geophysical Union.

poleward/radial electric field is a characteristic of the dark structures both during quiet and disturbed periods. Coincident Arecibo ISR radar measurement for the 17–18 February 1998 event show the presence of kilometer-scale striations [*Mathews et al.*, 2001], while the coherent scatter results of *Swartz et al.* [2000] show the presence of 3-m scale irregularities, both in the local hemisphere, suggesting that the instabilities generated are present over several decades of scale size.

Another recent event showing similar characteristics to the events observed over Puerto Rico occurred in the Pacific sector on 10–11 September 2005. Several frames from this event are presented in Figure 2. The structures are observed during a period of enhanced geomagnetic activity, the TEC levels derived from dual-frequency GPS receivers show a significant enhancement in electron content over quiet-time conditions, and the structures show the same reversal in drift direction and counterclockwise rotation before local midnight. An exciting additional data set is available from a 50-MHz coherent scatter radar on Christmas Island, magnetically conjugate to the imaging systems' fields of view. This instrument reveals the presence of 3-m irregularities at the magnetic equator along field lines connected to the edges on both sides of the structure observed in the images (J. J. Makela, M. C. Kelley, and R. T. Tsunoda, Observations of mid-latitude instabilities generating meter-scale waves at the equator, submitted to *Journal of Geophysical Research*, hereinafter referred to as Makela et al., submitted manuscript). As discussed below, this cannot easily be explained without considering coupling between the magnetic equator and midlatitudes.

3. DISCUSSION

Of the MSTID characteristics discussed above, the most striking feature is the alignment of the structures. Specifically, that they are typically aligned approximately 20° from the magnetic meridian. This is in stark contrast to equatorial depletions associated with equatorial spread *F*, which tend to be aligned with the magnetic meridian during their development, although they do tilt westward as the night progresses [see, e.g., *Makela*, 2006, and references therein]. The Perkins instability operating in the local ionosphere predicts this alignment, which is the primary reason it remains the most popular theory for explaining MSTIDs. In fact, this alignment seems to be a general characteristic of all of the nighttime ionospheric structure observed at midlatitudes, both during quiet and disturbed conditions. This is evident in all of the storm-time events detailed above, as is clear by careful consideration of the examples presented in Figures 1 and 2. Note that in each case, the structure, which begins early in the evening approximately 10° off the magnetic

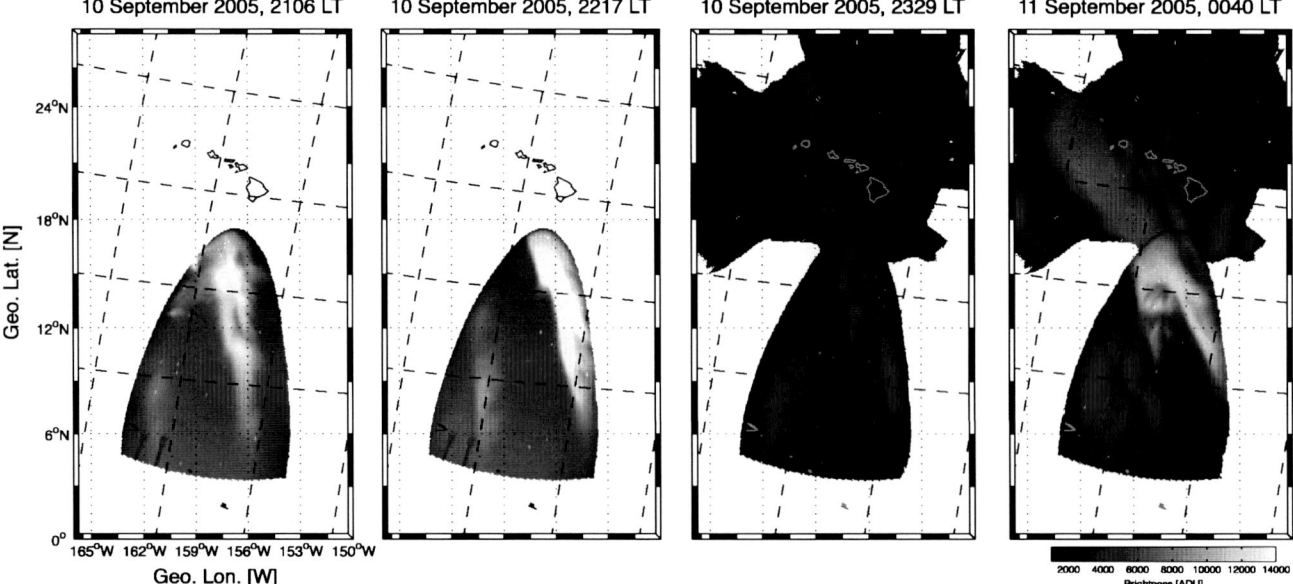

Figure 2. Observations of the 630.0-nm emission obtained from two imagers located on Maui, Hawaii, on 10–11 September 2005. Lines of constant magnetic latitude and longitude are reproduced on each image. Data from the all-sky imager were not available early in the evening due to the location of the moon.

meridian, rotates counterclockwise until it is between 20° and 30° off the meridian. For example, in the 10–11 September 2005 event, the structure rotates 15° over the span of an hour to become aligned with the "Perkins" direction. This rotation, in the cases presented here, takes place shortly before local midnight.

The origin of the storm-enhanced structures remains a topic of some controversy. That they only appear during times of enhanced geomagnetic activity is clear. However, there does not seem to be as clear of a correlation with geomagnetic indices (note that the Dst was only approximately −30 for the 17–18 February 1999 event, but much more disturbed for the other events) as there is for the development of storm-time equatorial irregularities [e.g., *Martinis et al.*, 2005]. A more complete study of the geophysical characteristics, including the effects of prompt penetrating and disturbance dynamo electric fields as well as the state of the interplanetary electric field, is needed to address this issue.

The structures observed over Hawaii and Puerto Rico could simply be colossal equatorial spread F bubbles that are generated at the magnetic equator, growing to several thousand kilometers in altitude. In this case, the electric fields generated internal to these structures map along the magnetic field lines where they would perturb the local ionosphere and generate the observed structure. The westward tilt (from the magnetic meridian) would then be consistent with the observed tilts with altitude that gradually develop with increasing altitude of the equatorial bubbles, as observed in radar data and optical data from both space- and ground-based instruments [see *Makela*, 2006, and references therein]. The westward drift, under this explanation would be because of a change in the equatorial/low-latitude drifts from east to west, also common during storm-time due to the development of a disturbance dynamo. Clearly, this explanation has its merits and explains many of the observed traits.

Indeed, from the Hawaii site, using both a narrowfield imager and an all-sky imager, unambiguous examples of such equatorial plasma bubbles reaching poleward of Hawaii (corresponding to an equatorial apex altitude or 2000 km) have been observed [*Kelley et al.*, 2002]. Such high equatorial plasma bubbles have also been observed in the South American [*Sahai et al.*, 1998; *Martinis and Mendillo*, 2007] and Japanese [*Otsuka et al.*, 2002] sector, reaching altitudes over 2000 km. To extend out of the poleward edge of the field of view from Puerto Rico, however, a bubble would have to reach an equatorial apex altitude of over 3250 km. Of course, the storm-enhanced events occur during disturbed conditions, when electric fields can be intensified, meaning it may be feasible for a bubble to grow to this altitude.

However, at least two characteristics of these structures argue against them being equatorial in origin. First, we consider the rapid growth of an equatorial depletion to reach the lower midlatitudes early in the evening. As an example, the first sign of structuring on 22–23 November 1997 occurs at approximately 2320 UT (1920 LT). As proposed by *Tsunoda* [1985], one prerequisite for the development of an equatorial depletion is that the E region footprints of the equatorial F region are in shade, so as not to short out the electric fields necessary for growth. On 22–23 November 1997, the southern E region footprint remains in sunlight until approximately the same time as the structure first appears in the imaging data. Each event studied suggests that this is a common trait (occurrence early in the evening) for these structures. Is it reasonable to have an equatorial plasma bubble develop and grow over 1000 km in altitude during the span of minutes? Second, unlike equatorial plasma bubbles, which invariably appear as depletions in electron density, these structures, upon closer examination, actually seem to be enhancements in electron content on an otherwise typical background. That is the electron content, as measured by GPS-derived TEC, appears "typical" (compared to quiet time values) in the dark bands and is enhanced several times over quiet time values in the bright regions. This suggests that the pertinent aspect of these events is the enhancement of the electron density and the associated gradients in electron content.

Regardless of the origin of the gradients in electron density, they will be subject to the local physics of the generalized $\mathbf{E} \times \mathbf{B}$ instability. For this instability mechanism to be effective, either the neutral wind must be oriented from high to low electron density or the $\mathbf{E} \times \mathbf{B}$ direction must be oriented from low to high electron density. Note, however, that for a given orientation of the wind or electric field, one side of a dark region in the airglow would be unstable, while the other would be stabilized. The Hawaii event shown in Figure 2, thus, highlights an interesting new facet of these structures hitherto unappreciated due to a lack of data. As described in more detail by Makela et al. (submitted manuscript), the coherent scatter radar at Christmas Island observed 3-m irregularities on field lines conjugate to the edges on both sides of the enhanced airglow. To satisfactorily explain this, one must invoke mechanisms operating in the equatorial plane (Rayleigh–Taylor) to explain the instabilities on the equatorward edge of the airglow enhancement and the generalized $\mathbf{E} \times \mathbf{B}$ mechanism operating in the off-equatorial plane to explain the instabilities on the poleward edge. The large-scale perturbation electric field generated off the equator must then map to the equatorial plane where it then generates instabilities at 3-m scale size to explain the complex 50-MHz echoes seen in the Christmas

Island radar. This event, to our knowledge, is the first example of what appears to be an instability process in the local off-equatorial ionosphere feeding irregularities at the magnetic equator.

To make progress in furthering our understanding of these fascinating structures, a new set of experiments needs to be carried out. Concentrating on the American sector to make use of the invaluable data obtained from the instrumentation at the Arecibo Observatory, it is logical to extend the portion of the ionosphere studied southward by installing new imaging systems in the southern Caribbean and/or northern South America, expanding on the conjugate observatories (Arecibo and El Leoncito, Argentina) already in place [e.g., *Martinis and Mendillo*, 2007]. Doing so would provide direct observations that could possibly resolve the outstanding issue of whether the storm-enhanced structure is generated in the local ionosphere poleward of the equatorial anomaly or if it is the signature of a colossal equatorial plasma bubble. Additional GPS-TEC monitors are needed in this region to track the electron content in better detail. Coherent scatter radar and GPS scintillation monitors would add valuable information on the occurrence and location of instabilities at various scale sizes within the structures or along their gradients. Conducting a new round of coordinated experiments seems warranted at this point, if we are to come to a better understanding of the disturbed midlatitude nighttime ionosphere.

Acknowledgments. This work was supported under grant ATM-0644654 from the National Science Foundation. The imager data directly used in this study were obtained in conjunction with Michael C. Kelley. Geophysical indices were obtained from the SPIDR database maintained by NOAA.

REFERENCES

Basu, S., S. Basu, S. Ganguly, and J. A. Klobuchar (1981), Generation of kilometer scale irregularities during the midnight collapse at Arecibo, *J. Geophys. Res.*, *86*(A9), 7607–7616.

Behnke, R. (1979), F-layer height bands in the nocturnal ionosphere over Arecibo, *J. Geophys. Res.*, *84*(A3), 974–978.

Garcia, F. J., M. C. Kelley, J. J. Makela, and C.-S. Huang (2000a), Airglow observations of mesoscale low-velocity traveling ionospheric disturbances at midlatitudes, *J. Geophys. Res.*, *105*(A8), 18,407–18,415.

Garcia, F. J., M. C. Kelley, J. J. Makela, P. J. Sultan, X. Pi, and S. Musman (2000b), Mesoscale structure of the midlatitude ionosphere during high geomagnetic activity: Airglow and GPS observations, *J. Geophys. Res.*, *105*(A8), 18,417–18,427.

Kelley, M. C., J. J. Makela, W. E. Swartz, S. C. Collins, S. Thonnard, N. Aponte, and C. A. Tepley (2000), Caribbean Ionosphere Campaign, Year One: Airglow and plasma observations during two intense mid-latitude spread-F events, *Geophys. Res. Lett.*, *27*(18), 2825–2828.

Kelley, M. C., J. J. Makela, B. M. Ledvina, and P. M. Kintner (2002), Observations of equatorial spread-F from Haleakala, Hawaii, *Geophys. Res. Lett.*, *29*(20), 2003, doi:10.1029/2002GL015509.

Kelley, M. C., W. E. Swartz, and J. J. Makela (2004), Mid-latitude ionospheric fluctuation spectra due to secondary E×B instabilities, *J. Atmos. Sol. Terr. Phys.*, *66*(17), 1559–1565, doi:10.1016/j.jastp.2004.07.004.

Makela, J. J. (2006), A review of imaging low-latitude ionospheric irregularity processes, *J. Atmos. Sol. Terr. Phys.*, *68*(13), 1441–1458, doi:10.1016/j.jastp.2005.04.014.

Makela, J. J., M. C. Kelley, J. J. Sojka, X. Pi, and A. J. Mannucci (2001), GPS normalization and preliminary modeling results of total electron content during a midlatitude space weather event, *Radio Sci.*, *36*(2), 351–361.

Martinis, C. R., and M. J. Mendillo (2007), Equatorial spread F-related airglow depletions at Arecibo and conjugate observations, *J. Geophys. Res.*, *112*, A10310, doi:10.1029/2007JA012403.

Martinis, C. R., M. J. Mendillo, and J. Aarons (2005), Toward a synthesis of equatorial spread F onset and suppression during geomagnetic storms, *J. Geophys. Res.*, *110*, A07306, doi:10.1029/2003JA010362.

Mathews, J. D., S. González, M. P. Sulzer, Q.-H. Zhou, J. Urbina, E. Kudeki, and S. Franke (2001), Kilometer-scale layered structures inside spread-F, *Geophys. Res. Lett.*, *28*(22), 4167–4170, doi:10.1029/2001GL013077.

Mendillo, M., J. Baumgardner, D. Nottingham, J. Aarons, B. Reinisch, J. Scali, and M. Kelley (1997), Investigations of thermospheric ionospheric dynamics with 6300-Å images from the Arecibo Observatory, *J. Geophys. Res.*, *102*(A4), 7331–7343.

Miller, C. A., W. E. Swartz, M. C. Kelley, M. Mendillo, D. Nottingham, J. Scali, and B. Reinisch (1997), Electrodynamics of mid-latitude spread F. 1. Observations of unstable, gravity wave-induced ionospheric electric fields at tropical latitudes, *J. Geophys. Res.*, *102*(A6), 11,521–11,532.

Ogawa, T., N. Balan, Y. Otsuka, K. Shiokawa, C. Ihara, T. Shimonmai, and A. Saito (2002), Observations and modeling of 630 nm airglow and total electron content associated with traveling ionospheric disturbances over Shigaraki, Japan, *Earth Planet. Space*, *54*, 45–56.

Otsuka, Y., K. Shiokawa, T. Ogawa, and P. Wilkinson (2002), Geomagnetic conjugate observations of equatorial airglow depletions, *Geophys. Res. Lett.*, *29*(15), 1753, doi:10.1029/2002GL015347.

Otsuka, Y., K. Shiokawa, T. Ogawa, and P. Wilkinson (2004), Geomagnetic conjugate observations of medium-scale traveling ionospheric disturbances at midlatitude using all-sky airglow imagers, *Geophys. Res. Lett.*, *31*, L15803, doi:10.1029/2004GL020262.

Perkins, F. (1973), Spread F and ionospheric currents, *J. Geophys. Res.*, *78*(1), 218–226.

Sahai, Y., P. R. Fagundes, J. A. Bittencourt, and M. A. Abdu (1998), Occurrence of large scale equatorial F-region plasma depletions during geo-magnetic disturbances, *J. Atmos. Sol. Terr. Phys.*, *60*(1), 1593–1604.

Saito, A., T. Iyemori, M. Sugiura, N. C. Maynard, T. L. Aggson, L. H. Brace, M. Takeda, and M. Yamamoto (1995), Conjugate occurrence of the electric field fluctuations in the nighttime mid-latitude ionosphere, *J. Geophys. Res.*, *100*, 21,439–21,451.

Shiokawa K., C. Ihara, Y. Otsuka, and T. Ogawa (2003), Statistical study of nighttime medium-scale traveling ionospheric disturbances using midlatitude airglow images, *J. Geophys. Res.*, *108* (A1), 1052, doi:10.1029/2002JA009491.

Swartz, W. E., M. C. Kelley, J. J. Makela, S. C. Collins, E. Kudeki, S. Franke, J. Urbina, N. Aponte, M. P. Sulzer, and S. A. Gonzalez (2000), Coherent and incoherent scatter radar observations during intense mid-latitude spread F, *Geophys. Res. Lett.*, *27*(18), 2829–2832.

Tsunoda, R. T. (1985), Control of the seasonal and longitudinal occurrence of equatorial scintillations by the longitudinal gradient in integrated E region Pedersen conductivity, *J. Geophys. Res.*, *90*(A1), 447–456.

J. J. Makela, Department of Electrical and Computer Engineering, University of Illinois at Urbana–Champaign, 316 Coordinated Science Lab, MC-228, 1406 West Green Street, Urbana, IL 61801-2918, USA. (jmakela@uiuc.edu)

Global-Scale Observations of the Limb and Disk (GOLD): New Observing Capabilities for the Ionosphere-Thermosphere

R. W. Eastes,[1] W. E. McClintock,[2] M. V. Codrescu,[3] A. Aksnes,[4] D. N. Anderson,[5] L. Andersson,[2] D. N. Baker,[2] A. G. Burns,[6] S. A. Budzien,[7] R. E. Daniell,[8] K. F. Dymond,[7] F. G. Eparvier,[2] J. E. Harvey,[9] T. J. Immel,[10] A. Krywonos,[1] M. R. Lankton,[2] J. D. Lumpe,[11] G. W. Prölss,[12] A. D. Richmond,[5] D. W. Rusch,[2] O. H. Siegmund,[10] S. C. Solomon,[6] D. J. Strickland,[11] and T. N. Woods[2]

The Global-scale Observations of the Limb and Disk (GOLD) mission of opportunity will greatly improve understanding of the Earth's thermosphere and ionosphere through measurements of the global-scale response to external and internal forces. GOLD will fly an UV imager on a geostationary satellite to measure densities and temperatures across almost an entire hemisphere in this poorly understood region of the Earth's upper atmosphere and lower space environment, at altitudes where temperatures are currently not well known. GOLD will provide the first global-scale observations of temperatures in the lower thermosphere (130–180 km), in addition to more familiar measurements such as aurora location and energy input, peak electron densities (N_mF_2) in the nighttime ionosphere, and atomic oxygen to molecular nitrogen column density ratios ($\Sigma O/N_2$) ratios. GOLD can provide nearly continuous real-time observations of one hemisphere. In addition to measurements on the disk of the Earth, GOLD can provide coincident measurements of molecular oxygen densities and the temperature profile in the lower thermosphere (150–250 km) from stellar occultations as well as exospheric temperatures from limb profiles of molecular nitrogen emissions. GOLD has two identical channels, each capable of all the measurements described. This allows GOLD to provide coincident measurements in any desired combination, e.g., disk temperatures and $\Sigma O/N_2$. Combined with the advanced models now available, measurements from GOLD will revolutionize our understanding of the global-scale

[1]Florida Space Institute, MS-FSI, Kennedy Space Center, Florida, USA.
[2]Laboratory for Atmospheric and Space Physics, University of Colorado, Boulder, Colorado, USA.
[3]NOAA-SWPC, Boulder, Colorado, USA.
[4]Universitetet i Bergen, Det matematisk-naturvitenskapelige fakultet, Bergen, Norway.
[5]Cooperative Institute for Research in Environmental Sciences, University of Colorado, Boulder, Colorado, USA.
[6]National Center for Atmospheric Research, Boulder, Colorado, USA.
[7]Naval Research Laboratory, Washington, D. C., USA.
[8]Ionospheric Physics, Stoughton, Massachusetts, USA.
[9]Center for Research and Education in Optics and Lasers, University of Central Florida, Orlando, Florida, USA.
[10]Space Sciences Laboratory, University of California, Berkeley, California, USA.
[11]Computational Physics, Inc., Springfield, Virginia, USA.
[12]Argelander Institut für Astronomie, Bonn, Germany.

Midlatitude Ionospheric Dynamics and Disturbances
Geophysical Monograph Series 181
Copyright 2008 by the American Geophysical Union.
10.1029/181GM29

response of the thermosphere and ionosphere to geomagnetic and solar forcing. The data and knowledge gained from GOLD will enhance space weather specification and forecasting capabilities.

1. INTRODUCTION

GOLD has completed a competitive phase A study as a possible mission of opportunity (MOO) to be flown in conjunction with the Living with a Star (LWS) programs Radiation Belt Storm Probes mission. Four features of the GOLD mission are prominent:

1. First global-scale measurements of temperatures in the Earth's upper atmosphere: In contrast to previous in situ measurements and remote sensing observations from low Earth orbit, which covered only localized regions of space, GOLD's geostationary perspective will provide unambiguous spatial and temporal measurements of temperatures at the upper levels of the Earth's atmosphere that are global in scale and have sufficient precision and cadence to permit observation of atmospheric tides. These new measurements will afford fresh insights into the global-scale response of solar and geomagnetic forcing on the thermosphere and ionosphere.

2. Simultaneous measurements of temperatures and density ratios of major species (O and N_2) to provide new constraints necessary for advancing space weather models: GOLD measurements will allow separation of the temporal and spatial evolution of column density ratios for the major species [atomic oxygen/molecular nitrogen, $\Sigma O/N_2$; *Meier et al.*, 2005] in the Earth's upper atmosphere. Simultaneous measurement of both the temperatures and the density ratios will provide the critical constraints essential for advancing the best models of the lower space environment. Previously, the Thermosphere–Ionosphere Mesosphere Energetics and Dynamics (TIMED) satellite has provided such data (as did the Polar BEAR satellite) and single wavelength observations from the DE 1 and IMAGE satellites have been used to infer O/N_2 column density ratios ($\Sigma O/N_2$).

3. First use of a commercial geostationary satellite for a NASA science payload: SES Americom, a company with decades of experience putting satellites into geostationary orbit and operating them, has committed to accommodating the GOLD UV imager as a MOO on one of its satellites in 2012. SES Americom will control the satellite from their existing mission operations centers. They will uplink commands for instrument control and transfer downlinked data to GOLD facilities for processing and distribution to the science community and other users. Commercial companies produce reliable satellites capable of operating for 15 years on orbit. Accommodation on commercial satellites has the potential to render significant cost savings, reduce financial risk, and provide numerous additional flight opportunities for future investigations in a variety of disciplines.

4. Real-time data availability: GOLD's geostationary orbit and operations plan provide real-time data availability throughout the continental United States. Since the data are broadcast as they are collected, near-real-time (15-minute latency) data will be available to space weather forecast centers, either through links to the planned receiving station or through a stand-alone station. In this way, GOLD's observations can be used to drive, test, and improve space weather forecasting capabilities.

GOLD's measurements of atmospheric temperatures across the disk will provide significant advances in remote sensing capabilities, and the $\Sigma O/N_2$ will extend the availability of measurements that are relied upon to track changes in the upper atmosphere. Except for IMAGE [*Mende et al.*, 2000a, 2000b, 2000c] and TIMED [*Paxton et al.*, 1999; *Christensen et al.*, 2003], FUV (~110–180 nm) imaging of the Earth by previous satellites—DE 1 [*Frank et al.*, 1985], Viking [*Murphree and Cogger*, 1988], Polar BEAR [*Del-Greco et al.*, 1989], POLAR [*Torr et al.*, 1995; *Brittnacher et al.*, 1997]— have typically been at substantially lower spectral resolution than the proposed measurements. The high-resolution measurements by IMAGE were focused on emissions from the geocorona (primarily hydrogen emissions). While the TIMED images are at low resolution, due to data downlink limitations, the instrument can measure spectra (but not images) at ~1-nm resolution.

Due to the combination of instrument capabilities and orbit that would be used, the $\Sigma O/N_2$ data provided by GOLD will provide a new perspective on this measurement. Previous missions have provided mid- and low-latitude observations primarily from low or medium Earth orbit (e.g., TIMED, Polar BEAR) or have concentrated on high latitudes through the use of highly elliptical orbits (e.g., DE 1, Viking, POLAR). A geostationary orbit provides a global-scale view of the subauroral regions, similar to that provided for the high latitudes by the auroral imaging missions such as Viking, DE 1, and POLAR. However, from a geostationary orbit, there would be no gaps in the observations due to passage through perigee, although there will be a gap (<1.5 h) near midnight during most seasons due to the Sun being near the field of view. Observing from a geostationary orbit allows separation of the spatial and temporal changes since the same region is observed throughout the day. Such separation of temporal and spatial changes is difficult from low

Earth orbit, where it takes ~12 h to revisit the same low–mid-latitude location (~24 h to revisit at the same time of day).

2. GOLD SCIENCE OBJECTIVES

The Sun, especially during geomagnetic storms, affects the Earth's thermosphere and ionosphere and, in turn, the space-based technologies we rely on. Efforts to determine the response of these regions and use that information to drive the next-generation space weather models has been hampered by a lack of essential temperature and composition information. Most images from GOLD will contain both dayside and nightside regions, and data from both regions will be available in such images. On the dayside, GOLD will simultaneously measure the $\Sigma O/N_2$ and the temperature of the neutral atmosphere at 150–180 km on a global scale with a 0.5- to 1-h cadence. The low resolution channel observations will continue scanning across the nightside, where peak electron densities (N_mF_2) and their distribution will be measured. The high-resolution channel is expected to concentrate on the dayside disk, where disk temperature measurements are possible, and on stellar occultations. This combination of neutral temperature, neutral composition, and peak electron density (N_mF_2) measurements will provide a new understanding of the effects of high-latitude forcing during geomagnetic storms, the temporal and spatial evolution of flare effects, and the temporal and spatial evolution of night-time ionospheric irregularities (bubbles) and will enable observations of thermospheric tides. GOLD observations will make it possible to understand the nonlinear feedback effects in space weather.

The GOLD mission is framed by four science questions (see Figure 1):

1. What is the global-scale response of the thermosphere and ionosphere to geomagnetic forcing? The main mechanisms responsible for thermosphere and ionosphere storm-time effects are known, but large uncertainties remain in their relative importance because there are relatively few observations of the global-scale response of $\Sigma O/N_2$ in the thermosphere and none of the temperature.

2. What is the global-scale response of the ionosphere and thermosphere to changing EUV radiation? Measuring the global-scale response will provide insights unavailable from localized measurements. It is unknown whether, on a global scale, the atmospheric and ionospheric ground state predicted by first principles models is internally consistent with solar EUV inputs. The absence of both local and global information about the relationship between neutral composition and the electron density limits our understanding of the relationship between these coupled parameters. GOLD can provide the neutral composition measurements needed for comparison with electron density measurements from coincident measurements or assimilated measurements (e.g.,

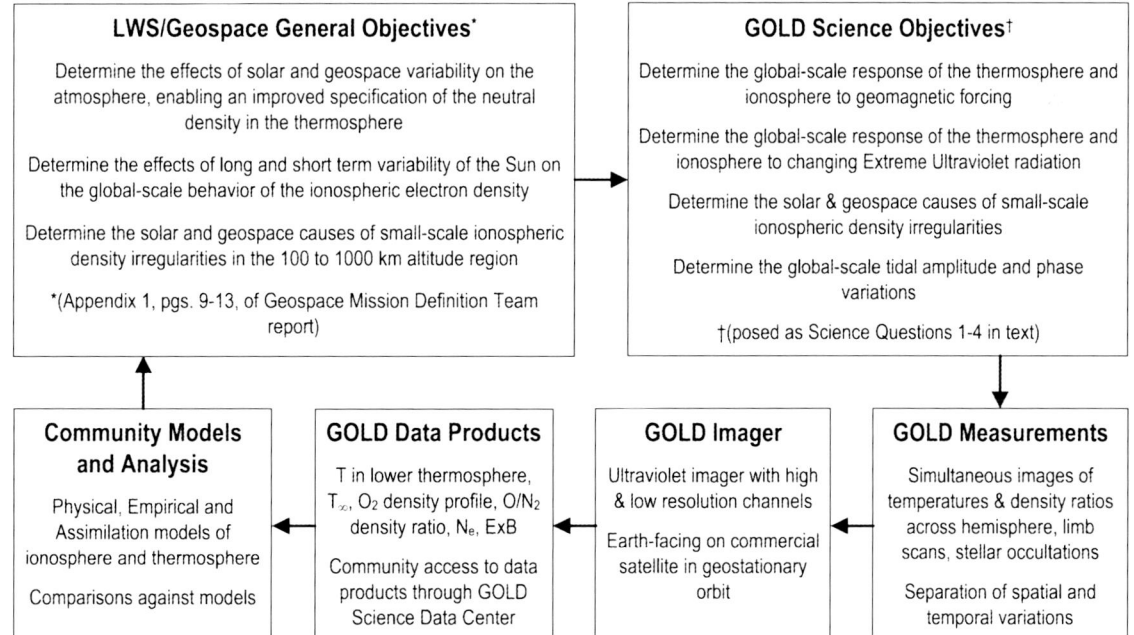

Figure 1. GOLD program science objectives address LWS/Geospace objectives.

using the global assimilation of ionospheric measurements model [*Schunk et al.*, 2004]).

3. What are the solar and geospace causes of small-scale ionospheric density irregularities? Our ability to characterize these irregularities and understand their driving mechanisms suffers from a lack of basic measurements. At low latitudes, our understanding is based on relatively limited geographic coverage, and even that understanding is not well-connected to the global scale variations and changes. Using low-latitude, nighttime electron densities (N_mF_2) as a function of latitude, longitude, and local time, GOLD will provide the first quantitative determination of the longitude dependence of low-latitude, nighttime E × B drift velocities and their coincidence with ionospheric density depletions.

4. What are the global-scale tidal amplitude and phase variations? Atmospheric tides are the dominant driver for wind and temperature variations in the low- and middle-latitude lower thermosphere, and they play a major role in the generation of ionospheric electric fields through ionospheric dynamo action. These same tides cause variations in the atmospheric temperatures, between 130 and 180 km, which will be measured by GOLD.

As summarized in Figure 1, these four questions play a central role in understanding how the Earth's upper atmosphere changes in response to solar and geospace variability, which is a critical objective of NASA's LWS program. GOLD also addresses two key elements of NASA's Heliophysics Research Objectives, which are to "understand the coupling between ionospheres and upper atmospheres mediated by strong ion-neutral interactions" and "determine changes in the Earth's magnetosphere, ionosphere, and upper atmosphere to enable specification, prediction, and mitigation of their effects." The global-scale view afforded by the geostationary vantage and coincident observations with the Solar Dynamics Observatory will yield crucial data for evaluating and improving predictive global models of the coupled thermosphere–ionosphere.

In order to answer the four questions listed above, the GOLD instrument will measure:

On the disk:

1. Neutral temperatures, column averaged, between 150 and 180 km (daytime)
2. Atomic oxygen/molecular nitrogen column density ratios ($\Sigma O/N_2$) (daytime)
3. Electron density (N_mF_2) variations in latitude and longitude (nighttime)

On the limb:

4. Molecular oxygen density profiles, between 150 and 250 km (daytime and nighttime)
5. Exospheric neutral temperatures (daytime)

As outlined in Figure 1, these GOLD measurements will be used with the most advanced space environment models available, e.g., assimilated mapping of ionospheric electrodynamics, Center for Integrated Space Weather Modeling, coupled thermosphere–ionosphere plasmasphere electrodynamics, National Center for Atmospheric Research thermosphere–ionosphere–electrodynamics general circulation model (TIE-GCM), global assimilation models of the ionosphere, and global airglow (GLOW), many of which have been developed by GOLD's coinvestigators. Testing these and other models against the GOLD observations will advance the next generation of models.

3. TECHNICAL APPROACH

A key feature of GOLD's technical approach is the use of two identical optical channels, each capable of performing any of the planned observations. Consequently, simultaneous observations of critical parameters, e.g., both $\Sigma O/N_2$ and temperatures in the lower thermosphere, are possible. The coincidence of these measurements, as well as the completeness of the temporal coverage available from the two channels, significantly increases the value of both measurements.

Instrument measurement performance and spacecraft accommodations have been carefully traced back to the science requirements. During phase A, detailed simulations were performed using realistic instrument performance models to demonstrate that GOLD measurements will meet the requirements for disk temperature precision, spatial resolution of bubbles, daytime stellar occultation measurements, and the retrieval of exospheric temperature from daytime limb measurements of N_2. The wavelength range and spectral resolution of the instrument allow the use of proven techniques for $\Sigma O/N_2$ determination [e.g., *Strickland et al.*, 1995, 1999] when operating a channel in the low-resolution mode. When operating in the high-resolution mode, the spectral resolution is sufficient to determine the N_2 LBH rotational temperature, which is equivalent to the temperature of the neutral atmosphere [*Aksnes et al.*, 2006, 2007]. Deriving O_2 densities and temperature profiles from stellar occultations [e.g., *Lumpe et al.*, 2002], deriving exospheric temperatures from limb profiles [e.g., *Meier et al.*, 2005; *Meier and Picone*, 1994], and identification of bubbles in nighttime disk images [e.g., *Immel et al.*, 2003] will rely on previously proven techniques and algorithms. The instrument design, which provides prudent margins for all critical performance parameters, uses only proven technologies currently in use by Laboratory for Atmospheric and Space Physics and its subcontractors. No new or external technologies are required to implement GOLD.

4. TEMPERATURE RETRIEVALS FROM N_2 DISK MAPS

When observing on the disk from a geostationary orbit, the temperature and emission rate will vary along the line of sight. Consequently, the observed spectrum is a weighted sum of emissions at the various temperatures along the line of sight, with altitudes near 150 km contributing most of the emission. Proof of the concept by both data analysis [*Aksnes et al.*, 2006, 2007] and model calculations have been performed.

Recent analysis of data from the high-resolution ionospheric and thermospheric spectrograph (HITS) instrument aboard the ARGOS [*Dymond et al.*, 1999] by *Aksnes et al.* [2006, 2007] show consistency with the models. While the HITS observations have more limited spectral coverage (144–154 nm), lower spectral resolution (0.13 nm), and more significant noise in the spectra than the proposed measurements, the retrieved temperatures are consistent with those from other sources.

To model these observations, volume emission rates were calculated for N_2, using temperatures and densities from the National Center for Atmospheric Research TIE-GCM model with parameterized *g* factors from the Atmospheric Ultraviolet Radiance Integrated Code model. The volume emission rates were used to weight the contributions of the corresponding N_2 spectra calculated using the temperature from the TIE-GCM model. Signals over the entire image were calculated, using a vertical viewing geometry and photoabsorption by O_2 in the calculation. Column emission rates for the 144- to 154-nm emissions of N_2 were 750 R at 70° solar zenith angle on the equator. The use of vertical viewing geometry for all locations gives brightnesses lower than would be expected for points away from nadir and approximately two times less than actual emissions near the limb. The resulting brightnesses were converted to the expected number of counts using an instrument model with sensitivities derated by a factor of 2 from the nominal instrument sensitivity and which includes the spectral resolution, instrumental line shape, and noise. Noise due to both the counting (Poisson) statistics and charged particle backgrounds equivalent to 100 counts/cm^2/s (and their production of X-rays during collisions with the instrument) was added.

Temperatures were then retrieved from the synthetic spectra for comparison with the original temperatures. The retrieval used a forward model and discrete inverse approach developed at NRL. The analysis used the 144- to 154-nm wavelength range, which is identical to that used earlier by *Aksnes et al.* [2006] when analyzing similar data. Calculations were performed for thousands of spectra, and the standard deviation from the average of the retrieved temperatures was calculated to determine the statistical uncertainties. The simulations included solar zenith angles of 70° and less. The average one standard deviation of the retrieved temperatures is significantly less than 15 K for observations at a 2-h cadence and 600 × 600 km^2 spatial scales and significantly less than 55 K for a 1-h cadence at 250 × 250 km^2. The offsets between input and retrieved temperatures were a few degrees for 2-h averages and a few tens of degrees for the 1-h averages. These offsets depend on the SNR and will be examined more closely in the near future. While these simulations did not include the 149.3-nm atomic nitrogen emission, scattered light from 149.3 nm is not problematic and the bands that are observed outside the 144- to 154-nm bandpass will more than double the signal (in counts) offsetting any bands lost.

The most important instrument parameters for these simulations are spectral resolution, band-integrated SNR (total counts in the spectrum), wavelength scale, and background. We find that accurate wavelength scale is important but that varying the resolution between 0.1 and 0.13 nm has little effect on the retrieved uncertainties. When accurate wavelength scales (good to 0.01 nm) are employed, SNR ~ 100 for the 144- to 154-nm emissions is sufficient to recover temperature to less than ±15 K, while SNR ~ 30 is adequate for ±55 K retrieval, even when backgrounds of up to 10 counts per spectral element (equivalent to 100 counts/cm^2/s) were included. If all of the molecular emissions (138–160 nm) to be observed by GOLD are used, the SNRs are significantly better, ~170 and ~50 rather than ~100 and ~30. This should allow the temperature uncertainties to be decreased even further.

Simulation results are shown in Plate 1. Plate 1 (left) shows the TIE-GCM disk temperatures and Plate 1 (right) shows the one standard deviation uncertainties in the retrieved temperatures. The retrieval shown in Plate 1 used 0.13 nm spectral resolution, SNR = 100. These results are consistent with earlier results from analysis of HITS observations [*Aksnes et al.*, 2006, 2007].

5. SUMMARY AND CONCLUSIONS

The GOLD MOO will greatly improve understanding of the Earth's thermosphere and ionosphere through measurements of the global-scale response to external and internal forces. The measurements provided by GOLD will allow scientists to answer the four main science questions that frame the GOLD mission. In addition, testing currently available space environment models against GOLD observations will help advance the next generation of models.

The disk temperatures provided by GOLD will be the first global-scale observations of temperatures in the lower

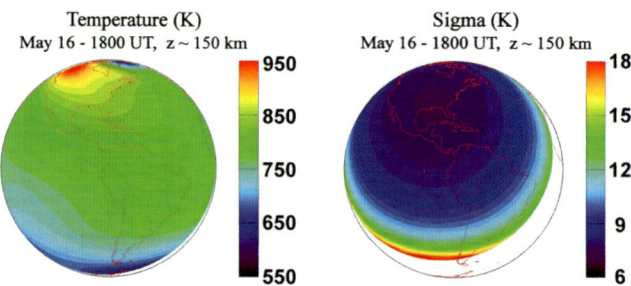

Plate 1. (left) TIE-GCM modeled temperatures near 150 km. (right) One standard deviation uncertainties in temperatures retrieved from simulated observations at solar zenith angles less than 70°, assuming averaging over 2 h and 600 × 600 km^2.

thermosphere. Proof of concept by both data analysis and model calculations have been performed. Modeling of temperature retrievals using GOLD instrument parameters and expected noise has shown that data from GOLD will allow temperatures to be recovered with a precision greater than ±15 K for a 2-h cadence and 600 × 600 km² spatial scales, while a precision greater than ±55 K can be obtained for a 1-h cadence and 250 × 250 km² spatial scales.

The GOLD instrument, with its two identical optical channels, each capable of performing any of the planned observations, allows simultaneous observations of critical parameters such as $\Sigma O/N_2$ and temperatures. The instrument design uses only proven technologies currently in use by Laboratory for Atmospheric and Space Physics and its subcontractors, and no new or external technologies are required to implement GOLD.

SES Americom, a company with decades of experience putting satellites into geostationary orbit and operating them, has committed to accommodating the GOLD UV imager as a MOO on one of its satellites in 2012. The data would be available within 15 minutes of the observations being made, making them ideally suited for space weather applications. If GOLD is selected, this would be the first use of a commercial geostationary satellite for a NASA science payload. Accommodation on commercial satellites has the potential to render significant cost savings, reduce financial risk, and provide numerous additional flight opportunities for future investigations in a variety of disciplines.

Acknowledgment. The work described in this article was supported by NASA contract NNG07EK01C to the University of Central Florida.

REFERENCES

Aksnes, A., R. Eastes, S. Budzien, and K. Dymond (2006), Neutral temperatures in the lower thermosphere from N_2 Lyman–Birge–Hopfield (LBH) band profiles, *Geophys. Res. Lett.*, *33*, L15103, doi:10.1029/2006GL026255.

Aksnes, A., R. Eastes, S. Budzien, and K. Dymond (2007), Dependence of neutral temperatures in the lower thermosphere on geomagnetic activity, *J. Geophys. Res.*, *112*, A06302, doi:10.1029/2006JA012214.

Brittnacher, M., J. Spann, G. Parks, and G. Germany (1997), Auroral observations by the polar ultraviolet imager UVI, *Adv. Space Res.*, *20*, 1037–1042.

Christensen, A. B., et al. (2003), Initial observations with the Global Ultraviolet Imager (GUVI) in the NASA TIMED satellite mission, *J. Geophys. Res.*, *108*(A12), 1451, doi:10.1029/2003JA009918.

Delgreco, F. P., R. W. Eastes, and R. E. Huffman (1989), UV ionospheric remote sensing with the Polar BEAR satellite, *Proc. SPIE, 1158*, 46–50.

Dymond, K. F., K. D. Wolfram, S. A. Budzien, C. B. Fortna, and R. P. McCoy (1999), The High-resolution Ionospheric and Thermospheric Spectrograph (HITS) on the Advanced Research and Global Observing Satellite (ARGOS): Quick look results, *Proc. SPIE, 3818*, 137–148.

Frank, L. A., J. D. Craven, and R. L. Rairden (1985), Images of the Earth's aurora and geocorona from the Dynamics Explorer Mission, *Adv. Space Res.*, *5*(4), 53–68.

Immel T. J., S. B. Mende, H. U. Frey, L. M. Peticolas, and E. Sagawa (2003), Determination of low latitude plasma drift speeds from FUV images, *Geophys. Res. Lett.*, *30*(18), 1945, doi:10.1029/2003GL017573.

Lumpe, J. D., R. M. Bevilacqua, K. W. Hoppel, and C. E. Randall (2002), POAM III retrieval algorithm and error analysis, *J. Geophys. Res.*, *107*(D21), 4575, doi:10.1029/2002JD002137.

Meier, R. R., and J. M. Picone (1994), Retrieval of absolute thermospheric concentrations from the far UV dayglow: An application of discrete inverse theory, *J. Geophys. Res.*, *99*(A4), 6307–6320.

Meier, R. R., G. Crowley, D. J. Strickland, A. B. Christensen, L. J. Paxton, D. Morrison, and C. L. Hackert (2005), First look at the 20 November 2003 superstorm with TIMED/GUVI: Comparisons with a thermospheric global circulation model, *J. Geophys. Res.*, *110*, A09S41, doi:10.1029/2004JA010990.

Mende, S. B., et al. (2000a), Far-ultraviolet imaging from the IMAGE spacecraft. 2. Wideband FUV imaging, *Space Sci. Rev.*, *91*, 271–285.

Mende, S. B., et al. (2000b), Far-ultraviolet imaging from the IMAGE spacecraft. 1. System design, *Space Sci. Rev.*, *91*, 243–270.

Mende, S. B., et al. (2000c), Far-ultraviolet imaging from the IMAGE spacecraft. 3. Spectral imaging of Lyman-alpha and OI 135.6 nm, *Space Sci. Rev.*, *91*, 287–318.

Murphree, J. S., and Cogger, L. L. (1988), The application of CCD detectors to UV imaging from a spinning satellite. *Proc. SPIE*, *932*, 42–49.

Paxton, L. J., et al. (1999), Global ultraviolet imager (GUVI): Measuring composition and energy inputs for the NASA Thermosphere Ionosphere Mesosphere Energetics and Dynamics (TIMED) mission, *Proc. SPIE, 3756*, 265–276.

Schunk, R. W., et al. (2004), Global assimilation of ionospheric measurements (GAIM), *Radio Sci.*, *39*, RS1S02, doi:10.1029/2002RS002794.

Strickland, D. J., J. S. Evans, and L. J. Paxton (1995), Satellite remote sensing of thermospheric O/N_2 and solar EUV, 1. Theory, *J. Geophys. Res.*, *100*, 12,217–12,226.

Strickland, D. J., R. J. Cox, R. R. Meier, and D. P. Drob (1999), Global O/N_2 derived from DE 1 FUV dayglow data: Technique and examples from two storm periods, *J. Geophys. Res.*, *104*, 4251–4266.

Torr, M. R., et al. (1995), A far ultraviolet imager for the International Solar Terrestrial Physics Mission, *Space Sci. Rev.*, *71*, 329–383.

A. Aksnes, Universitetet i Bergen, Det matematisk-naturvitenskapelige fakultet, Postboks 7803, N-5020 Bergen, Norway.

D. N. Anderson and A. D. Richmond, Cooperative Institute for Research in Environmental Sciences, University of Colorado, Boulder, CO 80303, USA.

L. Andersson, D. N. Baker, S. A. Budzien, K. F. Dymond, F. G. Eparvier, M. R. Lankton, W. E. McClintock, and D. W. Rusch, Naval Research Laboratory, Washington, DC 20375-5352, USA.

A. G. Burns and S. C. Solomon, National Center for Atmospheric Research, P.O. Box 3000, Boulder, CO 80307-3000, USA.

M. V. Codrescu, NOAA-SWPC, 325 Broadway, Boulder, CO 80303, USA.

R. E. Daniell, Ionospheric Physics, Stoughton, MA 02072-2226, USA.

R. W. Eastes and A. Krywonos, Florida Space Institute, MS-FSI, Kennedy Space Center, FL 32899, USA. (reastes@mail.ucf.edu)

J. E. Harvey, Center for Research and Education in Optics and Lasers, University of Central Florida, Orlando, FL 32816, USA.

T. J. Immel and O. H. Siegmund, Space Sciences Laboratory, University of California, Berkeley, CA 94720-7450, USA.

J. D. Lumpe and D. J. Strickland, Computational Physics, Inc., Suite 210, 8001 Braddock Road, Springfield, VA 22151, USA.

G. W. Prölss, Argelander Institut für Astronomie, Auf dem Hügel 71, Bonn D-53121, Germany.

T. N. Woods, Laboratory for Atmospheric and Space Physics, University of Colorado, Boulder, CO 80309, USA.

Index

airglow and aurora, 259, 319
atmospheric sciences, 221
current systems, 145
electric fields, 25, 121, 145, 157, 169, 179, 187, 201, 235
energetic particles: trapped, 179
equatorial ionosphere, 91, 99, 121, 145, 157, 297
impacts on technological systems, 35
instruments and techniques, 77, 311
interplanetary magnetic fields, 157
interplanetary shocks, 99
ionosphere, 99
ionosphere/atmosphere interactions, 201, 221, 259
ionosphere/magnetosphere interactions, 51, 77, 99, 135, 179, 235, 291
ionospheric disturbances, 9, 63, 83, 121, 271, 283, 311
ionospheric dynamics, 25, 51, 63, 83, 271, 283
ionospheric irregularities, 25, 283, 291, 297, 311
ionospheric storms, 9, 35, 51, 83, 91, 169, 247, 271, 297
magnetic storms, 135, 169, 247, 259

magnetic storms and substorms, 135, 169, 247, 259
magnetosphere: inner, 135, 179, 235
magnetosphere/ionosphere interactions, 51, 77, 99, 135, 179, 235, 291
midlatitude ionosphere, 9, 35, 51, 63, 77, 91, 121, 169, 187, 247, 271, 297, 311
modeling and forecasting, 25, 35, 83
plasma temperature and density, 9
plasma waves and instabilities, 283
radar atmospheric physics, 77
ring current, 135, 291
solar radiation and cosmic ray effects, 319
spectral analysis, 157
substorms, 145
theoretical modeling, 187
thermosphere: composition and chemistry, 221, 319
thermosphere: energy deposition, 63, 187, 201, 221, 235, 247, 259, 319
thermospheric dynamics, 63, 187, 201, 221, 235, 247, 259, 319
tomography and imaging, 91